TABLE 5.1 Power Supply Design

Given: R_S R_L $V_{LDC} = \dfrac{V_{LP} + V_{Lt1}}{2}$ $v_{LPP} = V_{LP} - V_{Lt1}$

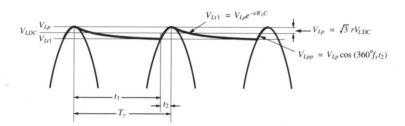

$V_{Lt1} = V_{Lp}e^{-t/R_LC}$

$V_{Lp} = \sqrt{3}\,rV_{LDC}$

$V_{Lpp} = V_{Lp}\cos(360° f_s t_2)$

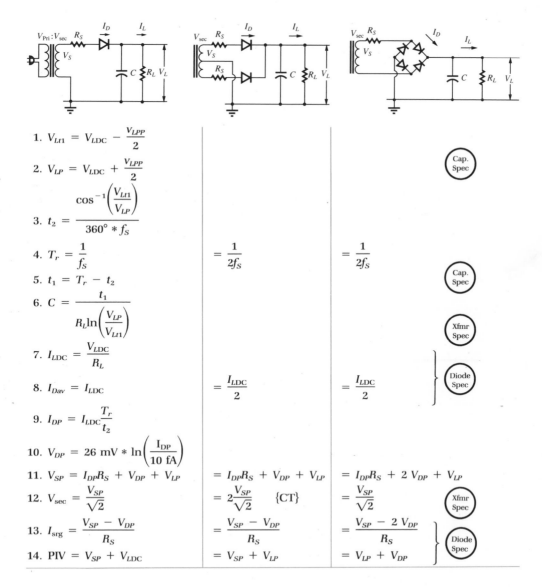

	Half-wave	Center-tapped	Bridge	Spec
1.	$V_{Lt1} = V_{LDC} - \dfrac{v_{LPP}}{2}$			
2.	$V_{LP} = V_{LDC} + \dfrac{v_{LPP}}{2}$			Cap. Spec
3.	$t_2 = \dfrac{\cos^{-1}\left(\dfrac{V_{Lt1}}{V_{LP}}\right)}{360° * f_S}$			
4.	$T_r = \dfrac{1}{f_S}$	$= \dfrac{1}{2f_S}$	$= \dfrac{1}{2f_S}$	Cap. Spec
5.	$t_1 = T_r - t_2$			
6.	$C = \dfrac{t_1}{R_L\ln\left(\dfrac{V_{LP}}{V_{Lt1}}\right)}$			Xfmr Spec
7.	$I_{LDC} = \dfrac{V_{LDC}}{R_L}$			
8.	$I_{Dav} = I_{LDC}$	$= \dfrac{I_{LDC}}{2}$	$= \dfrac{I_{LDC}}{2}$	Diode Spec
9.	$I_{DP} = I_{LDC}\dfrac{T_r}{t_2}$			
10.	$V_{DP} = 26\text{ mV} * \ln\left(\dfrac{I_{DP}}{10\text{ fA}}\right)$			
11.	$V_{SP} = I_{DP}R_S + V_{DP} + V_{LP}$	$= I_{DP}R_S + V_{DP} + V_{LP}$	$= I_{DP}R_S + 2V_{DP} + V_{LP}$	
12.	$V_{sec} = \dfrac{V_{SP}}{\sqrt{2}}$	$= 2\dfrac{V_{SP}}{\sqrt{2}}$ {CT}	$= \dfrac{V_{SP}}{\sqrt{2}}$	Xfmr Spec
13.	$I_{srg} = \dfrac{V_{SP} - V_{DP}}{R_S}$	$= \dfrac{V_{SP} - V_{DP}}{R_S}$	$= \dfrac{V_{SP} - 2V_{DP}}{R_S}$	Diode Spec
14.	PIV $= V_{SP} + V_{LDC}$	$= V_{SP} + V_{LP}$	$= V_{LP} + V_{DP}$	

INTRODUCTION TO ELECTRONIC DEVICES AND CIRCUITS

Robert R. Ludeman Andrews University

SAUNDERS COLLEGE PUBLISHING
A Division of Holt, Rinehart and Winston, Inc.
Philadelphia   Chicago   Fort Worth   San Francisco
Montreal   Toronto   London   Sydney   Tokyo

Text Typeface: Zapf Book Light
Compositor: The Clarinda Company
Acquisitions Editor: Barbara Gingery
Developmental Editor: Alexa R. Barnes
Managing Editor: Carol Field
Project Editor: Margaret Mary Anderson
Copy Editor: Elsie Testa
Manager of Art and Design: Carol Bleistine
Art Director: Christine Schueler
Art and Design Coordinator: Doris Bruey
Text Designer: Arlene Putterman
Cover Designer: Lawrence R. Didona
Text Artwork: Publication Services
Director of EDP: Tim Frelick
Production Manager: Charlene Squibb

Cover Credit: John Anderson / TSW-CLICK / CHICAGO

Printed in the United States of America

INTRODUCTION TO ELECTRONIC DEVICES AND CIRCUITS

0-03-009538-7

Library of Congress Catalog Card Number: 89-043108

9012   015   987654321

Dedication

To my students who have taught me so much.

Preface

INTRODUCTION TO ELECTRONIC DEVICES AND CIRCUITS is intended for the second semester of either a two year Electronics Technician course or a four year Electronics Engineering Technology program. The title of the book is indicative of its approach. Major emphasis is placed on the application of standard circuit analysis procedures to circuits containing electronic devices. The approach is very practical with frequent suggestions that the student should hook up the circuits discussed to see if their data agrees with the theory presented in the text. Many of the exercises in the accompanying laboratory manual use the same circuits as those in examples or problems in the text. The topics covered are sequential in that wherever it is possible, the behavior of a newly presented device is explained in terms of a previously studied device. Thus the student sees how individual devices fit into the broader picture that we call electronics and should be better equipped to understand unfamiliar devices when they are encountered.

Prerequisites: It is assumed that the student has a good background in DC/AC circuits. The ability to use such circuit analysis tools as Thevenin, Norton, KVL and KCL is assumed. They are used throughout the book.

For the most part the level of mathematics is limited to elementary algebra, but it is used extensively. Some trigonometric relationships are used and the student will need to know which calculator button to press to perform logarithmic and exponential calculations. Derivations that require calculus are placed in Appendix D.

Objectives: The overall objective of this book is to help students obtain a solid foundation of understanding on which they can build an enjoyable and successful career in electronics. The specific objectives are: (1) To build on the background in circuit analysis gained in the prerequisite electric circuits course and to apply those ideas to circuits containing active devices. (2) To provide an intuitively satisfying explanation of the operation of discrete semiconductor devices. (3) To introduce the student to the ideal operational amplifier and typical op amp circuits. (4) To introduce

v

simple timing circuits as a bridge to the study of digital circuits in more advanced courses.

Approach

Since understanding is always the goal, formulas are usually derived using basic circuit analysis procedures rather than being simply stated with no indication of their origin or limitations. The student is always urged to rely on fundamental ideas rather than the memorization of special purpose formulas which may or may not apply to the circuit under consideration.

The approach conforms as closely as possible to what the student will encounter in later class work and in industry after leaving school. Standard resistor values and engineering notation are used throughout the text. Practical circuits are used in design and analysis problems with heavy use of manufacturer's data sheets and information from commonly available data books. Since the preceding sources usually list transistor common emitter h parameters, a greatly simplified version of the common emitter h parameter model is used for the design and analysis of bipolar transistor circuits. This simplified model is no more difficult to use than the r parameter model used in some textbooks and it has the strong advantage of using the manufacturer's data and terminology. For the same reasons the analogous g parameter model is used for FETs. The ideal model is used exclusively in the study of op amp circuits and emphasis is placed on designing circuits so that the ideal model applies. The two-transistor model is used in the study of thyristors and UJTs.

Useful procedures are developed for both circuit design and circuit analysis. These procedures are then presented in "recipe" form which the student is encouraged to incorporate into a computer spreadsheet program.

Computer Usage

This book does not require the use of a personal computer, but since it is virtually universally available in the work place, the student is encouraged to use a spread sheet for circuit design and a commercially available circuit analysis program for circuit analysis. Examples of their use are included in Appendices G and H respectively.

The spreadsheet applications are sufficiently general so that any one of a number of spreadsheets could be used. SuperCalc was the program that was used for the design examples in Appendix G because of its power and its cost effectiveness. SuperCalc is a product of Sorcim/IUS, 2195 Fortune Drive, San Jose, CA 95131. PSpice was chosen for circuit analysis based on its power, its close relationship to the widely used mainframe program SPICE, and its wide availability thanks to the generosity of its developer, MicroSim Corporation, 23175 La Cadena Drive, Laguna Hills, CA 92653. Appendix H contains PSpice solutions to a number of examples as noted throughout the text. The student is strongly encouraged to become proficient in the use of tools of this type.

Learning Aids

Each chapter contains the following features:

- An opening photograph relevant to the chapter content.
- A chapter outline and a listing of special terms used in the chapter.
- A motivational introduction and objectives section which describes what the chapter covers, why it is important, and what understanding the student should gain from the chapter.
- Completely worked-out examples follow the development of almost all important ideas or derivations.
- Tables and figures wherever they will contribute to understanding.
- Color used in a functional way to focus the student's attention.
- A troubleshooting section with helpful suggestions for identifying circuit malfunctions.
- A concise summary of the ideas developed in the chapter. This section is intentionally made sufficiently brief so that the student is not tempted to study only this section of the chapter.
- A glossary of important terms keyed to the chapter section where they are discussed.
- End of chapter problems keyed to the chapter sections with the answers to most of the odd-numbered problems included at the end of the book.
- A direct conversational style of presentation intended to give the student a feeling of active involvement in the development of the subject matter.

Organization

Since the application of basic concepts is stressed throughout the text, new concepts are developed from previous understandings. Thus the topics in Parts I and II should be studied carefully and in the order they are presented. Parts III, IV, and V are dependent on Parts I and II, but they are more or less independent of one another so that they provide more flexibility in both the breadth and depth of study.

The following listing of the major parts of the text includes an estimate of the priority that should be given to each.

INTRODUCTION:

Chapter 1 is a very brief review of the basic circuit concepts that the student should know before entering the course. Assuming that the prerequisite electric circuits course has been mastered, this chapter can be skimmed.

PART I: DIODES

Since the P-N junction is an essential part of so many electronic devices, Chapters 2 and 3 should be studied carefully. Concepts developed in these chapters will be referred to repeatedly in later chapters. Chapters 4 through 7 are also important since power supplies are an essential part of most electronic devices.

PART II: BIPOLAR TRANSISTORS

The presentation of bipolar transistors in Chapter 8 builds directly on the student's understanding of the P-N junction from Chapters 2 and 3. Since bipolar transistors are used extensively in electronic circuits, and since the concept of the Q point and the load line (Chapter 9), biasing (Chapter 10), and equivalent circuits (Chapter 11), are used by other devices as well, an understanding of these chapters is crucial to the understanding of electronics.

Although Chapters 12 and 13 might be considered slightly less important, the concepts developed there have wide applicability and should not be ignored. Chapter 14 could be given lighter emphasis, although circuit efficiency is an important concept.

The complementary circuit presented in Chapter 15 is treated as an application of the emitter follower circuit presented in Chapter 13; it is important because of its uniqueness and its wide use in real devices.

PART III: FIELD EFFECT TRANSISTORS

Since field effect devices operate on a different principle than bipolar devices, this part of the book can be considered to be independent of what is to follow. Chapters 16 through 18 could be omitted without sacrificing understanding of the later parts of the book.

In the introduction of field effect devices in Chapters 16 and 18, an approach is taken that is similar to that used in connection with bipolar devices in Part II. Circuit design and analysis procedures are then developed using an approach very similar to that used for bipolar transistors.

PART IV: LINEAR INTEGRATED CIRCUITS

This part of the book is independent of Parts III and V. Chapter 19 examines briefly the internal circuitry of the 741 op amp and shows how its behavior can be understood in terms of bipolar transistor behavior studied in Part II. Chapters 20 through 22 examine typical op amp circuits. Finally Chapter 23 introduces the 555 timer as a bridge to the digital circuits the student will study in later courses.

PART V: MISCELLANEOUS SEMICONDUCTOR DEVICES

Chapters 24 and 25 are independent of Parts III and IV of the book. Models for the devices studied in Chapters 24 and 25 are based on the student's understanding of bipolar transistor behavior studied in Part II.

APPENDICES

A: Standard electronic symbols.

B: Standard resistor and capacitor values. In order to promote familiarity with standard values, students are encouraged to specify the nearest standard value whenever component values are calculated.

C: SI prefixes. The use of standard prefixes and engineering notation is encouraged throughout the text.

D: Calculus-based derivations. Since calculus is not required, these derivations are removed from the main body of the text. Students with sufficient mathematical background should refer to these derivations.

E: Algebra-based derivations. A derivation that is particularly long and tedious. Students with a strong interest in algebra should refer to this section.

F: Device parameters and data sheets. In order to encourage familiarity with commercially available devices and the data usually available for them, this listing of device parameters is frequently referred to.

G: SuperCalc examples. SuperCalc solution of examples identified in the text.

H: PSpice examples. PSpice solution of examples identified in the text.

I: r parameters *vs.* h parameters. A comparison of the r parameter transistor model with the simplified h model.

Ancillary Package

The following ancillaries are available with this textbook:

Laboratory Manual This manual contains a full set of student tested laboratory experiments which are closely correlated with the examples and problems in the textbook.

Instructor's Manual for the Laboratory Manual A laboratory manual containing expected student responses to the laboratory experiments. Supplied free of charge to instructors adopting the book.

Instructor's Manual with Transparency Masters and Test Bank Helpful suggestions for the teaching of each chapter, fully worked-out solutions to all end-of-chapter problems, 125 transparency masters which are duplicates of figures appearing in the text, test items and answer key. Supplied free of charge to instructors adopting the book.

Transparency Acetates A set of 25 transparency acetates of the most important figures in the textbook ready for projection. Supplied free of charge to instructors adopting the book.

Computerized Test Bank A computerized bank of the test items and their answers on floppy disk in IBM and compatible format. Items are keyed to the text by chapter and section number.

Acknowledgements

I wish to take this opportunity to thank the hard working staff at Saunders College Publishing/Holt, Rinehart and Winston, Inc., for having the patience and exerting the effort required to convert the ruminations of a teacher into a textbook: Barbara Gingery, Senior Acquisitions Editor, Margaret Mary Anderson, Project Editor, Alexa Barnes, Developmental Editor, and Laura Shur, Editorial Assistant. Without their support and encouragement, this book would never have seen the light of day. Further, I wish to thank the many individuals whose insightful criticisms were so helpful during the development process:

Ernest Abbott, Napa Valley College
William Barnes, County College of Morris
Robert I. Davis, Miami-Dade Community College—North Campus
Lucius Day, Metropolitan State College
Samuel Derman, City College of New York
Joseph Ennesser, DeVry Institute of Technology—Lombard Campus
George Granier, Pennsylvania State University—Capitol Campus
Ronald Johnson, Andrews University
David Longobardi, Antelope Valley College
David McDonald, Lake Superior State University
Edward Peterson, Arizona State University
Richard Sturtevant, Springfield Technical College
Raj Raghuram, Trenton State University
Peter Tampas, Michigan Technological University
John A. Triplett, Salt Lake City College
Andrew VanCamp, East Central College

Finally, I owe the greatest debt of gratitude to Nona who has taught me that love can surmount any obstacle, including electronics, computers, and even textbook writing.

Robert R. Ludeman
October 1989

Contents

P PSpice solution of examples in this section are in Appendix H.
S SuperCalc solution of examples in this section are in Appendix G.

Astronaut constructing a model of the moon.
(Courtesy of NASA.)

Chapter 1
Introduction

SPECIAL TERMS

Kirchoff's current law (KCL)
Kirchoff's voltage law (KVL)
Norton equivalent
Superposition
Thevenin equivalent
Voltage divider
PSpice
SuperCalc

1.1   Objectives

This book is intended to introduce you to the field of electronics as practiced by engineering technologists. We will not attempt to "cover" every device on the market. Such an encyclopedic coverage would necessarily be too shallow to provide any real understanding, and would become obsolete very quickly. Rather, we will try to "uncover" the basic principles of operation of electronic devices and show how your present understandings can be extended to include new devices.

Thus your objective as you study should be understanding. Try to understand not only what is happening, but why it is happening. You will find real understanding to be more satisfying than the memorization of many quickly forgotten facts. You will also find such understanding of greater value to you in later courses and in the job market after you leave school.

1.2   Prerequisite

The prerequisite for success in this course is a good understanding of elementary circuit ideas. We are going to USE the following concepts: Ohm's Law, KVL, KCL, Voltage dividers, Thevenin and Norton equivalents, Maximum power transfer, and Superposition. They are presented approximately in the order of their importance. We will make extensive use of the first four or five ideas. If your understanding of them is incomplete, you should set about correcting the situation. In any case, you will find it to your advantage to have the textbook and lecture notes from your circuits course nearby as you study this book.

1.3   Computer Usage

We will make occasional reference to the use of a personal computer as an aid in problem solving. Given its near universal availability and increasingly powerful software, in future courses and in your professional life after school you will find yourself handicapped if you are not able to use this time-saving tool.

Perhaps the most versatile program available for use on a personal computer is the spread sheet. If you learn to use it, you will find it to be a real time-saver. Several examples of its use in circuit design have been included in Appendix G. SuperCalc was used in these examples because of its power and its cost effectiveness but there are probably other programs that will work as well.

Although there are several good circuit analysis programs available, SPICE is probably most widely used on large computers. Many adaptations of SPICE for use on personal computers have been made. Maybe the "flavor" most widely known is PSpice, thanks to the generosity of its developer in making it available to schools. Appendix H contains a number of examples of the use of PSpice.

You are encouraged to check out these, or similar programs from your departmental library and use them for the solution of the problems in this book. As you gain facility with them, you will find that they make school work easier and that they will enhance your employability when you leave school.

1.4 Notation and Units

Throughout this book, when we speak of "current" we will mean positive current flow unless we specifically refer to negative carriers. Thus "current" will always flow from a point of higher $(+)$ voltage to a point of lower $(-)$ voltage. Capital letters will always be used to represent DC or time invariant quantities. Lower case letters will be used to represent AC or time variant quantities.

Circuit variables will be made specific through extensive use of subscripts such as R_1, I_5, etc. Voltage is unique in that we almost always refer to it in relative terms. We speak of the "voltage difference" or the "voltage lost" between two points. Thus voltage references should always have two subscripts. V_{AB} will refer to the reading of a voltmeter that has its positive (red) lead connected to point A in a circuit and its negative (black) lead connected to point B. In electronics we often have a "common ground" to which it is convenient to reference the circuit voltages. Often that common ground is the chassis of the device. When such a common ground reference is intended, a single subscript will be used for voltages. Thus V_A will refer to the reading of a voltmeter that has its positive lead connected to point A and its negative lead connected to ground. From the preceding discussion it is obvious that

$$V_{AB} = V_A - V_B \tag{1.1}$$

This is standard practice in electronics and you should have encountered it previously in your circuits class.

Numbers will never be used without appropriate units attached. Standard unit abbreviations will be used.

Engineering notation will be used throughout, with numbers usually rounded to three significant figures for convenience in writing. For your convenience, you are encouraged to set up your calculator to display in that format. At the same time, you should learn to use your calculator in such a way that, by storing intermediate results, you use as many decimal places as it is capable of. Thus, your calculator will say

$$\frac{8}{3} - \frac{4}{3} = 2.67 - 1.33 = 1.33$$

rather than

$$\frac{8}{3} - \frac{4}{3} = 2.67 - 1.33 = 1.34$$

Further, we will not make a big issue of the use of significant digits and the differences between 8, 8., and 8.00. Suffice it to say that if, for example, a resistor with ± 10 percent tolerance is used in an Ohm's Law calculation, there must be an uncertainty in the calculated voltage or current of at least ± 10 percent.

You should be familiar with engineering notation from your circuits class. If you have forgotten, you are strongly advised to refresh your memory. Engineering notation is virtually the universally used language of electronics.

1.5 The Concept of a Model

Typically, when we are confronted with something that is unfamiliar or complex our minds automatically make a model of that thing. When the astronauts returned from the moon they said, "The moon is like ———." They were making a model of the moon. In electronics we seldom think of a piece of wire as a crystal structure composed of myriads of protons and neutrons surrounded by clouds of electrons. (The thoughtful reader probably recognizes a second or a third level of modeling going on here.) Rather, we think of a piece of wire as—well, just as a piece of wire; a perfect lossless conductor or something of that sort. Of course, we know that our model is not perfectly accurate. The point is that it is a close enough approximation to reality to fill our needs—most of the time! And that is just what a model is all about. It is a simplification of reality that is accurate enough to be useful. When we begin searching for a model to describe circuits and semiconductor devices we should keep these conflicting requirements in mind: simplicity and accuracy.

A note of caution: It is important to remember that models have limitations. It is easy to document cases where money, time, and even lives have been lost when those limitations have been ignored.

1.6 Circuits Reviewed

The problems at the end of this chapter are of the type which you must be able to solve in connection with the study of electronics in the chapters which follow. If you find them easy to solve, you will find the following chapters also easy and interesting. If you find them difficult, you are strongly advised to review your circuits course.

You are encouraged to use a spread sheet or a circuit analysis program wherever it is appropriate. A star following a problem or example indicates that a SuperCalc solution is in Appendix G or that a PSpice solution is in Appendix H.

1.7 Troubleshooting

You will find troubleshooting tips throughout this book. Troubleshooting is much like a game of logic in which the circuit provides the troubleshooter with certain clues, usually in the form of voltages or waveforms, from which hypotheses must be formulated and tested. The most successful troubleshooter is usually the person who can recognize both normal and abnormal circuit behavior and who can make the most sound logical inferences based on that information. No one ever gets too much experience in troubleshooting.

1.8 Summary

In this chapter we have listed some of the prerequisite skills which we will assume that you have brought to this course and we have described the procedures and practices we will follow. If you incorporate these skills, procedures, and practices into your mind set, you will find your study of electronics to be interesting and rewarding.

PROBLEMS

1. Find the value of the current I_5 in the circuit of Figure 1.1.

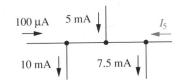

FIGURE 1.1 Problem 1.1.

2. Find the value of the current I_4 in the circuit of Figure 1.2.

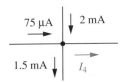

FIGURE 1.2 Problem 1.2.

3. Find the value of the voltage V_B in the circuit of Figure 1.3.

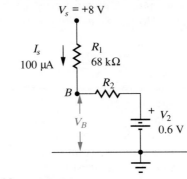

FIGURE 1.3   Problems 1.3, 1.4.

4. Find the value of the resistor R_2 in the circuit of Figure 1.3.
5. Find the value of the current I_L in the circuit of Figure 1.4.

FIGURE 1.4   Problems 1.5, 1.8.

6. Find the range of output voltage V_{out} available from the circuit of Figure 1.5.

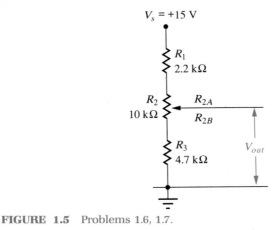

FIGURE 1.5   Problems 1.6, 1.7.

7. Find the setting of the 10 kΩ pot in Figure 1.5 (find the value of resistor R_{2B}) in order that $V_{out} = 7.5$ V.

8. Construct a Thevenin equivalent for the circuit "seen" by the load in Figure 1.4.

9. Construct a Thevenin equivalent for the circuit to the left of points A and B in Figure 1.6.

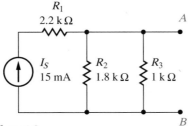

FIGURE 1.6   Problem 1.9.

10. Construct a Norton equivalent for the circuit "seen" by the load in Figure 1.4.

11. Construct a Norton equivalent for the circuit to the left of points A and B in Figure 1.6.

12. Find the value of the current I_5 in the circuit of Figure 1.7.

   ● (For a PSpice solution see Appendix H1.1) *Hint:* Simplify the circuit by constructing a Thevenin equivalent of the circuit to the left of resistor R_5 and constructing a Thevenin of the circuit to the right of resistor R_5.

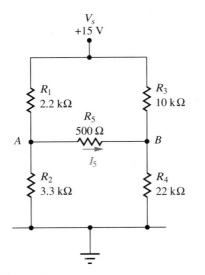

FIGURE 1.7   Problem 1.12.

13. Find the maximum power that can be delivered to the load in Figure 1.8.

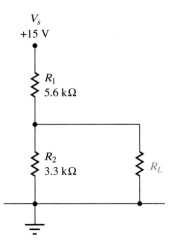

FIGURE 1.8   Problem 1.13.

14. Find the maximum power that can be drawn from terminals A and B in Figure 1.6.

15. Find the value of currents I_1 and I_2 in the circuit of Figure 1.9.

   ● (For a PSpice solution see Appendix H1.2)

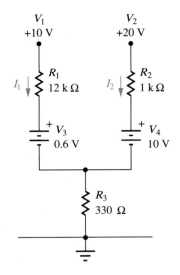

FIGURE 1.9   Problem 1.15.

16. Find the value of capacitor C in Figure 1.10 such that the voltage V_L will be no less than 14 volts 1/120 second after switch S is opened.

17. If a 600μF capacitor is used in Figure 1.10, find the value of the voltage V_L 1/60 second after switch S is opened.

   ● (For a PSpice solution see Appendix H)

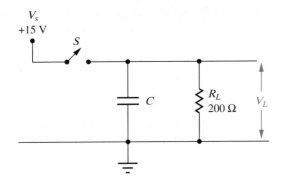

FIGURE 1.10   Problems 1.16, 1.17.

18. **Find the cutoff frequency of the circuit of Figure 1.11.**

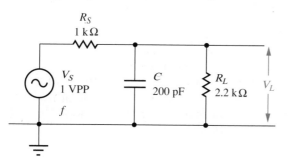

FIGURE 1.11   Problem 1.18.

19. **Figure 1.12 is a simplified version of the input to my stereo. What is its low frequency cutoff?**

● **(For a PSpice solution see Appendix H1.4)**

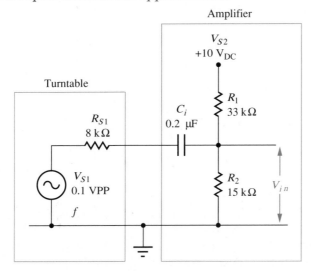

FIGURE 1.12   Problem 1.19.

20. At a frequency of 1 kHz, what will be the value of V_{in} for my stereo diagrammed in Figure 1.12 at the instant when the turntable voltage V_{S1} is:
    (a) At its positive peak value.
    (b) At its negative peak value.

Troubleshooting ▶

21. For the circuit of Figure 1.3, if V_S and V_2 are normal, what are the most likely faults (more than one fault could cause the indicated symptom) if:
    (a) $V_B = +8$ V
    (b) $V_B = +0.6$ V

22. Refine your answers to Problem 21 if:
    (a) I_s is very large.
    (b) $I_s = 0$ A

23. For the circuit of Figure 1.5, what are the most likely faults if $V_{out} = +15$ V regardless of the setting of the pot?

24. Refine your answer to Problem 23 if:
    (a) $+15$ V is read at both "top" and "bottom" of R_2.
    (b) $+15$ V is read at the "top" of R_2 and OV at the "bottom" of R_2.

25. Solve for V_A and V_B in Figure 1.7 if R_3 developed a "cold solder joint" and so became an open circuit.

PART I
DIODES

Physicists who invented the semiconductor: William
Shockley, Walter Brattain, and John Bardeen.
(Courtesy of UPI/Bettmann Newsphotos.)

A "clean room" where semiconductors are manufactured. (Courtesy of National Semiconductor.)

Chapter 2
The P-N Junction-A Model

SPECIAL TERMS

Periodic chart
Impurity atom
Hole or P carrier
Semiconductor
N type semiconductor
P type semiconductor
Depletion region
Majority carrier
Minority carrier
P-N junction
Barrier potential
Forward bias
Diode characteristic graph
Diode
Anode
Cathode
Reverse bias
Reverse saturation current (I_S)
Zener effect
Zener voltage

15

2.1 Introduction and Objectives

To fully describe the characteristics of the P-N junction one is tempted to talk about energy level diagrams, band gaps, and Fermi-Dirac statistics. The prerequisites for such a discussion would include an understanding of differential equations and quantum mechanics. Some of you may wish to take up those topics in more advanced courses, but that is not our objective here. In this chapter we will look at a very simplified (perhaps over simplified!), but intuitively appealing qualitative model to describe semiconductor behavior.

Since much of this chapter will be devoted to the development of a model, it might be well for you to return to Section 1.5 and reread what we said there about models. Since this is probably your first experience with semiconductors, our model will lean much more toward simplicity than toward accuracy. You will find the model presented here to be an aid to your understanding as we take up a more quantitative study of both diodes and transistors in later chapters.

After studying this chapter you will have a feeling for the behavior of a P-N junction and for the behavior of semiconductor devices in general and you should be able to:

- Describe in simplified form the structure of a silicon crystal and how both P type and N type semiconductors are formed.
- Describe the structure of a P-N junction and the distribution of carriers in a crystal containing a P-N junction under conditions of forward and reverse bias.
- Identify diode terminals and the symbol used to represent diodes.
- Recognize and use the diode characteristic graph.

2.2 Atoms and Crystals

Figure 2.1 contains a part of the periodic chart of the elements which is

I	II	III	IV	V	VI	VII	VIII
H Hydrogen							He Helium
Li Lithium	Be Beryllium	B Boron	C Carbon	N Nitrogen	O Oxygen	F Fluorine	Ne Neon
Na Sodium	Mg Magnesium	Al Aluminum	Si Silicon	P Phosphorus	S Sulfur	Cl Chlorine	Ar Argon
K Potassium	Ca Calcium	Ga Gallium	Ge Germanium	As Arsenic	Se Selenium	Br Bromine	Kr Krypton

FIGURE 2.1 A portion of the periodic chart of the chemical elements.

familiar to all chemistry students. The chemists would tell us that according to their model, atoms are composed of a positively charged nucleus surrounded by exactly the right number of electrons to make the atom electrically neutral. The electrons travel around the nucleus in carefully prescribed orbits. Our friends in chemistry would go on to tell us that those atoms which they place in the first column of their chart each have one electron in their outermost orbit. The atoms in column 2 each have two electrons in their outermost orbit, and so on until we reach the elements in column number 8 which have eight electrons in their outermost orbit.

The chemists would also tell us that since each of the elements in column 8 have eight electrons in their outermost orbit, they are chemically inert and do not tend to react with any other elements. Further, elements in the other columns tend to react with one another in ways which will result in eight outer electrons.

Thus, for example, ordinary table salt is formed when one atom of sodium (Na, column 1) combines with one atom of chlorine (Cl, column 7). The chlorine (Cl) atom takes the one electron from the sodium (Na) atom so that the chlorine atom will have eight outer electrons.

In general, atoms with more than four outer electrons tend to take electrons away from atoms that have less than four in order to develop the stable configuration of eight outer electrons. Any chemistry student can cite scores of examples of this tendency toward eight outer electrons.

Of more importance to us, let us look at what happens to the atoms of column 4. Do they tend to give away electrons or do they tend to pick up electrons? The answer is that they tend to share electrons. Using the chemists' model, Figure 2.2 is a two-dimensional representation of what a

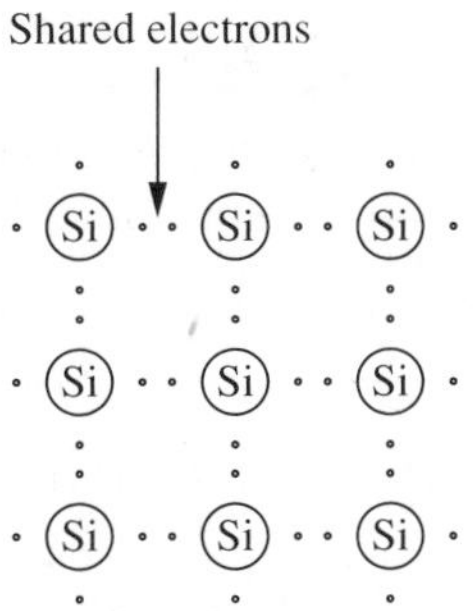

FIGURE 2.2 Silicon crystal structure demonstrating the sharing of electrons by atoms.

crystal of silicon (Si) would look like at the atomic level. The circles represent the nuclei of the atoms, and the dots represent the outer electrons. You will note that each silicon atom is shown with four outer electrons.

Since carbon (C) and germanium (Ge) are also in column 4, we could have used one of them for our example. In fact, germanium is used to make semiconductor devices. However, since silicon has better temperature stability, it is more commonly used today so we will use it in our example.

Note that although every atom diagrammed in Figure 2.2 has four outer electrons, the atom at the center of the diagram can share electrons with all four adjacent atoms so that it "thinks" that it has eight outer electrons. All of the atoms down inside the crystal will "feel" the same way. (We will not worry about the atoms at the surface of the crystal at this point.) It is this sharing of electrons that form the bonds which make a crystal of silicon a solid structure.

Since every electron is "tied down," there are no free electric charges that could conduct an electric current. For this reason a piece of pure silicon should be a perfect electrical insulator. Laboratory measurements show that this prediction is correct.

Of course, it is nice to be able to make insulators, but electronics would not be very interesting if that is all we could do. Let us look at what will happen when we introduce some "impurity atoms" into our crystal structure.

2.3 Holes and Electrons

Let us put our silicon crystal in an oven, heat it up to just below its melting temperature, and then introduce a vapor consisting of aluminum (Al) atoms. For our purposes, any element from column 3 could be used. The element the crystallographers actually use in this process is selected based on its compatibility with the rest of the crystal structure. The aluminum "impurity atoms" will now begin to diffuse into our crystal. We will stop the diffusion when our crystal looks like Figure 2.3.

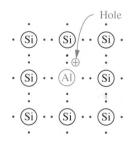

FIGURE 2.3 P type semiconductor showing an aluminum atom as the impurity atom which produces the positive carrier or "hole."

Notice that aluminum atoms have only three outer electrons (column 3) but our crystal structure demands four outer electrons to be shared with neighboring atoms. Thus the crystal structure is incomplete since it is short one electron. In Figure 2.3 this shortage is indicated by the $\oplus$ symbol. It is often called a "hole," or a "P carrier" for reasons that you will understand shortly.

It is important for you to realize that the $\oplus$ symbol does not represent an excess of positive charge in the crystal. We have not taken any electrons away from any atoms in the crystal so the net electrical charge is still zero. Rather, the $\oplus$ symbol indicates that the crystal structure is now incomplete. The silicon (Si) atoms in the crystal structure would "like" to see four bonding electrons but the electrically neutral aluminum (Al) atom can supply only three. Thus the "hole" or "P carrier."

Our crystal is no longer a perfect insulator. P carriers near the surface of the crystal can be "annihilated," or the holes can be "filled" by attaching a terminal to the crystal and forcing extra electrons into the crystal. In this way, the crystal becomes negatively charged. This negative charge soon prevents the addition of any more electrons unless we attach another terminal to the crystal and remove electrons. New P carriers will be created where the electrons are removed. Thus the P carriers appear to move through the crystal. The addition of the P carriers has made the crystal into an electrical conductor.

The crystal we have created is known as a "P type semiconductor" because its condition is due to positive P carriers or holes. When P type material conducts a current, we can think of the process as P carriers (holes) moving through the crystal in response to the applied electric field as shown in Figure 2.4.

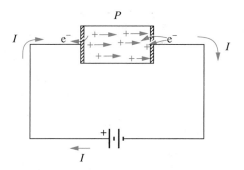

FIGURE 2.4   Positive carrier conduction within a P type semiconductor.

Since the P carriers are all that is of importance to us, the rest of the crystal structure has been omitted from Figure 2.4. Also, you must realize that atoms are very small. In the smallest crystal there are probably billions more carriers than could be drawn on Figure 2.4, but perhaps it will give you some idea of the processes taking place.

In spite of the belief, firmly held in some quarters, that there is no such thing as positive current flow, the physicists can perform experiments to show that positive carriers are "really" responsible for the condition in P type semiconductors—if there is such a thing as "really"!

In fact, the crystal is now a resistor whose resistance is determined by the size of the crystal and the amount of "impurity doping" it contains. This is one of the ways resistors are created in integrated circuits.

We have seen that introducing impurity atoms from column 3 of the periodic chart produces a P type semiconductor. Let us instead introduce impurity atoms from column 5. We will use phosphorus (P) for our example. Since phosphorus (P) has five outer electrons and only four of them are needed to form bonds with neighboring silicon atoms, we have one free electron. This free electron is represented by the ⊖ symbol in Figure 2.5.

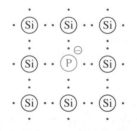

FIGURE 2.5 N type semiconductor showing a phosphorus atom as the impurity atom which supplies the negative carrier electron.

As was the case with the ⊕ symbol, again it is important to recognize that the ⊖ symbol does not represent an excess of negative charge, but rather an electron which is not part of the crystal structure and so is free to move about within the crystal. As you have probably guessed, this crystal is also a conductor and we have created an "N type semiconductor" in which conduction is due to N carriers (electrons).

2.4 The P-N Junction

Let us look at what will happen if P type carriers are present in one part of a crystal and N type carriers are present in an adjacent part of the crystal. We can visualize the creation of such a crystal thus: We will stand our pure silicon crystal on edge in the oven and introduce the aluminum (Al) vapor on one side while we introduce the phosphorus (P) vapor on the other side. The P and the N impurities will diffuse through the crystal toward each other. We will stop the process precisely when the impurities, diffusing into the crystal from opposite sides, meet in the center. The properties of the completed device will be critically dependent on our ability to accomplish this very difficult task exactly.

If we are very careful, we can form a sharply defined junction between the P and the N carriers. We have formed a "P-N junction." A P-N junction is diagrammed in Figure 2.6. As we mentioned before, Figure 2.6 can only

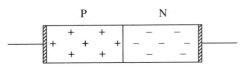

FIGURE 2.6 A P-N junction showing the P region on the right and the N region on the left.

give you a very rough idea of the distribution of the carriers within the crystal.

Again it is important to recognize that the crystal as pictured in Figure 2.6 is not carrying an electric charge. The + and − signs represent distribution of P and N carriers in the electrically neutral P and N regions of the crystal.

The P and the N carriers within our crystal will not be standing still. The physicists tell us that due to their temperature, they will be bouncing around in random fashion much like the billiard balls on a pool table after I shoot (I am not a very good shot!). What do you suppose will happen when an unlucky N carrier happens to stray into the P region and bump into a P carrier? You guessed it, unlike charges attract. They will combine. The result is the annihilation of the carriers nearest to the junction as shown in Figure 2.7. Since the formerly neutral P region has picked up

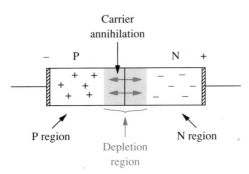

FIGURE 2.7 Depletion region near the P-N junction.

some negative carriers, it will become negatively charged. This overall negative charge will attract the remaining positive carriers in the P region back away from the junction.

Similarly, since the N region has lost some negative carriers, it becomes positively charged. This overall positive charge will attract the remaining negative carriers in the N region back away from the junction.

The net result of the above activity is that no carriers will remain near the junction. A "depletion region" will have been formed as shown in Figure 2.7. We know what that means: Wherever there are no carriers within the crystal, there can be no conduction. In the vicinity of the junction the crystal has become an insulator and no current can flow across the junction.

Obviously we have been describing the ideal situation, but the vast majority of carriers within our crystal must behave as we have described. In fact, they are known as "majority carriers." In the real world, "minority carriers" are also present. They are produced primarily by the thermal vibration of the atoms that make up the crystal so their effect is very temperature-dependent. As we shall see later, their effect is small.

2.5   Forward Bias

Let us hook our crystal up to a power supply and see what happens. First we will forward bias the junction. In other words, we will connect the positive battery terminal to the P region of the crystal and we will connect the negative battery terminal to the N region of the crystal. The circuit is shown in Figure 2.8.

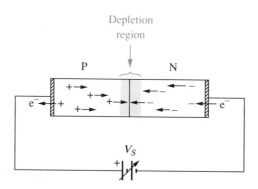

FIGURE 2.8   Forward biased P-N junction showing that forward bias tends to reduce the width of the depletion region.

We know that like charges repel and unlike charges attract. Thus, as we slowly increase the supply voltage V_S, within the P region, the P carriers will be repelled from the positive battery terminal and attracted toward the negative terminal. That is, they will be pushed toward the junction.

At the same time within the N region of the crystal, the N carriers will be repelled from the negative battery terminal and attracted toward the positive terminal. Thus they will also be pushed toward the junction. We are reducing the width of the depletion region.

Though a few carriers may be pushed from the battery into the crystal by the above procedure, no significant current will flow until sufficient voltage is applied to reduce the depletion region to zero width. At that point, carriers begin to diffuse across the junction, combinations begin to occur, and current begins to flow. As we continue to increase V_S, the current will increase very rapidly and nonlinearly. Figure 2.9 is a characteris-

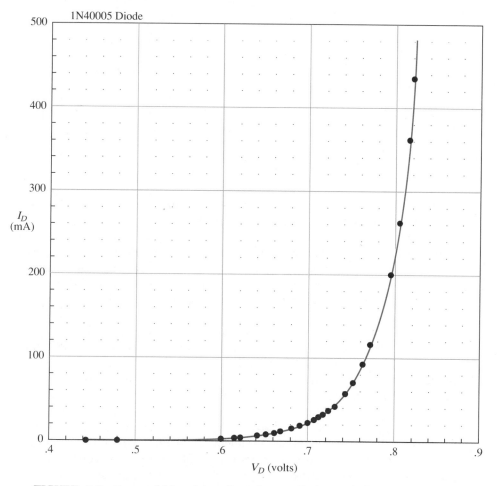

FIGURE 2.9   Forward biased junction characteristic graph for a silicon P-N junction.

tic graph drawn from experimental laboratory data. It shows the effect we have been talking about. Notice that almost no current flows until the voltage reaches nearly 0.6 V. Thus, the depletion region can be considered to be 0.6 V "wide," if we can measure that distance in terms of the voltage it represents.

If germanium had been used for the crystal structure, we would find the "width" of the depletion region to be about 0.2 V.

Notice how nonlinear the graph is. Between 0.6 V and 0.7 V the current flow only increases by about 20 mA, but between 0.7 V and 0.8 V, the current increases by something like 200 mA.

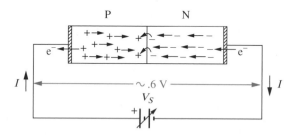

FIGURE 2.10   Conduction at the P-N junction within a forward biased silicon semiconductor.

Figure 2.10 illustrates how current flows under forward bias conditions. The external circuit removes electrons from the crystal structure at the terminal end of the P region leaving positive carriers (holes) behind. The positive carriers migrate through the P region toward the junction. At the junction they are annihilated by combining with negative carriers (electrons) which have migrated through the crystal from the terminal end of the N region where they entered the crystal from the external circuit.

The graph of Figure 2.9 tells us how much voltage you have to apply to force carriers back into the depletion region and so cause conduction to occur. This voltage is often called the "barrier potential." If you know the amount of current that is flowing through the crystal, you can read from the characteristic graph the amount of voltage that will be lost across the crystal.

Since graphs are hard to read and take up a lot of space, manufacturers rarely include them in their specification sheets. Usually they will provide you with one or two points on the graph and expect you to have a good enough mental picture of Figure 2.9 so that you can mentally create the rest of the graph. In Chapter 3 we will discover an equation that you can use in place of this graph in some common situations.

2.6 PSpice Diode Simulation

To be useful, a computer program for circuit analysis should be able to approximate the diode characteristic of Figure 2.9 fairly closely. Figure 2.11

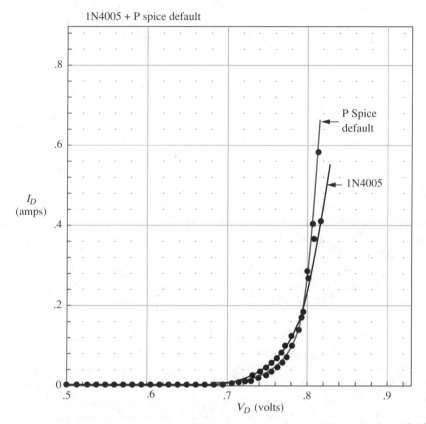

FIGURE 2.11 A comparison of laboratory data for a 1N4005 diode with the default PSpice diode model.

is the graph of Figure 2.9 with the default PSpice diode model superposed on it. Notice how closely PSpice models the real world. In Chapter 3 we will look at how such models are constructed. The rather simple PSpice program for producing Figure 2.11 is included in Appendix H.

2.7 Reverse Bias

Now let us examine the behavior of our crystal when we reverse bias it. In other words, we will reverse the power supply as shown in Figure 2.12.

As you remember from Section 2.4, even before we hooked our crystal

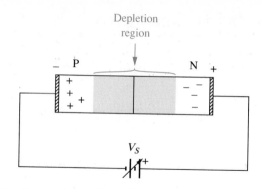

FIGURE 2.12   Reverse bias of a P-N junction.

to any external circuit, there was a nonuniform distribution of charge in the crystal due to the combinations that occurred at the junction when it was first formed. According to Figure 2.7, the P region became negatively charged and the N region became positively charged. That pulled the carriers back away from the junction causing the formation of the depletion region. Now we have hooked the negative terminal of the power supply to the P region. That will make the P region still more negatively charged. Similarly, the positive terminal of the power supply is connected to the N region, making it still more positively charged. Obviously, this will increase the width of the depletion region. Since there are no carriers in the depletion region, there can be no conduction; that is, ideally no conduction. Unfortunately, ideal conditions seldom exist. You may recall that toward the end of Section 2.4 we mentioned minority carriers. When the crystal is reverse biased, minority carriers cause a very small reverse current to flow. The graph of Figure 2.13 drawn from laboratory data shows that the reverse current is so very small that it can be ignored for most work. Notice that the vertical scale is in nanoamps. Also notice that the graph is nonlinear. A tangent to the curve (the dashed line in Figure 2.13) has an intercept of somewhere near $-8nA$ in this particular case. This current is known as the "reverse saturation current," I_S. As you would expect from the discussion in Section 2.4, the reverse saturation current is very temperature-dependent. In the next chapter you will find that although I_S is small, it is important in describing device behavior.

Figure 2.14 is a graph of the data from Figures 2.9 and 2.13 on the same scale. It will give you a better feel for how small the reverse current really is compared to the forward current. Notice that since the reverse current is measured in nanoamps, it is essentially zero on a graph scaled in milliamps.

We find that when we forward bias the junction, our crystal conducts

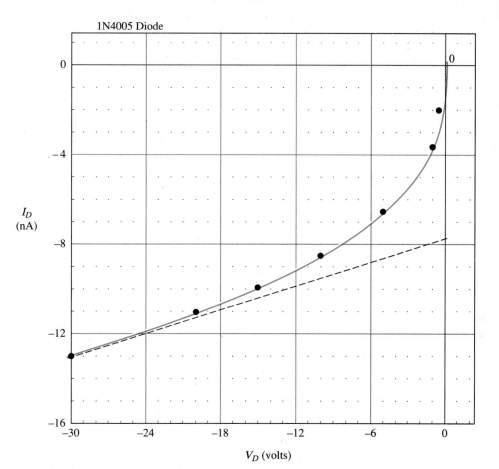

FIGURE 2.13 Reverse biased conduction characteristic for a 1N4005 diode. Note the vertical scale.

(with the loss of around 0.7 V) and when we reverse bias the junction, our crystal hardly conducts at all. We have created a diode. A diode does for electricity just what the plumber's check valve does for water. Current can flow one way but not the other.

2.8 Diode Identification

For your reference, Figure 2.15 shows first the P-N junction as we have been picturing it followed by the standard diode symbol used in circuit diagrams. Notice that positive current flow is in the direction of the arrow.

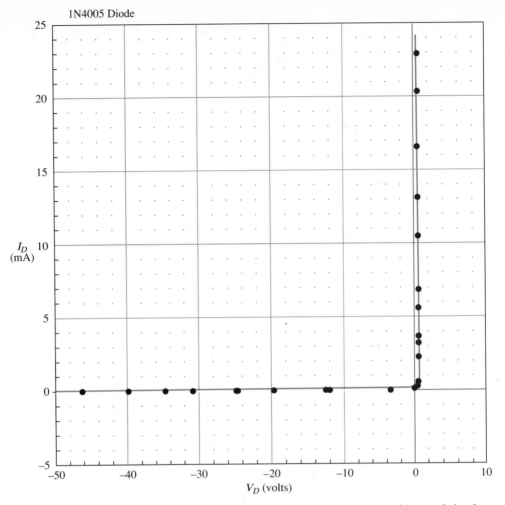

FIGURE 2.14 Junction characteristic showing both reverse bias and the forward bias on the same scale.

Using terminology borrowed from the chemists, the terminal connected to the P region is known as the "anode" and the terminal connected to the N region is known as the "cathode." Thus current flows through the diode from the anode to the cathode.

Included in Figure 2.15 is a diagram showing several common ways in which diodes are packaged. You will notice that in every case the cathode is identified in some way. It may be by a band of paint, even just a spot of paint, or a difference in shape, but the cathode is the end that is different.

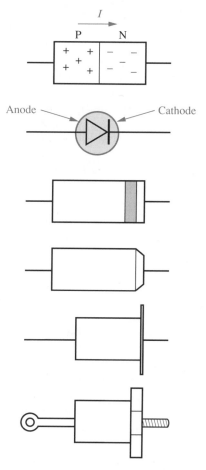

FIGURE 2.15 Diode symbols and their physical appearance.

2.9 Ohmmeter Tests

You should recall from your circuits class that an ohmmeter contains an internal DC power source which supplies current to the device under test. It measures the amount of voltage and current and it does an Ohm's Law calculation to tell you the resistance of the device. Since resistors are quite linear, it does not matter whether the ohmmeter applies a high voltage and reads a high current, or if it applies a low voltage and reads a low current. And it certainly does not matter about the polarity. The resistance will be the same.

As you can see from what we have said so far, diodes are quite different. If the positive ohmmeter lead is connected to the cathode lead of the diode and the negative lead is connected to the anode (see Figure 2.15), the junction will be reverse biased. Virtually no current will flow so the ohmmeter will read infinite (many megohms) resistance.

If the positive ohmmeter lead is connected to the anode lead of the diode, the junction will be forward biased. The amount of voltage supplied by the ohmmeter will determine whether the diode will conduct. If the ohmmeter power supply delivers 0.5 V or less, virtually no current will flow through a silicon diode. You will still get a reading of infinite resistance. However, if the ohmmeter power supply delivers a voltage of 1.5 V, the diode will drop nearly half of the supply voltage (around 0.7 V according to Figure 2.9) and so, almost independent of the scale you use, the meter will show a reading near midscale.

When testing circuits, sometimes it is desirable that any diodes or transistors in the circuit not conduct. Many modern ohmmeters have scales which use power supply voltages below the turn-on voltage of silicon diodes in order to accommodate this need. At other times you want any junctions in the circuit to conduct. To meet this need, many manufacturers have higher voltage "diode test" ranges built into their ohmmeters as well.

Incidentally, you should be aware that not all meters use the "right" polarity (red is +) on ohms scale. In order to simplify the circuit layout, some use a reverse polarity.

If you have paid attention to the details of the preceding discussion, it should be obvious to you that you can use an ohmmeter to make some simple diode tests. The better you become acquainted with the behavior of a given meter, the more sophisticated those tests can become.

2.10 Reading the Diode Characteristic Graph

The use of the diode characteristic graph is illustrated by the following example:

EXAMPLE 2.1
Using data from the diode characteristic graph of Figure 2.9, determine the power supply voltage V_S needed for a current I_D of (a) 20 mA (b) 300 mA in the circuit of Figure 2.16.

● (For a PSpice solution see Appendix H.)

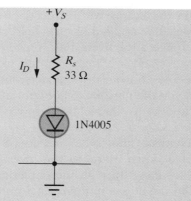

FIGURE 2.16   Example 2.1.

SOLUTION:

(a)   From Figure 2.9 for $I_D = 20$ mA, $V_D = 0.7$ V. Doing a KVL walk from V_S to ground in Figure 2.16 gives

$$V_S - I_D R_S - V_D = 0 \text{ V} \qquad (2.1)$$

or

$$V_S = I_D R_S + V_D \qquad (2.2)$$
$$= 20 \text{ mA} * 33 \ \Omega + 0.7 \text{ V} = 1.36 \text{ V}$$

(b)   From Figure 2.9 for $I_D = 300$ mA, $V_D = 0.81$ V. Substituting these values in Equation 2.2 gives

$$V_S = 300 \text{ mA} * 33 \ \Omega + 0.81 \text{ V} = 10.71 \text{ V}$$

Later we will find that the P-N junction is an important feature of transistors as well as diodes. Since it is so important, it will be well worth our time to study P-N junction characteristics more quantitatively. We will do that in the next chapter. But first let us look at what happens as we increase the reverse voltage across the diode.

2.11   Reverse Breakdown

As we have already seen, the higher the reverse voltage across the diode, the wider the depletion region becomes. Can this go on forever? Obviously not. The physical dimensions of the crystal place an upper limit on the width of the depletion region. What happens when the reverse voltage is pushed beyond the maximum value the diode can block? You are probably thinking, "Fire, smoke, and goodbye diode!" but that is not necessarily the case.

Let us look at the atoms within the depletion region as the reverse voltage is increased. The applied electric field causes the electrons within the atoms to be drawn away from their nuclei more and more strongly. If the reverse voltage reaches a high enough value, some of them finally break completely free and negative carriers suddenly appear throughout the crystal. The diode then begins to conduct in the reverse direction. This phenomenon is known as "Zener breakdown" named after a pioneer in the field.

Another mechanism known as "avalanche breakdown" also contributes to reverse breakdown. Here, as the reverse voltage is increased, the minority carriers traveling through the depletion region collide more and more violently with the atoms of the crystal until they finally begin knocking electrons loose from them. Those electrons knock yet other electrons loose and the avalanche begins. Since the avalanche is initiated by the minority carriers, you should expect it to be temperature-dependent, and it is. Once breakdown begins, a very small increase in reverse voltage causes a very large increase in reverse current. Perhaps you could view the situation much like ducks on a pond. When one flies, they all go. The voltage at which breakdown occurs is known as the *Zener voltage.*

Why does this breakdown not destroy the crystal? Remember that $P = VI$. So as long as the diode does not conduct any significant current, it will dissipate very little power even though the reverse voltage may be quite high. However, when breakdown occurs and current begins to flow, power is consumed and heat is produced. The heat must be dissipated by the diode and this causes its temperature to begin to rise.

If the current is so high that the diode cannot get rid of the heat fast enough, the temperature may rise so high that the crystal structure begins to break down. That is, the crystal begins to melt. Atoms begin to move around in the crystal and that nice sharp junction the manufacturer worked so hard to make begins to get fuzzy. That can drastically alter the characteristics of the P-N junction and render the diode useless.

On the other hand, if we limit the current, we can prevent the diode from overheating. Then when we return the reverse voltage to a value that is below the Zener voltage, the atoms collect their normal family of electrons again and, since none of the atoms have moved around within the crystal, the diode is unaffected by the experience.

In fact, there is a whole class of diodes that is specifically designed to be operated in the reverse breakdown region. As you may have guessed, they are known as *Zener diodes.* Figure 2.17 shows the characteristic graph for a typical Zener diode. If you compare the forward biased portion of this diode characteristic graph (the first quadrant) with that of the conventional diode characteristic graph shown in Figure 2.9, you will discover that in the forward biased direction, a Zener diode has the same sort of characteristic graph as any other diode although it is almost never operated in that region.

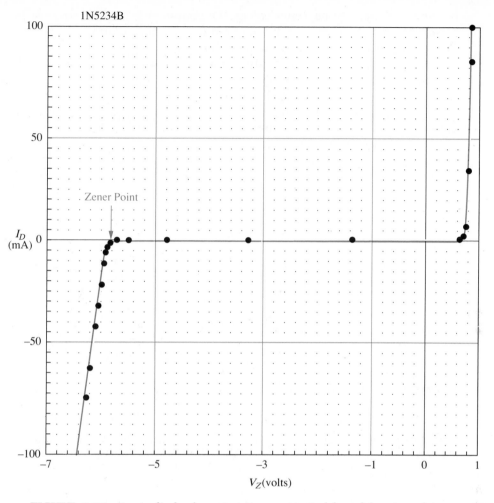

FIGURE 2.17 Zener diode characteristic constructed from laboratory data.

One difference between a Zener diode and a conventional diode is that in the Zener diode, the width of the crystal is carefully controlled so that the Zener diode will have exactly the desired Zener voltage. The manufacturers of conventional diodes do not try to control the exact value of the Zener voltage of their diodes. Their main concern is that the Zener voltage be well beyond the maximum reverse voltage that their diodes will ever encounter.

Another difference between a Zener diode and a conventional diode is that the Zener diode is specifically designed to have as near to a vertical characteristic graph in the reverse breakdown region as possible. This characteristic is important because ideally, whether the Zener is conducting a small or a large reverse current, the voltage loss across it will always be the same, namely its Zener voltage. Of course, this is of no concern to manufacturers of conventional diodes either since they do not expect their diodes to ever be operated in the Zener region.

As you might guess, due to the extra care needed in its manufacture, a Zener diode costs considerably more than a conventional diode.

Zener diodes are packaged in the same way as conventional diodes. Thus it is impossible to distinguish between them except by referring the part number printed on them to the manufacturer's data. The Zener diode symbol used in circuit diagrams is different from the conventional diode symbol however. In Figure 2.18 the "wings" on the cathode of the diode symbol identify it as a Zener.

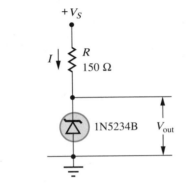

FIGURE 2.18   Example 2.2.

We will look more carefully at circuits containing Zener diodes in Chapter 7.

EXAMPLE 2.2

Using data from the Zener characteristic of Figure 2.17, determine over what range of voltages the power supply V_S of Figure 2.18 can vary if the minimum allowable voltage for V_{out} is 6 V and the maximum allowable V_{out} is 6.2 V.

SOLUTION:

MINIMUM V_{out}:
As usual the Zener is reverse biased so all Zener voltages and currents will be negative. Looking at Figure 2.17 we see that for the $V_Z = -6$ V, $I_Z = -20$ mA. To carry along a lot of negative signs is likely to be confusing so we will recognize that both V_{out} and I are positive values. Doing a KVL walk from V_S to ground gives

$$V_S - IR - V_{out} = 0 \text{ V} \tag{2.3}$$

or

$$V_S = IR + V_{out} \tag{2.4}$$

so

$$V_{Smin} = 20 \text{ mA} \cdot 150 \text{ } \Omega + 6 \text{ V} = 9 \text{ V}$$

MAXIMUM V_{out}:

Looking again at Figure 2.17 we see that for $V_Z = -6.2$ V $I_Z = -65$ mA. Substituting these values into Equation 2.4 gives

$$V_{Smax} = 65 \text{ mA} \cdot 150 \text{ } \Omega + 6.2 \text{ V} = 16.0 \text{ V}$$

As you can see, the voltage regulator circuit of Example 2.2 could be used to obtain a fairly constant 6 V from a rather variable voltage source such as your car's electrical system. Voltage regulator circuits will be looked at more carefully in Chapters 7, 13, and 21.

2.12 Troubleshooting

◄◄◄◄◄◄◄◄

From what we have said so far, you should recognize that unless it is carrying a very large current, if the forward voltage across a diode is much more than about 0.7 V, it is probably defective. Also, if the reverse current through a conventional diode is measurable, or if the reverse voltage across a Zener diode is not the value listed in the data books, it is probably defective. However, the in-circuit measurement of these quantities is not always easy to do. In the chapters that follow we will look more carefully at how such measurements can be made. In any case, if the condition of a diode is suspect, it is always best to free it from the circuit and measure its characteristics.

2.13 Summary

This chapter has introduced the ideas of crystal structure, impurity doping, semiconductor and P-N junction behavior. We have examined the properties of the P-N junction under conditions of both forward and reverse bias and we have examined the P-N junction characteristic graph. You should now be able to recognize diodes both by their appearance and by their circuit symbol and you should be able to determine their polarity using an ohmmeter. You will become better acquainted with diodes in Chapter 3.

GLOSSARY

Anode: The electrical terminal of a device which carries current from the external world into the device. (Section 2.8)

Barrier potential: The forward voltage that must be applied to a junction to cause conduction to be imminent. (Section 2.5)

Cathode: The electrical terminal of a device which carries current from the device out to the external circuit. (Section 2.8)

Depletion region: That region within a semiconductor where combinations have annihilated the carriers that would otherwise be present. (Section 2.4)

Diode: A two-terminal device which permits current flow in one direction but not in the other. (Section 2.7)

Diode characteristic graph: A graph of the relation that exists between diode current (vertical axis) and the voltage loss across the diode (horizontal axis). (Section 2.5)

Forward bias: The application of voltage to a device in the direction current would normally flow. (Section 2.5)

Hole or P carrier: A shortage of one electron in the crystal structure of a semiconductor. (Section 2.3)

Impurity atom: An atom with either three or five outer electrons which is introduced into a semiconductor crystal, changing the electrical characteristics of the crystal. (Section 2.2)

Majority carrier: A carrier that is in the "correct" part of the crystal; for example, a P carrier in the P region and an N carrier in the N region. (Section 2.4)

Minority carrier: A carrier that is in the "wrong" part of the crystal; for example, a P carrier in the N region and an N carrier in the P region. (Section 2.4)

N type semiconductor: A semiconductor whose conduction is due to the movement of free electrons within the semiconductor. (Section 2.3)

Periodic chart: A systematic way of displaying the name and properties of each of the chemical elements. The location of each element is determined by the configuration of its orbital electrons. (Section 2.2)

P-N junction: A place within a semiconductor where P type carriers and N type carriers interface. (Section 2.4)

P type semiconductor: A semiconductor whose conduction is due to "hole," or "P carrier" movement within the semiconductor. (Section 2.3)

Reverse bias: The application of voltage to a device in the direction opposite to the direction which current would normally flow. (Section 2.7)

Reverse saturation current (I_s):   The current that flows across a reverse biased P-N junction (Section 2.7)

Semiconductor: Material that has some of the electrical properties of a conductor and some of the properties of an insulator. (Section 2.3)

Zener effect: The combination of several processes whose effect is to cause a P-N junction to suddenly begin to conduct as its reverse bias is increased. (Section 2.11)

Zener voltage: The voltage required to cause a Zener diode to conduct in the reverse direction. (Section 2.11)

PROBLEMS

Section 2.5

1. From Figure 2.9, estimate I_D when V_D has the following values:
   (a) 0.5 V   (b) 0.6 V   (c) 0.7 V   (d) 0.8 V   (e) 0.9 V
2. According to Figure 2.9 what is the increase in diode current for a 1N4005 when V_D goes from:
   (a) $V_D = 0.6$ V to $V_D = 0.62$ V
   (b) $V_D = 0.8$ V to $V_D = 0.82$ V
3. Using data from the graph of Figure 2.9, calculate the value of resistor R in the circuit of Figure 2.19.

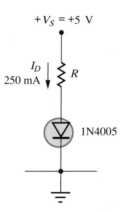

FIGURE 2.19   Problem 2.3.

4. In the circuit of Figure 2.19, how much power will be dissipated by:
   (a) The diode
   (b) The resistor.

Section 2.7

5. Using the graph of Figure 2.13 or Figure 2.14, determine the current that will flow in resistor R of Problem 2.3 if the polarity of the supply V_S is reversed.

6. According to Figure 2.13 or Figure 2.14 what is the increase in diode current for a 1N4005 when V_D goes from:
   (a) $V_D = -0.6$ V to $V_D = -0.62$ V
   (b) $V_D = -0.8$ V to $V_D = -0.82$ V.

Section 2.10

7. The maximum DC current rating for a 1N4005 diode is 1 amp. Using data from the graph of Figure 2.9, calculate the maximum DC power rating for the diode. You may have to do a little extrapolation!

8. Using data from the graph of Figure 2.17, find the value of V_{out} in Figure 2.20.

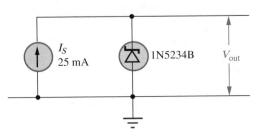

FIGURE 2.20   Problem 2.8.

9. Using data from the graph of Figure 2.17, find the supply voltage V_S required in Figure 2.21.

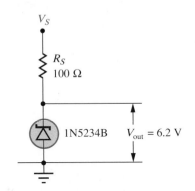

FIGURE 2.21   Problems 2.9, 2.10.

Section 2.11

10. Find the value of load resistor R_L that must be connected to the output of the circuit in Figure 2.21 in order to pull the output voltage V_{out} down to 6.0 V if the supply voltage V_S is 12.2 V.

Section 2.12

11. Using the results of Problem 3, estimate the value of I_D if the 1N4005 in Figure 2.19 became:
    (a) A short circuit.
    (b) An open circuit.
12. Using the results of Problem 5, estimate the voltage you would read across the 1N4005 in Figure 2.19 if the polarity of the power supply was reversed.
13. Estimate the value of V_{OUT} in Figure 2.20 if the 1N5234B became:
    (a) A short circuit.
    (b) An open circuit.
14. Using the results of Problem 9, determine the value of V_{OUT} in Figure 2.21 if the 1N5234B became:
    (a) A short circuit.
    (b) An open circuit.

Semiconductor devices mounted on a circuit board.
(Photograph by Chris Rogers/The Stock Market.)

Chapter 3
Diodes and Diode Models

SPECIAL TERMS

Load line
Q point
Iteration
Data sheet
Dynamic resistance (r_D)
Thermal runaway

3.1 Introduction and Objectives

In Chapter 2 we developed a rather crude qualitative model of the P-N junction. We also looked at typical diode forward and reverse characteristic graphs. Using those graphs we could answer some rather carefully chosen questions about how diode circuits behave. Did you notice that we did not try to answer questions like: If the supply voltage V_S of Figure 2.16 had a value of 2 V, what would be the diode voltage V_D and the diode current I_D? In this chapter you will learn several different ways to handle such practical problems and something about which approach is best in a given situation.

Although we will be talking about diode behavior in this chapter, you should be aware that what you learn here is applicable to transistors and other devices that contain P-N junctions.

At the conclusion of this chapter you should have a quantitative understanding of the behavior of the P-N junction and you should be able to:

- Solve practical circuits containing diodes using any one of the three or four models presented.
- Determine the important junction parameters from data contained on the manufacturer's data sheets.
- Determine the value of the dynamic diode resistance, r_D, both experimentally and theoretically.

3.2 Two Unknowns

Did you notice that there are two unknowns in the problem posed in the preceding section? Your algebra teacher would say that to solve for two unknowns you need two equations in V_D and I_D, or maybe two graphs

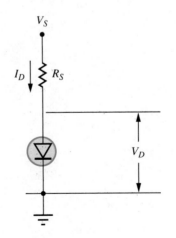

FIGURE 3.1 Diode circuit.

whose intersection will give V_D and I_D. Our knowledge of circuits will give us one of the relationships we need: Doing a KVL walk from V_S to ground in Figure 3.1 gives

$$V_S - I_D R_S - V_D = 0 \text{ V}$$

$$I_D = \frac{V_S - V_D}{R_s} \tag{3.1}$$

Equation 3.1 describes how the CIRCUIT functions in terms of V_D and I_D. This is one of the relationships we need. To obtain the second relationship we will look at how the DIODE functions in terms of V_D and I_D. This will involve developing a model that will describe the diode. We will look at four different models. The one that you should use in a given situation will be determined by how precisely you wish to describe the behavior of the diode.

3.3   Model 0: The Ideal Model

The simplest diode model is what we might call no model at all. That is, assume that the diode is a perfect conductor in the forward direction and a perfect insulator in the reverse direction. This takes care of one of the unknowns very easily, if not very accurately: $V_D = 0$ V. One look at the graph in Figure 2.9 tells you that this model is not very accurate in circuits where a few tenths of a volt are important. In circuits where you are dealing with voltages in the order of hundreds of volts, or where great accuracy is not important, this "quick and dirty" approach is not bad. All that is left is to solve the circuit equation for the current just as though the diode did not exist.

3.4   Model I: The Fixed Voltage Model

If you recall the shape of the diode characteristic graph in Figure 2.9 you will remember that for silicon diodes, as we have said before, the diode voltage is between 0.6 V and 0.8 V unless the current is very small or very large. Thus you will usually not make an error of much more than 0.1 V if you assume a fixed diode voltage of 0.7 V for a typical silicon diode. For a germanium diode the typical diode forward voltage is about 0.3 V. This fixed voltage assumption gives us a model that is more accurate than the ideal model, and yet it is very easy to use. It immediately gives us $V_D = 0.7$ V which is one of the unknowns we need to find. A KVL analysis of the circuit will provide us a circuit equation which we can solve to obtain the current. There are a lot of circuits for which this approximation is close enough.

EXAMPLE 3.1

Find the diode voltage and current in Figure 3.2.

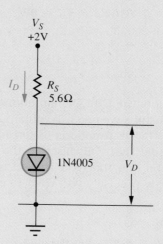

FIGURE 3.2   Example 3.1.

SOLUTION:

Obviously, the ideal diode model could not be used for this circuit because V_S is so small that to ignore V_D would introduce an unacceptably large error.

The fixed voltage model assumes the solution to the first part of the problem:

$$V_D = 0.7 \text{ V}$$

To find the current, we take a KVL walk from V_S down through the circuit to ground. This gives us

$$V_S - I_D R_S - V_D = 0 \text{ V}$$

Solving for current we have

$$I_D = \frac{V_S - V_D}{R_S} \tag{3.1}$$

$$= \frac{2 \text{ V} - 0.7 \text{ V}}{5.6 \ \Omega} = 232 \text{ mA}$$

3.5 Model II: The Graphical Model

Occasionally we would like to get a little better estimate of the diode voltage and current than the fixed voltage model provides. The most accurate way to describe any device is go to the lab, take data, and construct a graph of its behavior. Figure 2.9 contains such a graph for a typical 1N4005 diode. You might wonder if that graph fits YOUR "typical" 1N4005, but ignoring that question, the graph provides one relation between V_D and I_D. The KVL circuit equation provides the second relation. Thus we have enough information to solve for the two unknowns. The problem is, how do you solve an equation and a graph? The graphical model requires that we construct a graph of the circuit equation. The solution will then be the point of intersection of the two graphs.

EXAMPLE 3.2

Repeat Example 3.1 using the graphical model.

SOLUTION:

Since the circuit exclusive of the diode is composed of linear devices (V_S and R_S), the circuit equation, Equation 3.1 is also linear. In fact, we can easily write it in the form

$$y = (m)x + b$$
$$I_D = -\frac{1}{R_S}V_D + \frac{V_S}{R_S} \tag{3.2}$$

Thus the graph that describes the circuit operation will be a straight line. All we need in order to draw the graph is two points and a straight edge. From the circuit of Figure 3.2 we have the circuit equation

$$I_D = \frac{V_S - V_D}{R_S} \tag{3.1}$$

$$= \frac{2\text{ V} - V_D}{5.6\ \Omega} \tag{3.3}$$

Substituting any two values for V_D into Equation 3.3 and solving for the corresponding values for I_D will give us the pair of points we need.

Obviously, to achieve the best accuracy in drawing the straight line, the points should be chosen as far apart as possible. Since the horizontal axis of Figure 2.9 begins at $V_D = 0.4$ V let us use that value of V_D for our first point. Then solving for I_D we get

$$I_{D1} = \frac{2 \text{ V} - V_{D1}}{5.6 \ \Omega} \tag{3.3}$$

$$= \frac{2 \text{ V} - 0.4 \text{ V}}{5.6 \ \Omega} = 286 \text{ mA}$$

Figure 2.9 has been duplicated here as Figure 3.3 where the above point is identified as Point 1.

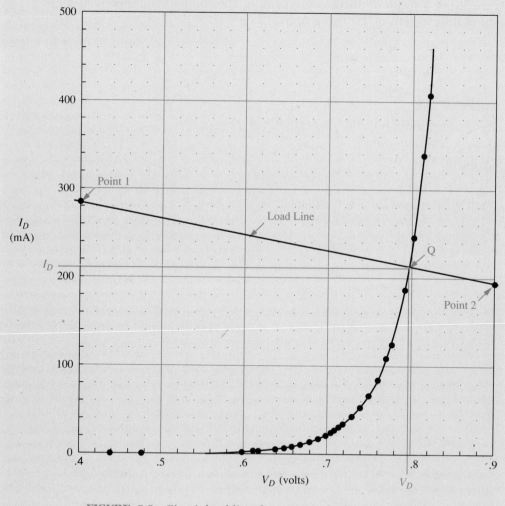

FIGURE 3.3 Circuit load line drawn on a diode characteristic graph.

Since the horizontal axis of Figure 3.3 only extends to $V_D = 0.9$ V, let us use that value for our second point. Again solving for the corresponding value of I_D we get

$$I_D = \frac{2 \text{ V} - V_{D2}}{5.6 \ \Omega} \tag{3.3}$$

$$= \frac{2 \text{ V} - 0.9 \text{ V}}{5.6 \ \Omega} = 196 \text{ mA}$$

The above point is identified in Figure 3.3 as Point 2. The straight line joining the above points is known as the "Load Line" for this circuit. Ohm's Law and the given circuit parameters V_S and R_S require that the circuit must always operate somewhere along the load line. The operation of almost all electronic devices can be described by a load line.

Notice that the circuit in effect says, "The operation must be somewhere along the load line." But the diode says, "No, the operation must be somewhere along the characteristic curve." Where can these two requirements both be met at the same time? Obviously at the intersection of the two lines. That point is known as the Operating Point, the "Quiescent Point" or the "Q Point." It is the quiescent point because everything is "quiet." That is, it is a DC condition. A familiar example would be when you have turned on your stereo but before you have put a record on the turntable. The circuits are all operating, current is flowing, but no signal is present. It is quiet (and peaceful!!). All of the transistors and signal diodes are operating at their Q points.

The Q point gives us the value for V_D and I_D that we seek. Reading from the graph of Figure 3.3, at the Q point, $V_D \approx 0.795$ V and $I_D \approx 215$ mA.

You might notice that the above values do not agree very well with those obtained from the fixed voltage model. This illustrates the fact that when a tenth of a volt is significant, the fixed voltage model is not the best choice.

The graphical model is probably as close to "reality" as we can get. The difficulty with its use is that you usually do not have a characteristic graph for the diode and to draw one is time-consuming. A curve tracer helps a great deal. It is a special purpose oscilloscope specifically designed to display device characteristic graphs quickly and easily. The serious circuit designer should not be without one.

3.6 Model III: The Junction Equation

Since graphs take up a lot of space and are expensive to construct and print, device manufacturers usually prefer to provide data about their devices in numerical form. Further, since calculators and computers work much better with equations than with graphs, designers often find it desirable to express device behavior in the form of an equation. Thus instead of converting the circuit equation into a load line as we did in Section 3.5 we will occasionally find it desirable to convert the characteristic graph of Figure 2.9 into an equation.

The solid state physicists tell us that theoretically the junction equation should have the form

$$I_D = I_S(e^{V_D/V_T} - 1)$$

For actual devices at forward voltages greater than about $+0.1$ V we can say

$$I_D \approx I_S e^{V_D/V_T} \tag{3.4}$$

A more useful form of the above equation is

$$V_D = V_T \ln\frac{I_D}{I_S} \tag{3.5}$$

V_T in the above equations represents the thermal energy possessed by the carriers within the diode crystal. It is given by

$$V_T = \frac{kT}{q} \tag{3.6}$$

where: k = Boltzmann's constant ($k = 1.38 \times 10^{-23}$ Joules/°K)
q = electron charge ($q = 1.6 \times 10^{-19}$ Coulomb)
T = absolute temperature

As you can see, V_T is determined by some of the most fundamental constants of nature. The trouble is, in order to have a number that we can use, we need to know the absolute temperature—the temperature of the carriers as they bounce around inside the crystal. Assuming that the carriers are at room temperature ($T = 300°K$) gives: $V_T = 25.9$ mV. It seems reasonable that the carriers inside the crystal might be considerably "hotter" than that as they are being pushed by the external electric field. Further, the equation assumes a uniform distribution of carriers on each side of the junction with a sharp discontinuity at the junction. That assumption is probably not too good for practical devices either. Thus V_T is almost always greater than 25.9 mV.

PSpice provides an "emission coefficient," N, by which the theoretical V_T is multiplied in order to account for these real world departures from ideality. Its value must be experimentally determined and is usually between 1 and 2. The default value used by PSpice is $N = 1$.

The I_S factor in Equation 3.5 is the "reverse saturation current" that results from the minority carriers which we first talked about in Section 2.4. Experiments show that if the theoretical value of $N = 1$ and $V_T = 25.9$ mV is used, it is reasonable to use a value of $I_S = 10$ fA (f = "femto-" = 10^{-15}). This is the default value used by PSpice. It is important to notice that if an empirical value is used for NV_T, then a corresponding empirical value for I_S must also be calculated. For approximations that are within about 0.1 V of typical experimentally observed values, it is safe to use

$$N = 1 \qquad V_T \approx 26 \text{ mV} \qquad I_S \approx 10 \text{ fA}$$

3.7 Determining NV_T and I_S

Occasionally it may be desirable to describe a specific device more precisely than the $NV_T = 26$ mV and $I_S = 10$ fA presented in Section 3.6. Due to the complexity of the situation it is practically impossible to determine meaningful values for NV_T and I_S from theory. Thus we must resort to laboratory measurement to obtain the parameters we need.

Since we must solve for two unknowns, we need two relationships. Fortunately, many manufacturers recognize this and so they provide at least two different sets of values for V_D and I_D. We will refer to them as (V_{D1}, I_{D1}) and (V_{D2}, I_{D2}). Inserting this information in the following equations gives us NV_T and I_S:

$$NV_T = \frac{V_{D2} - V_{D1}}{1n\dfrac{I_{D2}}{I_{D1}}} \tag{3.7}$$

$$I_S = \frac{I_D}{e^{V_D/(NV_T)}} \tag{3.8}$$

The above formulas are derived in Appendix E.

EXAMPLE 3.3
Determine the junction equation for a 1N4005 diode given the following voltage and current values: $V_{D1} = 0.6$ V, $I_{D1} = 2.3$ mA and $V_{D2} = 0.8$ V, $I_{D2} = 245$ mA.

SOLUTION:

Substituting in Equation 3.7

$$NV_T = \frac{V_{D2} - V_{D1}}{\ln\dfrac{I_{D2}}{I_{D1}}}$$

$$= \frac{0.8 \text{ V} - 0.6 \text{ V}}{\ln\dfrac{245 \text{ mA}}{2.3 \text{ mA}}} = 42.8 \text{ mV} \tag{3.7}$$

Substituting the above value for NV_T and the coordinates of either point 1 or point 2 in Equation 3.8 gives

$$I_S = \frac{I_D}{e^{V_D/(NV_T)}}$$

$$= \frac{245 \text{ mA}}{e^{0.8 \text{ V}/42.8 \text{ mV}}} = 1.9 \text{ nA} \tag{3.8}$$

Thus the empirical junction equation for the 1N4005 is:

$$V_D = NV_T \ln\frac{I_D}{I_S} \tag{3.5}$$

$$= 42.8 \text{ mV} * \ln\frac{I_D}{1.90 \text{ nA}} \tag{3.9}$$

Again, you should notice that the empirical values obtained for NV_T and I_S are strongly interdependent. Thus you cannot use a theoretical value for one and an empirical value for the other without getting into trouble.

In the event that the manufacturer fails to provide the necessary operating information you are left to guess at V_D or if your application is critical, get a handful of samples, go to the lab and get the information yourself.

Although an empirical equation like Equation 3.9 would probably come closer to "truth" for most diodes, the equation most commonly used, and therefore the equation with which you should become familiar is the theoretical junction equation

$$V_D = 26 \text{ mV} * \ln\frac{I_D}{10 \text{ fA}} \tag{3.10}$$

For typical germanium diodes, V_T is the same as for silicon, but $I_S \approx 1 \text{ μA}$.

3.8 Using the Junction Equation

EXAMPLE 3.4

Use the empirical junction equation, Equation 3.9 to calculate the diode voltage V_D for a 1N4005 when the diode current is:

(a) $I_D = 20$ mA

(b) $I_D = 300$ mA

● (For a PSpice solution, see Appendix H.)

SOLUTION:

(a) $I_D = 20$ ma

$$V_D = 42.8 \text{ mV} * 1\text{n}\frac{I_D}{1.90 \text{ nA}} \qquad (3.9)$$

$$= 42.8 \text{ mV} * 1\text{n}\frac{20 \text{ mA}}{1.90 \text{ nA}} = 0.692 \text{ V}$$

The theoretical equation, Equation 3.10, yields $V_D = 0.736$ V.

(b) $I_D = 300$ mA

$$V_D = 42.8 \text{ mV} * 1\text{n}\frac{300 \text{ mA}}{1.90 \text{ nA}} = 0.809 \text{ V}$$

The theoretical equation, Equation 3.10, yields $V_D = 0.807$ V.

The above calculated values agree fairly well with the values we have read from the graph in Figure 2.9 (See Example 2.1). As you can see, the empirical equation works very well for the 1N4005 diode and the theoretical equation works almost as well.

Let us return to the problem with which we opened this chapter: Given the circuit of Figure 3.2, find the diode voltage V_D and the diode current I_D. Now that we have the junction equation we should not have to resort to the graph of Figure 2.9. The two equations we must solve are the KVL equation which describes the circuit and the junction equation which describes the diode. They are

$$I_D = \frac{V_S - V_D}{R_S} \qquad (3.1)$$

$$= \frac{2\text{V} - V_D}{5.6 \ \Omega} \qquad (3.11)$$

and

$$V_D = V_T \ln \frac{I_D}{I_S} \qquad (3.5)$$

$$= 42.8 \text{ mV} * \ln \frac{I_D}{1.90 \text{ nA}} \qquad (3.9)$$

In principle we should be able to solve these two equations. The difficulty is that one of our unknowns is mixed up with a messy logarithm. Maybe you know a way to deal with that kind of problem, but if you do, you have to agree that it is messy. For our purpose it is much easier to use a procedure called "successive approximation" or "iteration." This approach is easy to implement using a spread sheet, and it is the procedure used by circuit analysis programs such as PSpice. The method consists of repeatedly guessing and testing the guess. We will proceed as follows:

First write the KVL equation for the circuit and solve it for the diode current (Equation 3.11 for Example 3.4). Next write the junction equation for the diode (Equation 3.9 for Example 3.4). Since we know that V_D must be about 0.7 V, carry out the following iterative procedure:

1. Guess $V_D = 0.7$ V
2. Put V_D in the circuit equation and solve for I_D.
3. Put I_D in the junction equation and solve for V_D.
4. GOTO 2.

Go around this loop until successive values agree to the number of decimal places you desire. Given the accuracy of the data in this case, to "desire" three decimal places is probably optimistic. Due to the exponential nature of the junction equation, the procedure converges rapidly and you will find that twice or three times around the loop is usually sufficient.

You can easily program your programmable calculator to speed up the task. A better approach would be to use a spread sheet such as SuperCalc to solve the problem. Perhaps the best way is to use a circuit analysis program such as PSpice.

EXAMPLE 3.5
Repeat Example 3.1 using the junction equation instead of the graph.

● (For a PSpice solution, see Appendix H.)

● (For a SuperCalc solution, see Appendix G.)

SOLUTION:
Guess:

1. $V_D = 0.7$ V

First iteration:

2. $I_D = \dfrac{V_S - V_D}{R_S}$ (3.1)

$= \dfrac{2\text{ V} - 0.7\text{ V}}{5.6\ \Omega} = 232$ mA

3. $V_D = NV_T \ln\dfrac{I_D}{I_S}$ (3.5)

$= 42.8\text{ mV} * \ln\dfrac{232\text{ mA}}{1.90\text{ nA}} = 0.797$ V

Second iteration:

2. $I_D = \dfrac{2\text{ V} - 0.797\text{ V}}{5.6\ \Omega} = 215$ mA

3. $V_D = 42.8\text{ mV} * \ln\dfrac{215\text{ mA}}{1.90\text{ nA}} = 0.794$ V

Third iteration:

2. $I_D = \dfrac{2\text{ V} - 0.794\text{ V}}{5.6\ \Omega} = 215$ mA

3. $V_D = 42.8\text{ mV} * \ln\dfrac{215\text{ mA}}{1.90\text{ nA}} = 0.794$ V

The above results compare well with the graphical result of Example 3.2.

If you have the program stored in your computer or in your calculator it should take less than a minute to go through the above example.

3.9 Reading Data Sheets

If you look at the manufacturer's diode data sheets in Appendix F you will find that although there is some variability, most contain the following kinds of information:

- Physical description. This section will include dimensions of the case and the leads.
- Absolute maximum ratings. This section will include temperature information, power dissipation limitations, and maximum voltage and current information. The reliability and lifetime of the device are threatened if these ratings are approached. Many conservative designers require

that the maximum ratings for the devices they specify must be at least 1.25 times the maximums they expect to exist in their application. Of course you can expect that devices with higher ratings will probably be more expensive.

- Electrical characteristics. This section contains information about the behavior of the device under typical operating conditions. Occasionally minimum and maximum values are given. If these values are given and typical values are omitted, it is safe to assume that typical values will be near the geometric average of these extremes as given by

$$\text{Average} \equiv \sqrt{(\text{min}) * (\text{max})} \tag{3.12}$$

If only minimum values, or only maximum values are given, you are left to guess at the operation of a typical device. Perhaps adding 20 percent to listed minimum values or subtracting 20 percent from listed maximum values is reasonable.

Also included in Appendix F is a sample of the kind of data table found in a typical data book. Data books usually contain the more important information from the manufacturers' data sheets in a very compact form.

EXAMPLE 3.6

Determine the junction equation for a BA216 diode.

SOLUTION:

From Appendix F for the BA216 $V_F = 0.56$ V at $I_F = 0.2$ mA and $V_F = 0.84$ V at $I_F = 15$ mA. Inserting these assumed typical values in Equations 3.7 and 3.8 gives:

$$NV_T = \frac{V_{D2} - V_{D1}}{1n\dfrac{I_{D2}}{I_{D1}}} \tag{3.7}$$

$$= \frac{0.84 \text{ V} - 0.56 \text{ V}}{1n\dfrac{15 \text{ mA}}{0.2 \text{ mA}}} = 64.9 \text{ mV}$$

$$I_S = \frac{I_D}{e^{V_D/(NV_T)}} \tag{3.8}$$

$$= \frac{15 \text{ mA}}{e^{0.84 \text{ V}/64.9 \text{ mV}}} = 35.6 \text{ nA}$$

Thus the diode equation is:

$$V_D = 64.9 \text{ mV} * 1n\frac{I_D}{35.6 \text{ nA}}$$

The constants in the above empirical junction equation look considerably different from those in the theoretical junction equation, Equation 3.10. It would be instructive for you to insert some typical values for I_D into the above equation and into Equation 3.10 and compare values of V_D that result. You would find that the current has to approach the maximum rated value for the diode before the results differ by 0.2 V. It would be even more instructive for you to take a couple of BA216s to your lab and see which equation comes closest to those encountered in practice.

3.10 Dynamic Diode Resistance r_D

You now know that if a diode is to conduct a current, then there will be a voltage loss across it. It will act like a resistor whose value is given by

$$R_D = \frac{V_D}{I_D} \qquad (3.13)$$

Notice that Equation 3.13 is really saying that R_D is the slope of a line drawn from the origin to the point (V_D, I_D) on the diode characteristic curve as shown in Figure 3.4. Well, not exactly. Inexplicably, we in elec-

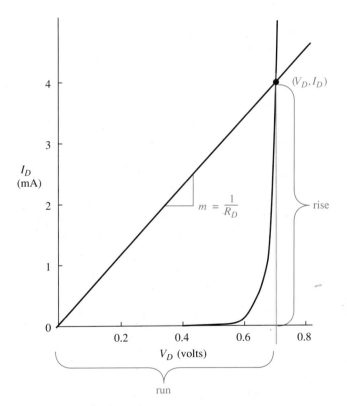

FIGURE 3.4 Static diode resistance, R_D.

tronics have chosen to draw our graphs with V on the horizontal axis and I on the vertical axis. That means that R_D is really the reciprocal of the slope of the above line. In terms that would make the mathematicians uncomfortable, R_D is the "run divided by the rise" as shown in Figure 3.4.

You should recognize that, due to the fact that diodes are very nonlinear, R_D will not be constant. It will decrease as I_D increases.

Now let us look at the behavior of a diode in a circuit that has a small AC signal current, i_D superposed on top of a DC Q point bias current, I_{DQ}. Such a circuit is shown in Figure 3.5.

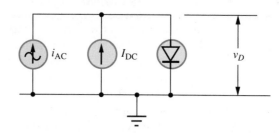

FIGURE 3.5   AC signal current, i_{AC} superposed on DC bias current, I_{DC}.

The positive peak diode current will be

$$I_{Dmax} = I_{DQ} + i_{DP} \tag{3.14}$$

The negative peak diode current will be

$$I_{Dmin} = I_{DQ} - i_{DP} \tag{3.15}$$

The resulting waveform is drawn on a typical diode characteristic graph in Figure 3.6. The corresponding value of V_{Dmax} and V_{Dmin} can be read from the characteristic graph, or they can be calculated using the junction equation. The peak to peak AC voltage across the diode, v_{DPP} is then given by

$$v_{DPP} \equiv V_{Dmax} - V_{Dmin} \tag{3.16}$$

Once v_{DPP} is known, Ohm's Law will tell us the diode's apparent resistance to AC. This is known as the "dynamic resistance," r_D, and is defined to be

$$r_D \equiv \frac{v_{DPP}}{i_{DPP}} \tag{3.17}$$

PSpice identifies r_D as REQ.

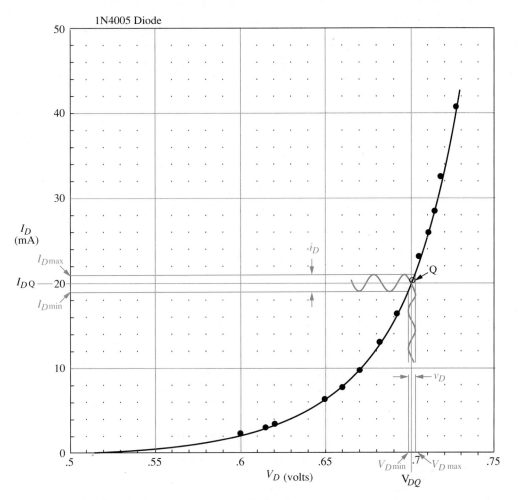

FIGURE 3.6 Diode response to an AC signal.

EXAMPLE 3.7

Determine the value of r_D for the circuit of Figure 3.7.

● (For a PSpice solution, see Appendix H.)

SOLUTION:

From Figure 3.7, $I_{DQ} = 20$ mADC and $i_D = 2$ mAPP $= 1$ mAP. Thus

$$I_{Dmax} = I_{DQ} + i_{DP} \tag{3.14}$$
$$= 20 \text{ mA} + 1 \text{ mA} = 21 \text{ mA}$$

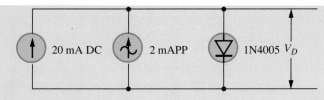

FIGURE 3.7   Example 3.7.

$$I_{D\min} = I_{DQ} - i_{DP} \tag{3.15}$$
$$= 20 \text{ mA} - 1 \text{ mA} = 19 \text{ mA}$$

The resulting waveforms are shown in Figure 3.6. Since it is a little difficult to read the required voltages from the graph, let us resort to the junction equation for the 1N4005. First, on the positive peak:

$$V_{D\max} = 42.8 \text{ mV} * \ln\frac{I_{D\max}}{1.90 \text{ nA}} \tag{3.9}$$

$$= 42.8 \text{ mV} * \ln\frac{21\text{mA}}{1.90 \text{ nA}} = 0.695 \text{ V}$$

(0.702 V read from the graph)

On the negative peak:

$$V_{D\min} = 42.8 \text{ mV} * \ln\frac{I_{D\min}}{1.90 \text{ nA}}$$

$$= 42.8 \text{ mV} * \ln\frac{19 \text{ mA}}{1.90 \text{ nA}} = 0.691 \text{ V}$$

(0.698 V read from the graph)

Thus the AC voltage required to push the 2 maPP current through the diode is:

$$v_{DPP} = V_{D\max} - V_{D\min} \tag{3.16}$$
$$= 0.695 \text{ V} - 0.691 \text{ V} = 4.29 \text{ mVPP}$$
$$(4 \text{ mVPP from the graph})$$

(This is a good illustration of the need to use all of the power of your calculator even though it is inconvenient to write down all of those digits.)

Substituting in Equation 3.17 gives

$$r_D \equiv \frac{v_{DPP}}{i_{DPP}} \tag{3.17}$$

$$= \frac{4.29 \text{ mVPP}}{2 \text{ mAPP}} = 2.14 \ \Omega$$

Note that the dynamic resistance is quite different from the static resistance. The dynamic resistance is what the AC signal sees as it moves up and down on the characteristic curve around the Q point. Thus, Equation 3.17 is NOT the same as Equation 3.13.

   If you examine Equation 3.17 and the preceding discussion carefully you will notice that Equation 3.17 is actually finding the slope (really 1/slope) of the diode characteristic graph at the Q point. Of course, we can find the above slope graphically by drawing a straight line tangent to the characteristic curve at the Q point and then calculating its slope (really 1/slope!). The procedure is illustrated in Figure 3.8. The tangent line has

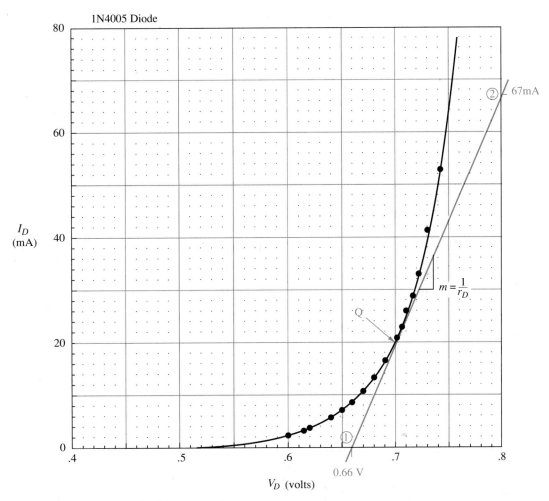

FIGURE 3.8   Diode dynamic resistance, r_D.

been extended as far as possible in order to make the result as accurate as possible. We can then write

$$r_D = \frac{1}{m} \tag{3.18}$$

Since the mathematicians would define slope to be

$$m \equiv \frac{\text{rise}}{\text{run}} = \frac{y_2 - y_1}{x_2 - x_1} \tag{3.19}$$

then

$$r_D = \frac{x_2 - x_1}{y_2 - y_1} = \frac{V_2 - V_1}{I_2 - I_1} \tag{3.20}$$

$$= \frac{0.8 \text{ V} - 0.66 \text{ V}}{67 \text{ mA} - 0 \text{ A}} = 2.09 \ \Omega$$

This graphical result agrees fairly well with the above calculated value of 2.14 Ω in Example 3.7.

If you know some calculus you know how to find the slope of a graph by finding the derivative of the function. Applying that approach to the junction equation we get the surprisingly simple relation:

$$r_D = \frac{N V_T}{I_D} \tag{3.21}$$

A derivation of the above Equation 3.21 appears in Appendix D.

When better information is not available, an approximate value for r_D can be obtained using the theoretical values $N = 1$ and $V_T = 26$ mV. This gives the approximation

$$r_D = \frac{26 \text{ mV}}{I_D} \tag{3.22}$$

Although the percentage of error obtained from using Equation 3.22 is large, the numerical value of the error is not too bad. Thus we will use it extensively, especially in connection with our study of transistors since it allows us to construct an AC equivalent circuit for a forward biased P-N junction which consists of a simple resistor to which we can apply Ohm's Law.

EXAMPLE 3.8

Calculate the AC component of the output v_{out} for the circuit in Figure 3.9a when $V_{SDC} = 5$ VDC.

● (See Appendix H for a PSpice solution.)

SOLUTION:

First we will obtain the junction equation for the diode: Looking in Appendix F, the data sheet for the 1N4305 lists the following forward voltages and currents:

$$V_D = 0.50 \text{ V at } I_D = 250 \text{ } \mu\text{A}$$
$$V_D = 0.70 \text{ V at } I_D = 10 \text{ mA}$$

Solving for the empirical values of NV_T and I_S gives

$$NV_T = \frac{V_{D2} - V_{D1}}{\ln\dfrac{I_{D2}}{I_{D1}}} \tag{3.7}$$

$$= \frac{0.7 \text{ V} - 0.50 \text{ V}}{\ln\dfrac{10 \text{ mA}}{250 \text{ } \mu\text{A}}} = 54.2 \text{ mV}$$

$$I_S = \frac{I_D}{e^{V_S/V_T}} \tag{3.8}$$

$$= \frac{10 \text{ mA}}{e^{0.7 \text{ V}/54.2 \text{ mV}}} = 24.7 \text{ nA}$$

Substituting the above values in Equation 3.5 yields the junction equation for the 1N4305:

$$V_D = 54.2 \text{ mV} * \ln\frac{I_D}{24.7 \text{ nA}} \tag{3.23}$$

Next we will solve for the Q point values of diode voltage and current. The KVL equation for the DC circuit is:

$$V_S - I_D R_S - V_D = 0 \text{ V} \tag{3.24}$$

Solving for I_D gives

$$I_D = \frac{V_S - V_D}{R_S}$$

$$= \frac{5 \text{ V} - V_D}{470 \text{ } \Omega} \tag{3.25}$$

Solving Equations 3.25 and 3.23 by iteration gives

$I_D = \dfrac{5 \text{ V} - V_D}{470 \text{ } \Omega}$	=	9.15	9.16	9.16	mA
$V_D = 54.2 \text{ mV} * \ln\dfrac{I_C}{24.7 \text{ nA}}$	=	0.695	0.695	0.695	V

Thus the Q point values are: $V_D = 0.695$ V and $I_D = 9.16$ mA.
Solving for the diode dynamic resistance r_D at $I_D = 9.16$ mA:

$$r_D = \frac{NV_T}{I_D} \tag{3.21}$$

$$= \frac{54.2 \text{ mV}}{9.16 \text{ mA}} = 5.92 \ \Omega$$

Finally, using the above value for r_D, the AC equivalent circuit of Figure 3.9b is constructed. Since R_S and r_D form a voltage divider, we have

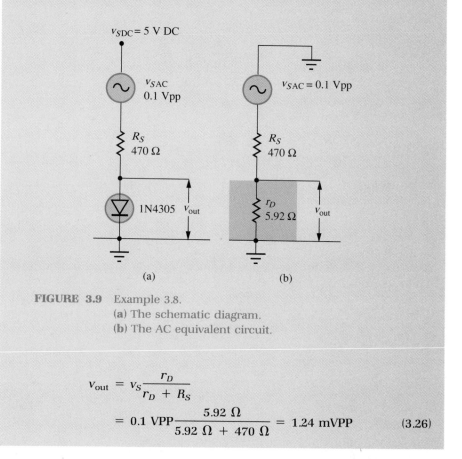

(a)   (b)

FIGURE 3.9   Example 3.8.
(a) The schematic diagram.
(b) The AC equivalent circuit.

$$V_{out} = V_S \frac{r_D}{r_D + R_S}$$

$$= 0.1 \text{ VPP} \frac{5.92 \ \Omega}{5.92 \ \Omega + 470 \ \Omega} = 1.24 \text{ mVPP} \tag{3.26}$$

From the above example you can see a good application for diodes: If the DC supply is at, or close to, zero the diode will not conduct and so virtually all of the AC signal is present at the output. When the DC supply is turned on, almost all of the AC signal is conducted to ground. You may have heard of diode switching circuits. Now you know how they work. We will study them in greater depth in Chapter 6.

3.11 Temperature Effects

You remember that the temperature of the crystal was important in determining the value of V_T in Equation 3.4. It is also a factor in determining I_S. In fact, borrowing some ideas from the chemists, we find that I_S approximately doubles for every 10°C increase in temperature.

An increase in V_T will tend to increase V_D but an increase in I_S will tend to decrease V_D. Since the latter is a much stronger effect, we find that V_D does, in fact, decrease with increases in temperature at a rate of about -2.5 mV/°C. Electronic thermometers and controllers often use specially designed diodes as their temperature-sensing element.

Looking at Figure 3.10 you can easily see that as long as V_S is constant, as the diode warms up, if V_D decreases significantly, then a larger portion

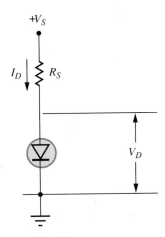

FIGURE 3.10   Circuit used to discuss temperature effects.

of V_S must be lost across R_S. That means that the current I_D will increase. An increase in I_D means the diode will get still hotter, and that will mean still less V_D and that means still more I_D and that means———!! Unless there is enough resistance in the circuit to prevent this "thermal runaway," the circuit can be destroyed. Fortunately, thermal runaway is not a problem in most simple circuits, but you should be aware that it can occur.

▶▶▶▶▶▶▶▶ **3.12 Troubleshooting**

The voltage loss across a forward biased diode that is functioning normally is rarely less than 0.6 V or more than 0.8 V. Thus diode operation outside of that range should bring questions to your mind.

Similarly, a normally functioning diode that is reverse biased conducts essentially no current. Thus you should question the performance of a diode that conducts in the reverse direction. The fact that there are diodes which violate the above "typical" conditions is what distinguishes technicians from technicians who think. We will look at some such diodes, and hopefully help you to think about them in later chapters of this book.

3.13 Summary

This chapter presented three, or perhaps four different diode models and methods for determining when each is appropriate. The greatest emphasis was placed on the junction equation since it is the most generally applicable. We discussed how to obtain the required parameters for the junction equation from both experimental data and the manufacturer's data sheets. Finally, we discussed the meaning of the dynamic diode resistance, r_D, how to determine its value, and how to use it.

GLOSSARY

Data sheet: A document prepared by the manufacturer of a device which contains the parameters which describe the behavior of the device and the specifications which the manufacturer guarantees that the device will meet. (Section 3.9)

Dynamic resistance (r): The resistance a device exhibits, typically at its Q point, to a very small AC signal. (Section 3.10)

Iteration: The repetition of a procedure. In our case, the procedure consists of guessing a solution, inserting the guess into the equation, and solving the equation to test the accuracy of the guess. (Section 3.8)

Load line: A line drawn on a graph which represents the behavior of a circuit. In electronics the graph is usually of current on the vertical axis and voltage on the horizontal axis. (Section 3.5)

Q point (quiescent point): The point on the load line of a circuit which represents its operation when no signal or outside stimulus is present. (Section 3.5)

Thermal runaway: The unstable behavior of a device which has a negative temperature coefficient which is characterized by increasing conduction as temperature rises, causing a marked deviation from the intended Q point. (Section 3.11)

PROBLEMS

Section 3.2

1. Find the voltage lost across the 1N4148 diode whose characteristic is graphed in Figure 3.11 when it is carrying 100 mA. Calculate the power dissipated by the diode.

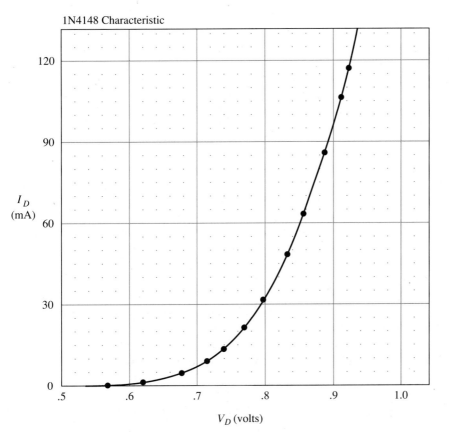

FIGURE 3.11   Problem 3.1.

2. Repeat Problem 1 for the 1N5819 diode whose characteristic is graphed in Figure 3.12.

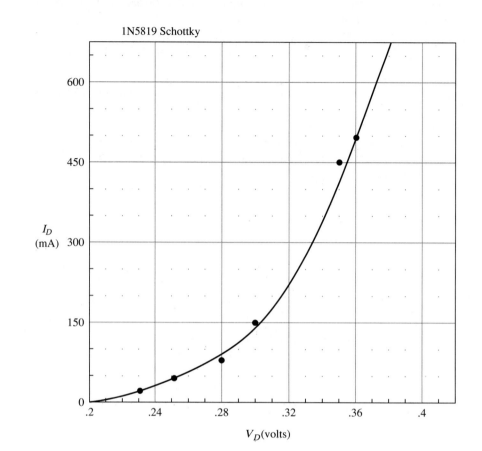

FIGURE 3.12   Problem 3.2.

Section 3.4

3. Calculate the supply voltage V_S required to deliver 300 V to the load R_L in Figure 3.13 if $R_L = 10$ k Ω.

(a)   Use Model 0, the ideal model for the diode.

(b)   Use Model I, the fixed voltage model for the silicon diode.

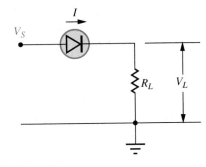

FIGURE 3.13 Problem 3.3.

4. Repeat Problem 3 if 3 V is to be delivered to a 100 Ω load.

Section 3.5

5. Repeat Problem 3 using Model II, the graphical model if the diode is the 1N4148 of Figure 3.11.
6. Repeat Problem 4 using Model II, the graphical model if the diode is the 1N4148 of Figure 3.11.
7. If the 1N4148 diode of Figure 3.14 has the characteristic of Figure 3.11, find the value of resistor R_S that will give a current of approximately I_D = 10 mA. Specify the nearest 10% resistor.

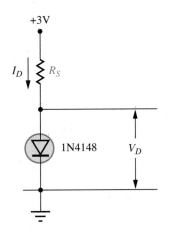

FIGURE 3.14 Problem 3.7.

8. Using the 1N4148 characteristic of Figure 3.11, find the required supply voltage, V_S, for the circuit of Figure 3.15.

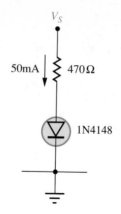

FIGURE 3.15   Problem 3.8.

9. Use the "fixed voltage" model to answer the question posed in the first paragraph of Section 3.1.
10. Find a graphical solution to the question posed in the first paragraph of Section 3.1.
11. Using the characteristic of Figure 3.11, find the current flowing in the circuit and the voltage across the 1N4148 in Figure 3.16.

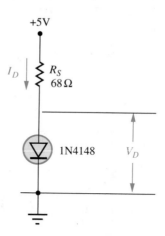

FIGURE 3.16   Problem 3.11.

Section 3.6

12. Find the error that was introduced when the -1 was ignored in the development of Equation 3.4. Assume $I_S = 10$ fA, $V_T = 26$ mV, and $V_D = 0.7$ V.

Section 3.7

13. Determine values for NV_T, N, and I_S from the characteristic of Figure 3.11. Write the junction equation for the 1N4148 diode.

14. Determine values for NV_T, N, and I_S from the 1N5819 characteristic of Figure 3.12. Write the junction equation for the 1N5819 Schottky diode. This is NOT a typical silicon diode.

15. Obtain the junction equation for a 1N4376 from the data contained in the manufacturer's data sheet in Appendix F.

Section 3.8

16. Use the junction equation developed in Problem 13 to determine the voltage lost across the 1N4148 when it is carrying 30 mA. Compare with the value read from the graph.

17. Use the junction equation developed in Problem 14 to determine the voltage lost across the 1N5819 Schottky diode when it is carrying 300 mA.

18. Use the junction equation developed in Example 3.3 to answer the question posed in the first paragraph of Section 3.1.

19. Use the "theoretical" silicon diode junction equation to answer the question posed in the first paragraph of Section 3.1.

20. Using the theoretical silicon diode junction equation rather than the characteristic graph, repeat Problem 11.

Section 3.10

21. Use Figure 3.11 to find the dynamic resistance, r_D, of the 1N4148 diode at a current of 10 mA. Compare with the result obtained from Equation 3.22.

22. Repeat Problem 21 for a current of 100 mA.

23. If the supply of Problem 11 had an AC ripple voltage of 0.1 V riding on top of the 5 VDC, determine the amount of AC ripple appearing across the diode.

24. Determine the value of V_{out} in Figure 3.17 if the diode has "typical" silicon diode parameters ($V_T = 26$ mV, $I_S = 10$ fA). *Hint:* Remember your circuits class—construct a Thevenin equivalent.

● (See Appendix H for a PSpice solution.)

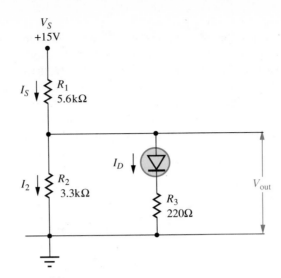

FIGURE 3.17   Problem 3.24.

25. Find the value of V_{out} in Figure 3.18. Use "typical" silicon junction parameters ($V_T = 26$ mV, $I_S = 10$ fA).

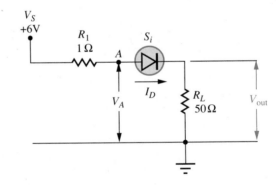

FIGURE 3.18   Problem 3.25.

Troubleshooting ▶   Section 3.12

26. Describe the effect on the circuit if the 68 Ω resistor in Figure 3.16 becomes:
    (a)   An open circuit
    (b)   A short circuit.

27. Determine the current and power dissipation in each circuit component in Figure 3.17 if resistor R_1 became:
    (a) An open circuit
    (b) A short circuit.
28. Determine the value of V_A and V_L in Figure 3.18 if:
    (a) The diode is hooked up backward
    (b) The diode is hooked up correctly but the power supply polarity is reversed.

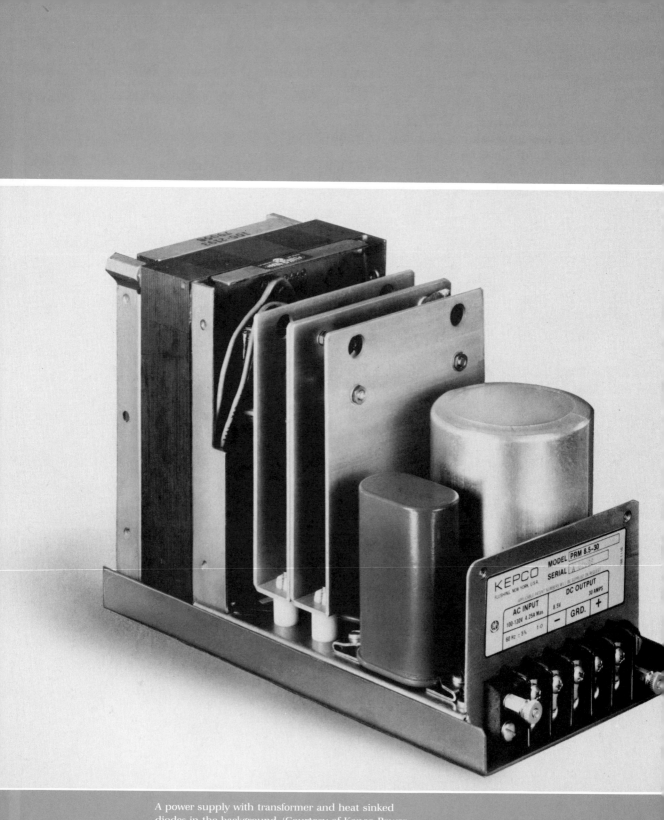

A power supply with transformer and heat sinked
diodes in the background. (Courtesy of Kepco Power
Supplies.)

Chapter 4
Power Supply Rectifiers

SPECIAL TERMS

Power supply
Rectifier
Isolation transformer
Half-wave rectifier
Peak inverse voltage (PIV)
Surge current (I_{surge})
Full-wave rectifier
Center-tapped transformer (CT transformer)
Phase splitting
Bridge circuit

4.1   Introduction and Objectives

In this chapter we will use what you already know about diodes to meet a practical need; the need to convert power from the AC power line into DC at any specified voltage. We will look at three circuits capable of accomplishing this task, and we will look at some of the advantages and disadvantages of each.

When you finish this chapter you should understand the concept of rectification and how common half-wave and full-wave rectifiers work. You should be able to:

• Analyze simple half-wave and full-wave rectifier circuits in terms of their output voltage, current, and waveforms.
• Design simple half-wave and full-wave rectifier circuits.

4.2   Power Supplies

It could be argued that the most important topic you will study in this book is that of the DC power supply. It is certainly the most often used circuit in electronics. Virtually all electronic devices require a DC power supply and when an electronic device fails, the fault is often traceable to the power supply.

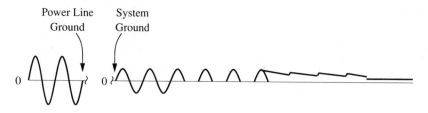

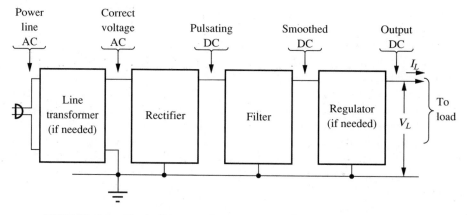

FIGURE 4.1   Block diagram of a power supply.

In Chapters 4 and 5 we will only be able to examine the simplest of power supply circuits. In fact, to explore this one topic fully would take more space than we have allotted for this entire book.

Figure 4.1 contains a block diagram for a typical power supply. Many power supplies use a transformer to isolate the circuit from the power line and to obtain the desired supply voltage. The rectifier converts the transformer's AC output to pulsating DC. The DC is then smoothed by the filter. Finally, a regulator may be used to maintain exactly the voltage required if the application demands such close control. In this chapter we will look at the first two of these components, and in Chapter 5 we will look at filters. Increasingly sophisticated regulator circuits will be discussed in Chapters 7, 13, and 21.

4.3 Safety Precautions

Since most power supplies obtain their power from the 60 Hz 120 V power line, and since that power line can supply more than enough power to severely damage equipment or people (you!) that are improperly connected to it, it should never be approached casually. You should THINK about the fact that one side of the power line is grounded and it is never safe to assume that you know which side. You should also remember that a lot of electronic test equipment is tied to earth ground. Thus, for example, you cannot with impunity connect the ground lead of your scope to one side of the power line. Neither can you, with impunity, hook it to the chassis of any device that is plugged into the power line since, as we shall see shortly, not all such devices are isolated from the power line. If you are unsure, you should plug the device you wish to test into an "isolation transformer" which is just a transformer with a 1:1 turns ratio which isolates the device from the power line. No well-equipped electronics bench should be without such a transformer.

4.4 Half-Wave Rectifier

If power line isolation is not mandatory for your circuit, and if the voltage delivered by the power company is adequate for your needs, the circuit in Figure 4.2a is all that is needed to construct a half-wave rectifier. When the AC line voltage V_S is positive the diode will be forward biased and, as we learned in Chapter 2, it will conduct. Then V_S (with the loss of around 0.7 V) will appear at the load. When the AC line voltage is negative, the diode will block almost all current flow and virtually no voltage will appear at the load. The associated waveforms are shown in Figures 4.2b and 4.2c.

The fixed supply voltage limits the usefulness of the circuit of Figure 4.2, but a more serious limitation is the fact that the load is NOT isolated

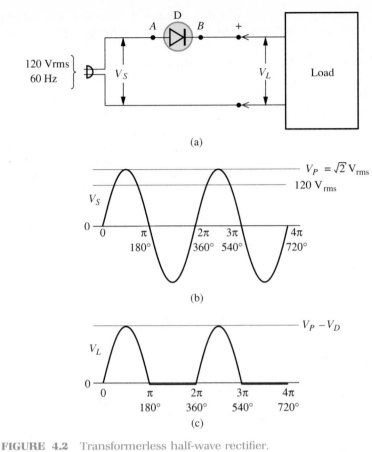

(a)

(b)

(c)

FIGURE 4.2   Transformerless half-wave rectifier.
(a) Circuit diagram.
(b) Supply Voltage waveform.
(c) Load voltage waveform.

from the power line. This presents a serious safety hazard which, as we
mentioned in Section 4.3, must be guarded against. The transformer coupled circuit of Figure 4.3 overcomes the above limitations. By specifying
the transformer the designer can obtain practically any desired output
voltage while obtaining isolation from the power line at the same time. Of
course, you pay a penalty in added weight, space, and cost for this luxury.
The output waveform will still be the same as that pictured in Figure 4.2c.

4.5 Half-Wave Rectifier Component Specification

In order to specify the components required for the half-wave transformer
coupled rectifier of Figure 4.3, we must find a relationship between the
supply voltage V_S and the average DC current I_{av} supplied to the load. Our

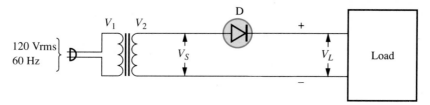

FIGURE 4.3 Transformer coupled half-wave rectifier.

task is complicated considerably by the fact that as you can see from Figure 4.2c, I_{av} is delivered in short pulses whose duration is less than 180 electrical degrees.

Since V_S is derived from the power line, you should remember that $V_{S\theta}$, the value of V_S at any particular conduction angle θ is given by

$$V_{S\theta} = V_{SP}\sin \theta \tag{4.1}$$

Also from Figure 4.2 it is obvious that current will begin to flow when V_S exceeds V_L by enough voltage to cause the diode to turn on. We will define this "turn-on voltage" to be V_{Son}. Similarly, we will define V_{Soff} to be the instantaneous supply voltage at turn-off.

We will define the conduction angles for the above instantaneous voltages as θ_{on} and θ_{off} respectively. These relationships are shown in Figure 4.4.

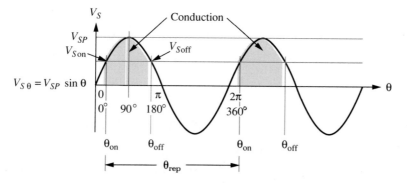

FIGURE 4.4 Conduction diagram for a half-wave rectifier showing the angle when conduction begins, θ_{on} and the angle when it ends, θ_{off}.

If you use a circuit analysis program, a current of something like 1 pA can usually be used to determine V_{Son} and V_{Soff}.

Using calculus (see Appendix D.2) we can show that for a sine wave

$$I_{av} = \frac{I_P}{\theta_{rep}}(\cos \theta_{on} - \cos \theta_{off}) \tag{4.2}$$

where

I_{av} = the DC average current

I_P = the peak value of the current. I_P can usually be calculated by applying KVL to the circuit when $V_S = V_{SP}$.

θ_{rep} = the angle between pulses, measured in radians.

θ_{on} = the angle where conduction begins.

θ_{off} = the angle where conduction ends.

The above relationships are shown in Figure 4.4.

As you can see from Figure 4.4, for a half-wave rectifier

$$\theta_{rep} = 2\pi \tag{4.3}$$

Also from Figure 4.4, for a simple unfiltered rectifier

$$V_{Soff} = V_{Son} \tag{4.4}$$

so that, as Figure 4.4 shows, if the angles are measured in degrees

$$\theta_{off} = 180° - \theta_{on} \tag{4.5}$$

Or, using a little trigonometry

$$\cos \theta_{off} = \cos(180° - \theta_{on})$$
$$= -\cos \theta_{on}$$

and so

$$\cos \theta_{on} - (\cos \theta_{off}) = 2 \cos \theta_{on} \tag{4.6}$$

Substituting Equations 4.3 and 4.6 in Equation 4.2 gives, for a half-wave rectifier when $V_{Soff} = V_{Son}$

$$I_{av} = \frac{I_P}{2\pi} 2 \cos \theta_{on}$$
$$= \frac{I_P}{\pi} \cos \theta_{on} \tag{4.7}$$

In the simplest case where the load is just a resistor, and if we ignore the diode turn-on voltage of around 0.6 V so that $V_{on} = 0$ V, conduction begins at $\theta_{on} = 0°$. Then Equation 4.7 reduces to

$$I_{av} = \frac{I_P}{\pi} \tag{4.8}$$

You should remember Equation 4.8 from your circuits class.

It is important to notice that everything that has been said here assumes that the circuit components are linear. Of course, you already know that diodes are not linear, but you have also seen that to assume a fairly constant diode forward voltage rarely introduces errors of ten percent.

That is the approach that is usually taken, and that is the approach we will take. We will assume a diode turn-on voltage of about $V_{Don} \approx 0.6$ V and a peak diode voltage of about $V_{DP} \approx 0.8$ V.

EXAMPLE 4.1

Suppose that you have available a transformer with a 12 Vrms secondary and an effective internal resistance of 0.3 Ω. If this transformer is used to build a half-wave rectifier battery charger for your car ($V_X = 13.2$ V; $R_L = 0.2\ \Omega$), find the average DC charging current.

● (For a PSpice solution, see Appendix H.)

SOLUTION:

The circuit is diagrammed in Figure 4.5. From our definition of turn-on, $I_{Don} = 0$ A and you can see from the circuit that

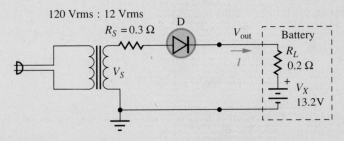

FIGURE 4.5   Example 4.1.

$$V_{Son} = V_{Don} + V_X \qquad (4.9)$$
$$= 0.6\ \text{V} + 13.2\ \text{V}$$
$$= 13.8\ \text{V}$$

The peak source voltage V_{SP} is given by

$$V_{SP} = \sqrt{2}\ V_{Srms} \qquad (4.10)$$
$$= \sqrt{2} * 12\ \text{Vrms}$$
$$= 17.0\ \text{VP}$$

Beginning at ground and taking a KVL walk around the circuit in a clockwise direction we have

$$V_S - IR_S - V_D - IR_L - V_X = 0\ \text{V} \qquad (4.11)$$

To obtain the peak current, we will solve Equation 4.11 for I_P

$$I_P = \frac{V_{SP} - V_X - V_{DP}}{R_S + R_L} \tag{4.12}$$

$$= \frac{17.0 \text{ V} - 13.2 \text{ V} - 0.8 \text{ V}}{0.3 \text{ }\Omega + 0.2 \text{ }\Omega} = 6 \text{ AP}$$

Of course instead of using $V_{DP} = 0.8$ V, you should recognize that we could have used Equation 3.10, the diode equation, and the iterative approach we used in Section 3.8. The required equations are:

$I_P = \dfrac{3.8 \text{ V} - V_{DP}}{0.5 \text{ }\Omega}$ =	6.20	5.83	5.83	A
$V_{DP} = 26 \text{ mV} \cdot \ln\dfrac{I_P}{10 \text{ fA}}$ =	0.886	0.884	0.884	V

You may think that these more "exact" values for I_D and V_{DP} would make a significant difference in the final result, but as we shall see, it affects the final result by only four parts out of 108 and so it is hardly worth the effort.

Solving Equation 4.1 for θ_{on} gives

$$\theta_{on} = \sin^{-1}\left(\frac{V_{Son}}{V_{SP}}\right) \tag{4.13}$$

$$= \sin^{-1}\left(\frac{13.8 \text{ V}}{17.0 \text{ V}}\right) = 54.3°$$

Since we now know I_P and θ_{on}, we can use Equation 4.7 to find the average current.

$$I_{av} = \frac{I_P}{\pi} \cos \theta_{on} \tag{4.7}$$

$$= \frac{6 \text{ A}}{\pi} \cos 54.3° = 1.12 \text{ A}$$

$$I_{av} = 1.08 \text{ A using the diode equation.}$$

If you went to a parts catalog to order the parts for the above battery charger, you would find a confusing array of transformers and diodes listed. Which ones should you order? To answer that question, you must specify the required characteristics of each part.

The essential specs for a transformer are the rms voltage rating for the primary, number of secondary windings and the rms voltage and current rating for each. Obviously, only one secondary winding is needed and its current rating can be assumed to be that of the average current delivered to the load.

Specifying the diode is not straightforward. One diode spec is the peak inverse voltage, PIV, that the diode must be capable of blocking. The circuit of Figure 4.6 shows the necessary relationships. For modern diodes

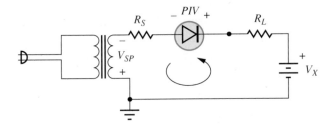

FIGURE 4.6   Diode peak inverse voltage, *PIV*, for a half-wave rectifier.

it is reasonable to assume that when the source in Figure 4.6 puts out its negative peak voltage, the reverse current is zero. Thus at that instant the voltage at the cathode of the diode is $+V_X$ and the voltage at the anode is $-V_{SP}$. Therefore

$$PIV = V_X - (-V_{SP})$$
$$= V_X + V_{SP} \qquad (4.14)$$

Another diode spec is the average forward current, which is determined by the long term ability of the diode to dissipate heat. For the half-wave rectifier this spec is just I_{AV}, and is given by Equation 4.7.

Two other diode current specs are usually given. One is the "repeated peak" current which is just I_P, given by Equation 4.12 for the half-wave rectifier. The other current spec is the "one cycle surge" current. We will find out why this is an important spec when we study filters in Chapter 5, but for now we could talk about what would happen to your battery charger if you should happen to let the leads touch. Of course, a good design would include a fuse to protect against such carelessness, but a fair amount of current will flow in the millisecond or two that it takes for the fuse to blow. Transformers are usually sturdy enough to stand that kind of abuse so they do not usually need a surge rating, but the surge current rating is usually specified for diodes. To estimate the surge current rating, I_{surge} needed for this circuit, assume that the supply voltage is at its peak, V_{SP} and with the leads shorted together, the only resistance in the circuit during the millisecond or two that it takes for the fuse to blow is the source resistance, R_S. Then

$$I_{\text{surge}} = \frac{V_{SP} - V_{DP}}{R_S} \qquad (4.15)$$

Of these current specs the most important are I_{AV} and I_{surge}. Usually if the surge current spec is met, the repeated peak rating will be no problem.

EXAMPLE 4.2

Specify both the transformer and the diode in Example 4.1 such that the average DC charging current $I_{AV} = 10$ A.

● (For a SuperCalc solution, see Appendix G.)

SOLUTION:

Again, assume $V_{Don} = 0.6$ V and $V_{DP} = 0.8$ V.

Therefore, as in the preceding example $V_{Son} = V_{Soff} = 13.8$ V. Thus Equation 4.7 is valid for this circuit

$$I_{AV} = \frac{I_P}{\pi} \cos \theta_{on} \tag{4.7}$$

or

$$I_P = I_{DC}\frac{\pi}{\cos \theta_{on}}$$

$$= 10 \text{ A}\frac{\pi}{\cos \theta_{on}} = \frac{31.4 \text{ A}}{\cos \theta_{on}} \tag{4.16}$$

From Equation 4.11, which is the KVL equation for the circuit of Figure 4.5, we have

$$V_{SP} = I_P(R_S + R_L) + V_{DP} + V_X \tag{4.17}$$

Substituting Equation 4.16 for I_P into Equation 4.17 gives

$$V_{SP} = \frac{31.4 \text{ A}}{\cos \theta_{on}}(0.3 \ \Omega + 0.2 \ \Omega) + 0.8 \text{ V} + 13.2 \text{ V}$$

$$= \frac{15.7 \text{ V}}{\cos \theta_{on}} + 14.0 \text{ V} \tag{4.18}$$

Since there are two unknowns in the above equation, another relationship is needed. Just as we did in Example 4.1, we will get the other relationship from the fact that the source waveform is a sine wave. Thus

$$\theta_{on} = \sin^{-1}\left(\frac{V_{Son}}{V_{SP}}\right) \tag{4.13}$$

Since the turn-on voltage V_{on} will be the same as it was in Example 4.1

$$\theta_{on} = \sin^{-1}\left(\frac{13.8 \text{ V}}{V_{SP}}\right) \tag{4.19}$$

In principle the problem is now solved. We should be able to combine Equations 4.18 and 4.19 to obtain V_{SP}. However, since messy trigonometric functions are involved, it will probably be easier to solve by iteration as we did when we were working with the diode junction equation in Section 3.8. A reasonable approach might be:

1. Guess at an initial value for θ_{on}.
2. Put θ_{on} in Equation 4.18 and solve for V_{SP}.
3. Put V_{SP} in Equation 4.19 and solve for θ_{on}.
4. Go to step 2. Go around this loop until successive values agree to the required accuracy.

Since we have no way of knowing where on the waveform of Figure 4.4 conduction will begin, we will guess right in the middle. That is half way between 0° and 90°. Thus:
Guess:

1. $\theta_{on} = 45°$

First approximation:

2. $V_{SP} = \dfrac{15.7 \text{ V}}{\cos \theta_{on}} + 14.0 \text{ V}$ (4.18)

$\quad = \dfrac{15.7 \text{ V}}{\cos 45°} + 14.0 \text{ V} = 36.2 \text{ V}$

3. $\theta_{on} = \sin^{-1}\left(\dfrac{13.8 \text{ V}}{V_{SP}}\right)$ (4.19)

$\quad = \sin^{-1}\left(\dfrac{13.8 \text{ V}}{36.2 \text{ V}}\right) = 22.4°$

Second approximation:

2. $V_{SP} = \dfrac{15.7 \text{ V}}{\cos 22.4°} + 14.0 \text{ V} = 31.0 \text{ V}$

3. $\theta_{on} = \sin^{-1}\left(\dfrac{13.8 \text{ V}}{31.0 \text{ V}}\right) = 26.5°$

Third approximation:

2. $V_{SP} = \dfrac{15.7 \text{ V}}{\cos 26.5°} + 14.0 \text{ V} = 31.5 \text{ V}$

3. $\theta_{on} = \sin^{-1}\left(\dfrac{13.8 \text{ V}}{31.5 \text{ V}}\right) = 26.0°$

Fourth approximation:

2. $V_{SP} = \dfrac{15.7 \text{ V}}{\cos 26°} + 14.0 \text{ V} = 31.5 \text{ V}$

3. $\theta_{on} = \sin^{-1}\left(\dfrac{13.8 \text{ V}}{31.5 \text{ V}}\right) = 26.0°$

You will note that these equations do not converge as quickly as the diode junction equation of Chapter 3, but iteration is probably still easier than trying to solve the equations analytically. Especially if you have a computer to do the work for you.

We can get the peak current I_P from either Equation 4.16 or 4.17. We will use Equation 4.16

$$I_P = \frac{31.4 \text{ A}}{\cos \theta_{on}} \tag{4.16}$$

$$= \frac{31.4 \text{ A}}{\cos 26°} = 34.9 \text{ A}$$

Now that I_P is known, to be really "accurate" about things, you may wish to go back to the diode junction equation and find out what V_{DP} "really" is.

$$V_{DP} = 26 \text{ mV} * \ln\left(\frac{I_P}{10 \text{ fA}}\right) \tag{3.10}$$

$$= 26 \text{ mV} * \ln\left(\frac{34.9 \text{ A}}{10 \text{ fA}}\right) = 0.931 \text{ VP}$$

To satisfy your own curiosity you may wish to put this value of V_{DP} back in Equation 4.17 just to see how much difference it makes. You will find that the change is not worth the effort.

Since transformer voltage ratings are always rms, we must convert the above calculated value of V_{SP} to rms. From your circuits class you should remember that

$$V_{Srms} = \frac{V_{SP}}{\sqrt{2}}$$

$$= \frac{31.5 \text{ V}}{\sqrt{2}} = 22.3 \text{ Vrms}$$

Since both the average DC current and the peak current are already known, the only diode specs we need to calculate are the peak inverse voltage, PIV, and the surge current I_{surge}.

From the circuit

$$\text{PIV} = V_X + V_{SP} \tag{4.14}$$

$$= 13.2 \text{ V} + 31.5 \text{ V} = 44.7 \text{ V}$$

Using the same argument used in Example 4.1, the diode surge current rating for this circuit will be

$$I_{\text{surge}} = \frac{V_{SP} - V_{DP}}{R_S} \tag{4.15}$$

$$= \frac{31.5 \text{ V} - 0.8 \text{ V}}{0.3 \ \Omega} = 102 \text{ A}$$

The rectifier components can now be specified as follows:

DIODE:

$\quad$ PIV > 44.7 V, $I_{av} > 10$ A, $I_P > 34.9$ AP, $I_{\text{surge}} > 102$ AP

TRANSFORMER:

$\quad V_{\text{pri}} = 120$ Vrms, $V_{\text{sec}} = 22.3$ Vrms, $I_{\text{sec}} > 10$ A

Due to the fact that the half-wave rectifier uses only half of the available power line waveform, it is not a very efficient circuit, especially when higher power is required. Further, you will find in Chapter 5 that if you need a nice smooth DC output, filtering the output of a half-wave rectifier is difficult due to the fact that the pulses are so far apart. The full-wave rectifier addresses both of these problems.

4.6 The CT Transformer Rectifier

In your electric circuits class you should have examined the behavior of the center-tapped (CT) transformer. It would be well for you to review that behavior at this point. Of greatest interest to us here is its phase-splitting capability. You should also recall that a label like "6.3 VCT" means that the whole transformer winding is rated at 6.3 Vrms. That means 3.15 Vrms on each side of the center tap.

$\quad$ Figure 4.7a contains the schematic diagram for a center-tapped transformer hooked up as a phase-splitter. Figure 4.7b shows the resulting waveforms at terminals A and B. It is important to note that V_A and V_B are equal in magnitude, but 180° out of phase with each other. When V_A is at its negative peak, V_B is at its positive peak. Thus, if diodes are placed as shown in Figure 4.8a, whenever Point A is positive, D_1 will conduct producing Pulse A at the load, and whenever Point B is positive, D_2 will conduct producing Pulse B at the load. Thus we have effectively two half-wave rectifiers that take turns supplying current to the load. Since their inputs are 180° out of phase, whenever one of them is reverse biased and so turned off, the other one is forward biased and supplying current to the load. Thus the load receives current every 180° (π radians) of the power line cycle and $\theta_{\text{rep}} = \pi$.

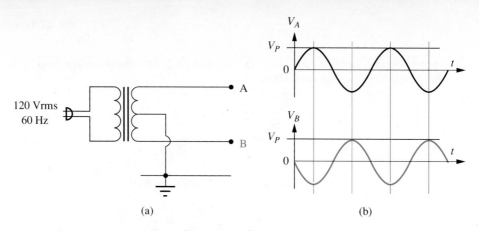

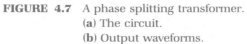

(a)

(b)

FIGURE 4.7 A phase splitting transformer.
(a) The circuit.
(b) Output waveforms.

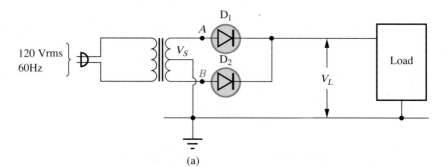

(a)

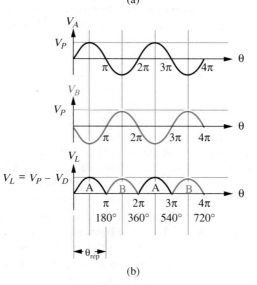

(b)

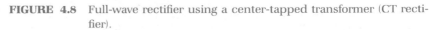

FIGURE 4.8 Full-wave rectifier using a center-tapped transformer (CT rectifier).
(a) The circuit.
(b) Resulting waveforms.

It is important to notice that this circuit requires a transformer with a center-tapped secondary winding. It is also important to notice that when D_1, for example, is forward biased, D_2 is all that prevents sudden destruction. If D_2 should conduct at the same time as D_1, there would then be nothing to prevent a rather large (infinite?) current from flowing, thus destroying the weakest link in the power chain between this circuit and the utility company. This sad state of affairs could occur if D_2 were defective, or if some benighted soul hooked it up backwards.

4.7 CT Transformer Rectifier Component Specification

A center-tapped transformer rectifier operates essentially the same as a half-wave rectifier. The equation for determining the average DC current delivered to the load, I_{AV} for this rectifier will be the same as that for the half-wave rectifier.

$$I_{av} = \frac{I_P}{\theta_{rep}}(\cos \theta_{on} - \cos \theta_{off}) \tag{4.2}$$

However, the value for the repetition angle, θ_{rep} is different. Instead of a repetition angle of 2π radians (360°), we will now have a repetition angle of 1π radians as indicated in Figure 4.8b. Thus, for the simple circuit where $V_{on} = V_{off}$, Equation 4.7 becomes

$$I_{av} = \frac{2I_P}{\pi} \cos \theta_{on} \tag{4.20}$$

Since the diodes conduct separately, for purposes of circuit analysis, we can ignore the diode that is not conducting. The KVL equation for this circuit is then identical to that of Section 4.5 for the half-wave rectifier.

$$V_{SP} = I_P(R_S + R_L) + V_{DP} + V_X \tag{4.17}$$

Circuit component specification is very similar to that for the half-wave rectifier with a few minor differences. You might at first think that since the transformer secondary is split at the center tap with each half carrying just half of the DC average current, the required transformer secondary current specification would be only half of the load current. However, since the two halves of the secondary winding are wound together on the same core, they share the same thermal environment. Therefore, the transformer secondary rms current specification will still be just the DC current to the load.

Since the diodes are usually physically separated from each other, their DC average current specification will be half the DC average current to the load. However, it is important to notice that each diode carries the peak current independent of the presence of the other diode in the circuit. Thus the procedure for the determination of the peak current and the surge current rating for the diodes will be the same as that for the half-wave rectifier.

Since the idea behind the peak inverse voltage rating PIV is still the same, we will use an approach similar to that of Section 4.5 to determine its value. However, the situation is a little different due to the fact that while we are looking for the PIV for one diode, the other diode is turned on and conducting. The circuit required to determine PIV is shown in Figure 4.9. At the instant shown, doing a KVL walk from the transformer through diode D_2 to the output we have

$$V_{SP} - I_P R_{S2} - V_{DP} = V_{LP} \qquad (4.21)$$

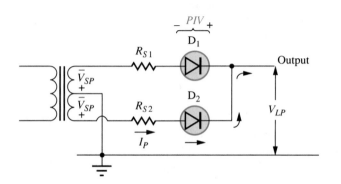

FIGURE 4.9   The *PIV* for a CT rectifier.

which is the voltage at the cathode of diode D_1. Since there is essentially no current flowing through R_{S1}, the voltage at the anode of D_1 is $-V_{SP}$. Thus

$$\text{PIV} = V_{LP} - (-V_{SP})$$
$$= V_{LP} + V_{SP}$$

or, substituting Equation 4.21 into the above equation,

$$\text{PIV} = 2 * V_{SP} - I_P R_S - V_{DP} \qquad (4.22)$$

EXAMPLE 4.3
Redesign the 10 ADC 13.2 VDC battery charger of Example 4.2 using a CT rectifier. Specify all circuit components.

SOLUTION:

Again assume that $V_{Don} = 0.6$ V, $V_{DP} = 0.8$ V, and the internal resistance of the battery is about $R_L = 0.2$ Ω. Further, assume that the transformer secondary has an effective internal resistance of about $R_S = 0.3$ Ω on each side of the center tap. Figure 4.10 contains the circuit.

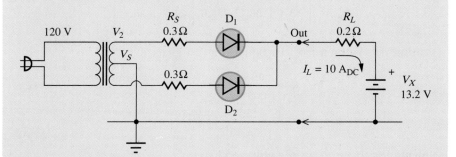

FIGURE 4.10 Example 4.3.

From the preceding examples we know that for this circuit

$$V_{Son} = V_{Soff} = 13.8 \text{ V}$$

Solving Equation 4.20 for I_P gives

$$
\begin{aligned}
I_P &= \frac{I_{av} * \pi}{2 \cos \theta_{on}} \\
&= \frac{10 \text{ A} * \pi}{2 \cos \theta_{on}} \\
&= \frac{15.7 \text{ A}}{\cos \theta_{on}}
\end{aligned}
\tag{4.23}
$$

Doing a KVL walk around the circuit of Figure 4.10, assuming only one diode conducting gives

$$V_{SP} - I_P R_S - V_{DP} - I_P R_L - V_X = 0 \text{ V}$$

Substituting Equation 4.23 into the above equation gives

$$
\begin{aligned}
V_{SP} &= \frac{15.7 \text{ A}}{\cos \theta_{on}} (0.3 \text{ Ω} + 0.2 \text{ Ω}) + 0.8 \text{ V} + 13.2 \text{ V} \\
&= \frac{7.85 \text{ V}}{\cos \theta_{on}} + 14.0 \text{ V}
\end{aligned}
\tag{4.24}
$$

From Example 4.2 we have

$$\theta_{on} = \sin^{-1}\left(\frac{13.8 \text{ V}}{V_{SP}}\right) \tag{4.19}$$

Using iteration on Equations 4.24 and 4.19 as we did in Example 4.2 yields:

Initial guess: $\theta_{on} = 45°$

$V_{SP} =$	25.1	23.4	23.7	23.7	23.7	V
$\theta_{on} =$	33.4	36.1	35.6	35.7	35.7	°

Putting this value for θ_{on} in Equation 4.23 gives

$$I_P = \frac{15.7 \text{ A}}{\cos \theta_{on}} \qquad (4.23)$$

$$= \frac{15.7 \text{ A}}{\cos 35.7°} = 19.3 \text{ AP}$$

Solving for the rms source voltage gives

$$V_{Srms} = \frac{V_{SP}}{\sqrt{2}}$$

$$= \frac{23.7 \text{ V}}{\sqrt{2}} = 16.8 \text{ V}_{rms}$$

Since the transformer is center-tapped, the transformer secondary rating, V_2, will be

$$V_2 = 2 \, V_{Srms}$$
$$= 2 * 16.8 \text{ V}_{rms} = 33.5 \text{ VCT}$$

The current rating for the transformer is the same as the current to the load which is 10 A. The average diode current rating is half the average load current, or 5 A. The peak diode current is $I_P = 19.3$ A. The diode surge current rating I_{surge} can be estimated using the argument first presented in Example 4.1. If the output is shorted, the KVL equation for the circuit of Figure 4.10 is

$$V_{SP} - I_{surge}R_S - V_{DP} = 0 \text{ V}$$

or

$$I_{surge} = \frac{V_{SP} - V_{DP}}{R_S}$$

$$= \frac{23.7 \text{ V} - 0.8 \text{ V}}{0.3 \text{ Ω}} = 76.3 \text{ A}$$

The PIV rating for the diode is

$$PIV = 2 * V_{SP} - I_P R_S - V_{DP} \qquad (4.22)$$
$$= 2 * 23.7 \text{ VP} - 19.3 \text{ AP} * 0.3 \text{ Ω} - 0.8 \text{ V} = 40.8 \text{ VP}$$

We can now specify the rectifier components as follows:

DIODES: (2 required)

$$PIV > 40.8 \text{ V}, I_{av} > 5 \text{ A}, I_P > 19.3 \text{ AP}, I_{surge} > 76.3 \text{ AP}$$

TRANSFORMER:

$$V_{pri} = 120 \text{ Vrms}, V_{sec} = 33.5 \text{ VCT}, I_{sec} > 10 \text{ A}$$

4.8 The Bridge Rectifier

Figure 4.11a contains the schematic for a bridge rectifier as you will usually see it drawn. Very often the four diodes are hooked up as shown in Figure 4.11a, encapsulated in a single package, and sold under a single part number.

This is probably the most common rectifier used in consumer products such as tape players, radios, and TVs. The AC input can be obtained from a transformer if isolation from the power line and/or a change in the input voltage is required. If these requirements can be avoided, the AC input can be obtained directly from the power line. The designer of lower-priced consumer products often takes advantage of the cost, weight, and space-saving offered by such a transformerless power supply. You should exercise all of the cautions described in Section 4.3 when working on these devices.

4.9 Bridge Rectifier Operation

To understand how the bridge rectifier works, let us first assume that at this particular instant, the source voltage is as shown in Figure 4.11b with Point A at a higher voltage than at Point B. Remember what that means: In the most elementary terms, Point A has an excess of positive charges which it would "like" to get rid of, and Point B has a shortage of positive charges and would "like" to pick up some charges. Thus current would like to flow away from Point A through diodes D_4 and D_1 and back to Point B. With no resistance in that path, the current would probably be quite large (infinite?). However, D_4 is headed the wrong way so that it is reverse biased and it blocks current flow. It has been circled in Figure 4.11b to remind you that it is blocking. Diode D_1 is forward biased and so it conducts. The current that flows through D_1 would also like to flow through D_2, and right back to the source at Point B. Again there is no resistance so the current would be large. This time it is D_2 that is headed the "wrong" way, saving us from catastrophe. D_2 has also been circled in Figure 4.11b to indicate that it is blocking.

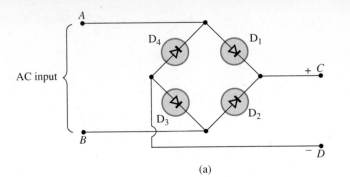

AC input

(a)

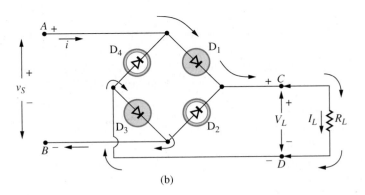

(b)

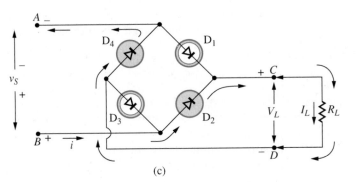

(c)

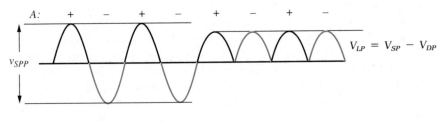

$$V_{LP} = V_{SP} - V_{DP}$$

(d)

FIGURE 4.11 Bridge rectifier.

(a) The circuit.

(b) Conduction path when terminal A is the more positive input terminal.

(c) Conduction path when terminal B is the more positive input terminal.

(d) Waveforms showing whether terminal A is the more positive or the more negative terminal.

Consequently current will flow from Point A through D_1 to Point C. Then from Point C through the load R_L to Point D. Will it now flow through diode D_4 and back to Point A, or will it flow through diode D_3 and on to Point B? Our very elementary discussion above was intended to answer that question. The fact that there is already an excess of positive charges at Point A makes it very unlikely that our positive current will want to go back to Point A. In fact, since there is a shortage of positive charges at Point B, our current will want to go there. Thus when Point A is positive, diodes D_1 and D_3 conduct and diodes D_2 and D_4 block so that current flows as indicated in Figure 4.11b creating a positive current flow in the load.

Now let us assume that it is one-half cycle later and the polarity of the AC source is reversed. Figure 4.11c contains the circuit for this situation. Again in very elementary terms, excess charge will be present at Point B and there will be a shortage of charge at Point A. D_3 will block and current will flow through D_2. Diode D_1 will block forcing the current to flow from Point C through the load to Point D. Then, attracted to Point A, current will flow back through D_4 as indicated in Figure 4.11c. Thus D_2 and D_4 conduct and D_1 and D_3 block so that current flows as indicated in Figure 4.11c again creating a positive current flow in the load.

The waveform that results from the above activities is drawn in Figure 4.11d.

Another way of looking at the bridge is to think of it as an intelligent switch. It looks at terminals A and B. It connects the one that is positive to terminal C and it connects the one that is negative to terminal D.

Obviously a "positive power supply" results if Point D is considered "ground" and Point C is considered "hot." A "negative power supply" results if Point C is considered "ground" and Point D is considered "hot."

For your safety as well as that of the equipment, it is important to recognize that Points A and B must NOT share a common ground with Points C and D.

4.10 Bridge Rectifier Component Specification

The bridge rectifier is a full-wave rectifier so conceptually it is much the same as the center tapped rectifier. The differences stem from the fact that two diodes conduct on each half of the power line cycle, and no center tap is needed on the transformer secondary winding, so the whole winding conducts on each half cycle.

EXAMPLE 4.4

Redesign the 10 A 13.2 VDC battery charger of Example 4.2 using a bridge rectifier. Specify all circuit components.

SOLUTION:

Again assume that $V_{Don} = 0.6$ V, $V_{DP} = 0.8$ V, $R_S = 0.3$ Ω, and $R_L = 0.2$ Ω. Figure 4.12 contains the circuit.

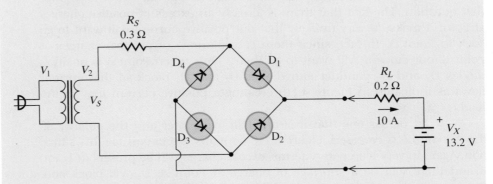

FIGURE 4.12   Example 4.4.

Since diodes D_1 and D_3 conduct separately from diodes D_2 and D_4, for purposes of circuit analysis we can ignore the diodes that are not conducting, but we must remember that on each half-cycle TWO diodes conduct. Therefore, the turn-on voltage V_{Son} becomes

$$V_{Son} = V_{Soff} = 2\,V_{Don} + V_X \qquad (4.25)$$
$$= 2 * 0.6\text{ V} + 13.2\text{ V} = 14.4\text{ V}$$

Since this full-wave rectifier is functionally similar to that of Example 4.3 we have

$$I_P = \frac{I_{av} * \pi}{2 \cos \theta_{on}}$$

$$= \frac{10\text{ A} * \pi}{2 \cos \theta_{on}}$$

$$= \frac{15.7\text{ A}}{\cos \theta_{on}} \qquad (4.23)$$

If we assume an instantaneous peak source voltage V_{SP} with polarity as shown in Figure 4.13, diodes D_1 and D_3 will be conducting and diodes D_2 and D_4 can be ignored. Taking a KVL walk clockwise around the circuit consisting of the transformer secondary, D_1, the battery being charged, and D_3 gives us

$$V_{SP} - I_P R_S - V_{DP} - I_P R_L - V_X - V_{DP} = 0\text{ V} \qquad (4.26)$$

Solving for the source voltage V_{SP} gives

$$V_{SP} = I_P(R_S + R_L) + 2\,V_{DP} + V_X \qquad (4.27)$$

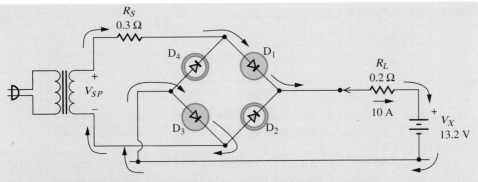

FIGURE 4.13 Current flow in Example 4.4.

Substituting Equation 4.23 into Equation 4.27 gives

$$V_{SP} = \frac{15.7\ A}{\cos\theta_{on}}(0.3\ \Omega + 0.2\ \Omega) + 2 * 0.8\ V + 13.2\ V \qquad (4.28)$$

$$= \frac{7.85\ V}{\cos\theta_{on}} + 14.8\ V$$

Again there are two unknowns so we call on our assumption that the waveform is a sine wave as we did in the previous examples. Using our new value for V_{Son} from Equation 4.25, we get

$$\theta_{on} = \sin^{-1}\left(\frac{V_{Son}}{V_{SP}}\right) \qquad (4.13)$$

$$= \sin^{-1}\left(\frac{14.4\ V}{V_{SP}}\right) \qquad (4.29)$$

Using iteration on Equations 4.28 and 4.29 as we did in previous examples, using as a first guess, $\theta_{on} = 45°$, we have

$V_{SP} =$	25.9	24.2	24.6	24.5	24.5	VP
$\theta_{on} =$	33.8	36.4	35.9	36.0	36.0	°

Putting the above value for θ_{on} in Equation 4.23 gives

$$I_P = \frac{15.7\ A}{\cos\theta_{on}} \qquad (4.23)$$

$$= \frac{15.7\ A}{\cos 36.0°} = 19.4\ AP$$

Solving for the rms source voltage gives

$$V_{Srms} = \frac{V_{SP}}{\sqrt{2}}$$

$$= \frac{24.5 \text{ VP}}{\sqrt{2}} = 17.3 \text{ Vrms}$$

In this case, the rms source voltage is the transformer secondary voltage rating. You should notice this important difference from the center-tapped circuit.

To determine the diode peak inverse voltage PIV for this circuit, referring to the circuit of Figure 4.13, we will take a KVL walk clockwise around the circuit consisting of the transformer secondary and either diodes D_1 and D_2 or diodes D_4 and D_3. The resulting equation is

$$V_{SP} - I_P R_S - V_{DP} - \text{PIV} = 0 \text{ V} \tag{4.30}$$

Solving for PIV

$$\text{PIV} = V_{SP} - I_P R_S - V_{DP}$$
$$= 24.5 \ V_P - 19.4 \text{ AP} * 0.3 \ \Omega - 0.8 \text{ V} = 17.9 \text{ VP}$$

Notice that the PIV requirement for the bridge circuit is about half what it was for the center-tapped circuit of Example 4.3.

Finally, using the argument presented in Section 4.5 to estimate the surge current I_{surge}, we write the KVL equation for the circuit assuming the output to be shorted

$$V_{SP} - I_{surge} R_S - 2V_{DP} = 0 \text{ V} \tag{4.29}$$

Solving for I_{surge}

$$I_{surge} = \frac{V_{SP} - 2V_{DP}}{R_S}$$

$$= \frac{24.5 \text{ V} - 2 * 0.8 \text{ V}}{0.3 \ \Omega} = 76.3 \text{ V}$$

We can now specify the components for this rectifier as follows:

DIODES: (4 required)
PIV > 17.9 V, $I_{av} > 5$ A if diodes are individually packaged and heat sinked, 10 A if diodes are packed together. $I_P > 19.4$ AP, $I_{surge} > 76.3$ AP

TRANSFORMER:
$V_{pri} = 120$ Vrms, $V_{sec} = 17.3$ Vrms, $I_{sec} > 10$ A

4.11 Rectifier Comparison

We have looked briefly at the three most common rectifier types. Table 4.1 provides an easy comparison of them under identical circuit conditions. You are encouraged to obtain a current electronics parts catalog and determine the parts cost for each of the rectifiers we designed. Obviously the half-wave rectifier is the simplest and probably the least expensive design. The fact that the current pulses are so far apart makes it also the least desirable. We will learn in Chapter 5 that this makes filtering more difficult. It also makes the peak current higher, which places greater stress on the circuit components. As the cost of diodes has declined and their quality and reliability have improved, the use of the half-wave rectifier has also declined. For most applications, the designer's choice is between the center-tapped rectifier and the bridge rectifier.

In terms of quality, the center-tapped circuit is very attractive. Since the center-tap can be grounded, it is very noise free and it is an easy and safe circuit to use. Since only two diodes are required, less heat sinking is required and the output voltage is more stable for changing loads.

The bridge circuit requires a less expensive transformer than the center tapped design since, according to Table 4.1, its secondary voltage rating is roughly half as high and it does not have to have a center tap. In fact, provided the voltage does not need to be stepped either up or down, a bridge rectifier can be connected directly to the power line without using a transformer at all. This makes the bridge circuit very attractive in applications where the cost, size, and weight of the transformer are significant.

Although more diodes are used in the bridge circuit, Table 4.1 indicates that their PIV rating needs to be only about half that of the center tapped design and, at least for low power applications they are often housed in a single package, thus reducing assembly costs.

The above factors contribute to make the bridge rectifier almost universally used as the input rectifier in switching power supplies and they are some of the factors the designer must weigh in deciding which type of rectifier to use for a specific application.

TABLE 4.1 Component Specification for a 10 A 13.2 V Battery Charger

| Rectifier Type | Diode/s | | | | Transformer |
	Number Required	PIV (Volts)	I_P (Amps)	I_{surge} (Amps)	V_2 (Volts)
Half-wave	1	44.7	34.9	102	22.3
CT	2	40.8	19.3	76.3	33.5 (CT)
Bridge	4	17.9	19.4	76.3	17.3

▶▶▶▶▶▶▶▶ **4.12   Troubleshooting**

By now you should be aware that if you are to understand the behavior of the circuits we have been dealing with, an oscilloscope is an almost indispensable tool. It is also an indispensable tool in troubleshooting. The amplitude, shape, and period of the waveform at various points in a circuit will tell the experienced technologist a great deal about the condition of the circuit.

The horizontal spacing of the pulses from a full-wave rectifier should be about 8.3 msec. If they are 16.6 msec apart, you know that the circuit is functioning as a half-wave rectifier so part of the rectifier is not working. The most likely suspect is one of the diodes. If alternate pulses from a full-wave rectifier have significantly reduced amplitude, again a problem exists, and the most likely cause is that one of the diodes is not conducting as well as it should.

Obviously if one of the diodes in a rectifier circuit short circuits, you have a more serious problem involving blown fuses or other more expensive damage.

4.13   Summary

We have looked at three types of rectifiers: the half-wave, the center tapped transformer, and the bridge type. We have looked at some of the advantages and disadvantages of each type for the design of a simple battery charger. You should now know the principles of operation of each type, how current flows in the circuit, and how rectification is accomplished.

Chapter 5 will continue the development of this chapter by adding the filtering required to construct a usable power supply.

GLOSSARY

Bridge circuit: A commonly used electric circuit consisting of two parallel voltage divider circuits usually connected to a source. The load or output is usually connected between the intermediate terminals of the voltage dividers. (Section 4.8)

Center-tapped transformer (CT transformer): A transformer having a "center tap" on at least one of its windings. (Section 4.6)

Full-wave rectifier: A rectifier that uses both halves of the AC waveform. (Section 4.6)

Half-wave rectifier: A rectifier that uses only one-half of the AC waveform. (Section 4.4)

Isolation transformer: A transformer intended to provide electrical isolation between electric circuits so that they no longer share a common ground. (Section 4.3)

Peak inverse voltage (PIV): The maximum reverse voltage that occurs across a diode in a circuit. (Section 4.5)

Phase splitting: The act of breaking, or "splitting" an AC waveform into two waveforms that are 180 electrical degrees out of phase with one another. (Section 4.6)

Power supply: Circuitry whose purpose is to supply electric power to a circuit. (Section 4.2)

Rectifier: A device or a circuit intended to convert alternating current into direct current. (Section 4.2)

Surge current (I_{surge}): Current that is much larger than normal and exists for a very short time. (Section 4.5)

PROBLEMS

Section 4.3

1. Assuming that your body is well grounded, draw a sketch showing current flow on each half of the AC cycle if you touched Point B of Figure 4.2a when the plug is connected to the 120 V power line such that:
   (a) Point A in Figure 4.2a is "hot."
   (b) Point A in Figure 4.2a is grounded.

Section 4.5

2. Find the current the battery charger of Example 4.1 will supply if your car's battery is really dead (V_x = 11 VDC).

3. Suppose your roommate (you would not do this!) tried to charge a 9 VDC battery (R_L = 1 Ω) using the battery charger of Example 4.1. Calculate the charging current.

4. Design a transformerless half-wave battery charger that will supply 100 mA to a 9 VDC battery. Assume R_L = 1 Ω. Specify all of the necessary components. *Hint:* Use a current limiting resistor.

5. Repeat Problem 4 using a transformer whose secondary resistance can be assumed to be 0.3 Ω.

Section 4.6

6. Repeat Problem 4 using a CT transformer full-wave battery charger. Assume that the resistance of the secondary is 0.3 Ω each side of the center tap.

Section 4.7

7. Draw the circuit for a CT rectifier that will put out negative voltage rather than positive voltage.

Section 4.8

8. Repeat Problem 4 using a transformer whose secondary resistance can be taken as 0.3 Ω, and a bridge rectifier.
9. Repeat Problem 4 using a transformerless bridge rectifier design.

Section 4.10

10. From your school library obtain a current electronics parts catalog and make out a parts order for the components needed to build the battery charger designed in
    (a)   Problem 4.   (b)   Problem 5.   (c)   Problem 6.
    (d)   Problem 7.   (e)   Problem 8.
11. Figure 4.14 shows another way to draw a bridge rectifier. Copy the

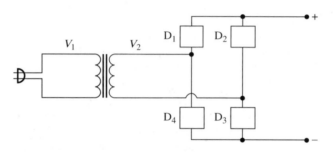

FIGURE 4.14   Problem 4.11.

circuit putting diode symbols on each of the diodes showing which way they are connected. Using different colored arrows on your diagram, show the direction of current flow on each half of the power line cycle.

12. Using the same transformer as was used in Example 4.1, redesign the battery charger using a bridge rectifier. Determine the new charging current.

Section 4.11

13. Design a "quick charger" that can supply 100 A to your car's battery ($V_X = 13.2$ V; $R_L = 0.1$ Ω). Use a CT rectifier design. Assume $R_S = 0.2$ Ω. Specify all components. How much power is dissipated by the transformer and the diodes?

Section 4.12

14. Describe the symptoms observed by the user (who does not have a scope or DMM) if the diode in the battery charger of Problem 4 became:

    (a)   A short circuit.

    (b)   An open circuit.

15. Repeat Problem 14 if one of the diodes in Problem 6 developed the indicated fault.

16. Being as quantitative as you can, draw the waveform you would observe at the battery terminals in

    (a)   Problem 6

    (b)   Problem 15 part b.

17. Suppose that after a couple of years of service, the garage mechanic to whom you sold the quick charger of Problem 13 brings it back to you with the complaint that it puts out much less current than it did when it was new. What is the most likely fault, and what tests will you perform to discriminate between that fault and other possible faults?

A power supply with filter capacitors on the right.
(Courtesy of Kepco Power Supplies.)

Chapter 5
Power Supply Filters

SPECIAL TERMS

Filter
Ripple factor (r)
Regulation (vr)
R-C filter
π filter
L-C filter

5.1   Introduction and Objectives

In Chapter 4 about the only use we found for the rectifiers we talked about was for the construction of battery chargers. The reason is that the output from a rectifier is not smooth. It supplies power in short pulses. In this chapter we will call on your knowledge of capacitors to develop a filter which will smooth the output from the rectifier and make it a viable substitute for battery power. We will then present a design recipe that can be used to design a usable power supply.

After having studied this chapter, you should understand the concept of power supply filtering and you should be able to:

- Use the design recipe we will develop to design simple power supplies such as those used to power many common electronic devices.
- Analyze the schematic diagram of a simple power supply in order to determine its operating parameters.

5.2   The Need for Filtering

Except as a very simple battery charger, the rectifiers that we studied in Chapter 4 usually are not very useful. The reason is that they deliver power in short pulses rather than at a constant rate as a battery would.

Figure 5.1 indicates that the output voltage from a rectifier can be considered as a DC voltage with an AC voltage riding on top of it. In fact,

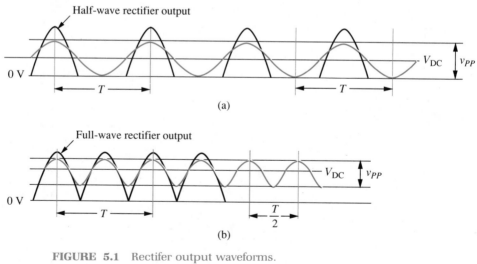

FIGURE 5.1   Rectifer output waveforms.
(a) Half-wave.
(b) Full-wave.

Equation 4.2 tells us what the magnitude of the average DC component of current will be. Just by looking at Figure 5.1a you can tell that for a half-wave rectifier, the amplitude of the AC component is about the same as the DC component. As you can see in Figure 5.1b, in the case of the full-wave rectifier the AC component is less, but it is still significant. Most electronic equipment demands that the AC component be negligible compared to the DC component. When it is not, you hear a low frequency rumble in your stereo (probably 120 Hz) or you see a wavy line moving slowly through the display on your computer screen. It is the effective elimination of the AC component that filtering is all about.

5.3 Ripple and Regulation

Two criteria are usually used to estimate the quality of a power supply. One is "ripple factor" or "percent ripple." Ripple factor, r, is defined to be the ratio of the RMS voltage present in a waveform divided by the DC voltage present. Since ripple factor is a unitless quantity, percent ripple $\%r$ is obtained by multiplying ripple factor by 100%. Thus

$$r \equiv \frac{V_{Lrms}}{V_{LDC}} \tag{5.1}$$

or

$$\%r \equiv \frac{V_{Lrms}}{V_{LDC}} \, 100\% \tag{5.2}$$

The other measure of power supply performance is its "regulation," vr, or "percent regulation," $\%vr$. Regulation is calculated by first finding the change in output voltage obtained by subtracting the output voltage when the circuit is fully loaded, V_L from the output voltage when the circuit has no load attached, V_O. Then to find vr, the above change is divided by the fully loaded voltage V_L. Regulation is also a unitless quantity so the percent regulation $\%vr$ is obtained by multiplying the regulation by 100%. Thus

$$vr \equiv \frac{V_{ODC} - V_{LDC}}{V_{LDC}} \tag{5.3}$$

or

$$\%vr \equiv \frac{V_{ODC} - V_{LDC}}{V_{LDC}} \, 100\% \tag{5.4}$$

You should notice that the smaller the ripple is, the better your power supply is. Also, the smaller the regulation, the better. This might seem strange. It is easy to think that if the regulation were 100 percent, that is as good as it could get. However, if you think of regulation as the amount the load can affect, or "regulate" the output voltage, you can see that the less the regulation the better. In fact, we would like for the output to remain exactly constant regardless of what load we placed on our power supply. That would be zero regulation, and that would be perfect!

EXAMPLE 5.1
The output of the power supply that powers the little clock-radio that wakes me up every morning is pictured in Figure 5.2a. When the wire

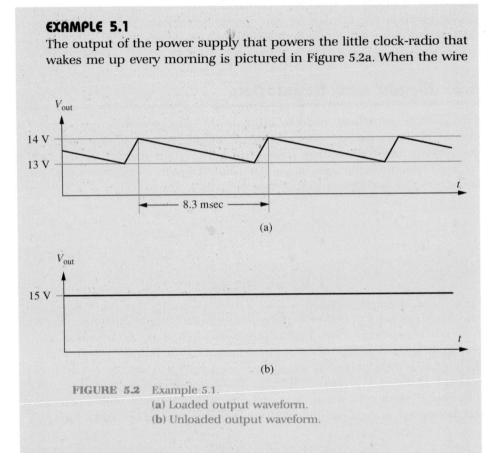

FIGURE 5.2   Example 5.1.
(a) Loaded output waveform.
(b) Unloaded output waveform.

that connects the power supply to the radio circuit is disconnected, the waveform is as shown in Figure 5.2b.

(a) Is it a half-wave or a full-wave rectifier?
(b) Find the percentage ripple.
(c) Find the percentage regulation.

SOLUTION:
(a) **Rectifier Type**   From Figure 5.2a the spacing between peaks on the waveform is $T_r = 8.3$ ms so the ripple frequency is

$$f_r = \frac{1}{T_r} \tag{5.5}$$

$$= \frac{1}{8.3 \text{ ms}} = 120 \text{ Hz}$$

Since the above ripple frequency is about twice the power line frequency, it is probably a full-wave rectifier.

(b) Ripple   From Figure 5.2a, the maximum voltage when the circuit is loaded is $V_{max} = 14$ V and the minimum value is $V_{min} = 13$ V. Thus the AC output voltage is

$$v_{PP} = V_{max} - V_{min} \tag{5.6}$$

$$= 14 \text{ V} - 13 \text{ V} = 1 \text{ } VPP$$

Since ripple is defined in terms of rms AC voltage, the above peak to peak value must be converted to rms. But be careful. This waveform is not a sine wave so we cannot use Equation 4.10 to obtain the rms value. The waveform looks more like a triangle. Appendix D.3 contains the derivation of a formula for obtaining the rms value of a triangle waveform. The formula is

$$v_{rms} = \frac{v_{PP}}{2\sqrt{3}} \tag{5.7}$$

$$v_{rms} = \frac{1 \text{ } VPP}{2\sqrt{3}} = 0.289 \text{ V}_{rms}$$

The DC voltage level will be the average value of the waveform shown in Figure 5.2, which is given by

$$V_{DC} = \frac{V_{max} + V_{min}}{2} \tag{5.8}$$

$$= \frac{14 \text{ V} + 13 \text{ V}}{2} = 13.5 \text{ VDC}$$

The percentage ripple will be

$$\%r \equiv \frac{V_{rms}}{V_{DC}} 100\% \tag{5.2}$$

$$= \frac{0.289 \text{ V}_{rms}}{13.5 \text{ V}_{DC}} 100\% = 2.14\%$$

(c) Regulation

$$\%vr \equiv \frac{V_{ODC} - V_{LDC}}{V_{LDC}} 100\% \tag{5.4}$$

$$= \frac{15 \text{ VDC} - 13.5 \text{ VDC}}{13.5 \text{ VDC}} 100\% = 11.1\% \tag{5.4}$$

5.4 Capacitor Filter—Light Loading

A simple capacitor filter is shown in Figure 5.3. The resulting output wave-form is shown in Figure 5.4. The rectifier circuit can be any of the types we have studied. During time interval t_2 the rectifier conducts charging

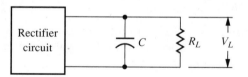

FIGURE 5.3 Capacitor filter circuit.

the capacitor to the peak voltage V_{LP}. During time interval t_1 the rectifier is reverse biased and so it is not conducting. The capacitor then supplies the load current by discharging into the load. As the capacitor discharges, the load voltage drops according to the capacitor discharge formula which you should remember from your electric circuits class

$$V_t = V_O e^{-t/\text{RC}} \tag{5.9}$$

At the end of the discharge time t_1 the voltage will be V_{Lt1}. Thus the load does not see a steady DC voltage. Rather there is an AC ripple riding on top of the DC. Its magnitude is

$$v_{LPP} = V_{LP} - V_{Lt1} \tag{5.10}$$

A quick look at Figure 5.4 indicates that the waveform of the ripple is not a sine wave. It is closer to a triangle wave such as we saw in Example 5.1.

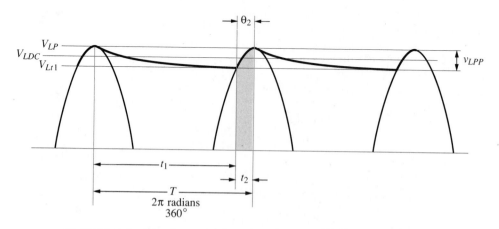

FIGURE 5.4 Waveform produced by a capacitor filtered rectifier circuit.

Thus, according to the derivation in Appendix D.3, to obtain the rms value of the ripple we should use

$$v_{rms} = \frac{V_{PP}}{2\sqrt{3}} \qquad (5.7)$$

The DC voltage to the load is approximately the arithmetic average of the waveform in Figure 5.4, which is given by

$$V_{LDC} = \frac{V_{LP} + V_{Lt1}}{2} \qquad (5.8)$$

The period of the ripple, T_r, is determined by the source frequency and the type of rectifier used. For a half-wave rectifier operated from the 60 Hz power line, $T_r \approx 16.7$ msec. For a switching power supply, T_r can be as short as 10 μsec.

You will notice that the charging portion of the waveform in Figure 5.4 is described by a trigonometric function and the discharging portion is described by a logarithmic function. When we are analyzing circuits, that will mean that we will want to solve the two functions simultaneously to obtain V_{t1} and t_1. As you might guess, an exact solution would be messy.

One way to avoid the above problem is to recognize that if you look at only a short segment of the capacitor discharge curve, as would be the case if R_L is large, C is large, or T_r is short, the curve is nearly a straight line. Further, with the aid of calculus we could show (see Appendix D.4) that the above straight line, which is line AB in Figure 5.5, will intersect the time axis at a time τ (Greek letter "tau") given by

$$\tau = RC \qquad (5.11)$$

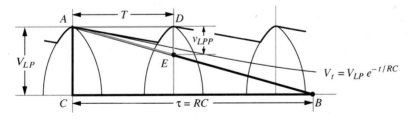

FIGURE 5.5 Linearized capacitor discharge curve.

You will notice that the above straight line in Figure 5.5 forms a pair of similar triangles, ABC and EAD. Since corresponding parts of similar triangles are proportional, we can write

$$\frac{BC}{AD} = \frac{AC}{DE} \qquad (5.12)$$

or approximately

- $$\frac{RC}{T_r} \approx \frac{V_{LP}}{V_{LPP}} \qquad (\% r < 5\%) \tag{5.13}$$

The above approximation is only good if the ripple is less than about 5 percent. Since many modern power supplies depend on regulator circuits to remove the ripple, larger ripple values are frequently acceptable. Equation 5.13 is not valid in such cases.

EXAMPLE 5.2

Assume that the clock-radio that produced the waveform in Figure 5.2 draws a current of 85 mADC. Find the value of the filter capacitor in its power supply.

SOLUTION:

The average DC voltage can be estimated from the waveform of Figure 5.2 to be

$$V_{DC} = \frac{V_{max} + V_{min}}{2} \tag{5.8}$$

$$= \frac{14\text{ V } + 13\text{ V}}{2} = 13.5\text{ VDC}$$

The equivalent circuit resistance of the radio is given by

$$R_L = \frac{V_{LDC}}{I_{LDC}}$$

$$= \frac{13.5\text{ VDC}}{85\text{ mADC}} = 159\ \Omega$$

Since it is known from Example 5.1 that the ripple is less than 5 percent, the approximate formula of Equation 5.13 applies. Solving for C gives

$$C = \frac{T_r V_{LP}}{R_L V_{LPP}}$$

$$= \frac{8.33\text{ msec} * 14.0\text{ VDC}}{159\ \Omega * 1\text{ VPP}} = 734\ \mu\text{F}$$

5.5   Capacitor Filter—Heavy Loading

If you looked very closely at Figure 5.5 you probably noticed that the discharge time is not really T and the ripple voltage will really be less than what is indicated in Figure 5.5. As you can see, the larger the allowable ripple, the larger these errors will be. To the extent that these errors exist, Equation 5.13 will overestimate the capacitor size needed for a given amount of filtering. The result will be a power supply that is bulkier and more expensive that it needs to be. That is poor engineering.

Let us now develop a more valid method of dealing with heavily loaded circuits.

In order to talk about the waveforms involved, you should remember from your circuits class the following relationship between the period of the waveform and the conduction angle for the waveform shown in Figure 5.4

$$\frac{\theta_2}{t_2} = \frac{2\pi}{T} = \frac{360°}{T} \tag{5.14}$$

or, substituting Equation 5.5 and solving for θ_2

$$\theta_2 = 2\pi f_S t_2 = 360°f_S t_2 \tag{5.15}$$

Further, we will find it more convenient to talk about the waveform output by the rectifier as starting from the peak voltage V_{LP} rather than at zero volts. The trigonometry student will recognize that this is easily done by calling the waveform a "cosine wave" rather than a sine wave. Thus during the time t_2 when the rectifier is conducting and the capacitor is being charged, we can write

$$V_{Lt1} = V_{LP} \cos \theta_2 \tag{5.16}$$

or, substituting Equation 5.15 into Equation 5.16

$$V_{Lt1} = V_{LP} \cos (2\pi f_S t_2) \tag{5.17}$$

Or, if you and your calculator think better in degrees than in radians

$$V_{Lt1} = V_{LP} \cos (360°f_S t_2) \tag{5.18}$$

Solving the above equation for t_2 gives

$$t_2 = \frac{\cos^{-1}\left(\dfrac{V_{Lt1}}{V_{LP}}\right)}{360°f_S} \tag{5.19}$$

It is important to notice that f_S in the above equation is the line frequency input to the rectifier, and is independent of the type of rectifier used. In the United States, f_S is usually 60 Hz unless you are dealing with the output rectifier of a switching power supply.

From Figure 5.4 it is evident that

$$t_1 = T_r - t_2 \tag{5.20}$$

where T_r is the repetition period for the rectifier. Its value will depend on the line frequency and the type of rectifier that is used. For the half-wave rectifier, T_r will usually be 1/60 sec. For a full-wave rectifier, T_r will usually be 1/120 sec.

Substituting Equation 5.19 into Equation 5.20 gives

$$t_1 = T_r - \frac{\cos^{-1}\left(\dfrac{V_{Lt1}}{V_{LP}}\right)}{360°f_S} \tag{5.21}$$

During the time interval t_1, when the capacitor is supplying the load current, the familiar capacitor discharge equation applies and we can write

$$V_{Lt1} = V_{LP}e^{-t1/RC} \tag{5.9}$$

or, solving the above equation for capacitance

$$C = -\frac{t_1}{R_L \ln\left(\dfrac{V_{Lt1}}{V_{LP}}\right)} \tag{5.22}$$

The above discussion contains all of the basic ideas we need to design or analyze a capacitor filter circuit.

EXAMPLE 5.3

Repeat Example 5.2 using the above more accurate method.

● (For a SuperCalc solution, see Appendix G.)

SOLUTION:
Using the voltages read from Figure 5.2, the conduction time t_2 is calculated

$$t_2 = \frac{\cos^{-1}\left(\dfrac{V_{Lt1}}{V_{LP}}\right)}{360°f_S} \tag{5.19}$$

$$= \frac{\cos^{-1}\left(\dfrac{13\text{ V}}{14\text{ V}}\right)}{360° * 60\text{ Hz}} = 1.01 \text{ msec}$$

Using the period read from Figure 5.2, the discharge time t_1 is

$$t_1 = T - t_2 \tag{5.20}$$

$$= 8.33 \text{ msec} - 1.01 \text{ msec} = 7.32 \text{ msec}$$

Using the above values and the effective load resistance calculated for Example 5.2, the capacitance C is given by

$$C = \frac{t_1}{R_L \, ln\left(\dfrac{V_{Lt1}}{V_{LP}}\right)} \tag{5.22}$$

$$= -\frac{7.32 \text{ msec}}{159 \, \Omega \, * \, ln\left(\dfrac{13 \text{ V}}{14 \text{ V}}\right)} = 622 \text{ μF}$$

If you compare this result with that of Example 5.2, you will see that the more approximate method of Section 5.4 does, in fact, overestimate the size of the filter capacitor needed.

EXAMPLE 5.4

Determine the DC output voltage, the percent ripple and the percent regulation for the power supply of Figure 5.6.

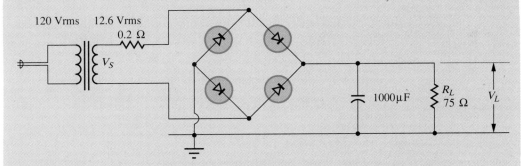

FIGURE 5.6   Example 5.4.

● (For a SuperCalc solution, see Appendix G. For a PSpice display of the waveform, see Appendix H.)

SOLUTION:

The output waveform will be similar to Figure 5.4, but since the recti-fier is a full-wave type, the period T_r will now be

$$T_r = \frac{1}{2f_S} \tag{5.23}$$

$$= \frac{1}{2 * 60 \text{ Hz}} = 8.33 \text{ msec}$$

The peak source voltage V_{SP} will be

$$V_{SP} = \sqrt{2} \, V_{Srms} \tag{4.10}$$

$$= \sqrt{2} * 12.6 \text{ Vrms} = 17.8 \text{ VP}$$

To calculate the peak voltage to the load, V_{LP}, assume that the peak diode voltage will be about 0.8 VP and recognize that the source resistance and the load form an unloaded voltage divider. If we assume that at the peak of the waveform, the capacitor will be fully charged, then

$$V_{LP} = (V_{SP} - 2\,V_{DP})\frac{R_L}{R_L + R_S} \tag{5.24}$$

$$= (17.8\text{ VP} - 2 * 0.8\text{ VP})\frac{75\ \Omega}{75\ \Omega + 0.2\ \Omega} = 16.2\text{ VP}$$

The equation for the minimum voltage to the load, V_{Lt1} can now be written.

$$V_{Lt1} = V_{LP}e^{-t1/RC} \tag{5.9}$$

$$= 16.2\text{ VP} * e^{-t1/(75\ \Omega\ *\ 1000\ \mu F)}$$

Since there are two unknowns in the above equation, another relation between the same two variables is required. We can get it from

$$t_1 = T_r - \frac{\cos^{-1}\left(\dfrac{V_{Lt1}}{V_{LP}}\right)}{360°f_S} \tag{5.21}$$

$$= 8.33\text{ msec} - \frac{\cos^{-1}\left(\dfrac{V_{Lt1}}{16.2\text{ VP}}\right)}{360° * 60\text{ Hz}}$$

By iterating Equations 5.9 and 5.21, V_{Lt1} and t_1 can be obtained. Looking at Figure 5.4, and remembering that $T_r = 8.33$ msec, it seems reasonable to use as our first guess $t_1 = 8$ msec. The result is

$V_{Lt1} = 16.2\text{ VP} * e^{-t1/75\ \Omega\ *\ 1000\ \mu F}$ =	14.6	14.7	14.7	V
$t_1 = 8.33\text{ msec} - \dfrac{\cos^{-1}\left(\dfrac{V_{Lt1}}{16.2\text{ V}}\right)}{360° * 60\text{ Hz}}$ =	7.13	7.19	7.19	msec

As you might expect by looking at Figure 5.4, these equations converge very rapidly. The peak to peak ripple voltage v_{LPP} is

$$v_{LPP} = V_{LP} - V_{Lt1} \tag{5.10}$$

$$= 16.2\text{ V}_P - 14.7\text{ V} = 1.48\text{ VPP}$$

Now the rms ripple voltage v_{Lrms} can be calculated remembering that the waveform is close to a TRIANGLE wave

$$v_{Lrms} = \frac{V_{LPP}}{2\sqrt{3}} \tag{5.7}$$

$$= \frac{1.48\ V_{PP}}{2\sqrt{3}} = 0.427\ \text{Vrms}$$

The DC output voltage V_{LDC} is

$$V_{LDC} = \frac{V_{LP} + V_{Lt1}}{2} \tag{5.8}$$

$$= \frac{16.2\ VP + 14.7\ V}{2} = 15.5\ \text{VDC}$$

The percent ripple $\%r$ can now be calculated:

$$\%r = \frac{V_{Lrms}}{V_{LDC}}\ 100\% \tag{5.2}$$

$$= \frac{0.427\ \text{Vrms}}{15.5\ \text{VDC}}\ 100\% = 2.77\%$$

To obtain the regulation we first calculate the open circuit voltage V_{open}. With the load removed, no current will flow and the output will go up to the peak source voltage V_{SP} less the turn-on voltage V_{Don} of two diodes. If we assume V_{Don} to be about 0.6 V we have

$$V_{Lopen} = V_{SP} - 2V_{Don}$$

$$= 17.8\ VP - 2 * 0.6\ V = 16.6\ V$$

The percent regulation $\%vr$ is then given by

$$\%vr = \frac{V_{Lopen} - V_{LDC}}{V_{LDC}}\ 100\% \tag{5.4}$$

$$= \frac{16.6\ V - 15.5\ V}{15.5\ V} = 7.07\%$$

5.6 Power Supply Design Recipe

In practice you will often have a need for a simple unregulated power supply. You usually know how much DC voltage, V_{LDC}, you need and you usually either know the effective resistance, R_L, that your load will present to the supply, or else you can easily calculate its value from the amount of current, I_{LDC}, the load will draw. In addition you usually have some idea how much ripple voltage v_{LPP} your application can tolerate. You seldom know beforehand the effective resistance of the secondary winding of the transformer you will use, but for most small transformers you will be fairly close if you assume about 0.3 Ω. Assuming that you have the above information available, Table 5.1 contains procedural recipes which can be used to design a workable power supply. Steps 1, 2, 3, 5, etc. apply to all three of the rectifier types we have studied. Steps 4, 8, etc. are different for the various rectifier types. For SuperCalc implementations of Table 5.1, see Appendix G. A description and explanation of each of the steps follows.

1. This expression for the minimum output voltage, V_{Lt1}, is obvious from an examination of the output waveform.
2. This expression for the peak output voltage, V_{LP}, is also obvious from an examination of the output waveform. An examination of the circuit indicates that the capacitor must be rated for at least this much voltage.
3. This expression for the conduction time t_2 is Equation 5.19 and was derived in Section 5.5.
4. The period of the ripple, T_r, is the time between output pulses from the rectifier. The value of T_r depends on the frequency of the source, f_S, and the rectifier type used.
5. This expression for the discharge time, t_1, is obvious from an examination of the output waveform. It is Equation 5.20 of Section 5.5.
6. This expression for the capacitor size, C, is Equation 5.22 and was derived in Section 5.5. It is one of the ratings needed to specify the capacitor.
7. The DC current to the load, I_{LDC}, is also the rms current supplied by the transformer secondary winding. It is one of the ratings needed to specify the transformer.
8. The average diode current, I_{Dav}, determines the long-term heat dissipating requirement of the diode. Its value is one of the ratings needed for diode specification.
9. The repeated peak diode current, I_{DP}, is the current that flows during the conduction time, t_2. It is one of the ratings needed for diode specification.
10. The peak diode voltage drop, V_{DP}, can be estimated by this expression which is Equation 3.10 from Section 3.7. If you use $V_{DP} = 0.8\ VP$ you will probably be within 0.1 V of "truth" for most diodes.
11. This expression for the peak source voltage, V_{SP}, is obtained from a KVL analysis of the circuit.
12. The transformer secondary voltage, V_S, is its rms rating and is needed to specify the transformer.
13. The surge current, I_{srg}, is the current that will flow when the circuit is initially turned on and the capacitor is totally discharged. This is the "one cycle surge current" rating needed for diode specification. The more sophisticated power supply designs often incorporate a thermistor or other protective circuitry in their rectifier to produce a "soft turn on" and so reduce the trauma when the power supply is turned on.

TABLE 5.1 Power Supply Design

Given: R_S R_L $V_{LDC} = \dfrac{V_{LP} + V_{Lt1}}{2}$ $v_{LPP} = V_{LP} - V_{Lt1}$

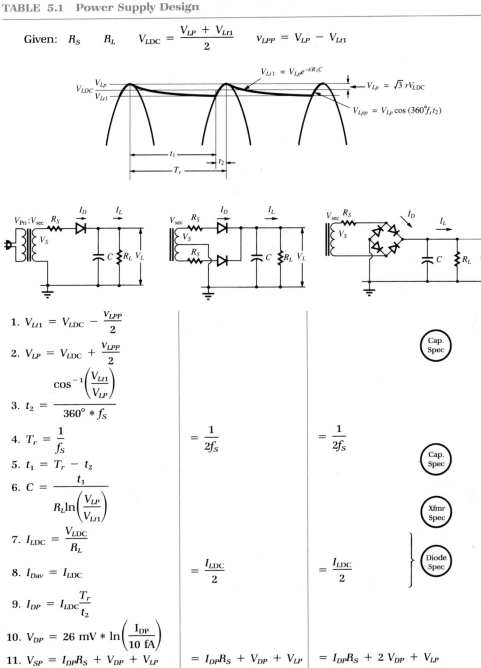

$V_{Lt1} = V_{Lp}e^{-t/R_LC}$

$V_{Lp} = \sqrt{3}\, rV_{LDC}$

$V_{Lpp} = V_{Lp}\cos(360°f_s t_2)$

1. $V_{Lt1} = V_{LDC} - \dfrac{v_{LPP}}{2}$

2. $V_{LP} = V_{LDC} + \dfrac{v_{LPP}}{2}$

3. $t_2 = \dfrac{\cos^{-1}\left(\dfrac{V_{Lt1}}{V_{LP}}\right)}{360° * f_S}$

4. $T_r = \dfrac{1}{f_S}$ $= \dfrac{1}{2f_S}$ $= \dfrac{1}{2f_S}$

5. $t_1 = T_r - t_2$

6. $C = \dfrac{t_1}{R_L\ln\left(\dfrac{V_{LP}}{V_{Lt1}}\right)}$

7. $I_{LDC} = \dfrac{V_{LDC}}{R_L}$

8. $I_{Dav} = I_{LDC}$ $= \dfrac{I_{LDC}}{2}$ $= \dfrac{I_{LDC}}{2}$

9. $I_{DP} = I_{LDC}\dfrac{T_r}{t_2}$

10. $V_{DP} = 26\ \text{mV} * \ln\left(\dfrac{I_{DP}}{10\ \text{fA}}\right)$

11. $V_{SP} = I_{DP}R_S + V_{DP} + V_{LP}$ $= I_{DP}R_S + V_{DP} + V_{LP}$ $= I_{DP}R_S + 2\,V_{DP} + V_{LP}$

12. $V_{sec} = \dfrac{V_{SP}}{\sqrt{2}}$ $= 2\dfrac{V_{SP}}{\sqrt{2}}\quad\{\text{CT}\}$ $= \dfrac{V_{SP}}{\sqrt{2}}$

13. $I_{srg} = \dfrac{V_{SP} - V_{DP}}{R_S}$ $= \dfrac{V_{SP} - V_{DP}}{R_S}$ $= \dfrac{V_{SP} - 2\,V_{DP}}{R_S}$

14. $\text{PIV} = V_{SP} + V_{LDC}$ $= V_{SP} + V_{LP}$ $= V_{LP} + V_{DP}$

Cap. Spec
Cap. Spec
Xfmr Spec
Diode Spec
Xfmr Spec
Diode Spec

14. The peak inverse voltage, PIV, is the maximum reverse voltage that appears across the diode. It is one of the ratings needed for diode specification.

The component ratings which must be specified for the design of a simple unregulated capacitor filtered power supply are:

TRANSFORMER:

V_S The rms secondary voltage rating.

I_S The maximum average current the transformer can supply continuously without overheating.

DIODES:

I_{av} The maximum average forward current the diode can pass continuously without overheating.

I_P The maximum recurrent peak current the diode can safely pass.

I_{srg} The maximum current the diode can conduct for a single power line cycle. This is the current that will flow if $V_S = V_{SP}$ at the instant when the power supply is first turned on and the filter capacitor is completely discharged. Usually if this specification is met, you do not need to worry about meeting the recurrent peak current specification.

PIV The maximum voltage the diode is capable of blocking.

CAPACITOR:

C The capacitance of the filter capacitor.

V_P The maximum voltage that can safely be placed across the capacitor.

As should be true of everything you do in electronics, you should not follow these steps blindly. They illustrate general principles whose specific application will vary from one situation to another. You should strive to understand the basic ideas and to apply sound circuit analysis procedures. These are the skills that will make you successful in electronics.

EXAMPLE 5.5

Design a center tapped transformer full-wave rectifier power supply that will deliver 3 VDC to a 33 Ω load with no more than 13 percent ripple.

SOLUTION:

Figure 5.7a contains a diagram of the circuit. The waveform at the load is shown in Figure 5.7b. From Equation 5.2 we can obtain the rms voltage at the load, v_{Lrms}.

$$v_{Lrms} = r_L V_{LDC} \tag{5.25}$$
$$= 0.13 * 3 \text{ VDC} = 0.390 \text{ Vrms}$$

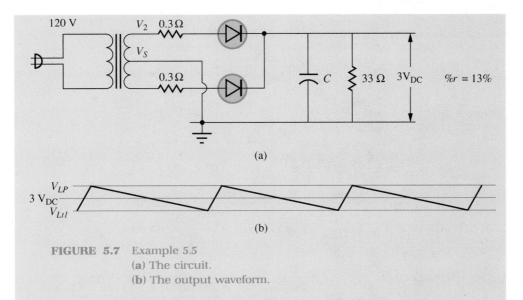

(a)

(b)

FIGURE 5.7   Example 5.5
(a) The circuit.
(b) The output waveform.

From Equation 5.7 we can obtain the peak to peak voltage at the load, v_{LPP}. Remember that the waveform as seen in Figure 5.7b is nearly a triangle wave so $\sqrt{3}$ is used rather than $\sqrt{2}$ as would be the case for a sine wave.

$$v_{LPP} = 2\sqrt{3}\, v_{Lrms} \tag{5.26}$$
$$= 2\sqrt{3} * 0.390 \text{ Vrms} = 1.35 \text{ VPP}$$

Now we can just plug in numbers into the design procedure of Table 5.1:

1. $V_{Lt1} = V_{LDC} - \dfrac{v_{LPP}}{2} = 3 \text{ VDC} - \dfrac{1.35 \text{ VPP}}{2} = 2.32 \text{ V}$

2. $V_{LP} = V_{LDC} + \dfrac{v_{LPP}}{2} = 3 \text{ VDC} + \dfrac{1.35 \text{ VPP}}{2} = 3.68 \text{ VP}$

3. $t_2 = \dfrac{\cos^{-1}\left(\dfrac{V_{Lt1}}{V_{LP}}\right)}{360°f_{line}} = \dfrac{\cos^{-1}\left(\dfrac{2.32 \text{ V}}{3.68 \text{ V}}\right)}{360° * 60 \text{ Hz}} = 2.35 \text{ msec}$

4. $T_r = \dfrac{1}{2f_S} = \dfrac{1}{2 * 60 \text{ Hz}} = 8.33 \text{ msec}$

5. $t_1 = T_r - t_2 = 8.33 \text{ msec} - 2.35 \text{ msec} = 5.98 \text{ msec}$

6. $C = \dfrac{t_1}{R_L \ln\left(\dfrac{V_{LP}}{V_{Lt1}}\right)} = \dfrac{5.98 \text{ msec}}{33 \text{ } \Omega * \ln\left(\dfrac{3.68 \text{ V}}{2.32 \text{ V}}\right)} = 393 \text{ } \mu\text{F}$

7. $I_{LDC} = \dfrac{V_{LDC}}{R_L} = \dfrac{3 \text{ VDC}}{33 \text{ } \Omega} = 90.9 \text{ mADC}$

8. $I_{Dav} = \dfrac{I_{LDC}}{2} = \dfrac{90.9 \text{ mA}}{2} = 45.4 \text{ mADC}$

9. $I_{DP} = I_{LDC}\dfrac{T_r}{t_2} = 90.9 \text{ mA}\dfrac{8.33 \text{ msec}}{2.35 \text{ msec}} = 321 \text{ mAP}$

10. $V_{DP} = 26 \text{ mV} * \ln\dfrac{I_{DP}}{10 \text{ fA}} = 26 \text{ mV} * \ln\dfrac{321 \text{ mA}}{10 \text{ fA}} = 0.809 \text{ VP}$

11. $V_{SP} = I_{DP}R_S + V_{DP} + V_{LP}$
$= 321 \text{ mA} * 0.3 \ \Omega + 0.809 \text{ VP} + 3.68 \text{ VP} = 4.58 \text{ VP}$

12. $V_{Srms} = 2\dfrac{V_{SP}}{\sqrt{2}} = 2\dfrac{4.58 \text{ VP}}{\sqrt{2}} = 6.48 \text{ VCT}$

13. $I_{srg} = \dfrac{V_{SP} - V_{DP}}{R_S} = \dfrac{4.58 \text{ VP} - 0.809 \text{ VP}}{0.3 \ \Omega} = 12.6 \text{ AP}$

14. $\text{PIV} = V_{SP} + V_{LP} = 4.58 \text{ VP} + 3.68 \text{ VP} = 8.26 \text{ VP}$

The specifications for the components are:

Transformer:

$$V_1 = 120 \text{ V}; V_S = 6.48 \text{ VCT}; I_S > 90.9 \text{ mA}$$

Diodes (2 required):

$$I_{Dav} > 45.5 \text{ mA}: I_{DP} > 321 \text{ mA}; I_{Dsrg} > 12.6 \text{ A}; \text{PIV} > 8.26 \text{ V}$$

Capacitor:

$$C > 393 \ \mu\text{F}; V_{rate} > 3.68 \text{ V}$$

If you have the opportunity, you should go to the lab and breadboard the above power supply to see how close these calculations agree with the real world. Also, if you want some good experience, you should obtain a current electronics parts catalog and check on the availability of the specified parts.

5.7 The R-C Filter

When additional filtering is needed, or when it is desirable to isolate one part of a circuit from another part, a simple R-C circuit such as that shown in Figure 5.8a is often used. The source, V_S, to which this filter is connected is usually the output from a capacitor filtered rectifier such as we have studied in the previous section. Thus the input to the R-C filter is usually the triangular waveform shown in Figure 5.8b.

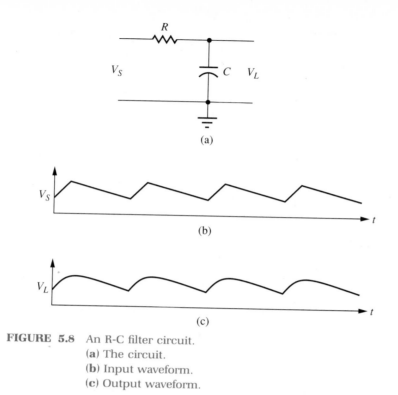

FIGURE 5.8 An R-C filter circuit.
(a) The circuit.
(b) Input waveform.
(c) Output waveform.

The basic idea behind the R-C filter is that the resistor R "decouples" the output, V_L, from the power supply while the capacitive reactance, X_C, of the capacitor, C, is so small that most of the ripple at V_L is conducted off to ground. Additionally, there is a sizable phase shift associated with the circuit which tends to smooth out the output waveform as pictured in Figure 5.8c. The result is an output waveform that contains much less ripple than the input. Also the output waveform more closely approximates a sine wave. Thus in our analysis of the circuit we may be at least partially justified in using the ideas of reactance and complex impedance which you should recall from your circuits class.

Often you are confronted with the need to design an RC filter to be connected to a power supply of known DC voltage, V_S, and AC ripple voltage, v_S. Usually the filter is to deliver a specified DC voltage, V_L, with no more than a specified ripple voltage, v_L, to a load whose equivalent resistance, R_L, is known or can be calculated. Your task is to decide on an appropriate resistor, R_F, and filter capacitor, C_F. One approach to the problem is:

The circuit is drawn as you will usually encounter it in Figure 5.9a. It is redrawn in Figure 5.9b to emphasize the fact that from a DC standpoint,

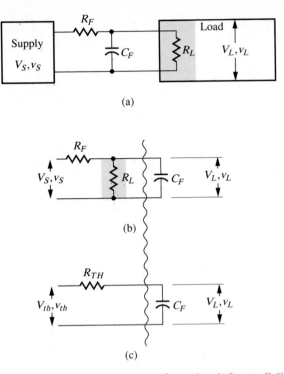

FIGURE 5.9 Construction of an equivalent circuit for an R-C filter circuit.
(a) The original circuit.
(b) Redraw of the original circuit to emphasize the Thevenin equivalent circuit.
(c) The Thevenin equivalent circuit.

the filter resistor, R_F, and the load resistor, R_L, form a voltage divider. The voltage divider is not loaded since, once the capacitor is charged, it will not draw any more DC current. Thus, from your circuits class you should remember that we can write

$$\frac{R_F}{V_S - V_L} = \frac{R_L}{V_L} \tag{5.27}$$

If we assume that you know what DC voltage V_L you need delivered to the load, we can solve for the value of R_F

$$R_F = R_L \frac{V_S - V_L}{V_L} \tag{5.28}$$

In order to find the value for C_F we will first construct a Thevenin equivalent for everything to the left of the wavy line in Figure 5.9b. The result is drawn in Figure 5.9c. From Figure 5.9b and your knowledge of Thevenin equivalents you know that

$$R_{Th} = R_L \parallel R_F \tag{5.29}$$

$$V_{Th} = V_S \frac{R_L}{R_L + R_F} \tag{5.30}$$

The resistors that make up the Thevenin equivalent respond to AC in the same way as they do to DC. Thus the ripple percentage of the Thevenin equivalent will be the same as it is for the supply. Applying Equation 5.25, we then have

$$v_{Thrms} = r_S V_{ThDC} \tag{5.31}$$

or, applying Equation 5.30 to the AC ripple

$$v_{Th} = v_S \frac{R_L}{R_L + R_F} \tag{5.32}$$

From Figure 5.9c, you can see that for the AC ripple, R_{Th} and C_F form an AC voltage divider whose total impedance is given by

$$Z = R_{Th} - jX_C \tag{5.33}$$

$$Z = \sqrt{R_{Th}^2 + X_C^2}$$

The voltage divider equation for the circuit can be written

$$\frac{Z}{v_{Th}} = \frac{X_C}{V_L} \tag{5.34}$$

or substituting Equation 5.33 into Equation 5.34

$$\frac{\sqrt{R_{Th}^2 + X_C^2}}{v_{Th}} = \frac{X_C}{V_L} \tag{5.35}$$

Solving Equation 5.34 for the capacitive reactance, X_C

$$R_{Th}^2 + X_C^2 = X_C^2 \frac{v_{Th}^2}{V_L^2}$$

$$X_C^2 \left(\frac{v_{Th}^2}{V_L^2} - 1 \right) = R_{Th}^2$$

$$X_C = \frac{R_{Th}}{\sqrt{\dfrac{v_{Th}^2}{V_L^2} - 1}} \tag{5.36}$$

Once the capacitive reactance is known, the capacitive reactance equation will give the size of the capacitor needed to provide the required filtering.

$$C_F = \frac{1}{2\pi f_r X_C} \tag{5.37}$$

Remember to use the ripple frequency, f_r, which is usually 60 Hz for a half-wave rectifier and 120 Hz for a full-wave rectifier.

EXAMPLE 5.6

A certain full-wave capacitor filtered power supply puts out 30 VDC with 10 percent ripple. Design an R-C circuit to be connected to the above supply that will deliver 20 mA at 10 VDC with no more than 0.5 percent ripple.

SOLUTION:

The circuit is drawn in Figure 5.10a. First the equivalent resistance of the load, R_L, is calculated

$$R_L = \frac{V_L}{I_L}$$

$$= \frac{10\ \text{V}}{20\ \text{mA}} = 500\ \Omega$$

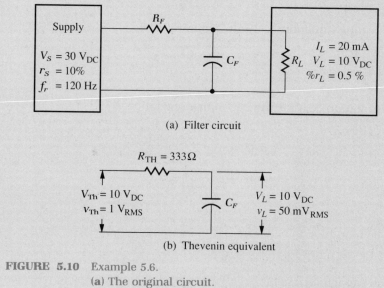

(a) Filter circuit

(b) Thevenin equivalent

FIGURE 5.10   Example 5.6.
(a) The original circuit.
(b) The Thevenin equivalent.

The value of the decoupling resistance is given by

$$R_F = R_L\frac{V_S - V_L}{V_L} \tag{5.28}$$

$$= 500\ \Omega\ \frac{30\ \text{V} - 10\ \text{V}}{10\ \text{V}} = 1\ \text{k}\ \Omega \qquad (\text{Use 1 k}\ \Omega)$$

The nearest 10% resistor is specified. Its power rating is

$$P_{RF} = I_{RF}^2 R_F$$

$$= (20\ \text{mA})^2 * 1\ \text{k}\ \Omega = 400\ \text{mW} \qquad (\text{use 1/2 Watt})$$

The Thevenin resistance as seen by the filter capacitor is

$$R_{\text{Th}} = R_L \parallel R_F \tag{5.29}$$

$$= 500\ \Omega \parallel 1\ \text{k}\ \Omega = 333\ \Omega$$

The AC voltage present in the Thevenin source is

$$v_{\text{Thrms}} = r_S V_{\text{ThDC}} \tag{5.31}$$

$$= 0.10 * 10\ \text{VDC} = 1\ \text{Vrms}$$

The maximum allowable ripple in the 10 VDC output is

$$v_{\text{Lrms}} = r_L V_{\text{LDC}} \tag{5.25}$$

$$= 0.005 * 10\ \text{VDC} = 50\ \text{mVrms}$$

The above values are summarized in Figure 5.10b.
    The capacitive reactance of the filter capacitor is given by

$$X_C = \frac{R_{\text{Th}}}{\sqrt{\dfrac{v_{\text{Th}}^2}{v_L^2} - 1}} \tag{5.36}$$

$$= \frac{333\ \Omega}{\sqrt{\dfrac{(1.0\ \text{Vrms})^2}{(50\ \text{mVrms})^2} - 1}} = 16.7\ \Omega$$

Using the capacitive reactance formula and recognizing that the ripple frequency for a full-wave rectifier is $fr = 120$ Hz

$$C_F = \frac{1}{2\pi f_r X_C} \tag{5.37}$$

$$= \frac{1}{2\pi * 120\ \text{Hz} * 16.7\ \Omega} = 79.6\ \mu\text{F} \qquad (\text{Use 80}\ \mu\text{F})$$

The nearest standard capacitor is specified.

You might notice in passing that when a capacitor filter has an R-C filter connected to it, as in the example above, and as drawn in Figure 5.11a, it is often called a "π filter," or more specifically a "π R-C filter." If you know the Greek alphabet and if you look at the hint in Figure 5.11b you might be able to guess the reason for the name: If you have a good imagination, the circuit configuration looks a little like the Greek letter π, don't you think?

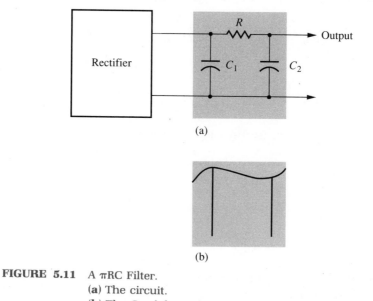

(a)

(b)

FIGURE 5.11   A πRC Filter.
(a) The circuit.
(b) The Greek letter from which the circuit gets its name.

As we have seen, the π R-C filter can do a good job of reducing the ripple in a power supply. Its biggest disadvantage is the fact that, as you may have guessed, it vastly increases the regulation of the power supply since the resistor R_F is part of the DC source resistance as seen at the output. In fact, it is usually the major part of the source resistance.

5.8 The L-C Filter

As indicated above, the resistor in the R-C filter is a problem. We need a large impedance to filter out the AC, but we need as small a resistor as possible in order that the DC can get through. Of course, those are exactly the characteristics of an inductor. An ideal inductor presents a high impedance to AC while exhibiting no resistance to DC. The reason that inductors are seldom used in simple low power filters is that they are expensive, bulky, and heavy. You can usually get more filtering for your money with the R-C circuit.

5.9 Troubleshooting ◀◀◀◀◀◀◀◀

When troubles develop in almost any electronic equipment, the problem is often traceable to the power supply. The reason is that the highest power levels and often the highest voltages are encountered in the power supply.

Total failure is probably the easiest symptom to troubleshoot. In that event, the first thing to check is the power line fuse. If it is blown, the best way to invite a call-back and a damaged reputation is to replace it and return it to the user. You need to satisfy yourself that you know why it blew, and that the underlying cause has been eliminated. Under no circumstances should you ever replace a blown fuse with a fuse with a higher rating than that specified by the manufacturer. To do so is to invite damage to more expensive circuit components.

One of the most frequent causes of blown fuses in a power supply is one or more shorted rectifier diodes. Another frequent problem is a very leaky, or a shorted filter capacitor. Also, you should not ignore the possibility that the load connected to the power supply is drawing excessive current.

The open circuit is perhaps a more difficult symptom. After ascertaining that the line fuse is good, the most likely suspect is usually the rectifier diodes. Open diodes are probably more common than shorted diodes. If you find open diodes, you need to be certain that they were not destroyed by a short circuit farther down stream.

A more difficult symptom to cure is that where total failure has not occurred. The symptom is most often excessive ripple or low output voltage. This symptom is often caused by a failure in one-half of a full-wave rectifier. This doubles the discharge time t_1 for the filter and so doubles the ripple voltage according to Equation 5.13. Looking at the ripple on an oscilloscope is the best way to detect this fault. If the pulses are 1/60 second apart (16.7 mSec) you have a half-wave rectifier. If they are 1/120 second apart (8.33 mSec) you have a full-wave rectifier.

Another source of excessive ripple is loss of capacity of the filter capacitor. As electrolytic capacitors age they are increasingly prone to this problem. Paralleling such a capacitor with a known good capacitor will result in a dramatic reduction in ripple.

5.10 Summary

In this chapter you should have learned the meaning of power supply ripple, how to evaluate it, and how to reduce it to any required value in the rectifiers studied in Chapter 4 by using either a simple capacitor filter or by adding either an R-C, or an L-C filter. The developments in both Chapter 4 and Chapter 5 are summarized in the design recipe presented in Chapter 5.

GLOSSARY

Filter: Circuitry intended to remove selected AC components from a waveform. (Section 5.2)

L-C filter: A filter consisting of an inductor in series with the load and a capacitor in parallel with the load. (Section 5.6)

π filter: A circuit configuration typically consisting of two capacitors to ground separated by another circuit component, usually either a resistor or an inductor. (Section 5.7)

R-C filter: A filter consisting of a resistor in series with the load and a capacitor in parallel with the load. (Section 5.7)

Regulation *(vr)*:   Usually voltage regulation. A measure of the ability of a source to maintain a constant output voltage in spite of variations in loading. It is the ratio of the change in output as the load is varied divided by the loaded output. It is usually expressed as a percent. (Section 5.3)

Ripple factor *(r)*:   A measure of the purity of a DC waveform. It is the ratio of the AC component (rms) present divided by to the DC component. It is usually expressed as a percent. The smaller its value, the more pure the DC waveform. (Section 5.3)

PROBLEMS

Section 5.3

1. Calculate the percent ripple for the battery charger of Example 4.1 when it is charging your car's battery.

2. My car's battery puts out 13.2 V when no current is being drawn. When I hit the starter, the voltage drops to 12 V. What is the percent regulation of the battery?

3. One of the power supplies in my computer uses a bridge rectifier and puts out a waveform like that shown in Figure 5.12. Find its percent ripple.

FIGURE 5.12   Problem 5.3.

4. Suppose that one of the diodes in the rectifier of Problem 3 became an open circuit. Describe the effect on the circuit operation and calculate the resulting percent ripple. (It made a wavy line appear on the screen.)

5. Suppose that the diode in Problem 4 had become shorted instead of becoming open. Describe the effect on the circuit.

Section 5.4

6. If the current drawn from the power supply of Problem 3 is 200 mA, calculate the value of the filter capacitor used in the circuit.

Section 5.5

7. Specify all of the components needed for the power supply of Problem 6.
8. Suppose that the power supply of Problem 6 is replaced by a switching power supply operating at 40 kHz. Calculate the size filter capacitor needed by this supply if the allowable ripple is unchanged.
9. In Problem 6, 50 mA of the output goes to supply a 3 VDC circuit that must have no more than 0.5 percent ripple. Design an R-C filter circuit that can be connected to the output of the circuit of Problem 3 that will meet these requirements.

Section 5.7

10. Design and specify all components for an independent power supply which gets its power directly from the 120 V power line that will meet the requirements of Problem 9. Use a center tapped transformer design with $R_S = 0.3 \ \Omega$.
11. Design and specify all components for a power supply that will replace the 9 V battery that powers a small radio receiver that draws 80 mA and can tolerate 10 percent ripple. Use a half-wave rectifier with a transformer having an effective secondary resistance of $R_S = 0.3 \ \Omega$.

Section 5.9

◀ Troubleshooting

12. Assume that my car's starter draws 100A. Using the data from Problem 3, determine:
    (a)   The internal resistance of my car's battery.
    (b)   The symptoms I would observe if the internal resistance became 1 Ω.
13. Describe the symptoms you would observe if one of the diodes in the circuit of Problem 9 became:
    (a)   An open circuit.
    (b)   A short circuit.
14. Suppose that the capacitance of the filter capacitor used in Problem 11 decreased to half its original value. Describe the symptoms that would be observed by:
    (a)   The customer (who has no scope or DMM).
    (b)   You, with all of your equipment.

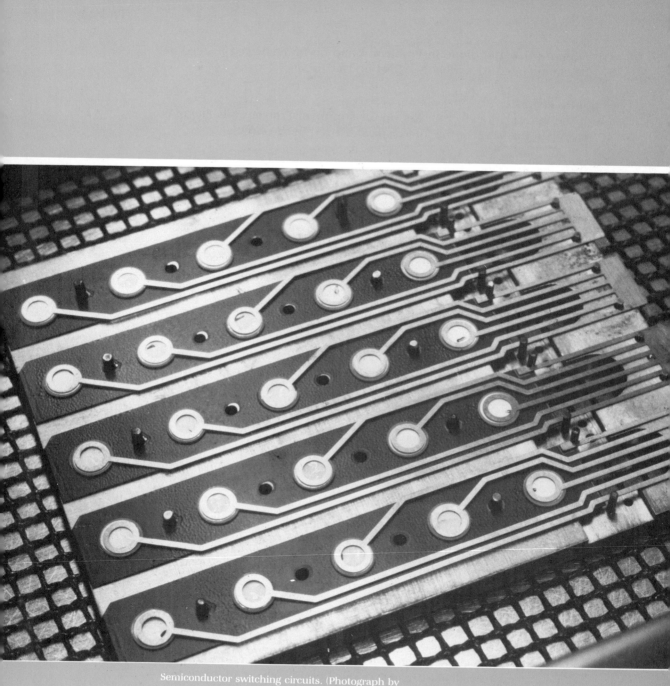

Semiconductor switching circuits. (Photograph by
Peter Vadnai/The Stock Market.)

Chapter 6
Other Typical Diode Circuits

SPECIAL TERMS

Clipping circuit
Distortion
Amplitude distortion
Harmonic distortion
Clamping circuit
Voltage multiplier circuit
Truth table

6.1   Introduction and Objectives

We could not possibly examine all of the circuits in which diodes are used. Consequently we will examine a few representative circuits in which diodes function in ways other than as simple rectifiers in power supplies. They are clippers, clamps, voltage multipliers, and logic circuits.

At the conclusion of this chapter you should have a deeper understanding of diode behavior in general and you should be more capable of figuring out how unfamiliar circuits function. You should be able to:

- Determine the output amplitude and draw the output waveform produced by diode clipping circuits.
- Design a simple clipping circuit that will clip at a specified voltage.
- Draw the output waveform for a diode clamping circuit.
- Draw the circuit and describe in quantitative terms the operation of a voltage multiplier circuit having any specified voltage multiplication factor.
- Describe the behavior of simple diode logic circuits.

6.2   Diode Clipping Circuits

Occasionally it is desired to limit the amplitude of a waveform so that it cannot be bigger than a certain predetermined value. A familiar example occurs in your local radio station. Federal regulations stipulate what their maximum radiated power can be. In order to have maximum coverage, the operator likes to push the signal level as close to that maximum as possible, but what do you suppose happens when the announcer gets exceptionally enthusiastic and yells into the microphone? A clipping circuit takes care of the extra enthusiasm.

A clipping circuit does exactly what its name implies. It clips the waveform. The half-wave rectifier presented in Section 4.4 can be considered to be a clipper of sorts. It clips off roughly half of the waveform.

Perhaps the simplest example of a peak clipper is shown in Figure 6.1. In order to understand how it works, you may wish to go back and review Sections 3.3 through 3.6. Remember that according to the graphs we drew

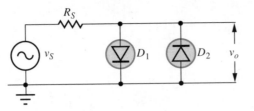

FIGURE 6.1   A peak clipper.

there, if diodes D_1 and D_2 of Figure 6.1 are assumed to be silicon diodes, then they will not begin to turn on until around $V_D = 0.6$ V. Thus if the signal generator output, v_S, is less than about 0.5 V_P then neither diode will conduct, no current will flow, and the circuit output voltage, v_o, will be equal to the circuit input voltage, v_S. This is the initial situation for the waveforms shown in the PSpice printout of Figure 6.2. When v_S, which

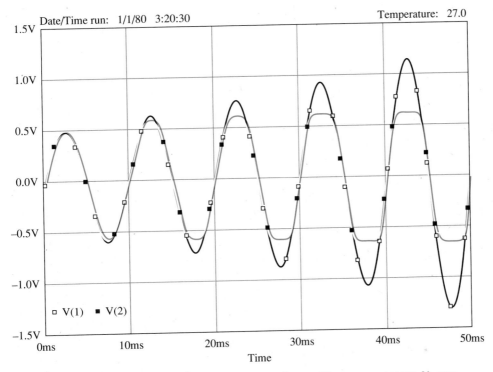

FIGURE 6.2 PSpice generated clipper waveforms. The open squares, V_1, represent the input signal waveform and the solid squares, V_2, represent the output signal waveform.

PSpice called "V(1)," exceeds about $+0.6$ V, D_1 in Figure 6.1 begins to conduct, current begins to flow through the 1 k Ω resistor, R_S, voltage begins to be lost across it, and the positive peak of v_o, which PSpice called "V(2)" in Figure 6.2, begins to reflect that loss. The result is that the output waveform is clipped at a peak voltage of near $+0.6$ V. On the negative half of the cycle the same activity occurs with respect to D_2. And the output is clipped at a negative peak of near -0.6 V.

We could quantitatively determine the value of v_o for any instantaneous value of v_S by using the successive approximation methods developed in Chapter 3, or we could request our favorite circuit analysis program to perform the chore for us as was done to produce Figure 6.2. See Appendix H for the PSpice program that was used.

The simple circuit we have examined does a fairly good job of clipping, especially if diodes are chosen that have a very steep characteristic graph (low dynamic resistance). One disadvantage of the circuit is that the user of the circuit does not have much control over the voltage at which clipping occurs. The circuit in Figure 6.3 is a clipper whose level is user adjustable. It is appropriately called a *biased* clipper.

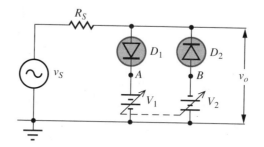

FIGURE 6.3   A biased clipper circuit.

The circuit functions in much the same way as the circuit of Figure 6.1 with the exception that on the positive half-cycle, since the cathode of D_1 has a DC bias voltage of $+V_1$, D_1 will not begin to conduct until v_o reaches about $V_1 + 0.6$ V. Thus we have raised the turn-on voltage by the amount of the bias voltage V_1. Since the circuit is symmetrical, everything we have said here about the positive half-cycle, D_1, and V_1 applies to the negative half-cycle, D_2, and V_2. You will notice that no current ever flows out of either supply V_1 or V_2. If any current flows it is always into V_1 or V_2.

It is important to notice that although clipping circuits serve a valuable function in electronics, they introduce at least two different kinds of distortion into the waveform. Any time the output waveform is not strictly proportional to the input waveform, that is distortion. Since the clipper circuit is intended to limit the output level, it will always produce amplitude distortion. Also, since the peaks of the output waveform are clipped, the shape of the waveform is changed. That is called *harmonic* distortion.

6.3 Diode Clamping Circuits

Pure AC is defined to be symmetrical about zero, but occasionally it is desirable that an AC signal be offset from zero. If a fixed DC offset is all that is desired, a circuit such as Figure 6.4 is sufficient. A KVL walk from the signal generator, v_S through the battery, V_X to the output, v_o gives

$$v_S + V_X = v_o \tag{6.1}$$

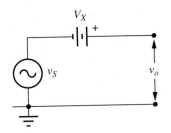

FIGURE 6.4 A circuit with a fixed DC offset.

If the signal amplitude, v_{SP}, is less than the offset voltage, V_X, then the output waveform will look like Figure 6.5a. if v_{SP} is equal to V_X then the output waveform will look like Figure 6.5b. If it is greater than V_X then it will look like Figure 6.5c.

Sometimes it is desired that the signal be offset a varying amount depending on the amplitude of the signal. For example, a common need is

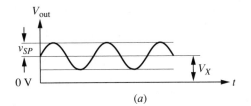

(a)

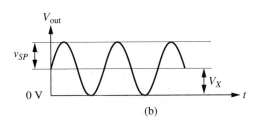

(b)

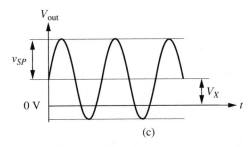

(c)

FIGURE 6.5 Fixed offset waveforms.
  (a) $v_{SP} < V_X$
  (b) $v_{SP} = V_X$
  (c) $v_{SP} > V_X$

that the negative peaks of the output waveform be "clamped" at some fixed value, like zero for example. That is, it is desired that the output waveform look like that of Figure 6.5b regardless of the amplitude. The circuit in Figure 6.6a is a simple form of such a clamping circuit. It clamps at about -0.6 V. It is called a "negative peak clamping circuit," or more simply, just a "negative clamp" because it is the negative peak of the input signal that is clamped at -0.6 VDC.

To understand how the clamping circuit in Figure 6.6a works, assume that the time constant, RC, for the circuit is much greater than the period T_S of the source. That means that once the capacitor, C, is charged, it will not discharge very quickly. Also assume that C is initially discharged and that the signal generator is putting out its negative peak instantaneous voltage, $-V_{SP}$, as shown in Figure 6.6b. The generator will then be attract-

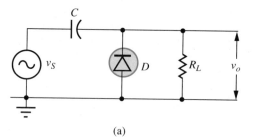

(a)

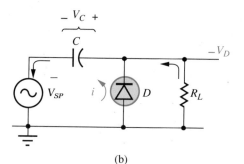

(b)

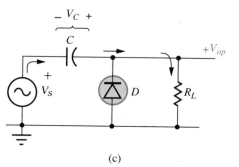

(c)

FIGURE 6.6   A negative peak clamping circuit.

ing current up through the diode, turning it on, and charging up the capacitor as shown.

Starting from ground in Figure 6.6b and doing a KVL walk in a clockwise direction against the indicated current flow around the loop containing the signal generator, the capacitor, and the diode gives

$$-V_{SP} + V_C + V_D = 0 \text{ V} \tag{6.2}$$

or

$$V_C = V_{SP} - V_D \tag{6.3}$$
$$\approx V_{SP} - 0.6 \text{ V}$$

If the time constant for the circuit is long, C will function much like the battery V_X in Figure 6.4. Thus, doing a KVL walk from the signal generator, v_S, through the capacitor to the output in Figure 6.6c gives

$$v_S + V_C = v_o \tag{6.4}$$

Substituting Equation 6.3 into Equation 6.4 gives

$$v_o = v_S + V_{SP} - 0.6 \text{ V} \tag{6.5}$$

When $v_S = +V_{SP}$ then $v_o = 2 V_{SP} - 0.6$ V.
When $v_S = 0$ V   then $v_o = V_{SP} - 0.6$ V.
When $v_S = -V_{SP}$ then $v_o = -0.6$ V.

The three waveforms of Figure 6.5 will now look like Figure 6.7. As we mentioned before, this circuit is called a "negative peak clamping circuit" or just a "negative clamp" because it is the negative peak of the input signal that is clamped at a fixed DC voltage, in this case -0.6 V.

You should notice that if the time constant, RC for the circuit is not long compared to the period of the waveform, then the clamping voltage becomes frequency-dependent and the circuit starts to look more like a high pass filter than a clamping circuit. On the other hand, if RC is too long, the output waveform tends to "float." That is, if the amplitude of the signal suddenly decreases, the waveform will briefly remain symmetrical about the same value as before and it will take some time for the waveform to sag down to the clamping voltage again. This effect is illustrated in Figure 6.8.

In order to determine the optimum value of the clamping capacitor for a given frequency and load resistor, R_L, we will use the statement made earlier that "the time constant, RC, is much greater than the period T of the waveform." You should know by now that in electronics, "much greater" means "at least 10 times." Thus we can say

$$RC \gg T$$

or

$$RC \geq 10T = \frac{10}{f} \tag{6.6}$$

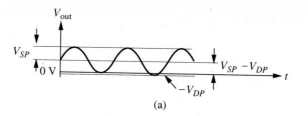

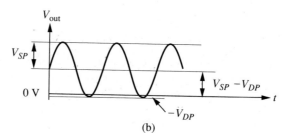

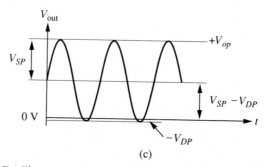

FIGURE 6.7 Clamper output waveform. Note that the clamping voltage is the same regardless of signal amplitude and that the midpoint of the waveform is always at $V_{SP} - V_{DP}$.

FIGURE 6.8 Floating in a clamped signal.

Solving for C gives the needed capacitor size.

$$C \geq \frac{10}{Rf} \qquad (6.7)$$

If the clamping diode is reversed, instead of clamping the negative peaks of the waveform, the positive peaks will be clamped. The circuit would then be called a "positive peak clamping circuit" or just a "positive clamp."

You should also recognize that by putting a fixed DC biasing voltage source in the circuit similar to what was done with the clipping circuit, the circuit can be designed to clamp at any given voltage.

EXAMPLE 6.1
Design a diode clamping circuit that will clamp the positive peaks of an input waveform at a voltage of +5 V. Use a load resistor of $R_L = 1$ M Ω and an operation frequency of 60 Hz.

SOLUTION:
Since we need a positive clamping circuit rather than the negative clamp of Figure 6.6, the diode will need to be reversed, and since the clamping voltage is to be +5 V, we will need a DC bias voltage of +4.4 V in addition to the 0.6 V we will get from the diode. This gives us the circuit of Figure 6.9. The capacitor size is given by

$$C = \frac{10}{fR} \qquad (6.7)$$

$$= \frac{10}{60 \text{ Hz} * 1 \text{ M } \Omega} = 0.167 \text{ μF}$$

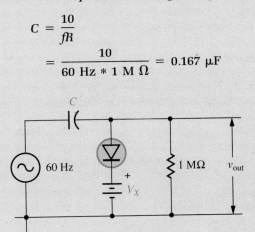

FIGURE 6.9 Example 6.1.

Since this is the minimum size for C, and since 0.167 μF is probably not a standard size, you would probably use a 0.2 μF capacitor.

6.4 A Diode Voltage Multiplier

The operation of some electronic circuits requires DC voltages on the order of a few thousand volts. While there are transformers and rectifiers capable of handling that kind of voltage, they are usually very bulky and very expensive. If only a small current is needed, a voltage multiplier circuit is sometimes a better choice. One such circuit is drawn in Figure 6.10.

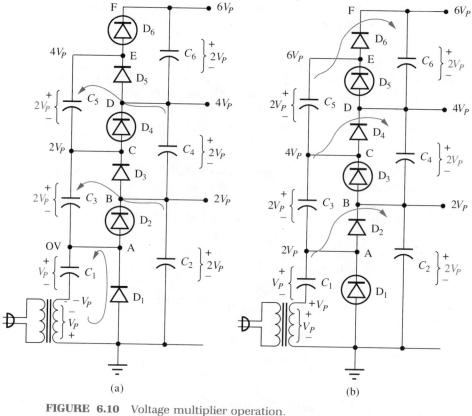

FIGURE 6.10 Voltage multiplier operation.
(a) The negative half-cycle.
(b) The positive half-cycle.

Note that the circuit is drawn so as to make the description of the circuit operation easier. There are many other ways to lay out this circuit. We will look at two later. In order to simplify things, we will ignore the 0.6 V or 0.8 V that it takes to turn on the diodes. That is not totally unrealistic since the voltages used in these kinds of circuits are usually many hundreds of volts so a few tenths of a volt is not too significant. For the same reason we will focus on only one part of the circuit at a time, and we will alternately look at what is happening on the negative half of the power line cycle using Figure 6.10a and then we will look at what is happening on the positive half of the power line cycle using Figure 6.10b.

To begin the analysis, focus on just the portion of the circuit containing the transformer, diode D_1, and capacitor C_1. You have seen this part of the circuit before. It is just the negative clamping circuit of Figure 6.6. You know how it works: On the negative half cycle (Figure 6.10a), current flows in the direction of the arrow, D_1 conducts, and $V_A = 0$ V, provided we ignore the 0.6 V lost across D_1. That means that C_1 gets charged up to V_P with the polarity as shown in Figure 6.10a. A quick check using KVL indicates that with the transformer putting out $-V_P$ and C_1 charged to $+V_P$, the voltage at Point A should be $V_A = 0$ V as we have said.

With C_1 now charged to $+V_P$ let us look at what happens on the positive half of the power line cycle which is shown in Figure 6.10b. We will now include diode D_2 and capacitor C_2 in the discussion. The transformer now puts out $+V_P$ which adds to the voltage $+V_P$ across C_1. That gives $V_A = +2\,V_P$. Diode D_1 is then reverse biased so it blocks current flow (D_1 is circled to indicate that it is blocking). At the same time D_2 is forward biased so current flows through it to charge C_2 bringing the voltage at Point B up to $2\,V_P$. Thus C_2 is charged up to a voltage of $2\,V_P$. If the output is taken from Point B we have a voltage doubler.

You may wish to argue that C_1 will be partially discharged in the process of charging C_2 and so the voltage will be less, but remember that the above activity is being very rapidly repeated so that the charge on C_1 is being continually replenished. As long as we do not draw a lot of current from the circuit, the voltage across C_2 will eventually (after perhaps a tenth of a second) approach $2\,V_P$.

With C_2 now charged to $2\,V_P$, let us return to Figure 6.10a and observe the next negative half cycle. We will now include diode D_3 and capacitor C_3 in the discussion. Again $V_A = 0$ V. Since $V_B = 2\,V_P$, D_2 will be reverse biased and will block. It is circled. Diode D_3 will be forward biased and will conduct charging C_3 to $2\,V_P$.

With C_3 now charged to $2\,V_P$, let us return to Figure 6.10b and observe the next positive half cycle. We will now include diode D_4 and capacitor C_4 in the discussion. V_A will again be $2\,V_P$ and will add to the $2\,V_P$ across C_3 so that the voltage at Point C will be $V_C = 4\,V_P$. Since $V_B = 2\,V_P$, D_3 will be reverse biased and will block (it is circled) and D_4 will conduct bringing the voltage at Point D up to $4\,V_P$. Note that this will mean only $2\,V_P$ across C_3 however, since Point B is already at $2\,V_P$. If we take the output from Point D we have a voltage quadrupler.

As you might guess, this process could be continued for as long as you wish. In practice, it is seldom carried beyond the quadrupler stage. One reason is that the capacitor size requirements quickly become ridiculous. You notice that C_2, C_4, etc., are in series and from your circuits class you know what that does to the total equivalent capacitance. Further, the effective internal resistance of the circuit gets progressively greater, all indicating that the circuit's ability to supply current approaches zero for very many stages of voltage multiplication.

Perhaps from the standpoint of understanding how the circuit works, Figure 6.10 is a good way to draw a voltage multiplier but it uses up a lot of paper. A more frequently used lay-out is shown in Figure 6.11. You

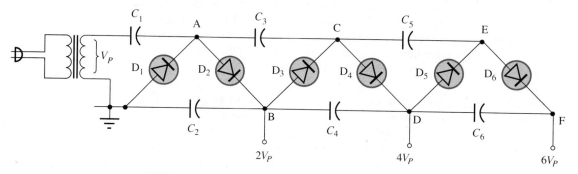

FIGURE 6.11   A "bridge girder" voltage multiplier.

should check the circuit carefully and convince yourself that it is in fact the same circuit as Figure 6.10. Perhaps you can see why this circuit is sometimes referred to as a "bridge girder voltage multiplier."

Figure 6.12 contains a different form of voltage multiplier circuit. It is sometimes referred to as a "stair case voltage multiplier." Can you see why?

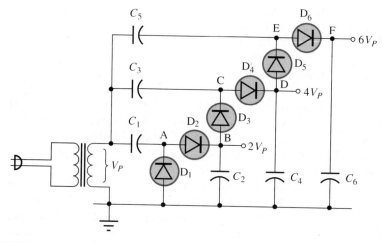

FIGURE 6.12   A "staircase" voltage multiplier.

6.5   Diode Logic Circuits

Logic circuits are binary circuits. That is, they are either on or off, high or low, one or zero. Two of the basic building blocks used to construct digital computers and other logic circuits are the AND gate and the OR gate. Both of these logic functions can be easily built using diodes.

The resistor and the two diodes that are within the logical AND symbol in Figure 6.13a form a two input AND gate. Obviously we could increase the number of inputs to as many as you might wish by adding more diodes. We will use switches S_A and S_B to illustrate the operation of the gate.

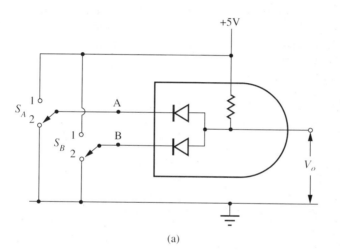

(a)

Inputs		Output
A	B	
0	0	0
1	0	0
0	1	0
1	1	1

(b)

FIGURE 6.13   A diode AND gate.
(a) Circuit.
(b) Truth table.

Note that both input A AND input B must be high (switch in position 1) in order for the output to be high. If either input is low (switch in position 2) the diode connected to that input will conduct pulling the output down to around 0.7 V which we will interpret as a logical zero.

Obviously, the name of the circuit is derived from the fact that both *A* AND *B* have to be high for the output to be high. The "truth table" that summarizes how the circuit responds to input signals is shown in Figure 6.13b.

The resistor and the two diodes that are within the logical OR symbol in Figure 6.14a form a two input OR gate. Again we could increase the number of inputs to as many as you might wish by adding more diodes. As we did for the AND gate, we will use switches S_A and S_B to illustrate the operation of the gate.

Note that if either input *A* OR input *B* is high (switch in position 1) the output will be high. If either input is high (switch in position 1) the diode connected to that input will conduct pulling the output up to around 4.3 V which we will interpret as a logical one. Obviously, the name of the circuit is derived from the fact that if either *A* OR *B* is high the output will be high. The truth table summarizing the behavior of this circuit is shown in Figure 6.14b.

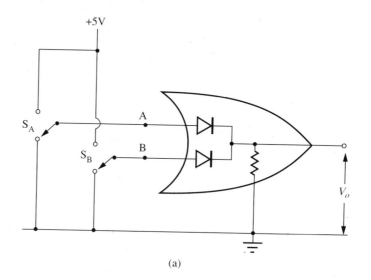

(a)

Inputs		Output
A	B	
0	0	0
1	0	1
0	1	1
1	1	1

(b)

FIGURE 6.14   A diode OR gate.
(a) Circuit.
(b) Truth Table.

In order to perform many of the functions of Boolean logic, we need the ability to perform the NOT function. Diodes do not lend themselves easily to the construction of this function. Later you will find that transistors can be used to do that job fairly easily.

6.6 Troubleshooting   ◀◀◀◀◀◀◀◀

In order to successfully troubleshoot any circuit you need a thorough understanding of both circuit analysis and the characteristics of the components that make up the circuit. Perhaps in the simplest of circuits such as those of previous chapters a memorized troubleshooting routine can be used in place of genuine understanding, but if you wish to advance to the more challenging and rewarding levels of technology, work toward real understanding.

6.7 Summary

Time and space have limited our examination of diode circuits to only a few examples, but if you have really understood what has been presented, you should be able to "feel your way through" some circuits that appear rather strange when you first look at them. After we get further into electronics we will encounter other interesting uses of diodes but in almost every case you will find that advantage is taken of either the diode's ability to conduct in only one direction, or else its very nonlinear turn-on characteristic.

GLOSSARY

Clamping circuit: A circuit that adds sufficient DC to a waveform to prevent the waveform from extending beyond a predetermined level. (Section 6.3)

Clipping circuit: A circuit that limits the amplitude of a waveform by clipping off the parts of the waveform that would otherwise extend beyond the predetermined limit. (Section 6.2)

Distortion: An unintended change in the waveform of a signal. (Section 6.2)

Amplitude distortion: An unintended change in the amplitude of a signal waveform. (Section 6.2)

Harmonic distortion: An unintended change in the harmonic content or the overtones present in a waveform. (Section 6.2)

Truth table: A table which displays in a systematic and compact form the response of a logic function to all possible combinations of inputs. (Section 6.5)

Voltage multiplier circuit: A circuit consisting of clamps and rectifiers that produce a multiplication of the input voltage. (Section 6.4)

PROBLEMS

Section 6.2

1. Using the fixed voltage model for silicon diodes introduced in Section 3.4, if the source in Figure 6.15a has the voltage waveform shown in Figure 6.15b, draw the output waveform superposed on a careful drawing of the source waveform.

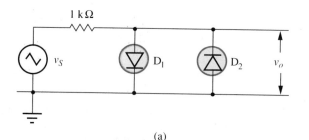

(a)

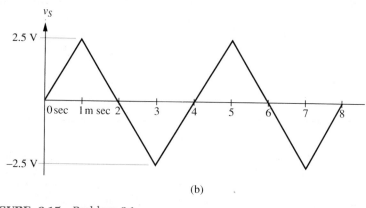

(b)

FIGURE 6.15   Problem 6.1.
(a) The circuit diagram.
(b) The input waveform, v_S.

2. On the waveform drawn in Problem 1 indicate the time interval over which diode D_1 is conducting and the time interval over which diode D_2 is conducting.

3. Using the diode equation introduced in Section 3.7 calculate the value of v_o in Problem 1 when $v_S = 0.75$ V. Explain why this result is different from that of Problem 6.1. *Hint:* Look at the graph of Figure 3.3.

4. Repeat Problem 3 when v_S is at its maximum value.

Section 6.3

5. Draw the circuit diagram for a diode clamping circuit that will provide negative peak clamping whose clamping level can be set to clamp at zero volts.

6. Repeat Problem 5 for a positive peak clamping circuit.

7. Determine the capacitor size for a clamping circuit designed to operate at 1 kHz if the load is a 1 k Ω resistor.

8. Approximately how long will the output from the clamping circuit of Example 6.1 float if the amplitude drops suddenly from 20 VPP to 10 VPP?

Section 6.4

9. Draw two copies of Figure 6.12, one for the negative half cycle and the other for the positive half cycle. On the diagrams indicate how each of the capacitors is charged in a way similar to what was done in Figure 6.10.

10. If the transformer secondary in Figure 6.12 is rated at 250 Vrms, list the voltage rating required for each of the diodes and capacitors in the circuit.

Section 6.5

11. Draw the circuit diagram for a 3 input diode AND gate.

12. Draw the circuit diagram for a 3 input diode OR gate.

13. Construct a truth table for the circuit of Problem 11.

14. Construct a truth table for the circuit of Problem 12.

Section 6.6

◀ Troubleshooting

15. Repeat Problem 1 if diode D_1 became:
    (a)   An open circuit.
    (b)   A short circuit.

16. Describe the effect on the output waveform of Figure 6.6 if:
    (a)   The diode became an open circuit.
    (b)   The resistor R_L became an open circuit.

17. Suppose that capacitor C_1 in the circuit of Figure 6.12 shorted and became a perfect conductor. Describe the most probable effect on:
    (a)   The other components in the circuit.
    (b)   The output voltage at points B, D, and F.

Commercial application of a digital readout.
(Photograph by Christopher Morrow/The Stock Market.)

Chapter 7
Special Purpose Diodes

SPECIAL TERMS

Light emitting diode (LED)
Dropping resistor
Seven segment display
Schottky diode
Varactor
Zener diode

149

7.1 Introduction and Objectives

In this chapter we will look very briefly at some of the special purpose diode types that are available. We will not try to examine all of the types that have been made. Rather, we will look at a few representative types and show how you can understand their behavior in terms of the basic principles with which you should already be familiar.

We will look briefly at the LED and the Schottky diode as representatives of the general class of special purpose diodes. Then, because it is probably the most important, we will look more carefully at the characteristics of the Zener diode and a simple Zener voltage regulator circuit. This should give you some idea how Zener diodes work and prepare you for the study of some more sophisticated regulator circuits in Chapter 13 and 21.

After having studied this chapter you should begin to understand the characteristics of the above-mentioned diodes and you should have a better understanding of the behavior of diodes in general. You should be able to:

- Design a simple circuit that will power a LED.
- Describe the behavior and use of LEDs, Schottky diodes, varactors, and Zener diodes.
- Design and analyze simple "brute force" Zener regulator circuits using the Zener data listed in common data books.

7.2 The Light Emitting Diode (LED)

The light emitting diode (LED) is a special purpose diode which is designed to produce light. The physicists tell us that whenever an electron falls from one energy level within an atom to a lower energy level, light is produced. Further, the frequency of the light is directly proportional to the voltage through which the electron falls. Of course, electrons are "falling through a voltage" when combinations occur at the junction within a diode. Thus all diodes produce light when they conduct electricity.

The junction voltage for silicon is so low that the light produced is not visible. The typical red LED is made from gallium arsenide whose junction voltage is about 1.4 V. Green LEDs are usually made from gallium phosphide which has a junction voltage of about 2.25 V.

Since an LED is a diode, as you might expect, if it is reverse biased, it will not conduct and so it gives off no light in that mode. Also, its turn-on characteristic is about as nonlinear as other diodes. Thus LEDS should be driven by a current source or, which is the same thing, an appropriate "dropping resistor" must be used in series with the device if it is to be powered by a voltage source. Many small LEDs are fairly well lit with a current of between 10 and 20 mA.

Obviously, the packaging for LEDs must be transparent and the junction must be shaped to the desired pattern but the packaging is similar to other diodes in that the cathode terminal (the terminal current flows out of) has something "funny" done to it. In the case of the familiar bullet-shaped LEDs commonly used as indicator lights, there is a flat spot on the rim of the bullet nearest to the cathode terminal.

The typical "seven segment" display usually contains at least seven LEDs. Usually, they share a common anode terminal and each segment has its own cathode terminal so that each segment can be lit separately or in combination with any other segments. Figure 7.1 shows some typical LEDs and the schematic symbol usually used for them.

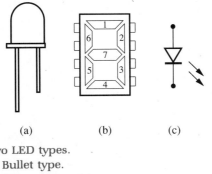

(a)   (b)   (c)

FIGURE 7.1 Two LED types.
(a) Bullet type.
(b) Seven segment type.
(c) Schematic symbol.

EXAMPLE 7.1

A model railroad enthusiast wishes to build a switch signal consisting of a green and a red LED. A 16 VDC supply is to be used. When the supply is positive, the green light is to come on and when it is negative, the red light is to come on. Design the circuit.

SOLUTION:

Since the turn-on voltages are slightly different for the diodes, separate dropping resistors will probably be needed for the lights and care will have to be taken that the polarity of the diodes is correct. Figure 7.2a contains the schematic. Doing a KVL walk from V_S to ground gives

$$V_S - I_D R_S - V_D = 0 \text{ V}$$

or

$$R_S = \frac{V_S - V_D}{I_D} \tag{7.1}$$

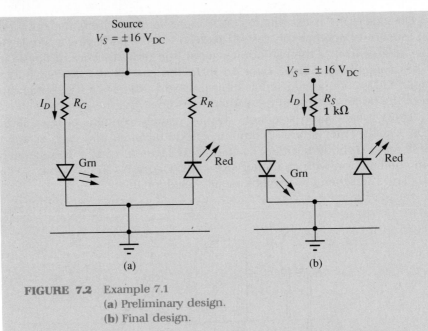

FIGURE 7.2 Example 7.1
(a) Preliminary design.
(b) Final design.

Using the approximations presented in this section

$$R_G = \frac{16 \text{ V} - 2.25 \text{ V}}{15 \text{ mA}}$$

$$= 917 \text{ } \Omega \qquad \text{(Use 1 k}\Omega\text{)}$$

$$R_R = \frac{-16 \text{ V} - (-1.4 \text{ V})}{-15 \text{ mA}}$$

$$= 973 \text{ } \Omega \qquad \text{(Use 1 k}\Omega\text{)}$$

Note that unless some very unnecessarily expensive resistors are used, the value of the dropping resistors is the same for each circuit. Obviously then, the circuit of Figure 7.2b should be used. The green light will not be quite as bright as the red one, but in view of the meanings usually attached to these colors, perhaps that is not all bad either.

7.3 The Schottky Diode

In some applications, such as when very small signal voltages are involved, the 0.6 V required to turn on a silicon diode is a disadvantage. As we mentioned in Section 2.5, only about 0.2 V is required to turn on a germanium diode, and it has been used in situations where low turn-on voltages are needed. However, germanium has other characteristics which limit its usefulness, notably it has poor temperature stability.

The Schottky barrier diode, or more simply the Schottky diode, which is also known as a "hot carrier" diode, is more commonly used in applications where low turn-on voltage is desirable. Perhaps more importantly, it is used in circuits where very high frequencies (above the megahertz range) are involved.

There is no difference in physical appearance between a Schottky diode and a conventional diode. The only way to distinguish them short of measuring their characteristics is to look up the part number in a data book. In a circuit diagram a Schottky diode is usually represented by the symbol in Figure 7.3.

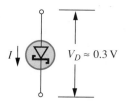

FIGURE 7.3 Schottky diode symbol.

The Schottky diode is formed by replacing the P type semiconductor material of the conventional diode with a metal electrode. This greatly reduces the width of the depletion region, and so reduces the turn-on voltage. The time it takes for carriers to cross the junction is also reduced. Thus the device will operate at a higher frequency. When you study digital electronics you will discover that the use of Schottky diodes in conjunction with other electronic devices will improve their frequency response as well.

7.4 Varactors

In your circuits class you probably visualized a capacitor as a pair of conducting plates separated by an insulator. If you stop to think about it, that is about the way we have described a reverse biased diode. You have P and N carriers (conductors) separated by a depletion region (insulator). Thus all diodes have some capacitance between their terminals. As you might expect, given the size of the "plates," diode capacitance is small, usually a few picofarads. However, in very high frequency applications, capacitance becomes an important diode parameter.

There is a whole class of devices designed to take advantage of the capacitor-like characteristic of diodes. They are called "varactors." Returning to your circuits class, you may remember that the capacitance of a capacitor is inversely proportional to the separation of the plates. The effective plate separation in a diode (i.e., the width of the depletion region) depends on the amount of reverse bias applied to the junction. Thus the capacitance of a varactor is electrically adjustable. This is the characteristic which makes a varactor a valuable device. Using it, a tuned circuit can be designed that will automatically adjust itself. There once was a day when you had to retune your TV every time you switched channels. That chore is now handled by a varactor.

The symbol customarily used for a varactor is shown in Figure 7.4.

FIGURE 7.4   Varactor symbol.

7.5 Zener Diode—Ideal Model

We first mentioned the Zener diode in Section 2.11. It may be worth your time to return to Chapter 2 and reread our discussion of its properties. According to what we said there, if we assume the Zener diode in Figure 7.5 to be ideal, as long as V_S is less than the Zener voltage, V_Z, the diode

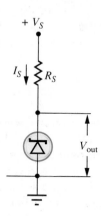

FIGURE 7.5   Zener diode regulator circuit.

will block all current flow so there will be no voltage lost across resistor R_S. The output voltage V_{out} will be exactly equal to V_S. That means we will be operating somewhere along the horizontal section of the ideal Zener diode characteristic graph of Figure 7.6 in the interval identified as "off."

Notice that due to the way the Zener is connected in the circuit, it is

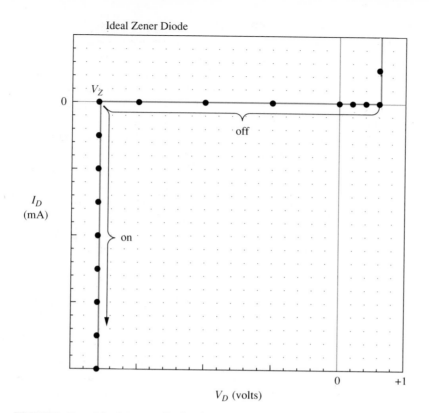

Ideal Zener Diode

FIGURE 7.6   Ideal Zener diode characteristic graph.

reverse biased when V_S is positive. This makes all Zener voltages and currents negative on the characteristic graph. For purposes of circuit analysis and in order to simplify equation writing both the Zener voltage and the Zener current will be considered positive from the circuit standpoint.

As soon as V_S in Figure 7.5 exceeds V_Z, the Zener begins to conduct current in the reverse direction. That means we will have passed the negative knee of the characteristic graph and will now be operating in the interval identified as "on" in Figure 7.6. V_{out} will now be equal to V_Z. Since we have assumed that the Zener is ideal, the slope of the graph in the "on" region is perfectly vertical. Thus V_Z will remain at exactly the same value regardless of the value of V_S as long as $V_S > V_Z$. As V_S increases, the current will increase to satisfy the KVL equation for the circuit

$$V_S - I_Z R_S - V_Z = 0 \text{ V} \tag{7.2}$$

Since the ideal characteristic in the "on" region is vertical, an increase in current does not mean an increase in voltage lost across the Zener diode. Thus ideally, the circuit in Figure 7.5 is a perfect voltage regulator circuit since the output is perfectly constant: $\%r = 0.0\%$ (see Section 5.3).

7.6 Zener Diode—Mathematical Model

The preceding model for a Zener diode is analogous to the fixed voltage model for a conventional forward biased diode. As was the case for conventional diodes, this model is adequate for a great deal of the work you will do with Zeners. Where large current variations are involved, or where small variations in V_Z are of concern, a more accurate model must be used. One such model would be a graphical model similar to the one presented in Section 3.5 for conventional diodes. Although that is probably the closest approximation to "truth" we could get, due to the problems with graphs that we mentioned in Section 3.5, a mathematical model is usually used. In order to develop such a model, we must first look at the data usually provided by the manufacturers of Zeners.

The most important data usually provided for Zener diodes are the Zener voltage rating, V_{ZT}, the test current at which the test voltage was measured, I_{ZT}, the Zener impedance Z_Z, and either the maximum power dissipation rating P_{Zrate} or the maximum current rating I_{Zrate}. Obviously, to specify one of the latter two ratings is to specify them both since they are related by the power equation

$$P_{Zrate} = V_Z I_{Zrate} \tag{7.3}$$

The value of P_{Zrate} is determined by the ability of the diode to dissipate heat. As we have mentioned for other such maximum ratings, when the device operation approaches this rating its reliability and life expectancy are diminished. It is good design practice to never allow a device to operate at more than about 90 percent of its rated power, or

$$\bullet \qquad I_{Zmax} = 0.9 * I_{Zrate} \tag{7.4}$$

In order to keep the Zener voltage as constant as possible, the Zener current should always be kept above about 10 percent of its rated value. Thus

$$\bullet \qquad I_{Zmin} = 0.1 * I_{Zrate} \tag{7.5}$$

You are already familiar with V_{ZT} which is usually specified at a stated test current I_{ZT} which is usually near 1/4 of I_{Zrate}.

The Zener impedance, Z_Z would be referred to as the dynamic resistance r_Z if Zener terminology were consistent with the terminology used for the rectifier diodes with which you are already familiar. Z_Z is determined by the slope (1/slope really) of the Zener diode characteristic curve when the Zener is being operated in its Zener region. The characteristic for most modern Zeners is nearly a straight line for currents greater than about 5 percent of I_{Zrate}. Thus, Z_Z can usually be considered constant in the "on" region. It may be calculated in the same way as r_D was calculated in Section 3.10. Its value is usually specified on manufacturers data sheets at the same test current as is employed to specify V_{ZT}.

Since it is the manufacturer's goal to make the diode as near to ideal as possible (Z_Z as small as possible), their data sheets often list only the maximum value for Z_Z. Experience shows that the typical value can be taken as approximately 80 percent of the specified maximum value or

●
$$\dot{Z}_Z \approx 0.8 * Z_{Z\max} \tag{7.6}$$

Since the manufacturer's data sheet usually specifies $Z_{Z\max}$, V_{ZT}, and I_{ZT}, these values can be used to obtain a Zener diode equation similar to the equation we developed in Section 3.6 for conventional diodes. Since, as we have mentioned, the characteristic for Zeners is typically nearly a straight line, we will be able to approximate it using a linear equation.

Since the above data contains the coordinates of a point, (V_{ZT}, I_{ZT}), and the slope (Z_Z), (1/slope, really!), we can plug these values into the familiar "slope, intercept" form of the equation of a straight line, $y = mx + b$, and write

$$V_{ZT} = Z_Z I_{ZT} + V_{Z0} \tag{7.7}$$

Solving for the intercept gives the value of V_{Z0} in terms of the manufacturer's published Zener diode parameters.

●
$$V_{Z0} = V_{ZT} - I_{ZT} Z_Z \tag{7.8}$$

Using the above value for V_{Z0} we can write the Zener diode equation

●
$$V_Z = I_Z Z_Z + V_{Z0} \tag{7.9}$$

These relationships are shown in Figure 7.7 and the resulting equivalent

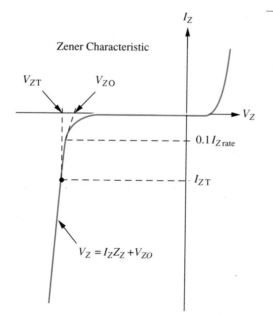

FIGURE 7.7 Graph for constructing a mathematical model of a Zener diode.

circuit is shown in Figure 7.8. It is important to remember that these relationships are only valid for diode reverse currents greater than about $0.1 * I_{Zrate}$.

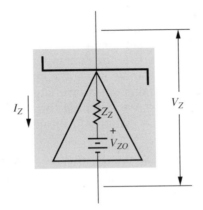

FIGURE 7.8   Equivalent circuit for a Zener diode.

EXAMPLE 7.2
Determine the Zener voltage for a 1N4730 Zener diode at 100 mA.

SOLUTION:
From Appendix F the Zener parameters for the 1N4730 are: $P_{DISS} = 1$ W, $V_{ZT} = 3.9$ V; $I_{ZT} = 64$ mA; and $Z_{Zmax} = 9$ Ω. A typical value for Z_Z will be assumed to be

$$Z_Z \approx 0.8 * Z_{Zmax} \tag{7.6}$$
$$= 0.8 * 9 \ \Omega = 7.20 \ \Omega$$

Substituting these values in Equation 7.8 gives

$$V_{Z0} = V_{ZT} - I_{ZT}Z_Z \tag{7.8}$$
$$= 3.9 \text{ V} - 64 \text{ mA} * 7.20 \ \Omega = 3.44 \text{ V}$$

Substituting values in Equation 7.9, the Zener diode equation for the 1N4730 is

$$V_Z = 3.44 \text{ V} + I_Z * 7.20 \ \Omega$$

Thus at $I_Z = 100$ mA:

$$V_Z = 3.44 \text{ V} + 100 \text{ mA} * 7.20 \ \Omega = 4.16 \text{ V}$$

7.7 Unloaded Zener Regulator Circuit Analysis

One of the more common applications of Zener diodes is to maintain a fixed output voltage in spite of a varying source voltage, and perhaps in spite of a varying load as well. First we will consider only fluctuations in V_S. These fluctuations could result from relatively long term variations in DC level, or they could result from the presence of AC ripple due to poor rectifier filtering. A typical circuit for this application is shown in Figure 7.9a. The Zener equivalent circuit is included in Figure 7.9b.

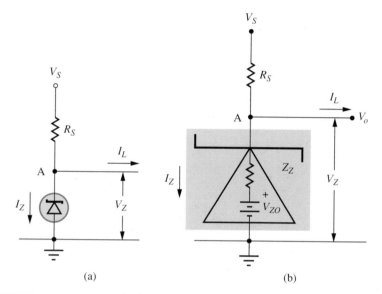

(a)   (b)

FIGURE 7.9 A Zener diode regulator circuit.
(a) Circuit diagram.
(b) Equivalent circuit.

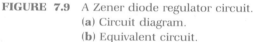

In order to see how V_Z will vary when V_S goes from V_{Smin} to V_{Smax}, we will first find I_Z for both extremes of V_S. Assuming that $I_L = 0$ A, a KVL walk from V_S to ground in Figure 7.9b gives

$$V_S - I_Z(R_S + Z_Z) - V_{Z0} = 0 \text{ V} \qquad (7.10)$$

or

$$I_Z = \frac{V_S - V_{Z0}}{R_S + Z_Z} \qquad (7.11)$$

Equation 7.11 can be solved for I_{Zmin} and for I_{Zmax}. These values can then be used in Equation 7.9 to obtain V_{Zmin} and V_{Zmax}, which is the variation in the output voltage which we seek.

In order for the analysis to be valid, as we mentioned in connection with Equation 7.5, I_{Zmin} must not be less than $0.1 * I_{Zrate}$.

EXAMPLE 7.3

The unloaded regulator circuit of Figure 7.10a is powered from a 12 V source that has 10 percent of sine wave ripple. What will be the ripple in the output?

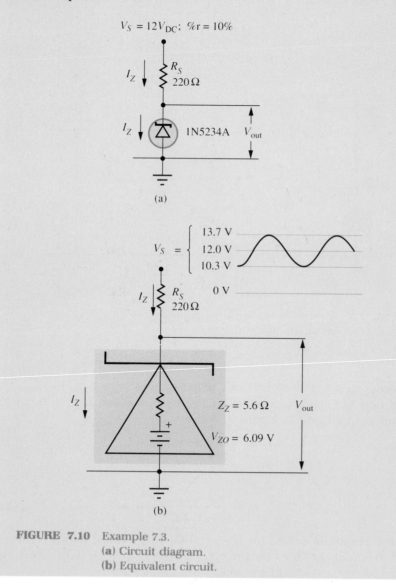

$V_S = 12V_{DC}; \quad \%r = 10\%$

R_S
220 Ω

I_Z

1N5234A V_{out}

(a)

$V_S = \begin{cases} 13.7 \text{ V} \\ 12.0 \text{ V} \\ 10.3 \text{ V} \end{cases}$

0 V

R_S
220 Ω

$Z_Z = 5.6 \, \Omega$ V_{out}

$V_{ZO} = 6.09 \text{ V}$

(b)

FIGURE 7.10   Example 7.3.
(a) Circuit diagram.
(b) Equivalent circuit.

SOLUTION:
From Appendix F the important data for the 1N5234A Zener diode are:
$P_{DISS} = 500$ mW, $V_{ZT} = 6.2$ V, $I_{ZT} = 20$ mA, $Z_{Zmax} = 7$ Ω.

The rated Zener current is given by

$$I_{Zrate} = \frac{P_{Zrate}}{V_Z} \tag{7.12}$$

$$= \frac{500 \text{ mW}}{6.2 \text{ V}} = 80.6 \text{ mA}$$

As a typical value of Z_Z for the 1N5234A we will use

$$Z_Z = 0.8 * Z_{Zmax} \tag{7.6}$$

$$= 0.8 * 7 \ \Omega = 5.6 \ \Omega$$

The value of V_{ZO} is then given by

$$V_{ZO} = V_{ZT} - I_{ZT}Z_Z \tag{7.8}$$

$$= 6.2 \text{ V} - 20 \text{ mA} * 5.6 \ \Omega = 6.09 \text{ V}$$

The above parameters are shown on the equivalent circuit in Figure 7.10b. The limits on the diode current in order for the above model to be valid are:

$$I_{Zmax} = 0.9 * I_{Zrate} \tag{7.4}$$

$$= 0.9 * 80.6 \text{ mA} = 72.6 \text{ mA} \tag{7.13}$$

and

$$I_{Zmin} = 0.1 * I_{Zrate} \tag{7.5}$$

$$= 0.1 * 80.6 \text{ mA} = 8.06 \text{ mA} \tag{7.14}$$

The Zener current, I_Z is given by doing a KVL walk through the circuit.

$$V_S - I_Z(R_S + Z_Z) - V_{ZO} = 0 \text{ V} \tag{7.15}$$

or

$$I_Z = \frac{V_S - V_{ZO}}{R_S + Z_Z}$$

$$= \frac{12 \text{ V} - 6.09 \text{ V}}{220 \ \Omega + 5.6 \ \Omega} = 26.2 \text{ mA}$$

Since the above current is well below the maximum of 72.6 mA indicated by Equation 7.13 and well above the minimum of 8.06 mA indicated by Equation 7.14, the linear model for the Zener which is shown in Figure 7.10b should apply for the postulated 10 percent variation in V_S.

The DC output voltage, V_{out} is then given by

$$V_Z = V_{ZO} + I_Z Z_Z \tag{7.9}$$
$$= 6.09 \text{ V} + 26.2 \text{ mA} * 5.6 \text{ } \Omega = 6.24 \text{ V}$$

From Section 5.3, our understanding of ripple indicates that the ripple voltage, v_S, in the supply is

$$v_{Srms} = r V_{SDC} \tag{5.25}$$
$$= 0.1 * 12 \text{ V} = 1.2 \text{ Vrms}$$

The principle of superposition allows us to consider the AC component of V_S separate from the DC component. From Figure 7.10b, resistors R_S and Z_Z form a voltage divider for the AC component. Thus

$$V_{out} = v_{Srms} \frac{Z_Z}{R_S + Z_Z} \tag{7.16}$$
$$= 1.2 \text{ Vrms} * \frac{5.6 \text{ } \Omega}{220 \text{ } \Omega + 5.6 \text{ } \Omega} = 29.8 \text{ mVrms}$$

Therefore the percent ripple in the output is

$$\%r \equiv \frac{V_{rms}}{V_{DC}} * 100\% \tag{5.2}$$
$$= \frac{29.8 \text{ mV}}{6.24 \text{ V}} * 100\% = 0.478\%$$

As you can see from the previous example, even the simplest of Zener regulator circuits does a reasonable job of ripple reduction.

7.8   Designing Zener Regulator Circuits

The designer of a regulator circuit such as that of Figure 7.11 usually knows the source voltage and the required load voltage and current. Based on that information, the Zener diode and the series resistor must be specified such that the Zener operation is always within the limits prescribed by Equations 7.4 and 7.5.

Obviously, the Zener voltage specification will be the required load voltage. That usually leaves two other variables to be specified. They are the Zener current rating, I_{Zrate}, and the series resistor value, R_S. Thus equations must be developed which relate these variables. The KVL equation for the circuit of Figure 7.11 is

$$V_S - I_S R_S - V_Z = 0 \text{ V} \tag{7.17}$$

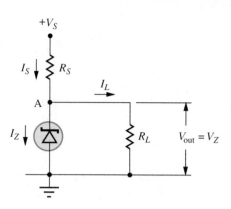

FIGURE 7.11   A loaded Zener regulator circuit.

The KCL equation for Point A is

$$I_S = I_Z + I_L \qquad (7.18)$$

Substituting Equation 7.18 into Equation 7.17 and solving for R_S gives

$$R_S = \frac{V_S - V_Z}{I_Z + I_L} \qquad (7.19)$$

Solving Equation 7.19 under minimum current and maximum current conditions will usually determine I_{Zrate} and R_S.

Minimum Zener current conditions can be determined as follows: Looking at Equation 7.18 you can see that the more current the load takes, the less there will be left to flow through the Zener. Also, from Figure 7.11, the smaller the supply voltage, the less the current flowing through the Zener. Or

$$\left. \begin{array}{l} V_S = V_{Smin} \\ I_L = I_{Lmax} \end{array} \right\} \longrightarrow I_Z = I_{Zmin} = 0.1\, I_{Zrate} \qquad (7.20)$$

Equation 7.19 then becomes

$$\bullet \qquad R_S = \frac{V_{Smin} - V_Z}{0.1\, I_{Zrate} + I_{Lmax}} \qquad (7.21)$$

Similarly, maximum Zener current conditions will be obtained when

$$\left. \begin{array}{l} V_S = V_{Smax} \\ I_L = I_{Lmin} \end{array} \right\} \longrightarrow I_Z = I_{Zmax} = 0.9\, I_{Zrate} \qquad (7.22)$$

Equation 7.19 then becomes

$$\bullet \qquad R_S = \frac{V_{Smax} - V_Z}{0.9\, I_{Zrate} + I_{Lmin}} \qquad (7.23)$$

The simultaneous solution of Equations 7.21 and 7.23 will provide the required value of I_{Zrate} and R_S.

EXAMPLE 7.4

Design and specify all components for a Zener regulator that will power a tape recorder rated at 6 VDC and 100 mA, from an automobile electrical system whose voltage varies from 15 VDC when the engine is running to 13 VDC when the engine is stopped.

SOLUTION:

We will design the circuit so that the maximum Zener current will be $I_{Zmax} = 0.9\,I_{Zrate}$. This will be the Zener current when the car's engine is running and the tape recorder is shut off. This situation is illustrated in Figure 7.12a. Equation 7.23 then becomes

$$R_S = \frac{V_{Smax} - V_Z}{0.9\,I_{Zrate} + I_{Lmin}} \qquad (7.23)$$

$$= \frac{15\text{ V} - 6\text{ V}}{0.9\,I_{Zrate} + 0\text{ A}} = \frac{10\text{ V}}{I_{Zrate}}$$

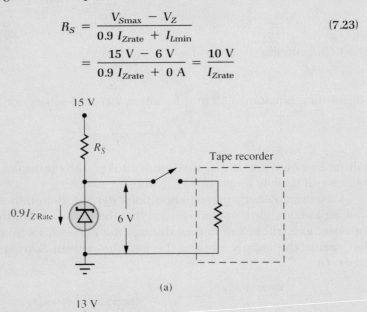

(a)

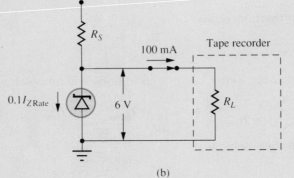

(b)

FIGURE 7.12   Example 7.4.
(a) When V_S is maximum and no load.
(b) When V_S is minimum and load is maximum.

We will also design for $I_{Zmin} = 0.1\, I_{Zrate}$ when the car's engine is off and the tape recorder is playing. This situation is illustrated in Figure 7.12b. Equation 7.21 then gives us

$$R_S = \frac{V_{Smin} - V_Z}{0.1\, I_{Zrate} + I_{Lmax}} \qquad (7.21)$$

$$= \frac{13\text{ V} - 6\text{ V}}{0.1\, I_{Zrate} + 100\text{ mA}} = \frac{7\text{ V}}{0.1\, I_{Zrate} + 100\text{ mA}}$$

Combining the results of Equation 7.23 and Equation 7.21 gives

$$\frac{10\text{ V}}{I_{Zrate}} = \frac{7\text{ V}}{0.1\, I_{Zrate} + 100\text{ mA}} \qquad (7.24)$$

Solving Equation 7.24 for I_{Zrate} gives $I_{Zrate} = 167$ mA.
The power dissipation rating for the Zener diode is

$$P_{rate} = V_Z I_{Zrate}$$

$$= 6\text{ V} * 167\text{ mA} = 1\text{ W} \qquad (7.3)$$

Looking in Appendix F, since a 1 W Zener would probably get too hot, a 1.5 W Zener should be used. Of the 1.5 W Zeners listed, the closest voltage rating would be given by the 1N5920A at $V_{ZT} = 6.2$ V. The other needed diode data are: $I_{ZT} = 60.5$ mA, $Z_{ZT} = 2\ \Omega$, $I_{Zrate} = 241$ mA.
In order that I_Z will never go below $0.1 * I_{Zrate}$ we will use

$$R_S = \frac{V_{Smin} - V_Z}{0.1\, I_{Zrate} + I_L} \qquad (7.21)$$

$$= \frac{13\text{ V} - 6.2\text{ V}}{0.1 * 241\text{ mA} + 100\text{ mA}}$$

$$= 54.8\ \Omega \qquad \text{(Use 47 Ohm)}$$

Notice that the diode spec values are used in determining R_S. Also, if the closest resistor size were used (56 Ω), the current would then drop below our minimum acceptable value.
In order to completely specify R_S, its maximum power dissipation must be given. It will be

$$P_{RSmax} = \frac{(V_{Smax} - V_Z)^2}{R_S} \qquad (7.25)$$

$$= \frac{(15\text{ V} - 6.2\text{ V})^2}{47\ \Omega}$$

$$= 1.65\text{ W} \qquad \text{(Use 2 W)}$$

7.9 Loaded Zener Regulator Circuit Analysis

When it is necessary to analyze a loaded Zener regulator circuit, it is usually best to construct a Thevenin equivalent for the circuit as seen by the Zener diode under conditions of both minimum Zener current and maximum Zener current. Then the same procedures can be applied to the equivalent circuit as those used for unloaded Zener circuits. Or, better, submit the circuit to your favorite circuit analysis program.

EXAMPLE 7.5

In the circuit of Example 7.4 a 1N5920A Zener is used. It has the following specifications: $V_{ZT} = 6.2$ V, $I_{ZT} = 60.5$ mA, $Z_{Zmax} = 2$ Ω, $P_{Zrate} = 1.5$ W. Determine the maximum and minimum voltage delivered to the tape recorder.

SOLUTION:

The loaded circuit is shown in Figure 7.13a. First we will construct equivalent circuits for both the tape recorder and the Zener diode. The equivalent resistance of the tape recorder is

$$R_L = \frac{V_L}{I_L} = \frac{6 \text{ V}}{100 \text{ mA}} = 60 \ \Omega$$

Next the diode equivalent circuit is obtained from the diode specs. Assuming as usual that

$$Z_Z = 0.8 * Z_{Zmax} \tag{7.6}$$
$$= 0.8 * 2 \ \Omega = 1.6 \ \Omega$$

then

$$V_{ZO} = V_{ZT} - I_{ZT}Z_Z \tag{7.8}$$
$$= 6.2 \text{ V} - 60.5 \text{ mA} * 1.6 \ \Omega = 6.10 \text{ V}$$

The Thevenin voltage seen by the Zener is given by

$$V_{Th} = V_S \frac{R_L}{R_L + R_S} \tag{7.26}$$

Thus

$$V_{Th \ max} = 15 \text{ V} \frac{60 \ \Omega}{60 \ \Omega + 47 \ \Omega} = 8.41 \text{ V}$$

$$V_{Th \ min} = 13 \text{ V} \frac{60 \ \Omega}{60 \ \Omega + 47 \ \Omega} = 7.29 \text{ V}$$

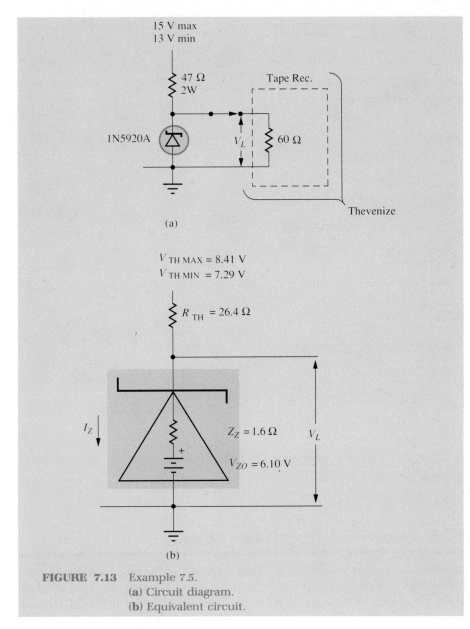

FIGURE 7.13 Example 7.5.
(a) Circuit diagram.
(b) Equivalent circuit.

The Thevenin resistance seen by the Zener is given by

$$R_{Th} = R_S \parallel R_L \tag{7.27}$$
$$= 47\ \Omega \parallel 60\ \Omega = 26.4\ \Omega$$

The overall equivalent circuit is shown in Figure 7.13b. Now the analysis procedure is the same as that of Example 7.3. Thus

$$I_Z = \frac{V_{Th} - V_{ZO}}{R_{Th} + Z_Z} \tag{7.11}$$

so

$$I_{Zmax} = \frac{8.41\ V - 6.10\ V}{26.4\ \Omega + 1.6\ \Omega} = 82.5\ mA$$
$$I_{Zmin} = \frac{7.29\ V - 6.10\ V}{26.4\ \Omega + 1.6\ \Omega} = 42.5\ mA$$

Substituting these values in the Zener equation gives

$$V_Z = V_{ZO} + I_Z Z_Z \tag{7.9}$$

so

$$V_{Zmax} = 6.10\ V + 82.5\ mA * 1.6\ \Omega = 6.23\ V$$

and

$$V_{Zmin} = 6.10\ V + 42.5\ mA * 1.6\ \Omega = 6.17\ V$$

As you can see, the regulator in the previous example does a fairly good job.

7.10 Improved Regulator Circuits

The simple regulator circuits we have studied so far are sometimes disparagingly referred to as "brute force regulators" because for a given value of V_S, the value of I_S is constant regardless of whether the load is connected or not. Thus the power supply is always fully loaded and the Zener must dissipate what ever power the load does not use. That means the power supply is going to run hot, the Zener is going to run hot, and a lot of heat sinking is going to be required.

In a more "intelligent" regulator, the Zener would be removed from the load circuit and replaced by some kind of variable resistor whose value could be adjusted automatically to maintain the output voltage exactly constant. The Zener circuit would then be used only as the reference to which the output voltage would be compared, and could therefore be a very low power circuit.

An even better approach would be for the regulator to control the supply voltage itself. If the load voltage dropped lower than the Zener reference voltage, the supply voltage would be adjusted to the correct value. Thus the source would not need to put out more voltage than the load required. This would provide the maximum efficiency.

This latter approach is the approach taken by the switching power supply. It derives its name from the fact that the power supply is switched on and off in response to the needs of the load. A nice byproduct of this approach is that as we mentioned in Section 5.3, the switching frequency is usually set at upwards of 40 kHz rather than using the 60 Hz power line frequency. Thus the filter capacitors need not be nearly as large.

Before we can discuss these "intelligent" regulators you must first learn about the transistors that are responsible for their "intelligence."

7.11 Troubleshooting ◀◀◀◀◀◀◀◀

It is possible to become so accustomed to seeing a forward diode voltage drop of around 0.7 V that you forget that the normal forward voltage for some diode types is quite different. You should also be aware that occasionally you will see schematic diagrams that use the standard diode symbol when a Schottky or other diode type is actually used.

When troubleshooting Zener regulator circuits, you should remember that if low voltage or excessive ripple is present in the supply such that the minimum supply voltage is less than the Zener voltage, the Zener will not be able to maintain the output voltage. Thus it is possible to blame the Zener diode for a problem that actually exists in its power supply.

7.12 Summary

We have briefly looked at some examples of diodes that have characteristics different from the usual silicon diode, the most important of which is the Zener diode. We have looked at an application of the Zener diode in a "brute force" voltage regulator circuit and we have found that if the diode characteristics are used, the solution of most diode circuits is rather straightforward, using conventional circuit analysis procedures.

GLOSSARY

Dropping resistor: A resistor placed in series with a power supply for the purpose of dropping the voltage, or limiting the amount of current delivered to a load. (Section 7.2)

Hot carrier diode: See Schottky diode.

Light emitting diode (LED):   A semiconductor diode which emits visible light when forward biased. (Section 7.2)

Schottky diode: A semiconductor diode formed by contact between a doped semiconductor layer and a metal coating, which has a nonlinear rectification characteristic. (Section 7.3)

Seven segment display: A display consisting of straight bar segments in such a configuration that by selectively activating the correct grouping of segments any numeral can be displayed. (Section 7.2)

Varactor: A solid state device whose capacitance can be varied by changing the voltage across it. (Section 7.4)

Zener diode: A semiconductor diode whose reverse breakdown voltage is carefully controlled and specified and that is designed to operate in its reverse breakdown mode. (Section 7.5)

PROBLEMS

Section 7.2

1. It is desired to hook a red LED across the telephone line so that whenever the phone rings the 48 V ring pulses will cause the LED to flash. Design a workable circuit using the approximations introduced in Section 7.2.

2. What should be the power rating of the 1 kΩ resistor used in Example 7.1?

3. How many times more current will it take for a seven segment LED display to show the number "2" than to show the number "1"?

4. What number should require the greatest amount of current for a seven segment LED display?

Section 7.3

5. Determine the approximate DC output voltage of the power supply of Figure 7.14 if the diodes are
   A. Conventional silicon diodes with $V_D = 0.7$ V.
   B. Schottky diodes with $V_D = 0.15$ V.

1.0 Vrms

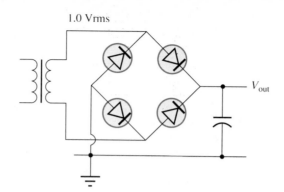

V_{out}

FIGURE 7.14   Problem 7.5.

Section 7.5

6. When building a half-wave rectifier, a student mistakenly used a 1N5240 Zener diode in place of the rectifier diode as shown in Figure 7.15. Discuss the effect of the error on the behavior of the circuit.
7. Repeat Problem 6 if the diode used was a 1N5226.

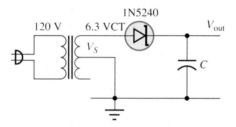

FIGURE 7.15   Problem 7.6.

Section 7.6

8. Using the specifications in Appendix F, write the Zener diode equation for a 1N4737 diode. Express the equation in simplest terms.
9. What should be the change in Zener voltage for the diode of Problem 8 if the diode current goes from 30 mA to 40 mA?
10. Repeat Problem 8 for a 1N5226 Zener diode.
11. Repeat Problem 9 for a 1N5226 Zener diode.
12. Using the graph in Figure 7.16, calculate the dynamic impedance Z_Z for the given 1N5234B Zener. Compare your result with the maximum value listed in Appendix F.

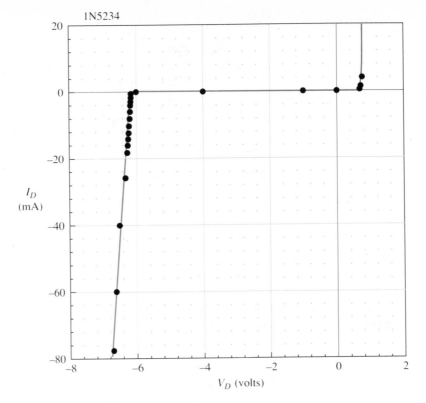

FIGURE 7.16   Problem 7.12.

Section 7.7

13. Suppose that you have a source that delivers 12 VDC with 10 percent ripple. Specify the resistor to be used with a 1N4733 Zener to make an unloaded regulator circuit to be connected to this source. Design the circuit to consume the minimum power possible using a standard 10 percent resistor.

14. Determine the amount of ripple present in the output of the regulator in Problem 13.

Section 7.9

15. Calculate the percentage of voltage regulation in the circuit of Examples 7.4 and 7.5 with the automobile engine running.

Troubleshooting ▶   ## Section 7.11

16. Describe the effect if the following faults developed in the circuit designed in Example 7.5. Consider each fault separately and consider the effect on both the circuit and the circuit output.

(a) Z_{Zmax} increases to 10 Ω.

(b)   The Zener becomes an open circuit.

(c)   The Zener becomes a short circuit.

(d) R_S becomes an open circuit.

(e) R_S becomes a short circuit.

17. Describe the effect on both the circuit and the load if someone accidentally hooked up the Zener diode backwards in Exampe 7.5.

18. Describe the effect on V_{out} in Problem 5 if one of the diodes became:

(a)   An open circuit.

(b)   A short circuit.

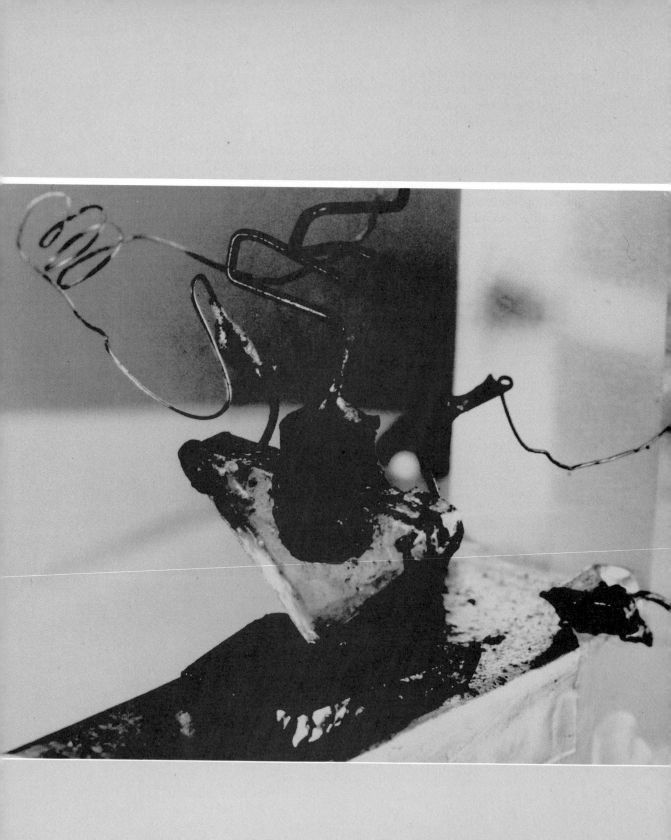

PART II
BIPOLAR TRANSISTORS

The first transistor. (Courtesy of Bell Telephone Laboratories.)

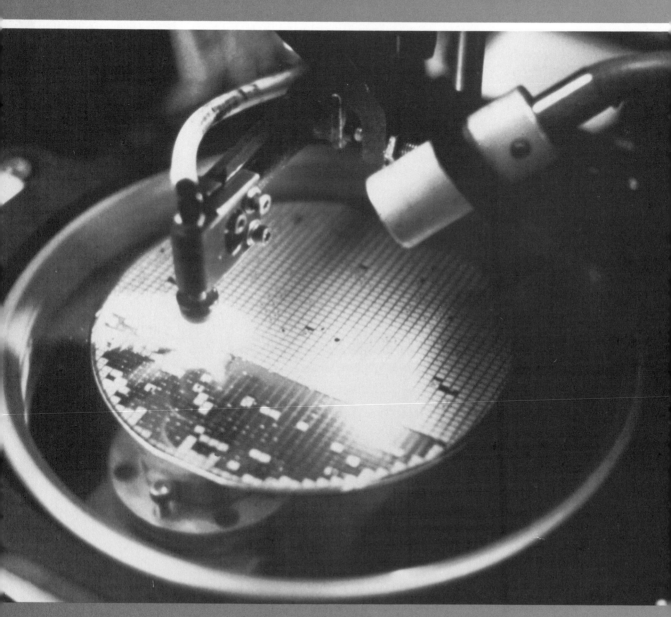

Cutting up a semiconductor wafer into individual
components circuits. (Courtesy of Motorola, Inc.)

Chapter 8
Transistor Theory

SPECIAL TERMS

Hybrid model
h parameters
h_{FE}
Beta
Gain
Active region
Cutoff
Saturation
Bias

8.1 Introduction and Objectives

In this chapter we will attempt to develop a model that will account for the major observed characteristics of bipolar transistors. As was the case in Chapter 2 where the P-N junction was first introduced, it is tempting to present a quantum mechanical description of what "really" goes on inside the device complete with energy level diagrams, band gaps, and the like. Again we will reject that approach in favor of an intuitive model which will be simple enough to understand and yet close enough to reality to be useful. Because bipolar transistors contain P-N junctions and because our approach will be similar, you will find it worth your time at this point to review what you learned in Chapters 2 and 3.

After studying this chapter you should understand how bipolar transistors function and how their behavior can be described using the common emitter characteristic graph. You will also begin to get a feel for the h parameter model commonly used to describe transistors. You should be able to:

- Identify transistor types and terminations from their circuit symbols.
- Apply P-N junction concepts to transistors.
- Use the definition of h_{FE} in circuit analysis situations.
- Recognize whether a transistor is saturated, cutoff, or operating in its active region.
- Construct and use the common emitter characteristic graph.
- Use transistor data commonly contained in data books to determine transistor behavior.

8.2 The DC h Parameter Model

The transistors we will be studying for the next several chapters are known as "bipolar" transistors because, as was the case with diodes, both positive charge carriers and negative charge carriers move around within the semiconductor crystal. Here we will be much more concerned with which type of carrier is involved and where it is going than was the case in our study of diodes.

The model we will begin to develop in this chapter and which we will complete in Chapter 11 for the description of transistors will be a "hybrid" model. It is a hybrid from the standpoint that it pictures the input as a Thevenin circuit composed of a voltage source in series with a resistor, and it pictures the output as a Norton circuit composed of a current source in parallel with a resistor. Because the model we will be using is known as a hybrid model, the parameters that we will be using to describe transistor behavior are called "hybrid parameters," or simply "h parameters." Although a dozen h parameters have been defined, we will find that we can adequately describe most transistor behavior using only two or three h parameters; the ones most often listed in manufacturers' data sheets.

8.3 P-N Junction Revisited

To begin, let us consider a P-N diode junction much like that of Sections 2.4 and 2.5. Again, if you have forgotten what we said there, what we are about to say will not make much sense unless you go back and review Sections 2.4 and 2.5.

Figure 8.1 shows a P-N junction in a circuit. You will notice that the terminal connected to the N region is grounded and is labeled "E." In transistor terminology it is called the "emitter." Perhaps this name comes from the fact that it "emits" negative carriers into the crystal. Since the emitter is grounded, the circuit we are developing is called a "common emitter" circuit.

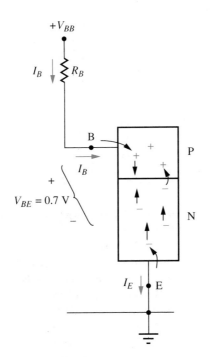

FIGURE 8.1 The base-emitter junction.

Again examining Figure 8.1 you will notice that the terminal connected to the P region is labeled "B." It is called the "base" for reasons which we will leave buried in history. The reason the terminal is on the side of the crystal rather than on the end of the crystal as we drew it in Figure 2.6 will become clear a little later. Notice that the base is connected through the "base biasing resistor" R_B to the "base bias supply" V_{BB}. It is standard practice to label power supplies using repeated subscripts such as V_{BB}, or V_{CC}.

Since the base supply is positive the "B-E junction," as this P-N junction is often called, is forward biased. Positive carriers generated at the base terminal will be moving through the P region of the crystal toward the junction. Negative carriers will be entering the N region of the crystal from the emitter terminal and also traveling toward the junction. Just as it was with diodes, any unlucky negative carrier that happens to stray across the junction will not survive long before it is captured by one of the positive carriers in the base region. This results in a positive "base current" I_B flowing into the base and a positive "emitter current" I_E flowing out of the emitter. The behavior of this forward biased junction can be described by a characteristic graph similar to that of Figure 2.9. As was the case with diodes, the typical voltage loss from base to emitter, V_{BE}, for a silicon transistor will be between 0.6 V and 0.8 V. As you might guess, for a closer approximation we can use the junction equation

$$V_{BE} = 26 \text{ mV} * \ln\left(\frac{I_E}{10 \text{ fA}}\right) \tag{8.1}$$

In case you have forgotten the origin of Equation 8.1, you should reread Sections 3.6 and 3.7.

You should notice that the current in Equation 8.1 is the EMITTER current, not the base current. By the time we finish constructing our transistor you will see that typically there is a big difference between the base current and the emitter current.

8.4 Adding Another Junction

In Figure 8.2a you will find that another N region has been added to the crystal. A terminal has been added to this second N region which is labeled "C." It is known as the "collector" perhaps because it "collects" the charges (negative charges in this case) that have traveled through the crystal. This completes the construction of our transistor. As you might guess, the transistor we have constructed is known as an NPN type transistor.

A collector resistor R_C connects the collector to the collector power supply V_{CC} (note the repeated subscript indicating a power supply). A standard NPN transistor symbol is drawn in Figure 8.2b. Notice that there is an arrow on the emitter lead which indicates the direction positive current will flow across the B-E junction.

Assume that the collector supply V_{CC} of Figure 8.2 is of sufficiently high positive voltage so that the collector to emitter voltage V_{CE} is reason-

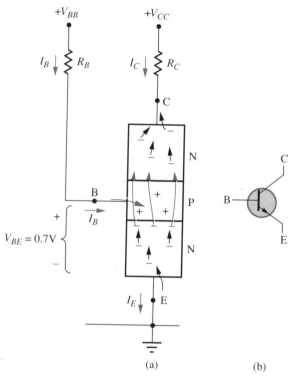

FIGURE 8.2 An NPN transistor
(a) Conceptual diagram.
(b) Standard symbol.

ably high, say around 5.0 V. Since the base to emitter voltage, V_{BE}, will be around 0.7 V for a silicon transistor, the collector to base junction will be reverse biased to about 4.3 V. Thus no current will flow across that junction.

Now let us look at how the collector circuit and the base circuit interact. In order to relate the situation to the world with which you are more familiar, we will ascribe to the carriers in the crystal some personality traits which they may not actually possess but which may help your understanding.

8.5 Current Gain (h_{FE})

Base current flowing from V_{BB} results in the injection of positive carriers into the base region. As you already know this causes the B-E junction to become forward biased. Carriers will then begin to diffuse across the junction just as was the case when we were studying diodes.

However, in this case, when a negative carrier from the emitter happens to wander across the B-E junction, it sees $+0.7$ V at the base but it also sees $+5$ V over at the collector. What would you do if you were that negative carrier, settle for $+0.7$ V, or "go for the gold"? Obviously, it heads for the collector and we begin to have collector current, I_C.

Our disappointed and snubbed positive carrier in the base region looks for another negative carrier crossing the B-E junction from the emitter only to see it head straight for the collector too. Can this process go on forever? Obviously not. Eventually some unlucky negative charge gets "caught" by a positive carrier in the base region before it can get through to the collector region. It combines with the positive charge that caught it. It is combinations of this type that constitute the base current.

You will notice that we have "gain." Injecting a single positive carrier into the base can result in several hundred negative carriers flowing to the collector, and that is gain; current gain. Bipolar transistors are current amplifying devices. A current of a few microamps flowing from V_{BB} into the base terminal can cause upwards of a milliamp of current to flow from V_{CC} into the collector. (Yes, it is really negative carriers going the other way, isn't it?) But it is now time to return from our brief excursion into what is "really" going on and again assume positive current flow in the circuit. Gain is an important parameter that describes transistor operation. You may occasionally hear the Greek letter beta (β) used to refer to the current gain of a transistor. PSpice uses the letters BF to stand for "forward beta." The term more commonly used by manufacturers and the term we will use is the h parameter h_{FE}. This h parameter is often referred to as the "forward current transfer ratio." This name grew from the idea that current appears to be transferred foreward from the base circuit to the collector circuit. The first subscript of this h parameter is derived from that idea. The second subscript refers to the fact that the emitter is the "common," or the "grounded" terminal of the transistor. h_{FE} is defined to be

$$h_{FE} \equiv \frac{I_C}{I_B} \qquad (8.2)$$

Sometimes distinctions are attempted between the DC value, h_{FE} and the AC value, h_{fe}. We will find that sample to sample variations are usually so great that such distinctions are of little practical value. Thus we will use h_{FE} and h_{fe} more or less interchangeably.

You will see in Section 8.7 that Equation 8.2 is only valid when the transistor is operated in its "active region." Obviously, if both collector current and base current are zero, you might wonder about the equation's validity. There are other conditions where the equation does not hold as well.

Notice that h_{FE} is a pure number having no units. It is a current divided by a current so the units will cancel out leaving a pure number. For most small transistors h_{FE} will have a value in the order of 100.

Our rather crude model suggests that the thinner the base region and the more lightly it is doped, the less chance a given negative carrier has of getting "captured" in its flight across the base region toward the collector. That should result in higher gain for the transistor. It turns out that in fact the gain of real transistors is determined by adjusting these factors during the manufacturing process.

Since both thickness and doping are difficult to control, there is a great deal of sample-to-sample variation in transistor gain. For many transistors you can expect variations of ± 50 percent or more. For example, if you look in Appendix F at the manufacturer's data sheet for a 2N4401 transistor you will see that for a collector current I_C of 10 mA h_{FE} can vary from 100 to as high as 300. One of the biggest challenges in transistor circuit design is to develop circuits that will meet the circuit design requirements regardless of this variation.

8.6 Relating V_{BE} and I_c

As we pointed out in Section 8.3, it is often adequate to use a value of $V_{BE} \approx 0.7$ V. For a more precise value, it is usually desirable to relate V_{BE} to I_C.

Figure 8.2 shows how positive current flows in the circuit in which the transistor is connected. The very small base current I_B flows from V_{BB} into the base. That allows the much larger current I_C to flow from V_{CC} into the collector. According to KCL, the emitter current is $I_E = I_C + I_B$. Since the gain of most transistors is upwards of 100 or more, I_C is so much larger than I_B that I_B can be ignored. Thus with an error of 1 percent or less,

$$I_C \approx I_E \tag{8.3}$$

Substituting the approximation of Equation 8.3 into Equation 8.1 gives us a very useful form of the junction equation applied to silicon transistors

• $$V_{BE} \approx 26 \text{ mV} * \ln\left(\frac{I_C}{10 \text{ fA}}\right) \quad \text{(Si)} \tag{8.4}$$

Equation 8.4 will estimate V_{BE} for most small silicon transistors with an error of usually less than 0.1 V. As was the case for diodes discussed in Section 3.7, for germanium transistors Equation 8.4 becomes

$$V_{BE} \approx 26 \text{ mV} * \ln\left(\frac{I_C}{1 \text{ }\mu\text{A}}\right) \quad \text{(Ge)} \tag{8.5}$$

For transistors PSpice uses $I_S = 0.1$ fA $= 10^{-16}$ A as its default value rather than the 10 fA used in Equation 8.4. You will find that your calculations will agree more closely with laboratory measurements if you specify $I_S = 10$ fA.

Most transistor circuits can be analyzed using nothing more than Equations 8.2, 8.4, and standard circuit analysis procedures.

EXAMPLE 8.1

The transistor of Figure 8.3 has an h_{FE} of 250. Determine the value of V_{BB} such that $I_C = 5$ mA.

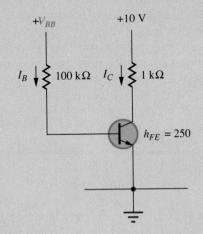

FIGURE 8.3 Example 8.1.

SOLUTION:

Since the emitter is grounded, Equation 8.4 will give us the voltage at the base

$$V_{BE} = 26 \text{ mV} * \ln\left(\frac{I_C}{10 \text{ fA}}\right) \tag{8.4}$$

$$= 26 \text{ mV} * \ln\left(\frac{5 \text{ mA}}{10 \text{ fA}}\right)$$

$$= 0.700 \text{ V}$$

Since I_C is given, Equation 8.2 will give us the base current

$$I_B = \frac{I_C}{h_{FE}} \tag{8.2}$$

$$= \frac{5 \text{ mA}}{250}$$

$$= 20 \text{ μA}$$

A KVL walk through the base circuit gives

$$V_{BB} - I_B R_B - V_{BE} = 0 \text{ V} \tag{8.6}$$

Solving for V_{BB} we have

$$\begin{aligned}
V_{BB} &= I_B R_B + V_{BE} \\
&= 20 \ \mu a * 100 \ k \ \Omega + 0.700 \text{ V} \\
&= 2.70 \text{ V} \tag{8.7}
\end{aligned}$$

8.7 Saturation, Cutoff, and Digital Operation

The above model requires that V_{CE} be a fairly high value. If V_{CE} gets down to a few tenths of a volt, we discover that to increase I_B does not cause an increase in I_C. The transistor is then said to be "saturated." That is, there are more than enough carriers at the base to supply the required collector current. Thus adding more carriers does not increase I_C. This is one of the conditions mentioned in Section 8.5 when Equation 8.2 is no longer valid. Under these conditions V_{CE} is very small, typically less than 0.5 V. Thus the transistor is acting like a switch that has been flipped to the "on" position.

At the other extreme, if there is no base current, there can be no collector current and the transistor is then said to be "cut off." This is obviously another condition when Equation 8.2 is no longer valid. The transistor is acting like a switch that has been flipped to the "off" position. In fact, many transistors are operated as switches. The circuit is so designed that the transistor is always either on or off, saturated or cut off. That is how all digital circuits operate such as the circuits in your calculator or your computer.

The region of transistor operation that is between saturation and cutoff is known as the "active" region. In the active region the transistor is conducting some current, but it is not fully turned on.

Sometimes the active region is referred to as the "linear region." Given the notorious nonlinearity of most transistors, "active region" is probably a better term.

Most of what we have said thus far in this chapter is only valid in the active region.

8.8 NPN vs PNP

The transistor described in the previous sections was an NPN type. That means that in order for the B-E junction to be forward biased, the base was at a more positive voltage than the emitter and the collector was at a still higher positive voltage. That required positive power supplies for both V_{BB} and V_{CC}.

We could have used a PNP transistor for our example. That would have necessitated reversing all carrier types, currents, and voltages, but the ideas would still be essentially the same. Of course, PNP transistors exist and are used in electronic circuits. Since positive carriers account for most of the current flow in such devices and since positive carriers are less mobile than negative carriers, the characteristics of PNP devices are harder to control. NPN devices are easier to work with and are more frequently used in circuits. In Chapter 15 we will study circuits in which NPN and PNP transistors are used together in what is known as a "complementary circuit." As you might guess, the name is derived from the fact that in the circuit the properties of the transistors complement each other.

8.9 Symbols, Appearance, and Ohmmeter Tests

Figure 8.4 shows the standard symbols for both an NPN transistor and a PNP transistor. Notice that in each case the emitter has an arrow attached to it which indicates the direction that positive current will flow through the B-E junction. By looking for the arrow on the schematic diagram you

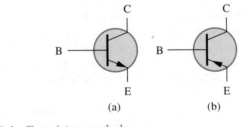

(a) (b)

FIGURE 8.4 Transistor symbols.
(a) NPN transistor.
(b) PNP transistor.

can always tell which terminal is the emitter and whether you are dealing with an NPN or a PNP transistor. Figure 8.5 shows how typical transistors are packaged. It is impossible from the packaging to distinguish between NPN and PNP transistor types, or even whether the device is a transistor or some other three terminal device for that matter.

Sometimes markings are placed on the transistor indicating which terminal is the emitter, which is the base, and which is the collector. In case you have to deal with an unknown and unmarked transistor, there are test instruments on the market which can determine which terminal is which and whether you have an NPN or a PNP type transistor. If you do

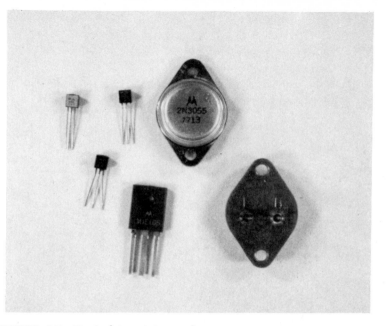

FIGURE 8.5 Typical transistor packages.

not have such an instrument, an ohmmeter and an understanding of the behavior of P-N junctions is usually sufficient to determine which terminal is the base and whether you have an NPN or a PNP transistor.

Remember that both the B-E junction and the B-C junction act very much like ordinary diode junctions. Thus the approach here is just an extension of the ideas developed in Section 2.9: With an ohmmeter set on its diode test scale, if the transistor is an NPN type, the meter will show conduction when the positive ohmmeter lead (+) is connected to the base (P type material) and the negative lead (−) is connected to either of the other transistor terminals (both are N type). If the transistor is a PNP type, the meter will show conduction when the negative ohmmeter lead (−) is connected to the base (N) and the positive ohmmeter lead (+) is connected to either of the other transistor terminals (both are P). Using this procedure it is not possible to distinguish between the emitter and the collector. CAUTION: As we mentioned in Chapter 2, some digital meters use reverse polarity for their resistance function.

EXAMPLE 8.2
Given the circuits in Figure 8.6 in each case tell whether the transistor is saturated, cut off, or active.

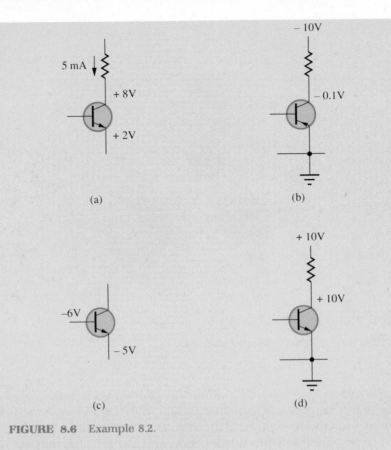

FIGURE 8.6   Example 8.2.

SOLUTION:

(a) Active. Current is flowing so the transistor is not cut off. V_{CE} is not near zero so the transistor is not saturated.

(b) Saturated. Current is flowing since there is a voltage drop across R_C. V_{CE} is nearly zero.

(c) Cut off. The B-E junction is reverse biased. Therefore, there can be no base current and therefore no collector current.

(d) Cut off. There is no voltage drop across R_C and so no current is flowing through it.

8.10   The Common Emitter Characteristic Graph

As was the case with diodes, the most accurate description of the behavior of a transistor is obtained by drawing a characteristic graph. In the case of the diode we had only two variables to deal with so by plotting I_D vs V_D

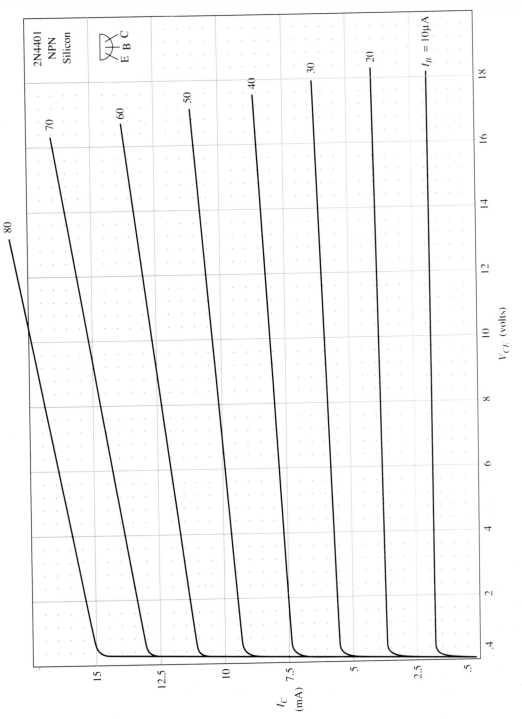

FIGURE 8.7 Common emitter characteristics for a 2N4401.

we had a reasonably complete picture of its behavior. We now have three variables to deal with. They are I_B, I_C, and V_{CE}. Figure 8.7 contains an example of the characteristic graph that is usually used to picture these three variables.

The graph of Figure 8.7 was plotted using a circuit such as the one shown in Figure 8.8. The 100 k Ω resistor will convert the base power supply V_{BB} into a reasonably good current source. The ammeter measuring I_C and the voltmeter measuring V_{CE} could very well be the inputs to an X-Y plotter, or to a more modern data collection system.

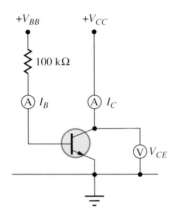

FIGURE 8.8   Circuit for constructing common emitter characteristic curves.

The data collection procedure would be to set I_B to a fixed value such as 10 μA in the case of Figure 8.7. Then you would take I_C vs V_{CE} data from $V_{CE} = 0$ V to the maximum value of interest. Then you would increment I_B to 20 μA and again take I_C vs V_{CE} data from $V_{CE} = 0$ V to maximum. You would repeat this procedure until you had a complete set of characteristic curves.

8.11   Reading the Characteristic Graph

According to Equation 8.2 the collector current I_C should depend on only I_B and h_{FE}. If that were exactly true, the characteristic curves in Figure 8.7 would be perfectly horizontal lines. Then you would get a fixed collector current for a given base current no matter what the value of V_{CE}. As you can see, that is not the case for a 2N4401. I_C depends to some extent on V_{CE} as well as on I_B. That is to say, a 2N4401 transistor is not exactly a constant current device. But it is not far from it. Since the variations in transistor properties are so notoriously large, it is hardly worth our effort to try to develop an equation that will show that added detail. Thus the equation we will use to relate I_C and I_B is Equation 8.2, the definition of h_{FE}. Further, as we have said before, it is probably not worth our time to make a distinction between an AC value for h_{fe} (note the lower case subscripts indicating AC) and a DC value for h_{FE}. We will return to this idea in Chapter 12 when we look at the transistor's ability to amplify AC signals.

EXAMPLE 8.3

(a) By referring to the characteristics of Figure 8.7 what should be the value of I_C when $I_B = 20$ µA for a 2N4401 transistor?
(b) Repeat (a) above for $I_B = 80$ µA.
(c) Calculate h_{FE} for the above values of I_C and I_B.

Compare with the value listed in the manufacturer's specification sheet in Appendix F.

SOLUTION:

(a) From Figure 8.7 you can see that for $I_B = 20$ µA, I_C ranges from about 3.5 mA to 4 mA depending on V_{CE}. A typical value might be $I_C = 3.8$ mA.
(b) The variation in I_C is worse at $I_B = 80$ µA, ranging from under 15 mA to perhaps 20 mA depending on how far you extend the graph. A typical value might be $I_C = 17$ mA.
(c) From (a) above we have

$$h_{FE} \equiv \frac{I_C}{I_B} \qquad (8.2)$$

$$= \frac{3.8 \text{ mA}}{20 \text{ µA}}$$

$$= 190$$

From (b.) above we have

$$h_{FE} \equiv \frac{I_C}{I_B} \qquad (8.2)$$

$$= \frac{17 \text{ mA}}{80 \text{ µA}}$$

$$= 213$$

According to the data listed in Appendix F, for currents near the values used in Example 8.3, h_{FE} should be from 100 to 300 with a typical value of 135. Thus the transistor used has a little higher gain than is typical, but it is within the limits set by the manufacturer.

8.12   Reading Data Sheets

If you examine some of the manufacturer's data sheets included in Appendix F you will see that there is some variation in the information provided depending to some extent on the bias of the manufacturer and the intended use of the device. There is also some variation in the symbols used but the following definitions are not uncommon:

P_{DISS} The maximum power the transistor can normally dissipate.

BV_{CEO} The maximum voltage that can exist between collector and emitter before breakdown will occur. Other maximum voltage ratings may also be given.

I_{CMAX} The maximum collector current the device can carry. The maximum base current is also often given.

h_{FE} The current gain of the transistor which may be specified at several different values of I_C.

h_{oe} The output conductance of the transistor is usually small enough that it can be ignored, but it is sometimes given.

C_{ob} The output capacitance of the transistor. Sometimes the base to emitter capacitance is also given.

If the transistor is intended to be used in switching applications, transit times and saturation voltages are also usually included.

▶▶▶▶▶▶▶   8.13   Troubleshooting

Since the B-E junction behaves much like a diode junction, V_{BE} is always near to 0.7 V for silicon transistors. If it is less than around 0.6 V, the transistor is probably cut off. If it is more than about 0.8 V, you can be reasonably sure that the transistor is defective.

8.14   Summary

In this chapter the rather intuitive ideas first presented in Chapter 2 relating to the behavior of the P-N junction were extended to "describe" transistor behavior. We showed that since the B-E junction is forward biased, the junction equation, which we first encountered in connection with diodes applies to transistors as well.

The h parameter h_{FE} was defined and the transistor common emitter characteristic graph was introduced. Both were applied to transistor circuits in which the transistor was operating within its active region.

GLOSSARY

Active region: The condition in which the output of an active device is approximately proportional to the input. (Section 8.7)

Beta: (β) See h_{FE}. (Section 8.5)

Bias: A nonsignal voltage or current applied to an active device for the purpose of rendering it responsive to an input signal. (Section 8.10)

Cutoff: The condition in which no carriers are present in an active device. (Section 8.7)

Gain: (A) The ratio of the output of a device compared to its input. (Section 8.5)

h parameters: Parameters used to define the behavior of an amplifier based on the hybrid model. (Section 8.2)

h_{FE}: Forward transfer ratio. The common emitter h parameter defined by the ratio of collector current to base current. (Section 8.5)

Hybrid model: A model which describes the behavior of a device. The model is obtained by applying two different circuit analysis procedures. In the case of the bipolar transistor, the base circuit is viewed as a Thevenin circuit and the collector is viewed as a Norton circuit. (Section 8.2)

Saturation: The condition in which there are more carriers present in an active device than are required to conduct the circuit current. (Section 8.7)

PROBLEMS

Section 8.3

1. Use Equation 8.1 to estimate the base to emitter voltage, V_{BE}, if the emitter current for the transistor in Figure 8.9 is   (a) I_E = 7.5 mA   (b) I_E = 75 mA   (c) I_E = 750 mA.

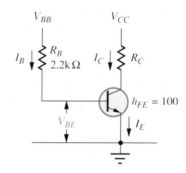

FIGURE 8.9 Problem 8.1.

2. If the emitter voltage for the transistor in Figure 8.10 is $V_E = 3$ V, find the base voltage, V_B.

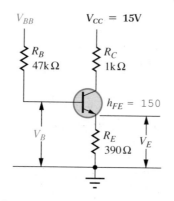

FIGURE 8.10 Problem 8.2.

Section 8.5

3. For a certain transistor, a base current of $I_B = 25$ μA results in an emitter current of $I_E = 4.5$ mA. Find the collector current.
4. Find the value of h_{FE} for the transistor in Problem 3.

Section 8.6

5. For Problem 1, calculate the required base power supply voltage, V_{BB}, for each value of I_E.
6. Find the value of base power supply voltage, V_{BB}, required for the circuit of Figure 8.11 if $I_C = 10$ mA.
7. Find the value of V_{BB} for Problem 2.
8. Find the value of V_{CE} for Problem 6.
9. Find the value of V_{CE} in Problem 2.

Section 8.7

10. For the circuit of Figure 8.11, estimate the approximate value of V_C, V_B, and V_E if the transistor is   (a)   cut off   (b)   saturated.
11. Repeat Problem 10 for Figure 8.10.
12. Estimate the value required for V_{BB} in Figure 8.11 if the transistor is (a)   cut off   (b)   barely saturated.
13. Repeat Problem 12 for Figure 8.10.

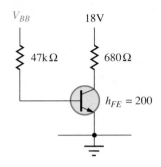

V_{BB} 18V

47kΩ 680Ω

$h_{FE} = 200$

FIGURE 8.11   Problem 8.6.

Section 8.11

14. Using the characteristics in Figure 8.7, estimate the value of h_{FE} for a 2N4401 transistor at   (a) $V_{CE} = 14$ V and $I_C = 3$ mA.   (b) $V_{CE} = 2$ V and $I_C = 13$ mA.

Section 8.12

15. By referring to the data in Appendix F, construct a transistor characteristic graph similar to Figure 8.7 for a 2N3055 transistor for $I_B = 1$ mA, 50 mA, and 150 mA. Use a range of $V_{CE} = 0$ V to $V_{CE} = 30$ V.

Section 8.13

◀ **Troubleshooting**

16. Assume that $V_{BB} = 5$V for the circuit of Figure 8.10. Calculate the resulting value of V_C, V_B, and V_E if the collector becomes an open circuit ($h_{FE} = 0$).
17. Repeat Problem 16 if instead of becoming open, the collector shorts to the emitter.
18. Repeat Problem 16 if instead of the above transistor defect, R_E becomes shorted ($R_E = 0$ Ω.)

At the Q point everything is very quiescent.
(Photograph by Christie Tito.)

Chapter 9
Load Line Construction and Use

SPECIAL TERMS

Load line
Q point
Coupling capacitor
Midband
Miller effect
Cutoff frequency
Low frequency cutoff
Bypass capacitor

9.1   Introduction and Objectives

In this chapter we will combine the behavior of the transistor with the behavior of the circuit of which it is a part. The common emitter characteristics describe the behavior of the transistor. We will draw a load line on the characteristics which will describe the behavior of the rest of the circuit. We first encountered the load line concept in connection with diodes in Chapter 3. Taken together, the characteristics and the load line will describe the total circuit behavior.

From this chapter you should gain an understanding of the meaning of the "Q point," how it is determined, its use, and its importance in understanding circuit behavior. You should then be able to:

- Construct both AC and DC load lines, interpret their meaning, and use them to predict circuit behavior.
- Draw the signal waveform produced by a transistor operating on a given load line at a given Q point showing when and why clipping occurs.
- Determine the location of the Q point for a given circuit.
- Optimize the Q point for a given circuit.

9.2   The DC Load Line

We first encountered the use of a load line in Section 3.5 when we were studying circuits containing diodes. If you have forgotten the load line concept, a review of that section would be worthwhile.

When a transistor is one of the components of an electric circuit, the circuit's behavior can be described by drawing a load line on the transistor's characteristic graph. The characteristic graph describes the transistor's behavior and the load line describes the behavior of the rest of the circuit. The circuit is constrained to operate somewhere along the load line. Exactly where is determined by the base current and the transistor's characteristics.

The situation here is very similar to that of Section 3.5 relating to diodes except that here we have the added freedom of varying the transistor's operation by varying the base current. In order to draw the load line, consider the circuit of Figure 9.1a. Taking a KVL walk through the collector circuit from V_{CC} to ground gives

$$V_{CC} - I_C R_C - V_{CE} = 0 \text{ V} \tag{9.1}$$

Solving Equation 9.1 for I_C, we have

$$I_C = \frac{V_{CC} - V_{CE}}{R_C}$$

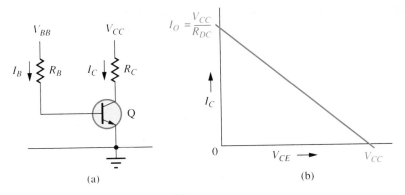

FIGURE 9.1 Drawing the load line.
(a) Circuit diagram.
(b) The load line.

It may be that in some cases you may encounter more resistors in the collector-emitter circuit than just R_C. You may even find resistors between the emitter and ground. For that reason perhaps we should replace R_C above with a more general term. We will use R_{DC} to stand for all of the DC resistances in the C-E circuit between V_{CC} and ground. We then have

$$I_C = \frac{V_{CC} - V_{CE}}{R_{DC}} \qquad (9.2)$$

This is the equation of the DC load line.

To be even more general in our approach, you will encounter circuits in which the emitter circuit is not terminated at ground as it is in Figure 9.1a. Rather an emitter power supply V_{EE} may be used. You should readily recognize that in that case Equation 9.1 becomes

$$V_{CC} - I_C R_C - V_{CE} = V_{EE} \qquad (9.3)$$

We will leave it up to you to alter Equation 9.2 and the rest of this section to include this additional generalization.

If you are mathematically inclined, it is obvious that if I_C is the dependent variable and V_{CE} is the independent variable, Equation 9.2 is linear with a slope of $-1/R_{DC}$, a vertical axis intercept of $I_0 = \dfrac{V_{CC}}{R_{DC}}$, and a horizontal axis intercept of V_{CC}.

On the other hand, if you would rather think in terms of what is going on in the circuit, by looking at Figure 9.1a you can see that if the transistor were totally turned off so that $I_C = 0$ A, then none of V_{CC} would be lost across R_{DC}. Thus

$$\text{IF} \quad I_C = 0 \text{ A} \quad \text{THEN} \quad V_{CE} = V_{CC} \qquad (9.4)$$

Equation 9.4 locates the horizontal axis intercept at V_{CC}. On the other hand, if the transistor were totally turned on so that $V_{CE} = 0$ V, then all of V_{CC} would be lost across R_{DC}. Thus

$$\text{IF} \qquad V_{CE} = 0 \text{ V} \qquad \text{THEN} \qquad I_0 = \frac{V_{CC}}{R_{DC}} \tag{9.5}$$

Equation 9.5 locates the vertical axis intercept at I_0. A straight line drawn between the above intercepts is the DC load line. The DC load line has been drawn on the characteristic graph in Figure 9.1b.

EXAMPLE 9.1

Draw a load line for the circuit of Figure 9.2a on the 2N4401 characteristics in Figure 9.2b and estimate from the graph the value of I_C and V_{CE} for (a) $I_B = 0$ A, (b) $I_B = 30$ μA, and (c) $I_B = 100$ μA.

SOLUTION:

According to the circuit of Figure 9.2a, $V_{CC} = 15$ V. Using the idea of Equation 9.4, V_{CC} has been located on the graph in Figure 9.2b. Substituting the data from the circuit of Figure 9.2a into Equation 9.5 gives

$$I_0 = \frac{V_{CC}}{R_{DC}} \tag{9.5}$$

$$= \frac{15 \text{ V}}{1.2 \text{ k}\Omega}$$

$$= 12.5 \text{ mA}$$

I_0 has been located on the graph of Figure 9.2b and the load line has been drawn connecting V_{CC} and I_0. The values of I_C and V_{CE} are then read from the load line for

(a) $I_B = 0$ A:

$I_C = 0$ A, $V_{CE} = 15$ V

In this case, the transistor is not conducting. It is operating at cutoff.

(b) $I_B = 30$ μA:

$I_C = 5.8$ mA, $V_{CE} = 8$ V

In this case, the transistor is operating in its active region.

(c) $I_B = 100$ μA:

$I_C = 12$ mA, $V_{CE} = 0.2$ V

In this case, the transistor is saturated.

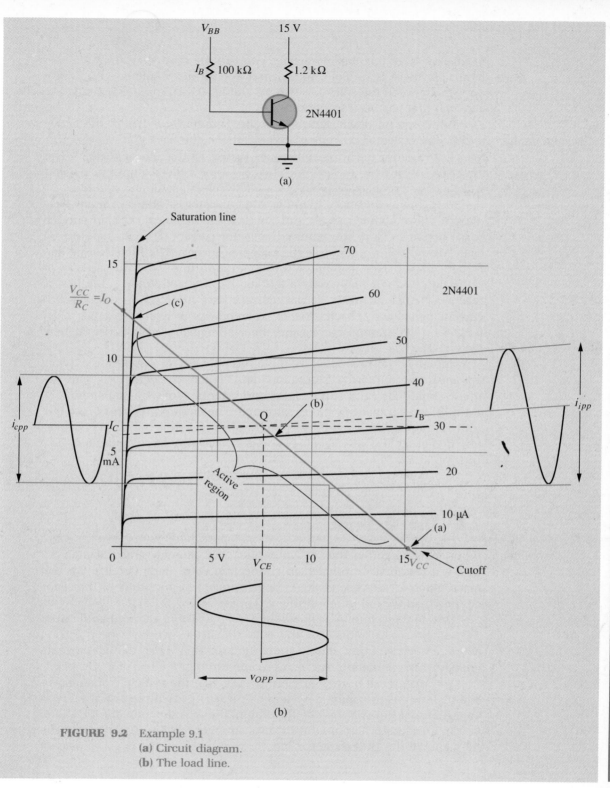

FIGURE 9.2 Example 9.1
(a) Circuit diagram.
(b) The load line.

From Part (a) of the above example, you can see that when the transistor is cut off, there is no collector current and so the whole supply voltage appears between the collector and the emitter. The transistor is acting like a switch that has been turned off.

From Part (b) of the above example, you can see that in the active region, the collector current is dependent on the base current. For this reason, the active region is sometimes known as the "linear region." The transistor is partially turned on. There is some collector current flowing and there is also a sizable voltage loss between collector and emitter.

Note that in the above example, the transistor is saturated at any base current above about $I_B = 65 \ \mu A$. There is, therefore, more than enough base current to carry the required collector current. Further increases in base current will not increase the collector current. The transistor is then turned on and is operating at the "saturation line." As you can see, at saturation, V_{CE} is usually only a few tenths of a volt. When saturation is reached further increases in base current have no effect on the collector current or voltage. The transistor is acting like a switch that has been turned on. Saturation, cutoff, and the active region are all identified in Figure 9.2b.

Hopefully the preceding example has shown you something of what was meant by the terms "saturation" and "cutoff" in Section 8.6. This example should also give you a better understanding of why Equation 8.2 is not valid for a transistor that is saturated. The fact that V_{CE} and I_C are not critically dependent on I_B once saturation is reached is one of the reasons why digital circuits are so reliable, and in fact, why large computers can function at all.

9.3 The Q Point

If you have forgotten the "Q Point" idea, you should reread Section 3.5 where it was first introduced in connection with diode circuits. We will make much greater use of the Q point concept in our study of transistor circuits than we did in our study of diodes.

Due to the nature of semiconductors, transistors cannot handle pure AC voltages or currents. For the NPN transistors we have been talking about, a negative input simply shuts the transistor off so that all currents are zero. This limitation can be overcome through the use of a "DC bias."

You may recall from your circuits class that the principle of superposition allows us to combine AC and DC. It tells us that the actual value of voltage or current will then be the sum of the AC value plus the DC value.

In a transistor circuit, the DC bias simply provides a DC offset to the AC signal. If the DC bias is sufficiently large, even on the negative peak of the AC signal, the net value of voltage and current seen by the transistor will still be positive. Thus the transistor will not reach cutoff. This is the purpose of DC bias in transistor circuits. The transistor voltages and currents when no signal is present are known as the *bias voltages* and *currents*. They determine the Q point.

In circuits such as that shown in Figure 9.2A, the transistor should be biased so that when the maximum expected AC signal current is added to the DC base bias current, the transistor will always be operating in the active region. The negative peak of the AC signal will not drive the transistor to cutoff and the positive peak of the AC signal will not drive the transistor into saturation. Since cutoff occurs at one end of the load line and saturation occurs near the other end, it seems reasonable that if you are designing an amplifier circuit, you should place the Q point at about the middle of the load line. Thus

$$I_C \approx \frac{I_0}{2} \tag{9.6}$$

You might argue that since saturation, I_{Csat}, is our concern, Equation 9.6 should read $I_C = I_{Csat}/2$. You would be correct. However, I_{Csat} is difficult to locate mathematically, and for most modern transistors there is very little difference between I_{Csat} and I_0 anyway, so we will use I_0 in Equation 9.6.

EXAMPLE 9.2

Locate the optimum Q point for the circuit in Figure 9.2a. Determine the expected transistor voltages and currents at the above Q point. Calculate the value of V_{BB} that will give the above Q point. Determine the AC signal voltage generated at the collector when an AC signal $i_i = 30$ μAPP is injected at the base of the transistor.

SOLUTION:

According to Example 9.1 $I_0 = 12.5$ mA for the circuit of Figure 9.2a. (Note the small difference between I_{csat} and I_0.) Thus

$$I_C = \frac{I_0}{2} \tag{9.6}$$

$$= \frac{12.5 \text{ mA}}{2} = 6.25 \text{ mA}$$

I_C locates the Q point on the load line drawn in Example 9.1 for this circuit. The Q point has been located on the graph of Figure 9.2b. We can guess, based on our knowledge of P-N junction behavior that V_{BE}

will be between 0.6 V and 0.8 V or we can estimate V_{BE} using the junction equation

$$V_{BE} = 26 \text{ mV} * \ln\left(\frac{I_C}{10 \text{ fA}}\right) \tag{8.4}$$

$$= 26 \text{ mV} * \ln\left(\frac{6.25 \text{ mA}}{10 \text{ fA}}\right) = 0.706 \text{ V}$$

The other Q point values V_{CE} and I_B can be read from the graph or they can be calculated mathematically. To obtain V_{CE} mathematically, we will first write the KVL equation for the circuit

$$V_{CC} - I_C R_C - V_{CE} = 0 \tag{9.1}$$

Solving for V_{CE}

$$V_{CE} = V_{CC} - I_C R_C$$
$$= 15 \text{ V} - 6.25 \text{ mA} * 1.2 \text{ k}\Omega = 7.5 \text{ V}$$

To obtain I_B mathematically we use the definition of h_{FE} from Section 8.5.

$$h_{FE} \equiv \frac{I_C}{I_B} \tag{8.2}$$

According to the values for h_{FE} calculated in Example 8.4 for this 2N4401, a typical value for h_{FE} might be $h_{FE} = 200$. Solving Equation 8.2 for I_B we have

$$I_B = \frac{I_C}{h_{FE}} \tag{9.7}$$

$$= \frac{6.25 \text{ mA}}{200} = 31.3 \text{ μA}$$

If you compare the above calculated values with those you can read from the graph in Figure 9.2b, you will see that they agree quite well.

To find the required value of V_{BB} we write the KVL equation for the base circuit

$$V_{BB} - I_B R_B - V_{BE} = 0 \text{ V} \tag{8.6}$$

Solving for V_{BB} gives

$$V_{BB} = I_B R_B + V_{BE}$$
$$= 31.3 \text{ μA} * 100 \text{ k}\Omega + 0.706 \text{ V} = 3.84 \text{ V}$$

It would be interesting for you to go to the laboratory and check to see how close the above calculations agree with "truth." Due to variations among components, do not expect agreement closer than 15 to 20 percent.

Adding the signal of $i_i = 30$ μAPP to the above base current using the principle of superposition gives

On the positive peak (+)

$$I_{B+} = I_B + \frac{i_{iPP}}{2}$$

$$= 31.3 \text{ μA} + \frac{30 \text{ μAPP}}{2} = 46.3 \text{ μA}$$

Solving for the positive peak collector current I_{C+} we have

$$I_{C+} = h_{FE}i_{B+}$$

$$= 200 * 46.3 \text{ μA} = 9.26 \text{ mA}$$

Solving for the collector voltage v_{CE+} gives

$$v_{CE+} = V_{CC} - I_{C+}R_C$$

$$= 15 \text{ V} - 9.26 \text{ mA} * 1.2 \text{ kΩ} = 3.89 \text{ V}$$

On the negative peak (−)

$$I_{B-} = I_B - \frac{i_{iPP}}{2}$$

$$= 31.3 \text{ μA} - \frac{30 \text{ μAPP}}{2} = 16.3 \text{ μA}$$

Solving for the negative peak collector current I_{C-} we have

$$I_{C-} = h_{FE}I_{B-}$$

$$= 200 * 16.3 \text{ μA} = 3.26 \text{ mA}$$

Solving for the collector voltage V_{CE-} we have

$$V_{CE-} = V_{CC} - I_{C-}R_C$$

$$= 15 \text{ V} - 3.26 \text{ mA} * 1.2 \text{ kΩ} = 11.1 \text{ V}$$

Solving for the AC output voltage v_o we have

$$v_o = V_{CE+} - V_{CE-}$$

$$= 3.89 \text{ V} - 11.1 \text{ V} = -7.21 \text{ VPP}$$

These results and the corresponding waveforms are drawn on the graph in Figure 9.2b.

Note that the negative value for v_o in the above example indicates phase inversion. When the base current increases, the collector voltage decreases. This phase inversion is an important characteristic of the common emitter amplifier.

If you look carefully you will see that there is not perfect agreement between the calculated results and the graph. This is typical behavior for transistors due to the fact that h_{FE} is not exactly constant over the entire active region.

9.4 Circuit Analysis

Often in electronics you are confronted with an existing circuit whose Q point you need to locate. That is, given the circuit, the transistor parameters, and the supply voltages, you need to find I_C, I_B, V_{CE}, and V_{BE}. The base circuit is usually the place to start in such an analysis. A reasonable procedure for locating the Q point is as follows:

Writing KVL for the base circuit we have

$$V_{BB} - I_B R_B - V_{BE} = 0 \text{ V} \tag{8.6}$$

From the definition of h_{FE} we can write

$$I_B = \frac{I_C}{h_{FE}} \tag{9.7}$$

Substituting Equation 9.7 in Equation 8.6 gives

$$V_{BB} - \frac{I_C R_B}{h_{FE}} - V_{BE} = 0 \text{ V} \tag{9.8}$$

It is important to notice that when base circuit components (R_B in this case) are referred to the collector circuit, they must be divided by h_{FE}. They look smaller because the base current is so much smaller. We will find that the reverse is also true: When collector circuit components are referred to the base circuit, they must be multiplied by h_{FE} because they are carrying proportionately larger currents. This phenomenon is known as the "Miller Effect." We will encounter it many times.

Solving Equation 9.8 for I_C gives

$$I_C = \frac{V_{BB} - V_{BE}}{\dfrac{R_B}{h_{FE}}} \tag{9.9}$$

Equation 9.9 is written to emphasize the fact that, as we said above, due to the gain of the transistor, R_B looks like it is much smaller than it actually is. We will use this idea often in the future.

Of course, there are two unknowns in Equation 9.9 but we have encountered that problem before and you probably already know what to do. We can simply assume that $V_{BE} \approx 0.7$ V, or for a little closer approximation, we can iterate Equation 9.9 with the junction equation, Equation 8.4. This will give both I_C and V_{BE}. Equation 9.7 will then give I_B and a KVL walk through the collector circuit will give V_{CE}.

EXAMPLE 9.3

For the circuit of Figure 9.2a, reproduced here as Figure 9.3, locate the Q point if $V_{BB} = 6$ V and estimate the appearance of the AC output voltage waveform v_o at the collector if a signal current of $i_i = 30$ μAPP is injected at the base of the transistor.

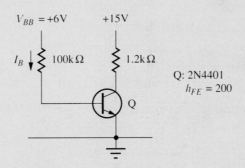

FIGURE 9.3 Example 9.3.

SOLUTION:

● (For a PSpice solution, see Appendix H.)

Substituting into Equation 9.9

$$I_C = \frac{V_{BB} - V_{BE}}{\dfrac{R_B}{h_{FE}}}$$

$$= \frac{6\text{ V} - V_{BE}}{\dfrac{100\text{ k}\Omega}{200}}$$

$$= \frac{6\text{ V} - V_{BE}}{500\ \Omega} \tag{9.9}$$

We will iterate the above equation with the junction equation

$$V_{BE} = 26\text{ mV} * \ln\!\left(\frac{I_C}{10\text{ fA}}\right) \tag{8.4}$$

Using $V_{BE} = 0.7$ V for the first guess, we have

	1	2	3	
$I_C = \dfrac{6.0 \text{ V} - V_{BE}}{500 \; \Omega}$	10.6	10.6	10.6	(mA)
$V_{BE} = 26 \text{ mV} * \ln\left(\dfrac{I_C}{10 \text{ fA}}\right)$	.720	.720	.720	(V)

Since the emitter is grounded, $V_B = V_{BE} = 0.720$ V. KVL applied to the collector tells us that

$$V_{CC} - I_C R_C - V_{CE} = 0 \text{ V}$$

or

$$V_{CE} = V_{CC} - I_C R_C$$
$$= 15 \text{ V} - 10.6 \text{ mA} * 1.2 \text{ k}\Omega = 2.33 \text{ V}$$

From the definition of h_{FE} we can find the base current

$$I_B = \frac{I_C}{h_{FE}} \tag{9.7}$$

$$= \frac{10.6 \text{ mA}}{200} = 53.0 \; \mu\text{A}$$

The above values for I_C, V_{CE}, and I_B agree with the values you can read from the graph in Figure 9.2b fairly well.

Adding the $i_{iPP} = 30 \; \mu\text{APP}$ signal to the above base current gives the following result:

On the positive peak:

$$I_{B+} = I_B + i_{iP} \tag{9.10}$$
$$= 53.0 \; \mu\text{A} + 15 \; \mu\text{A} = 68.0 \; \mu\text{A}$$

$$I_{C+} = h_{FE} I_{B+} \tag{9.11}$$
$$= 200 * 68 \; \mu\text{A} = 13.6 \text{ mA}$$

$$V_{C+} = V_{CC} - I_{C+} R_C \tag{9.12}$$
$$= 15 \text{ V} - 13.6 \text{ mA} * 1.2 \text{ k}\Omega = -1.32 \text{ V}$$

On the negative peak:

$$I_{B-} = I_B - i_{iP} \tag{9.13}$$
$$= 53.0 \; \mu\text{A} - 15 \; \mu\text{A} = 38.0 \; \mu\text{A}$$

$$I_{C-} = h_{FE} I_{B-} \tag{9.14}$$
$$= 200 * 38 \; \mu\text{A} = 7.60 \text{ mA}$$

$$V_{C-} = V_{CC} - I_{C-}R_C \tag{9.15}$$
$$= 15\ V - 7.60\ mA * 1.2\ k\Omega = 5.88\ V$$

If you have really understood what we have been doing, you should have all kinds of questions about the above calculated values for I_{C+} and V_{C+}. If you look at Figure 9.2b you will notice that neither of these values are on the load line. What is wrong? The problem is that the transistor saturates before it reaches these calculated values. The value of I_C is calculated from Equation 8.2, the definition of h_{FE}. You should remember from Section 8.5 that "Equation 8.2 is valid only when the transistor is operated in its active region." For the circuit of Figure 9.2a the collector current will never exceed 12.5 mA no matter what the base current is. You will save yourself a lot of difficulty if you learn to think about what circuits are doing; do not blindly plug numbers into formulas which may or may not apply.

Figure 9.4 is an attempt to picture what the resulting waveforms

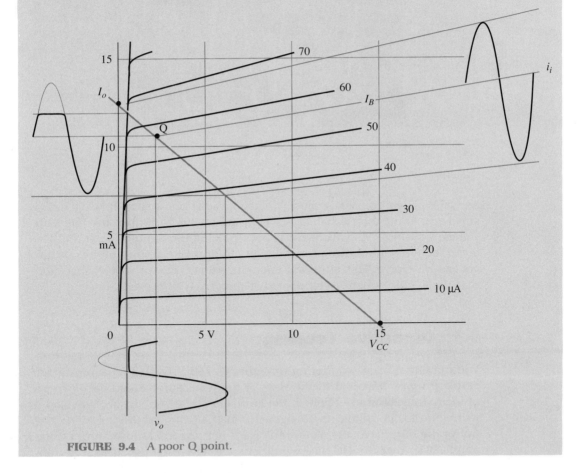

FIGURE 9.4 A poor Q point.

would look like. You notice that on the positive current peak the transistor saturates. The result is a clipped waveform. Figure 9.5 shows how the resulting collector voltage appears on a DC coupled oscilloscope. The vertical axis scale is 1 V per division with the bottom grid mark being 0 V. As you can see, clipping occurs at about +0.4 V as predicted in Figure 9.4.

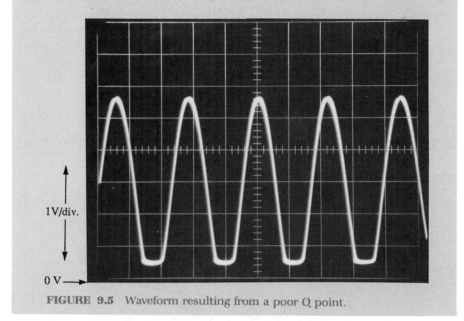

1 V/div.

0 V

FIGURE 9.5 Waveform resulting from a poor Q point.

As you can see from Example 9.3, placing the Q point such that V_{CE} is too small results in the probability of clipping at the bottom of the collector voltage waveform. As you might expect, placing the Q point such that V_{CE} is too large results in the probability of clipping at the top of the collector voltage waveform. This supports Equation 9.6 and our previous suggestion that the Q point ought to be placed at the center of the load line.

9.5 Capacitive Coupling

You may have been wondering how the AC signal we talked about in Example 9.3 was "injected at the base of the transistor." Since the purpose of many transistor amplifiers is the amplification of AC signals, we need a technique for coupling those signals into the input of the amplifier and out of the output. At the same time we must provide DC isolation for the amplifier so that the DC bias conditions we have established for the transistor will not be disturbed. You should be familiar with two common ways of accomplishing the above task. One way is to use a transformer. You remember that in order to induce a voltage in the secondary of a transformer a CHANGING magnetic field is required. That means an AC current is required in the primary. Any DC that is present in either the primary or the secondary windings is not transmitted by the transformer. These are exactly the characteristics we need. The transformer also offers some other attractive features such as the ability to match different impedances by specifying the turns ratio, and the ability to handle multiple inputs or outputs by specifying several different windings. Although they are used, transformers are expensive, bulky, and they have poor frequency characteristics. These disadvantages limit the usefulness of transformers as coupling devices.

The most common method of AC coupling is to use a capacitor. Figure 9.6 shows how AC signals are usually coupled into and out of a transistor amplifier. Resistor R_L represents the load to which the amplifier must supply power.

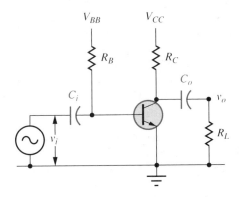

FIGURE 9.6   Capacitor coupling.

You should remember that when DC is applied to a capacitor the capacitor becomes charged and then no further current flows. Thus it provides good DC isolation. Of course, capacitors also have poor frequency characteristics since

$$X_C = \frac{1}{2\pi fC} \tag{9.16}$$

Thus at low frequency the capacitive reactance X_C becomes large. This results in a "low frequency cutoff," f_L. f_L is the frequency below which the capacitive reactance begins to "get in the way" of the AC signal being coupled into or out of the amplifier. We will have more to say about f_L in Chapter 12. Low frequency cutoff is a "single ended problem" because according to Equation 9.16 as frequency is increased a capacitor should become an increasingly good conductor of AC. According to Equation 9.16 the cutoff frequency will be dependent on the capacitance C of the capacitor. If the capacitance is made larger, the frequency at which cutoff occurs will be lower. We will look more closely at the capacitance values required for a given cutoff frequency in Chapter 12. For now we will assume that the capacitances and frequencies are high enough that X_C is negligible.

9.6   AC Load Line

Looking again at Figure 9.6, if we assume that the coupling capacitors are large enough to act as perfect conductors for AC, you can see that if the external load R_L is very small, all of the AC signal is going to get conducted straight to ground. The transistor will then see a dead short for AC. That will mean that there can be no AC signal voltage generated at the collector. The load lines we have been drawing such as Figure 9.2b and the resulting AC waveforms do not indicate this. The reason is that we have been drawing DC load lines. They have not been concerned with any AC load the transistor may be required to drive. We need a load line that will picture the AC load as seen by the transistor. We need an AC load line.

The transistor's operation will still be centered on the Q point and it will still be constrained to operate along the load line, but the load line along which it will operate is the AC load line, not the DC load line.

To construct the AC load line, you should remember two things:

1. Pure AC has an average value of zero. Thus, ideally, within the active region, the transistor's operation should be symmetrical about the Q point. Consequently, the Q point must remain the same regardless of any AC signals or loads.
2. The slope of the load line is determined by the resistance that the transistor sees in the circuit of which it is a part.

In the DC situation, according to Section 9.2 the transistor sees R_{DC} which is the sum of all of the DC resistances in the C-E circuit. By analogy, in the AC situation the transistor should see R_{AC}, the sum of all of the AC resistances in the C-E circuit.

Obviously, looking out of the emitter in Figure 9.6 the transistor sees zero resistance. Looking out of the collector, from an AC standpoint it sees two parallel paths. One path being through R_L to ground (remember that C_o is assumed to be a perfect conductor for AC). The other path is up through R_C to V_{CC}. V_{CC} looks like ground to the AC signal. The reason is that ideally, V_{CC} should remain exactly fixed regardless of any AC signal present. Thus, for the circuit of Figure 9.6

$$R_{AC} = R_C \parallel R_L \qquad (9.17)$$

Notice that Equation 9.17 gives us the slope of the AC load line, but not its position. To determine its position all we need to do is remember that the AC load line must pass through the Q point. From this information the mathematically inclined among us can draw the AC load line.

Probably the easiest way to draw the AC load line is to locate its horizontal intercept. We will call it v_{CC} so that it will correspond with V_{CC} of the DC load line. The lower case "v" will indicate AC. One way to locate v_{CC} is to remember that, due to the way we draw our graphs, resistance is really the negative reciprocal of the slope of the load line. Then, with reference to Figure 9.7 we can write

$$R_{AC} = \frac{\text{run}}{\text{rise}} = -\frac{V_2 - V_1}{I_2 - I_1}$$

$$= -\frac{v_{CC} - V_{CE}}{0\,A - I_C} = \frac{v_{CC} - V_{CE}}{I_C}$$

or

$$v_{CC} = V_{CE} + I_C R_{AC} \qquad (9.18)$$

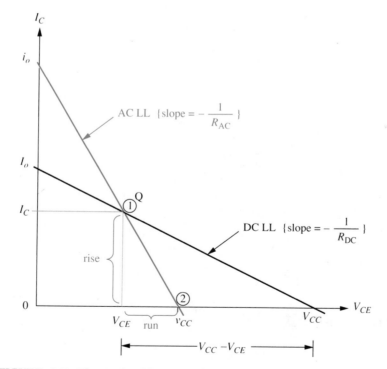

FIGURE 9.7 The AC load line.

EXAMPLE 9.4

Draw both the DC load line and the AC load line for the circuit of Figure 9.8a.

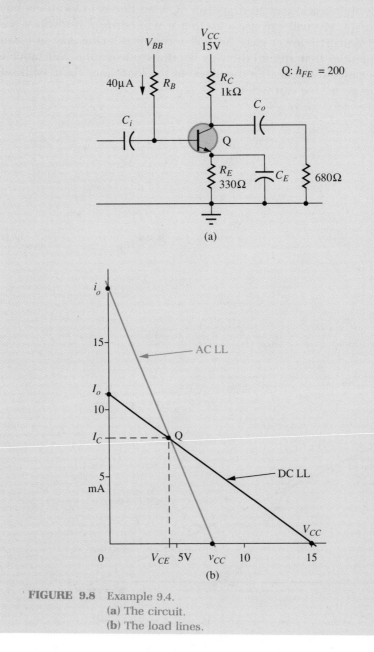

(a)

(b)

FIGURE 9.8   Example 9.4.
(a) The circuit.
(b) The load lines.

SOLUTION:

<u>DC LL:</u>

All capacitors are considered to be open circuits for purposes of DC analysis.

Notice that we have both a collector resistor and an emitter resistor in this circuit. According to Section 9.2

$$R_{DC} = R_C + R_E$$
$$= 1 \text{ k}\Omega + 330 \text{ }\Omega = 1.33 \text{ k}\Omega$$

Referring to the procedure detailed in Section 9.2, to locate the end points of the DC load line we have $V_{CC} = 15$ V and

$$I_0 = \frac{V_{CC}}{R_{DC}} \tag{9.5}$$

$$= \frac{15 \text{ V}}{1.33 \text{ k}\Omega} = 11.3 \text{ mA}$$

The DC load line is drawn from V_{CC} to I_0 in Figure 9.8b.

<u>Q Point:</u>

Since we know the value of I_B from Figure 9.8a, we can find I_C. From the definition of h_{FE} we can write

$$I_C = h_{FE}I_B \tag{9.19}$$
$$= 200 * 40 \text{ }\mu\text{A} = 8 \text{ mA}$$

This locates the Q point on the DC load line at $I_C = 8$ mA. The value of V_{CE} can be read from the graph or we can solve for it by taking a KVL walk down through the C-E circuit:

$$V_{CC} - I_C R_C - V_{CE} - I_C R_E = 0 \text{ V} \tag{9.20}$$

or

$$V_{CE} = V_{CC} - I_C(R_C + R_E)$$
$$= 15 \text{ V} - 8 \text{ mA}(1 \text{ k}\Omega + 330 \text{ }\Omega) = 4.36 \text{ V}$$

R_{AC}:

All capacitors are considered to be perfect conductors for purposes of AC analysis.

Looking out at the circuit from the emitter, we go directly to ground. Notice that R_E is "bypassed" by C_E. In the language of electronics C_E is the "emitter bypass capacitor."

Looking out from the collector we see $R_C \parallel R_L$. Thus we have

$$R_{AC} = R_C \parallel R_L \tag{9.17}$$
$$= 1 \text{ k}\Omega \parallel 680 \text{ }\Omega = 405 \text{ }\Omega$$

AC LL:

Two points are needed in order to draw the AC load line. The Q point gives us one point. v_{CC} will give us the other point. According to our discussion earlier in this section

$$v_{CC} = V_{CE} + I_C R_{AC} \tag{9.18}$$
$$= 4.36 \text{ V} + 8 \text{ mA} * 405 \text{ } \Omega = 7.60 \text{ V}$$

The location of v_{CC} is shown on the graph in Figure 9.8b. The AC load line has been drawn through v_{CC} and the Q point. Notice that the AC load line has been extended to the vertical axis intercept i_0 which corresponds to its DC counterpart, I_0.

9.7 Optimizing the Q Point

When the Q point was calculated in Example 9.4 above, you may have thought that it was not very well chosen since it was nowhere near the center of the DC load line. However, since AC operation is usually our concern, the Q point should be located at the center of the AC load line and not at the center of the DC load line. The AC load line is the load line along which the transistor actually operates.

For the circuits in Figure 9.2a and earlier, since no reactive components were present, the AC and the DC load lines were the same. For the more general case when reactive components may be present, Equation 9.6 should be rewritten as

$$I_C = \frac{i_0}{2} \tag{9.21}$$

and

$$V_{CE} = \frac{v_{CC}}{2} \tag{9.22}$$

In order to be able to design circuits so that the requirements of Equations 9.21 and 9.22 are met, let us return to Figure 9.7 and the ideas of Section 9.6. Looking at the DC load line, from the definition of slope we can write

$$R_{DC} = \frac{V_{CC} - V_{CE}}{I_C}$$

or

$$\frac{V_{CC}}{I_C} = R_{DC} + \frac{V_{CE}}{I_C} \tag{9.23}$$

Looking at the AC load line in Figure 9.7 we can write

$$R_{AC} = \frac{V_{CE}}{I_C} \qquad (9.24)$$

Substituting Equation 9.24 into Equation 9.23 gives

$$\frac{V_{CC}}{I_C} = R_{DC} + R_{AC}$$

solving for I_C gives the following circuit designers formula

••
$$i_C \overset{*}{=} \frac{V_{CC}}{R_{DC} + R_{AC}} \qquad (9.25)$$

Note Carefully:
The development in this section assumes that you are the designer and that you wish the Q point to be located at the center of the AC load line. If these assumptions are not met then Equations 9.24 and 9.25 are invalid and should not be used. A good example of a situation where Equation 9.25 is not valid is to use it in circuit analysis. Since the circuit analyst has no control over where the designer placed the Q point, his use of Equation 9.25 is invalid. The asterisk above the equal sign in Equation 9.25 is there to remind you that it is only valid for design purposes.

EXAMPLE 9.5
Determine the base current that will give the optimum Q point for the circuit of Figure 9.8a.

SOLUTION:
From Example 9.4 we know that $R_{DC} = 1.33 \text{ k}\Omega$ and $R_{AC} = 405 \text{ }\Omega$. Substituting these values in Equation 9.25 gives

$$I_C \overset{*}{=} \frac{V_{CC}}{R_{DC} + R_{AC}} \qquad (8.29)$$

$$= \frac{15 \text{ V}}{1.33 \text{ k}\Omega + 405 \text{ }\Omega}$$

$$= 8.65 \text{ mA}$$

From the definition of h_{FE} we have

$$I_B = \frac{I_C}{h_{FE}} \qquad (9.7)$$

$$= \frac{8.65 \text{ mA}}{200}$$

$$= 43.2 \text{ }\mu\text{A}$$

If you locate the above Q point and draw the AC load line carefully you will find that the Q point is exactly at the center of the AC load line. Equation 9.25 is widely applicable for obtaining "midpoint bias." All that is needed is to correctly identify R_{DC} and R_{AC}.

In the simplest of all cases, that of Example 9.2, since there are no reactive components in the circuit, it is obvious that the DC and the AC load lines are identical and $R_{AC} = R_{DC}$. Substituting in Equation 9.25 gives

$$I_C = \frac{V_{CC}}{R_{DC} + R_{AC}} \qquad (9.25)$$

$$= \frac{V_{CC}}{2R_{DC}}$$

which places the Q point exactly at the center of the load line—midpoint bias.

In Chapter 14 when we study transformer coupled circuits we will find that there too, even though R_{AC} may be many hundreds or even thousands of times larger than R_{DC}, Equation 9.25 is still applicable for midpoint bias.

▶▶▶▶▶▶▶ ## 9.8 Troubleshooting

Capacitively coupled circuits are reasonably easy to troubleshoot since the capacitors provide DC isolation for the amplifier. Injecting an AC signal into the circuit at the input coupling capacitor should result in an amplified AC signal with phase reversal at the collector. If it does not, troubleshooting is in order. As we have said before, the place to start is with a DC analysis of the circuit.

If the emitter is not grounded, care must be taken to obtain V_{BE} and V_{CE} from

$$V_{BE} = V_B - V_E \qquad (9.26)$$

and

$$V_{CE} = V_C - V_E \qquad (9.27)$$

As we have pointed out before, if V_{BE} is less than about 0.6 V the transistor should be cut off so V_{CE} should approach V_{CC}. If V_{BE} is more than about 0.8 V, the transistor should be saturated so V_{CE} should be near zero volts. If these conditions are not met, the transistor should be suspect and should be tested.

It is often a good practice to sketch the load line for the circuit you are troubleshooting and locate the Q point just to see if its location looks reasonable. If it does not, and the transistor is not the fault, then it is time to examine the other circuit components, including the possibility of leaky, shorted, or open capacitors.

9.9 Summary

In this chapter the DC load line was introduced as a way to represent the behavior of the DC circuit seen by the transistor, and the AC load line was shown to represent the behavior of the AC circuit. The Q point was shown to be at the intersection of the above lines. The designer's equation was shown to place the Q point at the center of the AC load line. The procedures introduced in this chapter will be extended and formalized in Chapter 10.

GLOSSARY

Bypass capacitor: A capacitor placed in parallel with other circuit components for the purpose of allowing AC to bypass those components. (Section 9.6)

Coupling capacitor: A capacitor placed in series with the signal path for the purpose of coupling a signal while blocking any DC voltage that may be present. (Section 9.5)

Cutoff frequency: The frequency at which the power output of a circuit has dropped to half of its midband power output. (Section 9.5)

Load line: A plot of a circuit's current response to variations in applied voltage. (Section 9.2)

Midband: That frequency range over which the output of a circuit is essentially independent of frequency. (Section 9.5)

Miller Effect: The phenomenon when the gain of a circuit influences the effective value of a circuit component. (Section 9.4)

Q point: The "quiescent point," or the condition of DC voltage and current in a circuit when no AC signal is present. (Section 9.3)

PROBLEMS

Section 9.2

1. Construct a load line for the circuit of Figure 9.9.
2. Construct a load line for the circuit of Figure 9.10.

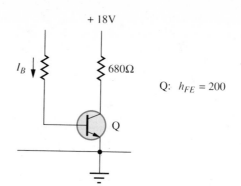

FIGURE 9.9   Problem 9.1.

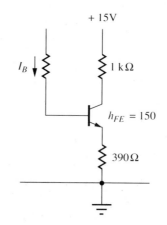

FIGURE 9.10   Problem 9.2.

Section 9.3

3. Determine V_C, I_C, and locate the Q point on the load line drawn in Problem 1 for each of the following values of I_B: (a) $I_B = 0$ A   (b) $I_B = 100$ μA   (c) $I_B = 150$ μA.

4. Determine V_C, I_C, and locate the Q point on the load line drawn in Problem 2 for each of the following values of I_B: (a) $I_B = 0$ A   (b) $I_B = 40$ μA   (c) $I_B = 80$ μA.

Section 9.4

5. Draw a schematic diagram consisting of a base power supply V_{BB}, a collector supply V_{CC}, a base resistor $R_B = 6.8$ kΩ, a collector resistor R_C, and a transistor that will operate as close to the load line and Q point shown in Figure 9.11 as possible using standard 10 percent resistor values (See Appendix B). Specify V_{BB}, V_{CC}, R_C, and h_{FE}.

6. Draw a schematic diagram consisting of a base power supply $V_{BB} = 10$ V, a collector power supply V_{CC}, a base resistor R_B, a collector resis-

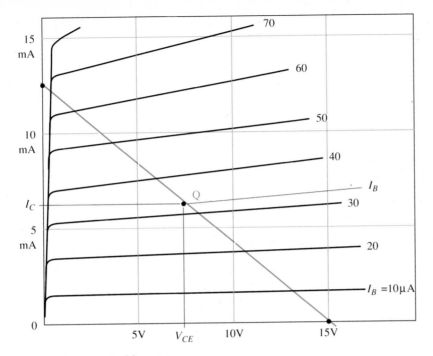

FIGURE 9.11   Problem 9.5.

tor R_C, an emitter resistor R_E = 220 Ω, and a transistor with h_{FE} = 250 that will operate as close to the load line and Q point shown in Figure 9.12 as possible using standard 10 percent resistor values (See Appendix B). Specify V_{CC}, R_C, and R_B.

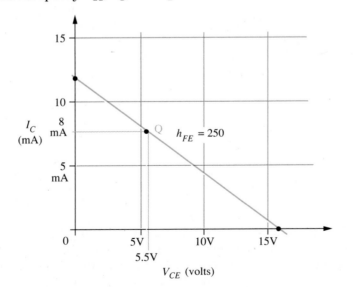

FIGURE 9.12   Problem 9.6.

7. Find the collector to emitter voltage, V_{CE} for the circuit of Figure 9.13 if the collector voltage is $V_C = 0$ V.

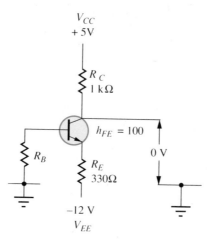

FIGURE 9.13   Problem 9.7.

Section 9.6

8. Find the base bias resistor R_B required to make $V_C = 0$ V in Figure 9.13. Specify the nearest 10 percent resistor (See Appendix B).
9. Draw the voltage waveform as it would be seen on a DC coupled oscilloscope connected to the collector of the transistor in Problem 5 if a 40 μAPP signal were injected at the base.
10. Draw the voltage waveform as it would be seen on a DC coupled oscilloscope connected to the collector of the transistor in Problem 6 if a 40 μAPP signal were injected at the base.
11. Assume that an AC load of $R_L = 1$ kΩ is coupled to the collector of the transistor in Problem 5 using a large coupling capacitor. Draw the AC load line for the resulting circuit. From the AC load line estimate the peak to peak value of the maximum unclipped voltage delivered to the load.

Section 9.7

12. Draw both AC and DC load lines for the circuit of Figure 9.14. From the AC load line estimate the peak to peak value of the maximum unclipped voltage delivered to the load.
13. Redesign the circuit of Figure 9.14 by respecifying I_B such that the Q point will be at the center of the AC load line. Estimate the peak to peak value of the maximum unclipped voltage the resulting circuit can deliver to the load. Determine the required value of V_{BB}.

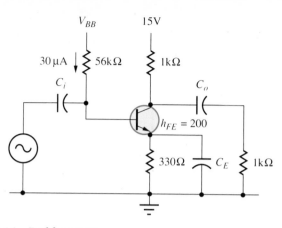

FIGURE 9.14   Problem 9.12.

14. Determine the value of R_B for the circuit of Figure 9.15 that will allow the maximum unclipped output to be delivered to the load. Specify the closest 10 percent resistor (see Appendix B).

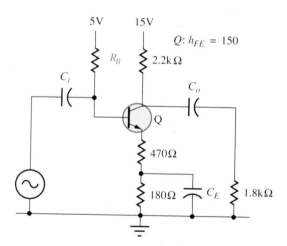

FIGURE 9.15   Problem 9.14.

Section 9.8

◀ Troubleshooting

15. Determine the Q point value of V_C, V_B, and V_E for the circuit of Figure 9.16.

16. Suppose that the B-E junction of the transistor in Figure 9.16 became an open circuit. Determine the values you would expect to see for V_C, V_B, and V_E.

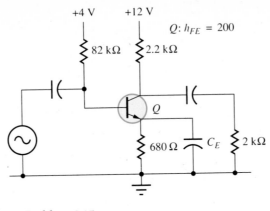

FIGURE 9.16   Problem 9.15.

17. Suppose that all of the transistor terminals in Figure 9.16 became shorted together. Determine the value you would expect to see for V_C, V_B, and V_E.

18. Suppose that the emitter bypass capacitor, C_E in Figure 9.16 became shorted. Determine the value you would expect to see for V_C, V_B, and V_E.

Troubleshooting a circuit board. (Photograph by
Lawrence Migdale/Photo Researchers, Inc.)

Chapter 10
Biasing Common Emitter Circuits

SPECIAL TERMS

Fixed bias
Stability factor (S)
Self bias
Emitter bias
Emitter swamping
Voltage divider bias
Tolerance stack-up

10.1   Introduction and Objectives

In this chapter we will study the most frequently used methods of biasing common emitter transistor amplifier circuits. At the conclusion of the chapter you should understand these circuits, the need for them, procedures for implementing them, the meaning of stability and how proper biasing produces stable circuits. You should then be able to:

- Determine the stability factor for a fixed bias, a self bias, or an emitter bias common emitter transistor amplifier circuit.
- Design a fully functional simple transistor amplifier stage.

10.2   Fixed Bias

Up to this point the circuits we have dealt with have had two power supplies, V_{CC} to supply the collector and V_{BB} to supply the base. Since power supplies are expensive, why not combine them? That is what "fixed bias" is all about. The circuit diagram then becomes that of Figure 10.1. Obviously, the power supply has to supply both V_{CC} and V_{BB}, but for simplicity we will just call it V_{CC}.

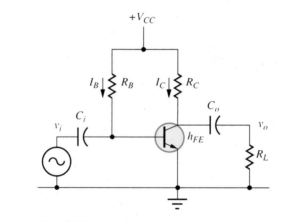

FIGURE 10.1   Fixed bias transistor circuit.

Often the value of V_{CC} and the external load resistor R_L are dictated by other circuit considerations and can be considered as given values by the designer of the circuit in Figure 10.1. In Chapter 12 you will discover that the value of R_C is dictated by the value of R_L (impedance matching). Thus all that is required for the design of the circuit of Figure 10.1 is the selection of the base biasing resistor, R_B. Our approach will not be a great deal different from that used in Chapter 9 except that here the value of V_{BB} will be dictated by the value of V_{CC}. We will proceed as follows:

If we assume that V_{CC}, R_C, and R_L are given, then Equation 9.17 gives the value of R_{AC} and the circuit designer's formula, Equation 9.25 gives the optimum value of I_C. Once I_C has been selected, Equation 9.7 gives the required value of I_B, and the junction equation, Equation 8.4 gives the value of V_{BE}. The required value of R_B can then be calculated by taking a KVL walk through the circuit of Figure 10.1 from V_{CC} through R_B and the base of the transistor to ground. Thus

$$V_{CC} - I_B R_B - V_{BE} = 0 \text{ V} \tag{10.1}$$

or

$$R_B = \frac{V_{CC} - V_{BE}}{I_B} \tag{10.2}$$

EXAMPLE 10.1

Determine the value of R_B that will give the optimum Q point for the circuit in Figure 10.2.

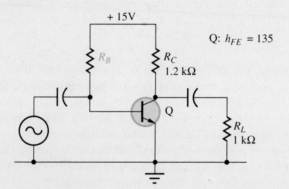

FIGURE 10.2 Example 10.1. Determining the value of R_B.

● (For a SuperCalc solution, see Appendix G.)

SOLUTION:

1. $R_{DC} = R_C$

   $= 1.2 \text{ k}\Omega$

2. $R_{AC} = R_C \parallel R_L$ (9.17)

   $= 1.2 \text{ k}\Omega \parallel 1 \text{ k}\Omega = 545 \text{ }\Omega$

3. $I_C \stackrel{*}{=} \dfrac{V_{CC}}{R_{DC} + R_{AC}}$ (9.25)

   $= \dfrac{15 \text{ V}}{1.2 \text{ k}\Omega + 545 \text{ }\Omega} = 8.59 \text{ mA}$

4. $V_{BE} = 26 \text{ mV} * \ln\dfrac{I_C}{10 \text{ fA}}$ (8.4)

     $= 26 \text{ mV} * \ln\dfrac{8.59 \text{ mA}}{10 \text{ fA}} = 0.714 \text{ V}$

5. $I_B = \dfrac{I_C}{h_{FE}}$ (9.7)

     $= \dfrac{8.59 \text{ mA}}{135} = 63.7 \text{ μA}$

6. $R_B = \dfrac{V_{CC} - V_{BE}}{I_B}$ (10.2)

     $= \dfrac{15 \text{ V} - 0.714 \text{ V}}{63.7 \text{ μA}} = 224 \text{ k}\Omega$ (Use 220 kΩ)

The nearest standard resistor with 10 percent tolerance is specified.

A quick look at Equation 10.1 shows why the circuit of Figure 10.2 is called a "fixed bias" circuit. Solving for the base current, we have

$$I_B = \frac{V_{CC} - V_{BE}}{R_B}$$

(10.3)

The only element on the right side of Equation 10.3 that is not exactly constant is V_{BE} and you know that V_{BE} will not vary more than about 0.2 V at the most. Thus the base bias current I_B is essentially fixed. A more descriptive name for the circuit would be to call it a "fixed base current circuit." "Fixed bias circuit" is the name usually used.

The analysis of a previously designed circuit requires the use of many of the same ideas as the design procedure, but *you must remember that the circuit designer's formula, Equation 9.25 cannot be used in any form for circuit analysis.* Our procedure will be almost the same as that used in Section 9.4: Do a KVL walk through the circuit, express the resulting equation in terms of I_C, and iterate the result with the junction equation.

The KVL equation for the base of the circuit of Figure 10.1 is Equation 10.1. From the definition of h_{FE},

$$I_B = \frac{I_C}{h_{FE}}$$

(9.7)

Substituting Equation 9.7 into Equation 10.1 and solving for I_C gives the needed equation.

$$I_C = \frac{V_{CC} - V_{BE}}{\dfrac{R_B}{h_{FE}}}$$

(10.4)

Notice the similarity between this equation and Equation 9.9.

Iterating Equation 10.4 with the junction equation gives I_C and V_{BE}. Substituting I_C into Equation 9.7 will give I_B and a KVL walk from V_{CC} down through R_C to the collector will then give the value of V_C.

$$V_{CC} - I_C R_C = V_C \qquad (10.5)$$

EXAMPLE 10.2

Analyze the circuit of Figure 10.3. That is, determine the Q point voltages and currents for the circuit in Figure 10.3. Any resemblance to Example 10.1 is intentional.

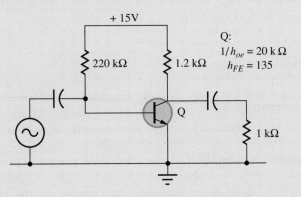

FIGURE 10.3   Example 10.2. Determining the Q point.

● (See Appendix H for a PSpice solution.)

SOLUTION:

$$I_C = \frac{V_{CC} - V_{BE}}{\dfrac{R_B}{h_{FE}}} \qquad (10.4)$$

$$= \frac{15\text{ V} - V_{BE}}{\dfrac{220\text{ k}\Omega}{135}} = \frac{15\text{ V} - V_{BE}}{1.63\text{ k}\Omega}$$

1. $I_C = \dfrac{15\text{ V} - V_{BE}}{1.63\text{ k}\Omega}$ =	8.77	8.76	8.76	mA
2. $V_{BE} = 26\text{ mV} * \ln\dfrac{I_C}{10\text{ fA}}$ =	0.715	0.715	0.715	V

$$3.\ I_B = \frac{I_C}{h_{FE}} \tag{9.7}$$

$$= \frac{8.76\ \text{mA}}{135} = 64.9\ \mu\text{A}$$

$$4.\ V_C = V_{CC} - I_C R_C \tag{10.5}$$

$$= 15\ \text{V} - 8.76\ \text{mA} * 1.2\ \text{k}\Omega = 4.48\ \text{V}$$

If you draw the AC load line carefully for Example 10.2 you will notice that the Q point is not exactly at the center of the AC load line. That is to be expected because we used the closest standard resistor for R_B in the design. If that were the only source of deviation from the intended behavior, Chapter 10 would end right here!

10.3   Stability Factor S

The stability factor S for a circuit is a measure of how far you can expect the actual circuit behavior to deviate from the design. There are a number of things that cause instability in transistor circuits, aging and temperature to name two. But all other sources of instability pale into—at least manageable proportions compared to sample-to-sample variability. If you look at the manufacturer's data sheets in Appendix F, for a 2N4401 transistor you will find that h_{FE} is more than slightly variable. In the current range used in Examples 10.1 and 10.2, a given transistor might have an h_{FE} as low as 40 or as high as 300 and still be labeled a 2N4401. The actual variability is probably less than that but it is not small. If you can learn to design stable circuits in the face of such sample-to-sample variability, all other sources of instability will be taken care of in the process.

The stability factor, S, is a measure of amplifier circuit quality much as voltage regulation was a measure of power supply circuit quality in Chapter 5. The value of S varies from zero, which means perfect stability, to one, which means no stability at all.

We would like for our circuits to be very stable. That means their Q points would always be exactly where they were designed to be. However, as you will see shortly, a perfectly stable circuit will not amplify signals at all. Thus in amplifier design we must seek a compromise between amplification and stability.

Incidentally, that is another advantage of digital circuits. Since they are not concerned with amplification as such, digital circuits can be made perfectly stable.

Whenever variations in one thing cause variations in another, the easiest way to talk about the situation is to use the language of calculus. Using that form, the stability factor S of a circuit is defined to be

$$S \equiv \frac{\% \, dI_C}{\% \, dh_{FE}} \tag{10.6}$$

where $\% \, dI_C$ is the percentage change in I_C caused by $\% \, dh_{FE}$, the percentage change in h_{FE}.

Equation 10.3 shows us that for the fixed bias circuit, I_B is virtually constant. Since $I_C = h_{FE} I_B$, we can see that I_C is directly proportional to h_{FE}. Consequently, if h_{FE} changes by a given percentage, I_C will change by the same percentage. Obviously then, for the fixed bias circuit, $S = 1$. That means no stability at all. The collector current, and therefore the Q point are determined as much by h_{FE} as by the judgment of the designer.

A calculus based derivation of the above value for S is included in Appendix D.

EXAMPLE 10.3

Find the new collector voltage V_C^* (the "hot" value for V_C) for the circuit in Figure 10.3 if h_{FE} were to double so that $h_{FE}^* = 270$ (the "hot" value for h_{FE}).

SOLUTION:
According to the preceding discussion, $S = 1$. Thus doubling h_{FE} results in doubling I_C. Therefore

$$I_C^* = 2 \, I_C$$
$$= 2 * 8.76 \text{ mA} = 17.5 \text{ mA}$$

Solving for the collector voltage gives

$$V_C^* = V_{CC} - I_C R_C$$
$$= 15 \text{ V} - 17.5 \text{ mA} * 1.2 \text{ k}\Omega = -6.00 \text{ V}$$

We saw this kind of absurd result in connection with the AC output waveform in Example 9.3. The problem here is the same as it was there: The transistor has long since reached saturation and so the equations we have been using are not valid.

At saturation:

$$V_{CE\text{sat}} \approx 0 \text{ V}$$

and

$$I_{C\text{sat}} \approx I_0$$
$$= \frac{V_{CC}}{R_{DC}} \tag{9.5}$$
$$= \frac{15 \text{ V}}{1.2 \text{ k}\Omega} = 12.5 \text{ mA}$$

Examples 10.2 and 10.3 illustrate the fact that for the fixed bias circuit, the designer has very little control over where the Q point will be. If the h_{FE} for a given transistor happens to be low, the Q point will be near cutoff. If it happens to be high, the transistor may be saturated. Due to this poor stability the fixed bias circuit is rarely used in actual practice.

You might at first think that since

$$I_C = h_{FE}I_B \qquad (9.19)$$

any change in h_{FE} must of necessity result in the same percentage change in I_C with the result that the stability must always be $S = 1$. Of course, you are right for the fixed bias circuit. Equation 9.19 indicates that the only way to get stability is to somehow obtain an inverse relationship between I_B and h_{FE}. If a given increase in h_{FE} resulted in exactly the same percentage decrease in I_B then according to Equation 9.19, I_C would not change at all and we would have perfect stability. The remainder of this chapter will be devoted to the study of circuits which attempt to approach this goal while at the same time preserving the ability of the amplifier to amplify signals.

10.4   Self Bias

One way to reduce I_B when h_{FE} begins to increase is to use for the base bias supply a voltage source whose value decreases as I_C increases. As we saw in Examples 10.2 and 10.3, the collector is such a source. The circuit of Figure 10.4 uses what is often called "collector bias" or "self bias." Because it is simple and yet it is fairly stable, you will see various forms of the circuit in Figure 10.4 used in practice.

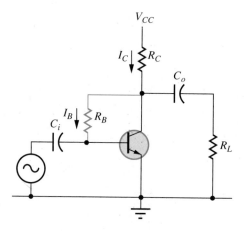

FIGURE 10.4   Self biased circuit.

You might at first think that the design and analysis of the self biased circuit would be quite complicated due to the fact that R_C carries both I_C and I_B. Fortunately, I_B is so small compared to I_C that we can assume that R_C carries only I_C. With this simplification the circuit design and the circuit analysis procedures are but very little different from what we have done before.

In order to design the circuit of Figure 10.4, assuming as we did before, that V_{CC}, R_C, and R_L are given, the circuit designer's equation, the definition of h_{FE}, and the junction equation will give I_C, I_B, and V_{BE} respectively, just as they did for the fixed bias circuit. The value required for R_B can then be determined by taking a KVL walk from V_{CC} in Figure 10.4 down through R_C, over through R_B and through the base to ground. The result is

$$V_{CC} - I_C R_C - I_B R_B - V_{BE} = 0 \text{ V} \qquad (10.7)$$

or

$$R_B = \frac{V_{CC} - I_C R_C - V_{BE}}{I_B} \qquad (10.8)$$

EXAMPLE 10.4
Redesign the circuit of Example 10.1 to be a self biased circuit. The new circuit is shown in Figure 10.5.

● (See Appendix G for a SuperCalc solution.)

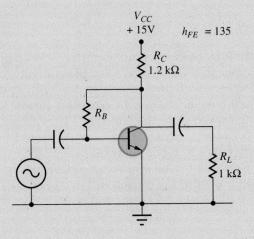

FIGURE 10.5 Example 10.4. Determining the value of R_B.

SOLUTION:

Under the assumption that $I_B \ll I_C$, steps 1 through 5 of Example 10.1 remain unchanged. Thus $I_C = 8.59$ mA, $V_{BE} = 0.714$ V, and $I_B = 63.7$ μA. Step 6 must be altered as follows:

$$6.\ R_B = \frac{V_{CC} - I_C R_C - V_{BE}}{I_B} \tag{10.8}$$

$$= \frac{15\,V - 8.59\,mA * 1.2\,k\Omega - 0.714\,V}{63.7\,\mu A} = 62.4\,k\Omega \qquad (\text{Use } 68\,k\Omega)$$

The nearest standard 10 percent resistor is specified.

The analysis of an existing self biased circuit proceeds much the same as the fixed bias circuit with the slight complication that Equation 10.7 introduces. Again our approach will be to obtain a KVL equation for the base, express the equation in terms of I_C, and iterate the result with the junction equation. Equation 10.7 is the required KVL equation. Substituting

$$I_B = \frac{I_C}{h_{FE}} \tag{9.7}$$

into Equation 10.7 gives

$$V_{CC} - I_C R_C - \frac{I_C}{h_{FE}} R_B - V_{BE} = 0\ V$$

or

$$I_C = \frac{V_{CC} - V_{BE}}{R_C + \dfrac{R_B}{h_{FE}}} \tag{10.9}$$

Using Equation 10.9 in place of Equation 10.4, the analysis proceeds in exactly the same way as it did for the fixed bias circuit in Example 10.2.

Notice that, as we mentioned in connection with Equation 9.9, R_B appears to be a much smaller resistor in Equation 10.9 (it is divided by h_{FE}) because it is being viewed from the collector—the Miller Effect again.

The first step in the analysis of any circuit should be an estimate of the circuit stability factor S. If you know some calculus you can derive an equation for S for the self biased circuit by substituting Equation 10.9 into Equation 10.6. The procedure is carried out in Appendix D. The result is

$$S = \frac{R_B}{R_B + h_{FE}R_C} \tag{10.10}$$

It is interesting to note that, just as we pointed out in Section 9.4 in connection with the development of Equation 9.9, looking from the base circuit the resistors in the collector circuit appear to be h_{FE} times larger than they really are, and looking from the collector circuit, resistors in the base circuit appear to be divided by h_{FE}—the Miller Effect again! Equations 10.9 and 10.10 illustrate this idea. We will see many more examples of this effect before we are through.

EXAMPLE 10.5

Analyze the circuit in Figure 10.6.

- (See Appendix H for a PSpice solution.)

SOLUTION:

1. $S = \dfrac{R_B}{R_B + h_{FE}R_C}$ $\hspace{3cm}$ **(10.10)**

$$= \frac{68 \text{ k}\Omega}{68 \text{ k}\Omega + 135 * 1.2 \text{ k}\Omega} = 0.296 = 29.6\%$$

$$I_C = \frac{V_{CC} - V_{BE}}{R_C + \dfrac{R_B}{h_{FE}}} \hspace{3cm} \text{(10.9)}$$

$$= \frac{15 \text{ V} - V_{BE}}{1.2 \text{ k}\Omega + \dfrac{68 \text{ k}\Omega}{135}} = \frac{15 \text{ V} - V_{BE}}{1.70 \text{ k}\Omega}$$

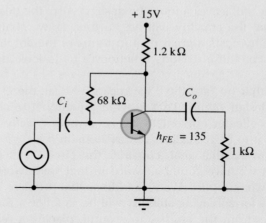

FIGURE 10.6   Example 10.5. Location of Q point.

2. $I_C = \dfrac{15\ V - V_{BE}}{1.70\ k\Omega}$

=	8.41	8.40	8.40	mA

3. $V_{BE} = 26\ mV * \ln\dfrac{I_C}{10\ fA}$

=	0.714	0.714	0.714	V

4. $I_B = \dfrac{I_C}{h_{FE}}$ (9.7)

$\quad = \dfrac{8.4\ mA}{135} = 62.2\ \mu A$

5. $V_C = V_{CC} - I_C R_C$ (10.5)

$\quad = 15\ V - 8.40\ mA * 1.2\ k\Omega = 4.92\ V$

If you replace $h_{FE} = 135$ in Example 10.5 with $h_{FE}{}^* = 270$ as we did for the fixed bias circuit in Example 10.3, you will find that $V_C{}^* = 3.20\ V$. This is certainly not an insignificant deviation from the expected value of 4.92 V which we got using the typical value for h_{FE}, but at least the transistor does not saturate as it did in the fixed bias circuit of Example 10.3. This should give you some feeling for what $S = 29.6$ percent means.

10.5   Emitter Bias

The self biased circuit of Section 10.4 certainly gave us a significant improvement in stability over the simple fixed bias circuit of Section 10.2. However, the circuit has some serious shortcomings. For many applications the stability still leaves a lot to be desired. Also, the designer has little control over what the stability will be. Further, as we shall see in Chapter 12, that stability costs a great deal in terms of the amplifier's ability to amplify signals. For these reasons it behooves us to look at other ways to achieve stability.

Remember that by "stability" we mean how far the Q point can deviate from the design values. That is what Equation 10.6 says. We would like for the DC collector current to always remain exactly where the designer intended it to be regardless of variations in h_{FE}.

To move toward that goal, consider the circuit in Figure 10.7. You might think that we have gone backward in that we have returned to the need for two power supplies. We will address that problem in Section 10.6.

Our strategy for achieving stability will be as follows: Since the collector current flows from the emitter to ground, placing a resistor between the emitter and ground in the circuit of Figure 10.7 will cause the voltage at the emitter to change if the collector current changes. Since V_{BE} is vir-

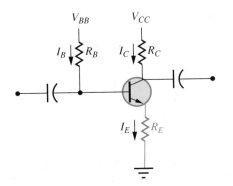

FIGURE 10.7 Emitter biased circuit.

tually constant, any change in emitter voltage requires a similar change in the base voltage. Thus if we can keep the base voltage constant, I_C must necessarily also remain constant. That should give us the stability we seek.

To calculate the stability factor for the circuit of Figure 10.7 we will first apply KVL to the base circuit:

$$V_{BB} - I_B R_B - V_{BE} - I_E R_E = 0 \text{ V} \tag{10.11}$$

Remember that since I_B is much smaller than I_C,

$$I_E \approx I_C \tag{8.3}$$

Also, from the definition of h_{FE},

$$I_B = \frac{I_C}{h_{FE}} \tag{9.7}$$

Substituting Equations 8.3 and 9.7 into Equation 10.11 and solving for I_C gives

$$I_C = \frac{V_{BB} - V_{BE}}{\dfrac{R_B}{h_{FE}} + R_E} \tag{10.12}$$

In a fashion similar to that used to derive Equation 10.10, if you know some calculus you can substitute Equation 10.12 into Equation 10.6 as is done in Appendix D to obtain a stability equation for the emitter biased circuit of Figure 10.7. The result is

$$S = \frac{R_B}{R_B + h_{FE} R_E} \tag{10.13}$$

Notice that the effect of R_E in Equation 10.13 is multiplied by h_{FE}. It looks like a bigger resistor because it is being viewed from the base circuit. This is the same effect as we saw in connection with Equations 9.9 and 10.9— Miller again!

Equation 10.13 does not tell us how big either R_B or R_E should be. It merely tells us something about their relative size. If we knew one of them, Equation 10.13 would tell us the value to use for the other one in order to achieve a given stability.

Some authors call R_E the "emitter swamping resistor" because it is supposed to "swamp out" any variations in V_{BE}. Since, as we have mentioned before, V_{BE} will almost always be between 0.6 V and 0.8 V, it varies no more than about 0.2 V. Thus they argue that R_E should be chosen so that $V_E \gg 0.2$ V. Thus let $V_E = 2$ V. This rule will usually lead to a stable design, but it is a bit too rigid. Suppose that you wish to design the amplifier for a hearing aid that has to operate on a single 1.4 V battery. The 2 V rule is a little hard to implement in that case.

Perhaps a better approach is as follows: First recognize that

$$R_{DC} = R_C + R_E \tag{10.14}$$

Thus if R_{DC} is to remain fixed, R_C must be decreased by whatever amount you put in R_E. Since the AC output signal appears across R_C, any reduction in R_C is going to reduce the amplifier's ability to amplify signals. As we have said before, we must seek a compromise between stability and amplification. A reasonable compromise is to let

$$R_E = \frac{R_{DC}}{4} = \frac{R_C}{3} \tag{10.15}$$

Once R_E is determined, solving Equation 10.13 for R_B gives

$$R_B = \frac{S}{1 - S} h_{FE} R_E \tag{10.16}$$

A reasonably good compromise between amplification and stability is usually achieved if a stability factor of $S \approx 10$ percent is used, or, it is not uncommon to use

$$R_B \approx \frac{h_{FE} R_E}{10} \tag{10.17}$$

Next the circuit voltages can be calculated by first recognizing that

$$I_E \approx I_C \tag{8.3}$$

then

$$V_E = I_E R_E = I_C R_E \tag{10.18}$$

Equation 8.4, the junction equation will give the value of V_{BE}. From the circuit of Figure 10.7, you can see that since the emitter is no longer grounded, the base voltage, V_B, is given by

$$V_B = V_{BE} + V_E \tag{10.19}$$

Also from the circuit of Figure 10.7, you can see that

$$V_{BB} = V_B + I_B R_B \qquad (10.20)$$

Finally, the value of R_{AC} must be considered. In order to calculate R_{AC} for previous circuits, since the emitter was grounded, it was only necessary to visualize what the transistor "saw" as it "looked at" what was connected to its collector terminal. In the circuit of Figure 10.7, we have a resistor in the emitter circuit as well. In order to determine R_{AC} for this circuit, it is necessary to recognize that the transistor is in series with both the collector and the emitter circuits. Thus Equation 9.17 now becomes

$$R_{AC} = R_C \parallel R_L + R_E \qquad (10.21)$$

EXAMPLE 10.6

Redesign the circuit in Figure 10.2 as the emitter biased circuit shown in Figure 10.8.

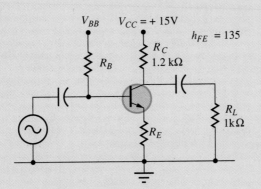

FIGURE 10.8   Example 10.6. Circuit design.

● (For a SuperCalc solution, see Appendix G.)

SOLUTION:

1. $R_E = \dfrac{R_C}{3}$ \hfill (10.15)

$= \dfrac{1.2 \text{ k}\Omega}{3} = 400 \ \Omega \qquad$ (Use 390 Ω)

2. $R_{DC} = R_C + R_E$ \hfill (10.14)

$= 1.2 \text{ k}\Omega + 390 \ \Omega = 1.59 \text{ k}\Omega$

Note that R_C and R_E are real resistors that you will have to get out of a box and hook up if you are to build this circuit. Thus the values specified and used in all later calculations are the nearest standard sizes. However, R_{DC} is not a resistor that you will hook up. Thus its value is not standardized.

3. $R_{AC} = R_C \parallel R_L + R_E$ (10.21)

$\qquad = 1.2 \text{ k}\Omega \parallel 1 \text{ k}\Omega + 390 \ \Omega = 935 \ \Omega$

Again, this is not a resistor you will hook up so do NOT standardize it.

4. $I_C = \dfrac{V_{CC}}{R_{DC} + R_{AC}}$ (9.25)

$\qquad = \dfrac{15 \text{ V}}{1.59 \text{ k}\Omega + 935 \ \Omega} = 5.94 \text{ mA}$

5. $V_E = I_C R_E$ (10.18)

$\qquad = 5.94 \text{ mA} * 390 \ \Omega = 2.32 \text{ V}$

6. $V_{BE} = 26 \text{ mV} * \ln\dfrac{I_C}{10 \text{ fA}}$ (8.4)

$\qquad = 26 \text{ mV} * \ln\dfrac{5.94 \text{ mA}}{10 \text{ fA}} = 0.705 \text{ V}$

7. $V_B = V_{BE} + V_E$ (10.19)

$\qquad = 0.705 \text{ V} + 232 \text{ V} = 3.02 \text{ V}$

8. $I_B = \dfrac{I_C}{h_{FE}}$ (9.7)

$\qquad = \dfrac{5.94 \text{ mA}}{135} = 44 \ \mu\text{A}$

9. $R_B = \dfrac{h_{FE} R_E}{10}$ (10.17)

$\qquad = \dfrac{135 * 390 \ \Omega}{10} = 5.27 \text{ k}\Omega \qquad \text{(Use 5.6 k}\Omega\text{)}$

For this circuit, R_B is a real resistor so a standard size is specified. This situation will change in Section 10.6.

10. $V_{BB} = V_B + I_B R_B$ (10.20)

$\qquad = 3.02 \text{ V} + 44 \ \mu\text{A} * 5.6 \text{ k}\Omega = 3.27 \text{ V}$

Except for the fact that the emitter is no longer connected directly to ground, the circuit analysis procedure for the emitter biased circuit is essentially the same as it was for the fixed bias circuit of Section 10.2.

EXAMPLE 10.7
Analyze the circuit of Figure 10.9.

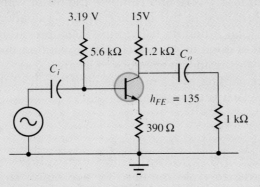

FIGURE 10.9 Example 10.7. Circuit analysis.

● (For a PSpice solution, see Appendix H.)

SOLUTION:

1. $S = \dfrac{R_B}{R_B + h_{FE}R_E}$ (10.13)

  $= \dfrac{5.6\ \text{k}\Omega}{5.6\ \text{k}\Omega + 135 * 390\ \Omega} = 0.0961 = 9.61\%$

  $I_C = \dfrac{V_{BB} - V_{BE}}{\dfrac{R_B}{h_{FE}} + R_E}$ (10.12)

  $= \dfrac{3.19\ \text{V} - V_{BE}}{\dfrac{5.6\ \text{k}\Omega}{135} + 390\ \Omega} = \dfrac{3.19\ \text{V} - V_{BE}}{431\ \Omega}$

2. $I_C = \dfrac{3.19\ \text{V} - V_{BE}}{431\ \Omega}$

= 5.77	5.76	5.76	mA

3. $V_{BE} = 26\ \text{mV} * \ln\dfrac{I_C}{10\ \text{fA}}$

= 0.704	0.704	0.704	V

4. $V_E = I_C R_E$ (10.18)

  $= 5.76\ \text{mA} * 390\ \Omega = 2.25\ \text{V}$

5. $V_B = V_{BE} + V_E$ (10.19)

  $= 0.704\ \text{V} + 2.25\ \text{V} = 2.95\ \text{V}$

6. $V_C = V_{CC} - I_C R_C$ (10.5)

  $= 15\ \text{V} - 5.76\ \text{mA} * 1.2\ \text{k}\Omega = 8.09\ \text{V}$

Steps 4, 5, and 6 provide the required Q point voltages.

In order to get a feel for the meaning of the kind of variation in Q point values you might expect for a circuit with a stability factor of near 10 percent, it would be instructive for you to repeat the analysis of the circuit in Example 10.7 for $h_{FE}^* = 270$ and compare the result with that of the previous circuits discussed in this chapter. This reevaluation can be accomplished in less than a minute using the PSpice procedure developed for Example 10.7. It is done in Appendix H. You will find that V_C changes by less than 0.5 V. This is adequate stability for most work.

The only remaining problem is the elimination of the need for two power supplies. That is the next topic.

10.6 Voltage Divider Bias

Obviously, in order to avoid the need for two power supplies in Section 10.5, we could have adapted the procedures of the fixed bias circuit of Section 10.2. That would necessitate the use of a large value for R_B and, according to Equation 10.13, that would result in poor stability—again. What is needed is a procedure for reducing the value of V_{CC} to the value dictated by Equation 10.20. A voltage divider will do the job for us.

A quick review of your electric circuits class will remind you that the Thevenin equivalent of the circuit in Figure 10.10a looks like Figure 10.10b where

$$V_{BB} = V_{CC}\frac{R_2}{R_2 + R_1} \tag{10.22}$$

and

$$R_B = R_1 \parallel R_2 \tag{10.23}$$

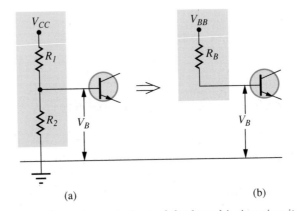

(a)   (b)

FIGURE 10.10 Thevenin equivalent of the base biasing circuit.
(a) Divider biasing circuit.
(b) Thevenin equivalent circuit.

Although circuit considerations may dictate the ratio of R_1 to R_2, the resistor values used to obtain that ratio are still under the designer's control so any desired level of stability within the limits imposed by Equation 10.23 can be achieved.

Since the determination of V_{BB} and R_B were a part of Section 10.5, the addition of the voltage divider in the base circuit adds only a couple of steps to the design procedure presented there. In order to determine R_1, we will start with Equation 10.22 and multiply both sides by R_1

$$V_{BB}R_1 = V_{CC}\frac{R_2R_1}{R_2 + R_1} \tag{10.24}$$

You may recognize the right side of Equation 10.24 as the method of paralleling resistors known as the "product over the sum" method. Thus

$$V_{BB}R_1 = V_{CC}R_1 \| R_2 \tag{10.25}$$

Substituting Equation 10.23 into Equation 10.25 gives

$$V_{BB}R_1 = V_{CC}R_B \tag{10.26}$$

or

$$R_1 = R_B\frac{V_{CC}}{V_{BB}} \tag{10.27}$$

Once R_1 is determined, R_2 can be determined by using the voltage divider equation

$$\frac{R_2}{V_{BB}} = \frac{R_1}{V_{CC} - V_{BB}} \tag{10.28}$$

or

$$R_2 = R_1\frac{V_{BB}}{V_{CC} - V_{BB}} \tag{10.29}$$

EXAMPLE 10.8

Modify the circuit of Example 10.6 to include voltage divider base biasing.

● (For a SuperCalc solution, see Appendix G.)

SOLUTION:

The voltage divider biased circuit is shown in Figure 10.11. Since almost all of the design procedure used in Example 10.6 applies to the voltage divider biased circuit, we will assume that the first nine steps

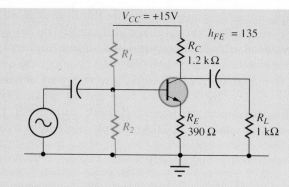

FIGURE 10.11   Example 10.8. Divider bias determination.

of Example 10.6 have been completed. Thus, $V_{CC} = 15$ V, $V_B = 3.02$ V, $I_B = 44$ μA, and $R_B = 5.27$ kΩ. Notice that it would now be incorrect to use a standardized value for R_B. Now R_B is no longer a real resistor which you would have to find in a resistor box if you were to hook up the circuit. Rather it is the parallel combination of R_1 and R_2. Repeating step 10 of the design procedure in Example 10.6 we have

10. $V_{BB} = V_B + I_B R_B$ (10.20)

$\quad\quad = 3.02$ V $+ 44$ μA $* 5.27$ kΩ $= 3.25$ V

11. $R_1 = R_B \dfrac{V_{CC}}{V_{BB}}$ (10.27)

$\quad\quad = 5.27$ kΩ $\dfrac{15 \text{ V}}{3.25 \text{ V}} = 24.3$ kΩ   (Use 27 kΩ)

Since R_1 is a real resistor that you will have to get from a resistor box, the nearest standard 10 percent resistor is specified.

12. $R_2 = R_1 \dfrac{V_{BB}}{V_{CC} - V_{BB}}$ (10.29)

$\quad\quad = 27$ kΩ $\dfrac{3.25 \text{ V}}{15 \text{ V} - 3.25 \text{ V}} = 7.47$ kΩ   (Use 6.8 kΩ)

Again R_2 is a real resistor you will be getting from a box so the nearest standard resistor is specified.

You will notice that the incorporation of standard resistor values at every step of the design procedure compensates for the fact that standard values must be used when the circuit is hooked up. This avoids what is known as "tolerance stack up." The last resistor, R_2, is the only one whose approximation cannot be compensated for.

The analysis of an existing voltage divider biased circuit requires constructing a Thevenin equivalent for the base voltage divider. From that point on, the analysis follows exactly the same procedure as that used in Example 10.7.

EXAMPLE 10.9
Analyze the circuit of Figure 10.12a.

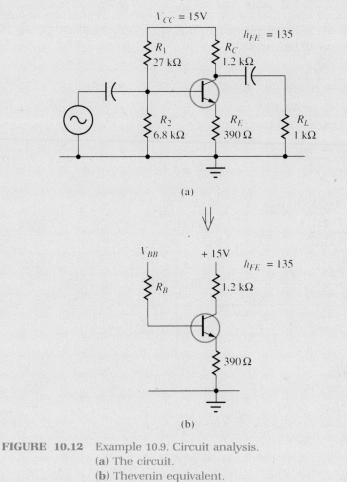

(a)

(b)

FIGURE 10.12 Example 10.9. Circuit analysis.
(a) The circuit.
(b) Thevenin equivalent.

● (For a PSpice solution, see Appendix H.)

SOLUTION:
Constructing a Thevenin equivalent of the base circuit of Figure 10.12 gives

1. $V_{BB} = V_{CC} \dfrac{R_2}{R_2 + R_1}$ (10.22)

$= 15 \text{ V} \dfrac{6.8 \text{ k}\Omega}{6.8 \text{ k}\Omega + 27 \text{ k}\Omega} = 3.02 \text{ V}$

2. $R_B = R_1 \| R_2$ (10.23)

$= 27 \text{ k}\Omega \| 6.8 \text{ k}\Omega = 5.43 \text{ k}\Omega$

The circuit is now that of Figure 10.12b which is just the separately biased circuit of Example 10.7. Thus the procedure which we introduced there will be used.

3. $S = \dfrac{R_B}{R_B + h_{FE}R_E}$ (10.13)

$= \dfrac{5.43 \text{ k}\Omega}{5.43 \text{ k}\Omega + 135 * 390 \text{ }\Omega} = 0.0935 = 9.35\%$

$I_C = \dfrac{V_{BB} - V_{BE}}{\dfrac{R_B}{h_{FE}} + R_E}$ (10.12)

$= \dfrac{3.02 \text{ V} - V_{BE}}{\dfrac{5.43 \text{ k}\Omega}{135} + 390 \text{ }\Omega} = \dfrac{3.02 \text{ V} - V_{BE}}{430 \text{ }\Omega}$

4. $I_C = \dfrac{3.02 \text{ V} - V_{BE}}{430 \text{ }\Omega} =$	5.39	5.38	5.38	mA
5. $V_{BE} = 26 \text{ mV} * \ln\dfrac{I_C}{10 \text{ fA}} =$	0.702	0.702	0.702	V

6. $V_E = I_C R_E$

$= 5.38 \text{ mA} * 390 \text{ }\Omega = 2.10 \text{ V}$

7. $V_B = V_{BE} + V_E$

$= 0.702 \text{ V} + 2.10 \text{ V} = 2.80 \text{ V}$

8. $V_C = V_{CC} - I_C R_C$ (10.5)

$= 15 \text{ V} - 5.38 \text{ mA} * 1.2 \text{ k}\Omega = 8.54 \text{ V}$

9. $I_B = \dfrac{I_C}{h_{FE}}$ (9.7)

$= \dfrac{5.38 \text{ mA}}{135} = 39.9 \text{ }\mu\text{A}$

Steps 6, 7, 8, and 9 provide the required Q point voltages and currents.

Notice that this circuit has all of the advantages of the circuit of Section 10.5 and in addition, it requires only a single power supply. It is the most commonly used transistor amplifier circuit. You should practice working with it until you are comfortable with it. We will help you toward that end in the next chapter.

10.7   Troubleshooting ◀◀◀◀◀◀◀◀

As we have said so often before, the place to start troubleshooting is with a visual inspection for such things as broken wires and suspicious looking solder joints. Next should come a Q point check.

If $V_C - V_E$ is near zero, the transistor is probably saturated. What is causing the saturation? A fault in the base circuit could be holding V_B too high (R_1 shorted? R_2 open?). A fault in the emitter circuit could be pulling V_E too low (R_E shorted?). Either condition will give excessive base current and so tend to saturate the transistor. A fault in the collector circuit could be causing V_C to be low (R_C open?). Of course, the transistor could be shorted, in which case V_{BE} will probably be less than 0.6 V.

If $V_C - V_E$ is excessively high, the transistor is probably cut off. Again ask yourself what could logically cause the situation. Apply logic similar to that in the above paragraph.

10.8   Summary

In this chapter we extended the ideas presented in Chapter 9. We studied various methods of biasing common emitter circuits and we looked at the kind of stability achieved by each method. At this point you should be able to do the DC design and analysis of simple but fully functional bipolar transistor amplifier circuits. This chapter is really just an intermediate step along the way to the development of usable circuits. The next step is to examine the AC characteristics of the circuit. This is the topic of the following chapters.

GLOSSARY

Emitter bias: A technique for biasing a transistor circuit which uses an "emitter swamping" resistor. (Section 10.5)

Emitter swamping: A biasing procedure requiring that the bias voltage applied to the emitter is so large that any variations in base to emitter voltage will be negligible. (Section 10.5)

Fixed bias: An abbreviated name for the "fixed current base bias" technique for biasing the base circuit of a transistor amplifier circuit. (Section 10.2)

Self bias: A technique for biasing the base circuit of a transistor amplifier in which the bias is obtained from the collector. (Section 10.4)

Stability factor (S): A measure of how stable the collector current is in a given circuit. S has a range of from 0 for a very stable circuit and 1 for a very unstable circuit. (Section 10.3)

Tolerance stack up: Circuit error which results from the cumulative effect of uncompensated component deviation from intended design values. (Section 10.6)

Voltage divider bias: A technique for biasing the base circuit of a transistor amplifier which permits the designer a measure of control over the size of the base biasing resistance through the use of a voltage divider. (Section 10.6)

PROBLEMS

Section 10.2

1. Given the circuit of Figure 10.13, find the value of resistor R_B that will give the optimum Q point. Specify the nearest 10 percent resistor value.

2. Find the Q point voltages for the circuit designed in Problem 1.

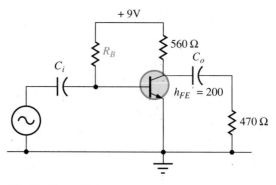

FIGURE 10.13   Problem 10.1

Section 10.3

3. Find the collector current I_C for the circuit designed in Problem 1. Find the new value of collector current, I_C^* if h_{FE} increases by 10 percent to

$h_{FE}{}^* = 220$. Using Equation 10.6, calculate the stability of the circuit. Compare with theory.

Section 10.3

4. Repeat Example 10.3 if h_{FE} were to drop to the minimum value for a 2N4401 given in the manufacturer's data sheets in Appendix F.

Section 10.4

5. Specify the nearest standard 10 percent resistor for R_B in Figure 10.14 that will give the optimum Q point.

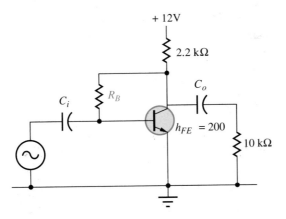

FIGURE 10.14   Problem 10.5.

6. Determine the stability factor and the Q point voltages for the circuit designed in Problem 5.
7. Find the collector current I_C for the circuit designed in Problem 5. Using the circuit analysis procedure presented in Section 10.4, find the new value of collector current, $I_C{}^*$ if h_{FE} increases by 10 percent to $h_{FE}{}^* = 220$. Using Equation 10.6, calculate the stability of the circuit. Compare with the value obtained from Equation 10.10.
8. Calculate the new value for V_C in Example 10.5 if $h_{FE}{}^* = 270$. Compare the resulting change in V_C with that of Example 10.3.
9. Estimate the approximate percentage error introduced in V_C by ignoring the fact that in the self biased circuit R_C actually carries I_B in addition to I_C. Assume that $h_{FE} = 100$.

Section 10.5

10. Use the procedure of Section 10.5 to design the amplifier in Figure 10.15.

11. Calculate the stability factor and the Q point voltages for the circuit designed in Problem 10.

12. Calculate the new value for V_C in Example 10.7 if $h_{FE}{}^* = 270$. Compare the resulting change in V_C with that of Problem 8.

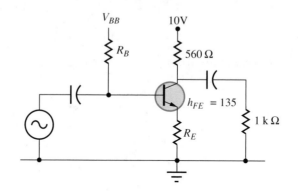

FIGURE 10.15   Problem 10.10.

Section 10.6

13. Redesign the circuit of Problem 10 to use a single power supply and voltage divider biasing.

14. Determine the stability factor and the Q point voltages for the circuit of Figure 10.16.

15. Determine the Q point voltages for the circuit of Figure 10.16 if the h_{FE} of the transistor is changed to $h_{FE} = 100$.

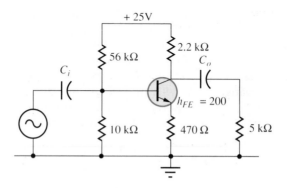

FIGURE 10.16   Problem 10.14.

Section 10.7

16. For the circuit of Figure 10.16, calculate the new Q point voltages if the 10 kΩ resistor became an open circuit.

17. For the circuit of Figure 10.16, calculate the new Q point voltages if the 470 Ω resistor became a short circuit.

18. For the circuit of Figure 10.16, calculate the new Q point voltages if:
    (a) The collector became an open circuit and the base became shorted to the emitter.
    (b) The collector became shorted to the emitter and the base became an open circuit.

The term "hybrid" is used in electronics and in other fields as well. (Charlton Photos.)

Chapter 11
Common Emitter
AC Equivalent Circuits

SPECIAL TERMS

h parameters
h_{ie}
$1/h_{oe}$
h_{re}
h_{fe}

11.1   Introduction and Objectives

In this chapter we will examine the reasons for using an equivalent circuit. We will look very briefly at the r parameter transistor model as an example of the other models which you may encounter. Since h parameters are more commonly found in the manufacturer's data sheets, we will study them in greater depth. Although a dozen h parameters have been defined for the description of a transistor, we will use no more than three. You will find that they describe transistor behavior in sufficient detail to meet your needs as an electronics technologist.

At the conclusion of this chapter you should have a reasonable understanding of what an equivalent circuit is and how to use the h parameter AC model as an aid in circuit design and analysis. You should be able to:

- Define h_{ie}, h_{fe}, and $1/h_{oe}$.
- Determine the value of the above h parameters for a given Q point and common emitter characteristic graph or for a given Q point and information supplied by typical data books.
- Draw a simplified h parameter equivalent circuit for a bipolar junction transistor.
- Use the above equivalent circuit and h parameters to perform circuit analysis using standard circuit analysis procedures.

11.2   Choosing a Model

In Section 1.5 we discussed the concept of a model. Since this whole chapter will be devoted to modeling, you may wish to reread that section at this point.

The circuit analysis techniques with which you are familiar, KVL, KCL, Ohm's Law, etc., can deal with only voltages, currents, resistors, and reactances. Transistors are not included. Thus our goal will be to construct an equivalent circuit which represents the transistor as though it were some combination of those familiar components. We can then apply the rules of circuit analysis with which you are already familiar to circuits which contain transistors.

An "exact" AC equivalent circuit requires the use of a dozen or more components including voltage and current sources, inductors, capacitors and resistors in various series and parallel configurations. As you discovered in Chapter 10, sample-to-sample variability is a problem with transistors. Therefore, it makes little sense to develop a very complicated model for the description of a device whose parameters may vary by 50 percent to 100 percent and more. The results obviously cannot be more accurate than the data from which they were drawn. This is just another way of stating that elegant dictum, "garbage in, garbage out"! Thus a simpler model will probably yield equally good results.

A number of simplifications of the "exact" model have been proposed. Most suffer from the same weakness: The manufacturers have not elected to list the required parameters on their data sheets. They have elected to list h parameters. Specifically, they have chosen to list common emitter h parameters. Thus we will use a very simplified version of the common emitter h parameter model for AC circuit analysis. We will find that it is accurate enough for our purposes. You should be aware that if you become involved with very high frequencies, high gains, or certain special purpose circuits, a more sophisticated model must be used.

11.3 The AC h Parameter Model

The low frequency AC common emitter h parameter equivalent circuit for a transistor is shown in Figure 11.1. You will notice that all of the variables and subscripts are lower case indicating AC values. DC bias considerations are totally ignored by this model. You will also notice that the model suggests another reason for the name "hybrid." The base circuit looks suspiciously like a Thevenin circuit and the collector looks like a Norton circuit. The equivalent is a voltage source at the input and a current source at the output; a hybrid.

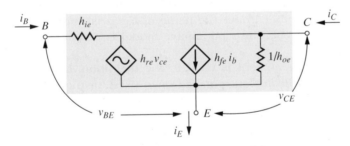

FIGURE 11.1 Common emitter h parameter model.

All of what you already know about the AC behavior of transistors is evident in the model: The emitter current is the sum of the base and the collector currents. The collector current is h_{fe} times the base current— ALMOST! Notice that some current is going to flow through the resistor labeled $1/h_{oe}$.

If you remember how current sources work, you remember that the current $h_{fe}i_b$ will be unaffected by the collector voltage. This would indicate that the transistor characteristics should be perfectly horizontal lines.

However, the higher the collector voltage, the more current will flow through $1/h_{oe}$. That will account for the slope of the transistor characteristic curves. You will learn more about Figure 11.1 in the remaining sections of this chapter.

Since the h parameter model is an AC model, all of the parameters are defined in terms of changing conditions. As we have said before, the language of calculus is ideally suited to talk about changing conditions. Accordingly, all of the h parameters are defined in terms of derivatives.

11.4   Input Resistance (h_{ie})

h_{ie} is defined to be

$$h_{ie} \equiv \left. \frac{dV_{BE}}{dI_B} \right|_{V_{CE}} \tag{11.1}$$

In case you are uncomfortable with the derivative in Equation 11.1, all the equation is really saying is, "While keeping V_{CE} constant, find the slope (of course, it is $1/m$) of the B-E junction graph." Obviously, since h_{ie} is defined as a voltage divided by a current, it will have units of resistance. That is why it is pictured as a resistance in Figure 11.1.

In fact, h_{ie} includes the bulk resistance of the base region of the transistor as well as the bulk resistance of the emitter region. Some models would call them r_b and r_e, respectively. You will find these r parameters used by some authors. Since we cannot separate them it seems reasonable to lump them together under the name of h_{ie} as is done on most manufacturer's data sheets. The r parameter model is discussed more fully in Appendix I.

Since the base current is used in the definition, h_{ie} is referenced to the base. If the collector current had been used, or more correctly, if the emitter current had been used, h_{ie} would be referenced to the emitter.

EXAMPLE 11.1
Find h_{ie} for a 2N3904 transistor at $I_B = 50\ \mu A$ using the graph in Figure 11.2

SOLUTION:
According to Equation 11.1, we need to find the slope of the graph in Figure 11.2 where $I_B = 50\ \mu A$. Therefore, a line has been drawn tangent to the graph at $I_B = 50\ \mu A$. For the sake of accuracy, it has been extended as far as possible and its end points have been identified as

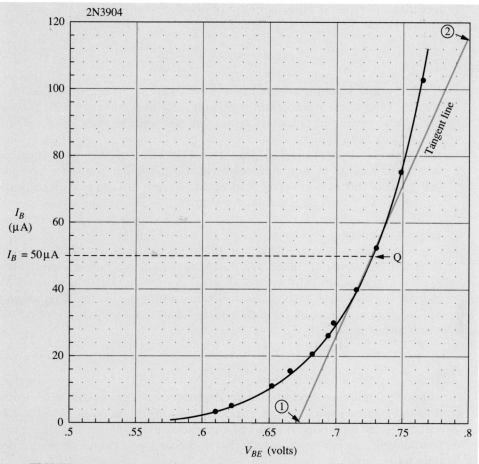

FIGURE 11.2 Base-emitter junction characteristic and the determination of h_{ie}.

points 1 and 2. Then from Equation 11.1, the idea of a derivative, and the definition of slope we have:

$$h_{ie} = \frac{V_{BE2} - V_{BE1}}{I_{B2} - I_{B1}} \tag{11.2}$$

$$= \frac{0.8 \text{ V} - 0.672 \text{ V}}{113 \text{ }\mu\text{A} - 0 \text{ A}} = 1.13 \text{ k}\Omega$$

In Section 8.3 we pointed out that the B-E junction is simply a diode junction. You will notice that the shape of the graph in Figure 11.2 looks much like the diode characteristic graphs we saw in Chapter 3. We also pointed out that the equation that comes reasonably close to fitting the graph of Figure 11.2 is the junction equation

$$V_{BE} = 26 \text{ mV} * \ln\frac{I_C}{10 \text{ fA}} \tag{8.4}$$

If you assume something like $h_{FE} = 250$, which is not unreasonable for a 2N3904, plug some numbers into Equation 8.4 and compare with the experimental results of Figure 11.2; you will see what kind of accuracy you can expect from Equation 8.4. You will find an error of around 0.1 V for the particular 2N3904 used to draw Figure 11.2. It would be nice to calculate the constants in Equation 8.4 to fit the particular transistor you are using as we did in Section 3.7, but again, sample-to-sample variability diminishes the value of the effort. For that reason you might argue for just using some reasonable figure like 0.6 V or 0.7 V for V_{BE} and ignore Equation 8.4 completely.

   If you have reviewed what we did in Section 3.10, you may be wondering if the dynamic diode resistance r_D which we talked about there applies to transistors. The answer is, yes it does. As you may have guessed, that is what we were talking about when we introduced the h parameter h_{ie}. As you have probably already guessed from your knowledge of r_D, we can approximate h_{ie} by

●

$$h_{ie} \approx \frac{26 \text{ mV}}{I_B} \tag{11.3}$$

Although the PSpice circuit analysis program uses a somewhat more complicated model, PSpice will yield results that are quite close to what you will get using Equations 8.4 and 11.3 provided that you set IS = IE − 14 in the .MODEL statement. You will also find that these values fit laboratory data fairly well.

EXAMPLE 11.2

Repeat Example 11.1 using Equation 11.2, to estimate the value of h_{ie}.

SOLUTION:

$$h_{ie} \approx \frac{26 \text{ mV}}{I_B} \tag{11.3}$$

$$= \frac{26 \text{ mV}}{50 \text{ }\mu\text{A}} = 520 \text{ }\Omega$$

Again, as you should expect by now, the agreement with experiment is not as good as we might like. You will find some 2N3904s whose h_{ie} is less than that predicted by Equation 11.2 and many whose h_{ie} is greater. This kind of variability among transistors is a fact of life that you will have to learn to live with.

11.5 Reverse Transfer Ratio (h_{re})

According to that rather unlikely allegory you were told in Section 8.5, it is the voltage at the collector that causes the carriers coming from the emitter to prefer the collector over the base. This should lead you to expect that changes in voltage at the collector should influence the base circuit to some extent. That is, there should be some reverse transfer of collector signal back to the base circuit. h_{re} provides for that possibility in the h parameter model. Look again at the h parameter equivalent circuit in Figure 11.1. Notice that there is a signal source in the base circuit; it is $h_{re}v_{ce}$. Thus, if you monitor the base terminal, with a large AC signal coupled into the collector you should see a small fraction of that signal at the base. h_{re} is that small fraction.

For almost all transistors h_{re} is very small. It is usually in the order of 10^{-4}. Since it is so small, for most work and for the remainder of this book, it can be ignored. The h parameter model for a transistor then becomes the circuit in Figure 11.3.

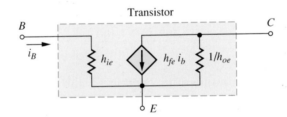

FIGURE 11.3 Simplified h parameter model.

11.6 Forward Transfer Ratio (h_{fe})

In Section 8.5 we have already defined the DC quantity h_{FE} to be

$$h_{FE} \equiv \frac{I_C}{I_B} \qquad (8.2)$$

As you might guess, since AC deals with change, the language of calculus should be used in defining the AC quantity h_{fe}. Thus we have

$$h_{fe} \equiv \frac{dI_C}{dI_B} \bigg|_{V_{CE}} \qquad (11.4)$$

dI_C is just the change in I_C produced by dI_B, a change in I_B. The statement $|V_{CE}$ means that during the above changes V_{CE} must be kept constant. What the definition of Equation 11.4 really says is that h_{fe} is the vertical spacing of the transistor's collector characteristic curves.

We should point out as we did in Section 8.5 that h_{fe} is a unitless quantity since it is a current divided by a current.

EXAMPLE 11.3

Using the 2N4401 transistor characteristics in Figure 11.4, calculate h_{fe} at $I_C = 10$ mA and $V_{CE} = 10$ V. Compare with h_{FE} at the same point.

SOLUTION:

The mathematicians would tell us that dI_B should be an infinitesimally small change, but since h_{fe} probably does not change a great deal in the region of the designated point, the interval between the plotted characeristic curves nearest to the designated point will be used to determine dI_B. Thus, with respect to Figure 11.4 we can say

$$h_{fe} \approx \frac{I_{C2} - I_{C1}}{I_{B2} - I_{B1}} \tag{11.5}$$

$$= \frac{10.3 \text{ mA} - 8.1 \text{ mA}}{50 \text{ μA} - 40 \text{ μA}} = 220$$

Using the procedure outlined in Sections 8.5 and 8.6 to determine h_{FE} we have

$$h_{FE} \equiv \frac{I_C}{I_B} \tag{8.2}$$

$$= \frac{10 \text{ mA}}{49 \text{ μA}} = 204$$

In Section 8.2 we argued that due to its variability it is probably not worth our time to try to distinguish between the DC value h_{FE} and the AC value h_{fe}. Example 11.4 bears that out. If we could be confident that either the h_{FE} or the h_{fe} of a given 2N4401 was between 204 and 220 we would be very pleased indeed. Thus we will use h_{FE} and h_{fe} more or less interchangeably in our discussion.

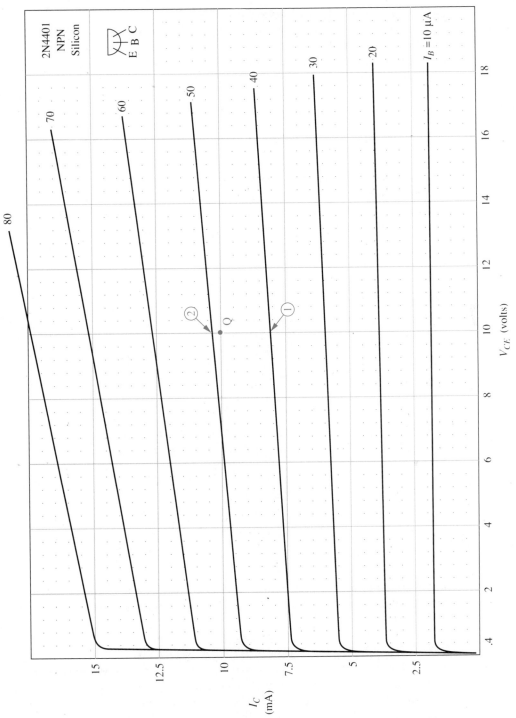

FIGURE 11.4 Common emitter characteristics and the determination of h_{fe}.

11.7 Output Conductance (h_{oe})

The term h_{oe} is probably traceable back to the days of the slide rule and the vaccum tube. Since we usually think more in terms of resistance than conductance, a more understandable term would be $1/h_{oe}$. As you can see in the equivalent circuit of Figure 11.3, the output resistance of a transistor is $1/h_{oe}$. Again, using the language of calculus we can define $1/h_{oe}$ as

$$1/h_{oe} \equiv \frac{dV_{CE}}{dI_C} \bigg|_{I_B} \tag{11.6}$$

What we are really looking for here is the slope (really 1/slope!!) of the collector characteristic curve at the point of interest. Since most modern transistors are fairly good current sources, their characteristic curves are not far from horizontal, and that means their output resistance is very high.

EXAMPLE 11.4

Using the 2N4401 transistor characteristics in Figure 11.5, calculate $1/h_{oe}$ at $I_C = 10$ mA and $V_{CE} = 10$ V.

SOLUTION:

Since the point of interest is not on one of the characteristic curves in Figure 11.5, the closest curve will be used. A line is drawn that is tangent to the characteristic at its closest approach to the point of interest. In order to improve accuracy, that line is extended as far as possible. Using the definition of Equation 11.6 we can write

$$1/h_{oe} = \frac{V_{CE2} - V_{CE1}}{I_{C2} - I_{C1}} \tag{11.7}$$

$$= \frac{20 \text{ V} - 0 \text{ V}}{11.4 \text{ mA} - 9.2 \text{ mA}} = 9.09 \text{ k}\Omega$$

You will notice that the characteristics have less slope for smaller collector currents. That means that $1/h_{oe}$ has a higher value for smaller currents.

11.8 Simplified h Parameter Model

For most of the work you are likely to do with transistors Figure 11.3 is a good enough model. In fact, for many modern transistors, $1/h_{oe}$ is so large that if the AC resistance in the collector-emitter circuit is no more than a

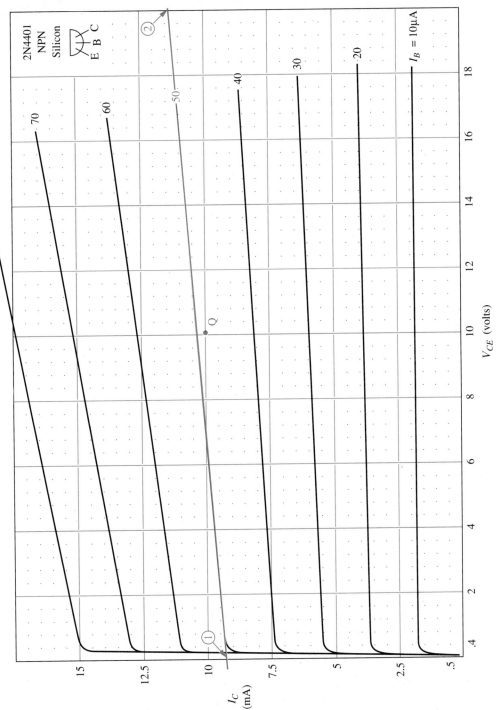

FIGURE 11.5 Common emitter characteristics and the determination of $1/h_{oe}$.

few thousand ohms, $1/h_{oe}$ can be ignored with very little loss in accuracy. In that case, the only value we really must have to work with a given transistor is its current gain h_{FE} or h_{fe}. This simplified model is essentially identical to the r parameter model described in Appendix I.

11.9 Total Circuit AC Equivalent

The goal in constructing an AC equivalent of a transistor amplifier circuit is to represent the circuit in terms of resistive and reactive components which follow the rules of circuit analysis with which we are already familiar. The h parameter model allows us to represent a transistor in terms of such components. All that remains is to combine that model with the external circuit components.

In order to simplify our AC equivalent circuit, we will make two assumptions. First we will assume that the capacitors in the circuit are so

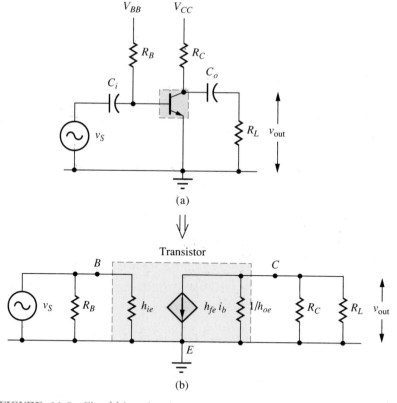

FIGURE 11.6   Fixed bias circuit.
(**a**) Circuit diagram.
(**b**) AC equivalent circuit.

chosen that at the frequencies of interest they act like perfect conductors of AC. In Chapter 12 we will address the question of how those capacitor sizes are determined. Second, we will assume that the AC impedance of the power supply is so small that it acts like ground to AC. You should be warned that failure to reevaluate this assumption when very high gain multistage amplifiers are involved can lead to disastrous oscillation.

The above two assumptions and the h parameter model of Figure 11.3 are used to obtain AC equivalent circuits in Figures 11.6, 11.7, and 11.8. Notice that since the capacitors are assumed to be perfect conductors, they do not appear in the AC equivalent circuit. Resistor R_E, or R_{E2} which is bypassed by capacitor C_E does not appear either since the capacitor acts as an AC short circuit. Also since power supplies are assumed to act as AC grounds, resistors R_B, R_1, and R_C which are connected to them are shown as grounded in the AC equivalent circuit.

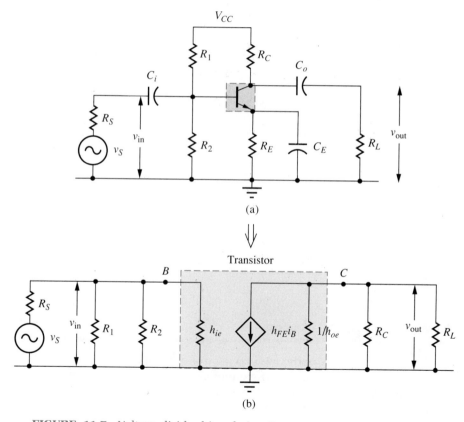

(a)

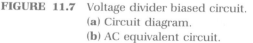

(b)

FIGURE 11.7 Voltage divider biased circuit.
(a) Circuit diagram.
(b) AC equivalent circuit.

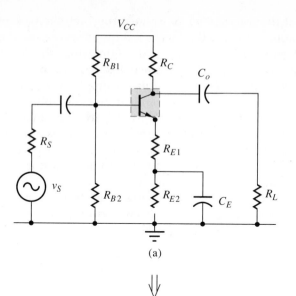

(a)

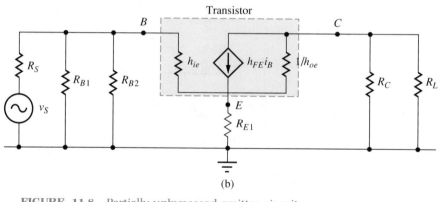

(b)

FIGURE 11.8 Partially unbypassed emitter circuit.
(a) Circuit diagram.
(b) AC equivalent circuit.

11.10 A PSpice Model

The PSpice transistor-model, which is quite complicated and includes such things as temperature and frequency effects, can be adjusted to fit the behavior of a given transistor almost exactly. As we mentioned in Section 8.6 and again in Section 11.4, if the PSpice default model is changed from $I_S = 0.1$ fA to $I_S = 10$ fA, it will agree remarkably well with the simple model we have developed. It will also almost always fit the behavior of the devices themselves within the tolerance of the manufacturer's specs. This value along with the value of h_{FE} (PSpice uses "BF") for the transistor you are using can be easily set in the PSpice ".MODEL" statement which will have the following form for a transistor named "Q1" that is an NPN type with an h_{FE} of 135:

.MODEL Q1 NPN (IS = 1E − 14 BF = 135)

11.11 Summary

Chapter 10 provided the information needed to establish the DC environment in which the transistor functions in a circuit. This chapter introduced the h parameters and the h parameter equivalent circuit needed to allow conventional circuit analysis procedures to be used to determine the circuit's response to an AC signal. In Chapter 12 these procedures will be applied.

GLOSSARY

h parameters: An abbreviation for the term "hybrid parameters." A set of four parameters used to describe two port devices in terms of their input and their output characteristics. The parameters used here are:

h_{fe} The forward transfer ratio. This parameter describes the current gain of the transistor (Section 11.6)

h_{ie} The input impedance. This parameter describes the input resistance as seen looking into the base of the transistor. (Section 11.4)

$1/h_{oe}$ The output impedance. This parameter describes the output impedance looking into the collector of the transistor. (Section 11.7)

h_{re} The reverse transfer ratio. This parameter is not used in this book. (Section 11.5)

PROBLEMS

Section 11.4

1. Using the graph in Figure 11.2, find the value of h_{ie} for the given 2N3904 transistor at   (a) I_B = 20 μA   (b) I_B = 100 μA. Compare with the approximation of Equation 11.3.

2. For the circuit of Figure 11.9, estimate the value of h_{ie}.

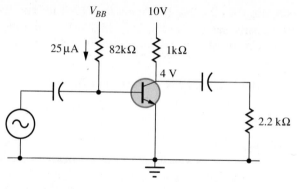

FIGURE 11.9   Problem 2.

Section 11.6

3. For the circuit of Figure 11.9, estimate the value of h_{fe}.
4. Using the transistor characteristics in Figure 11.10, calculate h_{fe} at the point where   (a) $V_{CE} = 5$ V and $I_C = 5$ mA.   (b) $V_{CE} = 15$ V and $I_C = 15$ mA.

Section 11.7

5. Determine $1/h_{oe}$ for the transistor and the Q points specified in Problem 4.
6. Determine $1/h_{oe}$ for the 2N4401 transistor in Figure 11.5 when $I_B = 10$ μA.

Section 11.8

7. Using the characteristics in Figure 11.11, determine as many of the h parameters as possible for the transistor in the circuit of Figure 11.12. *Hint:* You may have to do a little iterating.

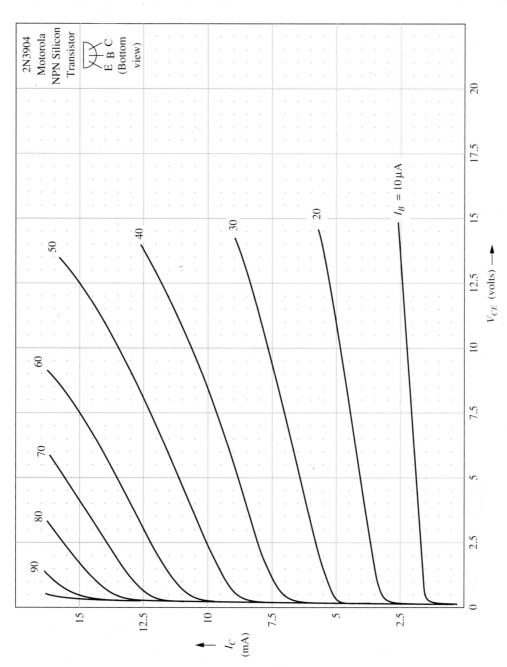

FIGURE 11.10   Problem 4.

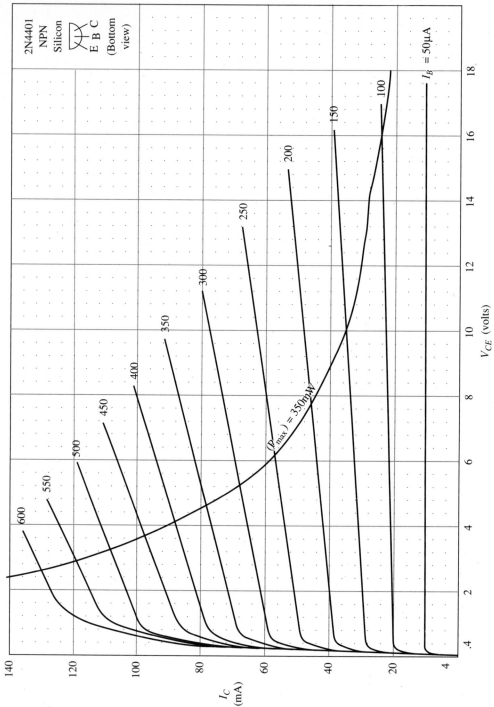

FIGURE 11.11 Problem 7.

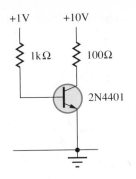

+1V   +10V

1kΩ   100Ω

2N4401

FIGURE 11.12   Problem 7.

Section 11.9

8. Draw an AC equivalent circuit for the circuit of Figure 11.13.

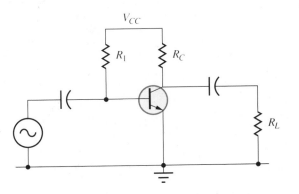

V_{CC}

R_1 R_C

R_L

FIGURE 11.13   Problem 8.

9. Draw the AC equivalent circuit for the circuit in Figure 11.14.

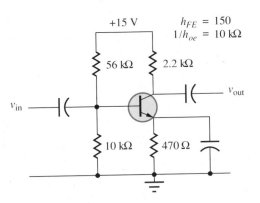

+15 V $h_{FE} = 150$
             $1/h_{oe} = 10\ k\Omega$

56 kΩ   2.2 kΩ

v_{out}

v_{in}

10 kΩ   470 Ω

FIGURE 11.14   Problem 9.

10. Draw the AC equivalent circuit for the circuit in Figure 11.15.

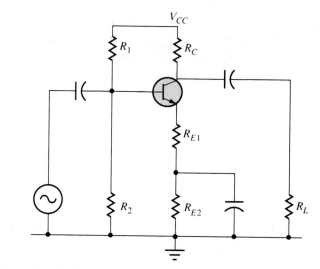

FIGURE 11.15   Problem 10.

Video section of a computer circuit board.
(Photograph by Martin Dohrn/Photo Researchers,
Inc.)

Chapter 12
AC Characteristics of Common Emitter Circuits

SPECIAL TERMS

Impedance
Gain
Decibel (dB)
Bode plot

12.1   Introduction and Objectives

For the past several chapters we have been avoiding, with more or less success, the issue of the behavior of transistor circuits in the presence of AC signals. In this chapter we will confront that issue. You will learn how a common emitter amplifier interacts with the source which supplies the AC signal to it, and how it interacts with the load it must drive. Also we will stop indulging in the fiction that capacitors are perfect conductors of AC. Rather, you will learn to calculate the effect capacitors have on circuit behavior and how transistor internal capacitance affects the circuit. Further, you will learn how to calculate amplifier gain and how its frequency dependence is typically described.

After having studied this chapter you should have a better understanding of the role of the h parameter model and how it is applied in circuit design and analysis and you should be able to:

- Perform a circuit analysis on a common emitter circuit, including the determination of all Q point DC voltages and currents, circuit response to an AC signal including input and output impedances, circuit gain, and cutoff frequencies.
- Design and specify all circuit components for a common emitter amplifier given the DC supply voltage, the source and load resistances, and the required voltage gain and low frequency cutoff.
- Construct a Bode plot for an amplifier.

12.2   Circuit Analysis

Since the rules of circuit analysis cannot handle transistors, in order to perform an AC circuit analysis on circuits such as that of Figure 12.1a, we must first construct an AC equivalent of the circuit as shown in Figure 12.1b. In Chapter 11 you should have learned how to perform this task.

In order to be assured that the circuit will behave as the AC equivalent circuit predicts, and in order to estimate the value of h_{ie} in the equivalent circuit, the place to begin any circuit analysis is with a DC analysis of the circuit. In Chapter 10 you should have learned how to perform this task.

Thus the analysis of any transistor amplifier circuit should involve these steps:

1. Perform a DC circuit analysis.
2. Construct an h parameter AC equivalent circuit.
3. Perform an AC circuit analysis.

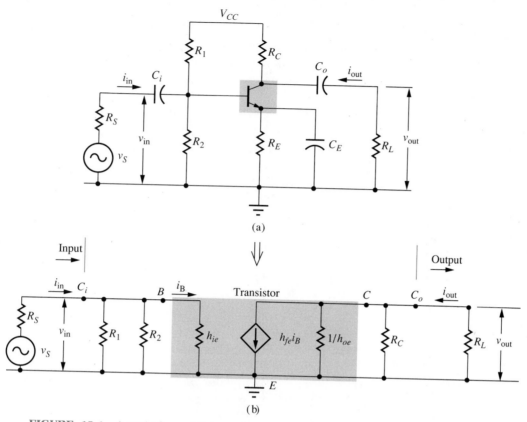

FIGURE 12.1 A typical amplifier stage.
(a) Circuit diagram.
(b) AC equivalent circuit.

12.3 Input Impedance (R_{in})

From the equivalent circuit of Figure 12.1b, you can see that as the AC source "looks into the amplifier circuit" at C_i, it sees

$$R_{in} = R_1 \parallel R_2 \parallel h_{ie} \qquad (12.1)$$

Note that R_{in} is an AC quantity. We would use the letter Z_{in} for the input impedance, except that we are dealing with a purely resistive quantity if we assume that C_i is large. Thus we will use R_{in} to represent the AC input impedance.

EXAMPLE 12.1

Find the input signal voltage, v_{in} that is applied to the transistor in Figure 12.2a if the signal source has an open circuit output voltage of $v_S = 10$ mVPP and an internal resistance of $R_S = 600$ Ω.

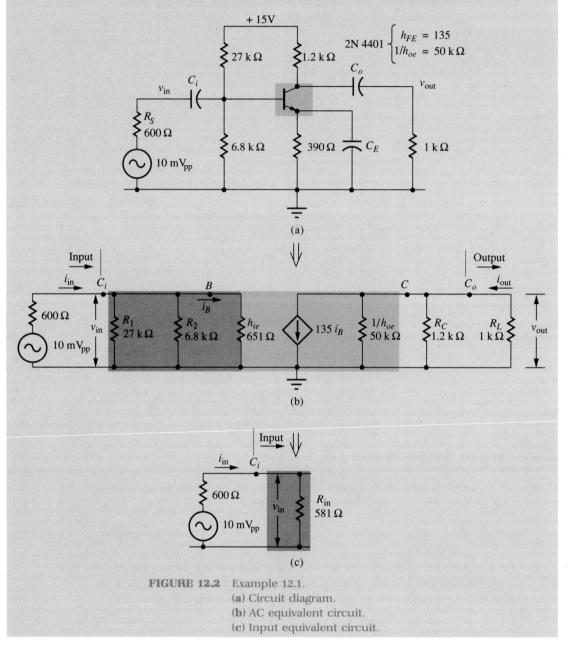

(a)

(b)

(c)

FIGURE 12.2 Example 12.1.
(a) Circuit diagram.
(b) AC equivalent circuit.
(c) Input equivalent circuit.

● (For a PSpice solution, see Appendix H.)

SOLUTION:
From a DC standpoint the circuit in Figure 12.2a is identical to the circuit of Example 10.9. Since the DC analysis of the circuit was performed in Example 10.9, we will not repeat it here. The results were $I_C = 5.38$ mA, $I_B = 39.9$ μA, $V_E = 2.10$ V, $V_B = 2.80$ V, and $V_C = 8.54$ V. Since V_{CE} is a bit over 6 V, the transistor is not cut off, and it is not saturated. Thus the h parameter AC equivalent circuit should be valid.

Applying the approach of Section 11.4, our best estimate of h_{ie} is

$$h_{ie} \approx \frac{26 \text{ mV}}{I_B} \tag{11.3}$$

$$= \frac{26 \text{ mV}}{39.9 \text{ μA}} = 651 \text{ } \Omega$$

The AC equivalent circuit is drawn in Figure 12.2b. Looking into the input of that circuit we can write

$$R_{in} = R_1 \parallel R_2 \parallel h_{ie} \tag{12.1}$$
$$= 27 \text{ k}\Omega \parallel 6.8 \text{ k}\Omega \parallel 651 \text{ } \Omega = 581 \text{ } \Omega$$

As shown in Figure 12.2c, assuming that the signal frequency is high enough so that all of the capacitors act like perfect conductors, the signal source impedance R_S and the circuit input impedance R_{in} form a voltage divider, so

$$v_{in} = v_S \frac{R_{in}}{R_{in} + R_S} \tag{12.2}$$

$$= 10 \text{ mVPP} \frac{581 \text{ } \Omega}{581 \text{ } \Omega + 600 \text{ } \Omega} = 4.92 \text{ mVPP}$$

Notice that h_{ie} is typically the most significant part of R_{in}. Thus our knowledge of R_{in} is no better than our knowledge of h_{ie}.

12.4 Input Coupling Capacitor C_i

In order to talk about the effect of the input coupling capacitor, we need to stop thinking of it as a perfect conductor and include it in the circuit of Figure 12.2c. That has been done in Figure 12.3. You should recognize that the circuit is, in fact, just a high pass filter. If you have forgotten how circuits containing capacitors respond to AC signals, you should return to your circuits class for a review. Specifically, you need to know about fre-

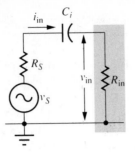

FIGURE 12.3   Input circuit used to determine C_i.

quency response, low frequency cutoff for high pass filters, and high frequency cutoff for low pass filters.

Remember that we define low frequency cutoff, f_L, as the frequency at which

$$R_{ckt} = X_C = \frac{1}{2\pi f_L C} \qquad (12.3)$$

Equation 12.3 can be applied to any circuit. Solving for C we have

$$C = \frac{1}{2\pi f_L R_{ckt}} \qquad (12.4)$$

Usually your application, or perhaps the marketing department of the company where you work, dictates the value for f_L. Since we do not want the coupling capacitor to interfere with the AC signal, you should use a frequency for f_L that is at least one decimal place lower than most of the frequencies the amplifier will be called upon to amplify. Thus if you wish to amplify a signal whose frequency is 1 kHz, choose $f_L = 100$ Hz or less.

If you are designing a cheap "boom box" to be sold at the local discount store, probably $f_L = 100$ Hz is adequate, but to meet the definition of "high fidelity" requires $f_L = 20$ Hz.

From Equation 12.4 it is obvious that the lower you wish f_L to be (and the better you want your stereo to sound), the bigger (and the more expensive) C_i must be.

We will define R_{ckt} to be the resistance that the capacitor "sees" as it looks at the circuit of which it is a part. As you can see in Figure 12.3, when C_i looks to its left it sees the source resistance R_S and when it looks to its right it sees R_{in}. Obviously, this is a series circuit. Thus for the circuit of Figure 12.3

$$R_{ckt} = R_S + R_{in} \qquad (12.5)$$

Combining Equations 12.4 and 12.5 we have for the circuit of Figure 12.3

$$C_i = \frac{1}{2\pi f_L (R_S + R_{in})} \qquad (12.6)$$

EXAMPLE 12.2

Calculate the value of C_i for the circuit of Example 12.1 that will give a cutoff frequency of about 35 Hz. Specify the nearest standard capacitor.

SOLUTION:

From Example 12.1 we have R_{in} = 581 Ω. Substituting values from Figure 12.2c into Equation 12.6 we have

$$C_i = \frac{1}{2\pi f_L (R_S + R_{in})} \tag{12.6}$$

$$= \frac{1}{2\pi * 35 \text{ Hz}(600 \text{ Ω} + 581 \text{ Ω})} = 3.85 \text{ μF} \quad \text{(Use 3.9 μF)}$$

The closest standard capacitor size from Appendix B is specified.

12.5 R_{out} and Choosing R_C

Looking at the equivalent circuit of Figure 12.1b, you can see that as the AC load "looks into the amplifier circuit," the resistance it sees is

$$R_{out} = R_C \parallel 1/h_{oe} \tag{12.7}$$

Again, because we are dealing with AC we are tempted to use the letter Z_{out} for this quantity but, assuming that C_o is appropriately chosen, the circuit output impedance is virtually a pure resistive quantity so we will use the letter R.

EXAMPLE 12.3

Calculate the output impedance of the circuit in Example 12.1.

SOLUTION:

Substituting the values from Figure 12.2 into Equation 12.7 we have

$$R_{out} = R_C \parallel 1/h_{oe} \tag{12.7}$$
$$= 1.2 \text{ kΩ} \parallel 50 \text{ kΩ} = 1.17 \text{ kΩ}$$

From the above example, you can see that, as we said in Section 11.7, $1/h_{oe}$ does not have a great effect on R_{out}. If we just ignored it and said

$$R_{out} \approx R_C \tag{12.8}$$

we would typically introduce an error of less than 10 percent.

You might at first think that in order to get the maximum power delivered to the load, you should try to make the output impedance of your amplifier as low as possible. In fact, if you could make it zero ohms, the amplifier should be able to supply a rather large current! The problem is that if R_{out} is to be small, according to Equation 12.7, R_C must be small. That will mean a near vertical load line, and consequently a very small output signal voltage. In fact, if you think about the load line, you will recognize that the larger you make R_C, the more signal output voltage you will have! Of course, with a large R_C and the resulting flat load line, the signal amplitude that the circuit can handle without clipping is small. Also a flat load line means a small base current. You may recall that the junction characteristic gets very nonlinear near cutoff so distortion becomes a problem. In order to balance these factors and select the optimum value for R_C in a given situation requires sound judgment, experience, and maybe even some trial and error.

Perhaps a reasonable first approach to the selection of R_C might be to match the output impedance of the amplifier to the impedance of R_L, the load it is required to drive. Substituting this idea into Equation 12.7 gives

$$R_L = R_C \parallel 1/h_{oe} \qquad (12.9)$$

Since R_L is usually dictated by external considerations, the circuit designer is usually required to solve Equation 12.9 for R_C. We can do algebra on Equation 12.9 if we remember that Equation 12.9 is just a shorthand method of saying

$$\frac{1}{R_L} = \frac{1}{R_C} + \frac{1}{1/h_{oe}} \qquad (12.10)$$

Solving Equation 12.10 for the R_C term we have

$$\frac{1}{R_C} = \frac{1}{R_L} - \frac{1}{1/h_{oe}} \qquad (12.11)$$

Notice that in Equation 12.11, instead of adding the second reciprocal as you have become accustomed to doing when you see the " $\parallel$ " symbol, we need to subtract it. Perhaps it will not introduce too much confusion if we write Equation 12.11 thus

$$R_C = R_L \parallel (-1/h_{oe}) \qquad (12.12)$$

If you have a program in your calculator to solve parallel resistors (and I hope that you do!) you will find that it will solve Equation 12.12 provided you enter $1/h_{oe}$ as a negative quantity.

EXAMPLE 12.4

Suggest a better value for the collector resistor in the circuit of Example 12.1.

SOLUTION:

Substituting values from Figure 12.2b into Equation 12.12 gives

$$R_C = R_L \,\|\, (-1/h_{oe}) \tag{12.12}$$
$$= 1 \text{ k}\Omega \,\|\, (-50 \text{ k}\Omega) = 1.02 \text{ k}\Omega \quad (\text{Use } 1 \text{ k}\Omega)$$

The nearest standard resistor is specified.

Again we notice that $1/h_{oe}$ does not contribute much.

12.6   Output Coupling Capacitor C_o

In order to calculate the value of C_o, we will again make use of the coupling capacitor equation, Equation 12.3. Again we must look at the capacitor in relation to the circuit of which it is a part, just as we did in Section 12.4. Figure 12.4 contains the essential part of the AC circuit. Although the

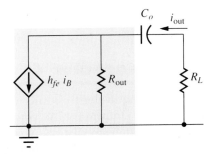

FIGURE 12.4   Output circuit used to determine C_o.

circuit contains a current source rather than a voltage source as in Figure 12.3, if you remember how Norton circuits work, you realize that from the capacitor's standpoint, the circuits are similar. Thus, using the same reasoning used in Section 12.4, we have

$$C_o = \frac{1}{2\pi f_L (R_{out} + R_L)} \tag{12.13}$$

Good circuit design practice requires that the designer use the same f_L for C_o as was used for C_i. It makes no sense, for example, to use a big (and expensive) input capacitor so that the input lets those beautiful low frequency bass notes get into your amplifier only to prevent them from getting out to the loudspeaker by using a little (and cheap) output capacitor. Obviously, a small input capacitor and a big output capacitor makes no sense either.

EXAMPLE 12.5

Calculate the size of the output coupling capacitor in the circuit of Example 12.1 for $f_L = 35$ Hz.

SOLUTION:

In Example 12.3 we calculated the output impedance of the circuit of Example 12.1 to be $R_{\text{out}} = 1.17$ kΩ. Inserting this value into Equation 12.13 gives

$$C_o = \frac{1}{2\pi f_L(R_{\text{out}} + R_L)} \qquad (12.13)$$

$$= \frac{1}{2\pi * 35 \text{ Hz}(1.17 \text{ k}\Omega + 1 \text{ k}\Omega)} = 2.1 \ \mu\text{F} \qquad (\text{Use } 2.2 \ \mu\text{F})$$

12.7   Emitter Bypass Capacitor C_E

In Figure 12.2b and all similar equivalent circuits we have pictured the emitter as grounded from an AC standpoint. As you might guess, that assumption simplifies the circuit analysis. Also, as you will discover in Section 12.14, if the emitter is at AC ground, signal amplification is maximized. In order to calculate the size of C_E we again apply the coupling capacitor equation, Equation 12.3. Again we must evaluate the circuit of which the capacitor is a part. The relevant equivalent circuit is drawn in Figure 12.5.

To visualize the circuit of which C_E is a part, picture yourself as a charge on the positive plate of C_E in Figure 12.5. Think about how you can get from the positive plate around through the circuit to the negative plate of C_E. Obviously, one way would be to go through R_E. Many students, and even textbook authors forget that that is not the only way. You could enter the emitter terminal of the transistor. Then you have to choose whether to go out of the base terminal or out of the collector terminal. If you choose the base, looking at the circuit of Figure 12.5 you can see that you must go through h_{ie} and on through either R_2, R_1, or R_S. If you choose the

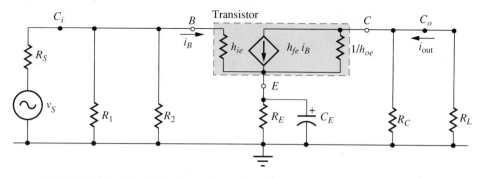

FIGURE 12.5 Equivalent circuit used to determine C_E.

collector, you must go through $1/h_{oe}$ and on through R_C or R_L in order to get to the negative plate of C_E.

As you might guess, there is a formidable amount of calculator button-pushing coming up. We can simplify our task somewhat by recognizing that $1/h_{oe}$ is quite large. Thus we can probably ignore the path out through the collector. Can we ignore the path out through the base? You should recall that we have argued on numerous previous occasions that when you look from the emitter back to the base circuit, since the currents are all much smaller, the resistors look as though they are much smaller. Again, we meet the Miller effect. Their values all have to be divided by h_{fe}. Since the base circuit resistance will be divided by h_{fe}, we certainly cannot ignore that path. including it in parallel with R_E gives

$$R_{\text{ckt}} = R_E \left\| \frac{h_{ie} + R_1 \| R_2 \| R_S}{h_{fe}} \right. \tag{12.14}$$

Once R_{ckt} is known, the result can be substituted in the coupling capacitor equation which can then be solved for C_E. The result is

$$C_E = \frac{1}{2\pi f_L \left(R_E \left\| \dfrac{h_{ie} + R_1 \| R_2 \| R_S}{h_{fe}} \right.\right)} \tag{12.15}$$

EXAMPLE 12.6

Calculate the size of C_E for the circuit of Example 12.1 for $f_L = 35$ Hz.

SOLUTION:

Substituting the numbers from Figure 12.2 into Equation 12.15 we have:

$$C_E = \frac{1}{2\pi f_L \left(R_E \left\| \dfrac{h_{ie} + R_1 \| R_2 \| R_S}{h_{fe}} \right. \right)} \tag{12.15}$$

$$C_E = \frac{1}{2\pi * 35\,\text{Hz} \left(390\,\Omega \left\| \dfrac{651\,\Omega + 27\,\text{k}\Omega \| 6.8\,\text{k}\Omega \| 600\,\Omega}{135} \right. \right)} = 527\,\mu\text{F}$$

(Use 500 μF)

The closest standard capacitor size is specified.

If you worked the preceding example on your own calculator, you may have noticed that since the resistance in the base circuit is divided by h_{fe}, it is by far the most important resistance in Equation 12.15. Further, the source resistance is often the most important resistance in the base circuit. In fact, contrary to your intuition when you first look at the circuit of Figure 12.5, in many cases R_E is irrelevant in determining the value of C_E.

12.8 Signal Gain (A)

Gain, A (the symbol used probably has its roots in the word Amplification) is defined to be

$$A \equiv \frac{\text{output}}{\text{input}} \tag{12.16}$$

Notice that the units for the output and the input must always be the same. Thus the gain is always a unitless quantity. In the specific case of voltage gain, A_V we have

$$A_V \equiv \frac{v_{\text{out}}}{v_{\text{in}}} \tag{12.17}$$

You should also note that, as we mentioned in Section 9.3, for the common emitter amplifier v_{out} is a negative quantity, indicating that it is out of phase with v_{in}. This can be seen from the direction of the current arrows in Figure 12.6. Thus A_V will be a negative quantity, the negative sign simply indicating that the output is out of phase with the input.

The most straightforward procedure for finding A_V is to first do a DC analysis of the circuit as we did in Chapter 10. Then draw the AC equivalent circuit, express both v_{out} and v_{in} in terms of the transistor's AC base current, i_B, and substitute the results into Equation 12.17. Then i_B will appear in both the numerator and denominator. It can then be canceled out of the equation which can then be solved for A_V.

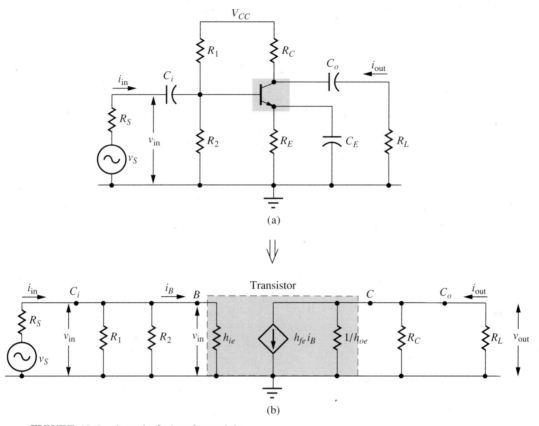

FIGURE 12.6 A typical signal amplifier.
(a) Schematic circuit diagram.
(b) AC equivalent circuit.

The equivalent circuit for the typical common emitter circuit of Figure 12.6a is shown in Figure 12.6b. Notice that R_E does not appear in Figure 12.6b since we assume that C_E is correctly chosen.

For the circuit of Figure 12.6b, Ohm   us that

$$v_{out} = -h_{fe}i_B(1/h_{oe} \parallel R_C \parallel R_L) \qquad (12.18)$$

Applying Ohm's Law to the input we have

$$v_{in} = i_B h_{ie} \qquad (12.19)$$

Substituting Equations 12.18 and 12.19 into Equation 12.17 and canceling i_B which appears in both numerator and denominator, we have

$$A_v \equiv \frac{v_{out}}{v_{in}} \qquad (12.17)$$

$$= -\frac{h_{fe}(1/h_{oe} \parallel R_C \parallel R_L)}{h_{ie}} \qquad (12.20)$$

By referring to Equation 12.16 we can write the equation for the current gain A_i

$$A_i \equiv \frac{i_{out}}{i_{in}} \tag{12.21}$$

Applying Ohm's Law to the output circuit of Figure 12.6b we have

$$i_{out} = \frac{v_{out}}{R_L} \tag{12.22}$$

Looking at the input circuit we have

$$i_{in} = \frac{v_{in}}{R_{in}} \tag{12.23}$$

or, substituting Equation 12.1 into Equation 12.23

$$i_{in} = \frac{v_{in}}{R_1 \parallel R_2 \parallel h_{ie}}$$

Substituting Equations 12.22 and 12.23 into Equation 12.21 we can write

$$A_i \equiv \frac{i_{out}}{i_{in}} \tag{12.21}$$

$$= \frac{\dfrac{v_{out}}{R_L}}{\dfrac{v_{in}}{R_{in}}} = \left(\frac{v_{out}}{v_{in}}\right)\left(\frac{R_{in}}{R_L}\right) \tag{12.24}$$

Substituting Equation 12.17 into Equation 12.24 gives

$$A_i = A_v \frac{R_{in}}{R_L} \tag{12.25}$$

By referring to Equation 12.16 we can write the equation for the circuit power gain as

$$A_P \equiv \frac{P_{out}}{P_{in}} \tag{12.26}$$

but you know that

$$P = VI \tag{12.27}$$

Substituting Equation 12.27 into Equation 12.26 gives

$$A_P = \frac{v_{out}i_{out}}{v_{in}i_{in}} \tag{12.28}$$

Substituting Equation 12.17 and 12.21 into Equation 12.28 gives

$$A_P = A_v A_i \tag{12.29}$$

EXAMPLE 12.7

Calculate the voltage gain, the current gain, and the power gain for the circuit of Example 12.1.

SOLUTION:

The circuit of Example 12.1 is reproduced in Figure 12.7a and the equivalent circuit is shown in Figure 12.7b. The voltage gain is given by

$$A_v = -\frac{h_{fe}(1/h_{oe} \parallel R_C \parallel R_L)}{h_{ie}} \quad (12.20)$$

$$= -\frac{135(50 \text{ k}\Omega \parallel 1.2 \text{ k}\Omega \parallel 1 \text{ k}\Omega)}{651 \ \Omega} = -112$$

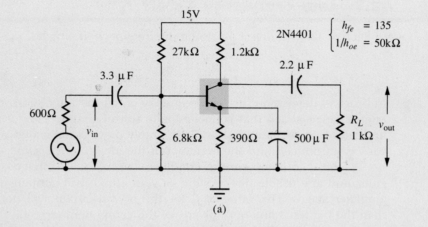

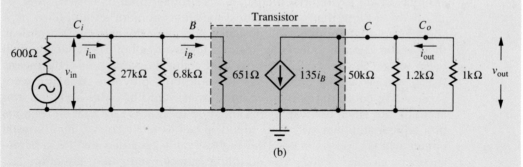

FIGURE 12.7   Example 12.8.
(a) Circuit diagram.
(b) Equivalent circuit.

The current gain is given by

$$A_i = A_v \frac{R_{in}}{R_L} \quad (12.25)$$

From Example 12.1 we have $R_{in} = 581 \ \Omega$, so

$$A_i = -112 \ \frac{581 \ \Omega}{1 \ k\Omega} = -65.0$$

The power gain is given by

$$A_P = A_v A_i \qquad (12.29)$$
$$= (-112)(-65.0) = 7280$$
$$= 7280$$

12.9 Low Frequency Cutoff f_L

In Section 12.4 we defined f_L to be the frequency at which

$$R_{ckt} = X_C = \frac{1}{2\pi f_L C} \qquad (12.3)$$

In order to determine the low frequency cutoff for any circuit which contains a capacitor, all that is needed is to solve Equation 12.3 for f_L. We will find later that the situation is considerably more complicated when feedback is introduced into the system, but the ideas are the same.

A typical common emitter amplifier stage contains three capacitor circuits, and any usable device such as your stereo will contain many such amplifier stages. The value of f_L for the overall circuit will not be lower than the highest value of f_L for any of the component circuits. That is just another way of saying that a chain is no stronger than its weakest link.

If we assume that f_L is identical for all sections of an amplifier, as it should be if the circuit is well-designed and all of the capacitors are ideal, then as the signal frequency is lowered toward f_L, all of the capacitors will begin to "get in the way" of the signal at about the same time. This means that f_L will be slightly higher than it would be if only one capacitor were present. We could develop an equation that would take this situation into account, but since real capacitors must be used, it is rare that all sections of a typical amplifier will have identical values of f_L. Thus f_L for the total circuit will be approximately the highest value obtained for the separate capacitor circuits. Solving the coupling capacitor equation, Equation 12.3 for f_L gives the required equation

$$f_L = \frac{1}{2\pi C R_{ckt}} \qquad (12.30)$$

For the typical common emitter circuit, R_{ckt} is obtained in the same way as it was determined when we calculated C_i, C_o and C_E in Sections 12.4, 12.6, and 12.7 respectively. Thus, solving Equation 12.6 for f_L we can write

$$f_{Lin} = \frac{1}{2\pi C_i (R_S + R_{in})} \tag{12.31}$$

Similarly, solving Equation 12.13 for f_L we have

$$f_{Lout} = \frac{1}{2\pi C_o (R_{out} + R_L)} \tag{12.32}$$

And, solving Equation 12.15 for f_L gives

$$f_{Le} = \frac{1}{2\pi C_E \left(R_E \, \middle\| \, \dfrac{h_{ie} + R_1 \| R_2 \| R_S}{h_{fe}} \right)} \tag{12.33}$$

As we mentioned before, the observed value of f_L for the circuit will be approximated by whichever of the above equations gives the highest value.

EXAMPLE 12.8

Find the low frequency cutoff for the circuit of Example 12.7.

SOLUTION:

Figure 12.7 contains the circuit as it has been developed. In Example 12.1, R_{in} for this circuit was found to be 581 Ω. Thus

$$f_{Lin} = \frac{1}{2\pi C_I (R_S + R_{in})} \tag{12.31}$$

$$= \frac{1}{2\pi \; 3.9 \; \mu F (600 \; \Omega + 581 \; \Omega)} = 34.6 \text{ Hz}$$

In Example 12.3, R_{out} for this circuit was found to be 1.17 kΩ. Thus

$$f_{Lout} = \frac{1}{2\pi C_o (R_{out} + R_L)} \tag{12.32}$$

$$= \frac{1}{2\pi \; 2.2 \; \mu F (1.17 \; k\Omega + 1 \; k\Omega)} = 33.3 \text{ Hz}$$

For the emitter circuit we have

$$f_{Le} = \frac{1}{2\pi C_E \left(R_E \, \middle\| \, \dfrac{h_{ie} + R_1 \| R_2 \| R_S}{h_{fe}} \right)} \tag{12.33}$$

$$= \frac{1}{2\pi \; 500 \; \mu F \left(390 \; \Omega \, \middle\| \, \dfrac{651 \; \Omega + 27 \, k\Omega \| 6.8 \, k\Omega \| 600 \, \Omega}{135} \right)} = 36.9 \text{ Hz}$$

As you can see, although the other circuits would still be conducting fairly well, the circuit that will tend to stop the amplifier from amplifying is the emitter circuit. The low frequency cutoff will not be lower than about $f_L = 36.9$ Hz.

As is the case in Example 12.8, due to the fact that C_E must be so very large, it is not uncommon for the emitter circuit to be the one that determines f_L.

The most important circuit characteristics at cutoff will be summarized at the end of Section 12.12.

12.10   High Frequency Cutoff f_H

In Section 7.4 we talked about the fact that a P-N junction acts like a capacitor. The junction capacitance tends to place an upper limit on the frequency at which a transistor can operate.

You should remember from your circuits class that the capacitance of a capacitor is directly related to plate area. Since transistors are rather small, it should come as no surprise that junction capacitances are small. Further, you should remember that capacitance is inversely related to plate separation. Therefore, you should not be surprised by the fact that the base-emitter capacitance C_{ib} is larger than the collector-base capacitance C_{ob}. After all, the base junction is forward biased and so you should expect the carriers that form the "plates" of C_{ib} to be closer together than the carriers that form the plates of the reverse biased collector junction, C_{ob}.

Figure 12.8 represents the location of C_{ib} and C_{ob} within the transistor. Figure 12.9b indicates where they appear in the equivalent circuit of the typical common emitter amplifier of Figure 12.9a. A quick look at some of

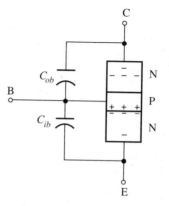

FIGURE 12.8   Transistor junction capacitances.

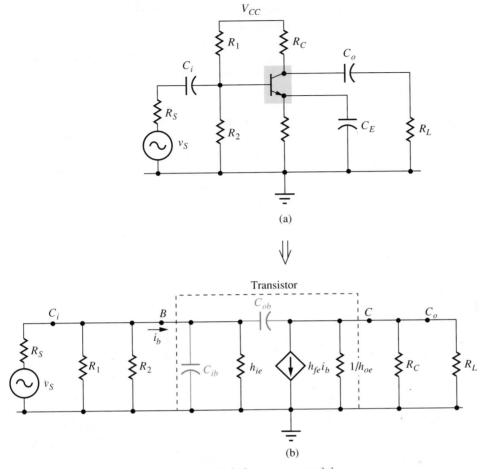

FIGURE 12.9 Common emitter high frequency model.
(a) Circuit diagram.
(b) AC equivalent circuit.

the data in Appendix F will indicate the values you can expect for typical transistors.

From Figure 12.9b, it is easy to see why C_{ib} tends to diminish the amplifier's ability to amplify high frequency signals. It makes the input circuit into a low pass filter. It will conduct high frequencies off to ground. This will produce a high frequency cutoff f_H which will place an upper limit on the frequency that can be amplified in much the same way as f_L places a lower limit on an amplifier.

Adapting the reasoning used in connection with f_L to the circuit of Figure 12.9, and considering only C_{ib}, we have

$$f_{Hi} = \frac{1}{2\pi\, C_{ib}(h_{ie} \parallel R_2 \parallel R_1 \parallel R_S)} \tag{12.34}$$

The effect of C_{ob} is not so readily apparent. To understand the effect of C_{ob}, you need to remember that, according to Section 12.8 the voltage gain of a common emitter amplifier is large. Thus a signal of a millivolt or two at the base may produce a signal of over 100 mV at the collector. Since C_{ob} exists between the base and the collector and since the amount of charge a capacitor will store depends on the voltage difference between its plates, the effective capacitance of C_{ob} is actually C_{ob} multiplied by the voltage gain of the amplifier. This is the Miller effect, showing up again.

Once the effect of C_{ob} is understood, the rest of the analysis of the effect of C_{ob} on f_H proceeds in a similar fashion as the determination of f_L in Section 12.9. From Figure 12.9b you can see that when C_{ob} looks to the left, it sees h_{ie}, R_2, R_1, and R_S all in parallel, and when it looks to the right it sees $1/h_{oe}$, R_C, and R_L all in parallel. Thus we have

$$f_{Ho} = \frac{1}{2\pi \, A_v C_{ob}(h_{ie} \parallel R_2 \parallel R_1 \parallel R_S \, + \, 1/h_{oe} \parallel R_C \parallel R_L)} \tag{12.35}$$

EXAMPLE 12.9

Find the high frequency cutoff f_H for the circuit of Example 12.8 if $C_{ib} = 22$ pF and $C_{ob} = 4$ pF.

SOLUTION:

Figure 12.10b contains the relevant circuit values obtained from the previous examples in this chapter. Looking first at the effect of C_{ib}, we substitute the circuit values in Equation 12.34

$$f_{Hi} = \frac{1}{2\pi \, C_{ib}(h_{ie} \parallel R_2 \parallel R_1 \parallel R_S)} \tag{12.34}$$

$$= \frac{1}{2\pi \, 22 \text{ pF } (651 \, \Omega \parallel 6.8 \text{ k}\Omega \parallel 27 \text{ k}\Omega \parallel 600 \, \Omega)} = 24.5 \text{ MHz}$$

Looking at the effect of C_{ob}, since the voltage gain was calculated for this circuit in Example 12.7 to be $A_v = 112$, and substituting circuit values in Equation 12.35 we have

$$f_{Ho} = \frac{1}{2\pi \, A_v C_{ob}(h_{ie} \parallel R_2 \parallel R_1 \parallel R_S \, + \, 1/h_{oe} \parallel R_C \parallel R_L)} \tag{12.35}$$

$$= \frac{1}{2\pi * 112 * 4 \text{ pF } (651 \, \Omega \parallel 6.8 \text{ k}\Omega \parallel 27 \text{ k}\Omega \parallel 600 \, \Omega + 50 \text{ k}\Omega \parallel 1.2 \text{ k}\Omega \parallel 1 \text{ k}\Omega)}$$
$$= 426 \text{ kHz}$$

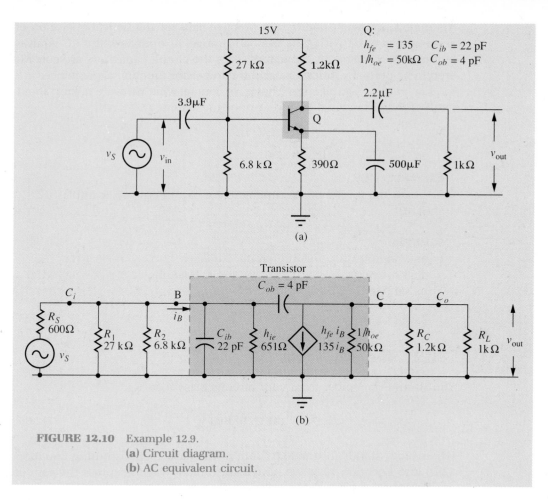

FIGURE 12.10 Example 12.9.
(a) Circuit diagram.
(b) AC equivalent circuit.

Using reasoning similar to that used in determining the overall circuit low frequency cutoff, the overall circuit high frequency cutoff, f_H will be no higher than the lowest value of f_{Hi} or f_{Ho}. Thus, f_H for the circuit of Example 12.8 can be no higher than $f_H = 426$ kHz. In fact, it will probably be somewhat lower since we did not take into account the frequency-limiting characteristics of other circuit components such as wire, resistors, etc.

Due to the fact that its effect is multiplied by A_v, C_{ob} is usually the transistor capacitance that has the most to do with f_H for the common emitter circuit.

12.11 Midband

The range of frequencies that extend from one order of magnitude (one decimal place) above f_L to one order of magnitude below f_H is known as the *midband range*. Good design practice calls for the typical operation of an amplifier to be within its midband range. At midband, the AC equivalent circuit contains no capacitors since the circuit capacitors are considered to be perfect conductors, and the transistor junction capacitances are too low to be of significance. That is, of course, what we were talking about in all of the circuits we looked at prior to Chapter 12.

EXAMPLE 12.9
Estimate the width of the midband for the circuit of Examples 12.8 and 12.9.

SOLUTION:
From Example 12.8, for the circuit of Figure 12.10, $f_L = 36.9$ Hz. Thus midband begins at 369 Hz (10 f_L). From Example 12.9, $f_H = 426$ kHz. Thus midband extends to 42.6 kHz (0.1 f_H).

12.12   The dB Scale

You should remember from your circuits class that

$$dB \equiv 10 \log\left(\frac{P_2}{P_1}\right) \tag{12.36}$$

where "log" stands for the logarithm to the base 10. Substituting Equation 12.26 into Equation 12.36 we have a convenient way to express the gain of an amplifier

$$dB = 10 \log(A_P) \tag{12.37}$$

Thus the dB gain of the amplifier in Example 12.7 is

$$dB = 10 \log (A_P) \tag{12.37}$$
$$dB = 10 * \log (7280) = 38.6 \text{ dB}$$

Of course, the above result is really a unitless quantity, but it is common to attach the "dB" identifier to the number.

Since the dB ratio is the ratio of two power values, it is also used to compare the power output of an amplifier to its output under some standard or controlled condition.

For example, if you look at the VU meters on a high quality tape recorder, you will discover that the scale begins at about -20 dB and extends to about $+3$ dB. In this case, zero dB indicates the amplifier output power that fully magnetizes the tape. In the case of a radio station, zero dB is the maximum allowable power output permitted by the licensing agency. Amplifier frequency response is conveniently represented on the same kind of scale. Zero dB is the power output at midband, P_{mid}, against which the power output at any other frequency, P_f, is compared. In this case we can write

$$dB = 10 \log\left(\frac{P_f}{P_{mid}}\right) \tag{12.38}$$

It is usually more convenient to use voltages rather than power in making such comparisons. To that end, substituting the power equation

$$P = \frac{v^2}{R}$$

into Equation 12.38, if we assume that the circuit resistance is constant, it will cancel out and we have

$$dB = 10 \log\left(\frac{v_f^2}{V_{mid}^2}\right)$$

or, if you know how to handle logarithms

$$\bullet \qquad dB = 20 \log\left(\frac{v_f}{V_{mid}}\right) \tag{12.39}$$

If you had trouble following that last step, remember that the exponent of a number becomes the coefficient of its logarithm.

We could show that at frequencies far from the midband, the output voltage, what we have been calling v_f, approaches a linear function of frequency. Then, from Equation 12.39, a plot of dB vs the log of frequency will approach a straight line with a slope of "20 dB per decade." A "decade" refers to multiplying or dividing the frequency by 10. The reason for this slope can be seen if we write

$$dB = 20 \log\left(\frac{v_f}{V_{mid}}\right) \tag{12.39}$$

$$y = m * x$$

Thus the slope should be 20.

EXAMPLE 12.11

Express the circuit gain at f_L and f_H for the circuit in Examples 12.8, 12.9, and 12.10 in dB.

SOLUTION:

In Section 12.9 we argued that at low frequencies, the amplifier acts like a high pass filter, and in Section 12.10 we argued that at high frequencies it acts like a low pass filter. If you know about such filters, you know that in either case at the cutoff frequency

$$R_{ckt} = X_C = \frac{1}{\sqrt{2}} Z_{ckt} \qquad (12.40)$$

The above relationships are shown in the impedance triangle in Figure 12.11a. From the voltage triangle in Figure 12.11b

$$v_{Rckt} = v_{XC} = \frac{1}{\sqrt{2}} v_S \qquad (12.41)$$

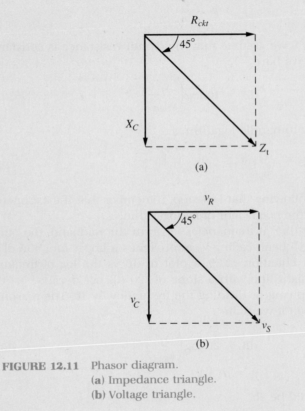

(a)

(b)

FIGURE 12.11   Phasor diagram.
(a) Impedance triangle.
(b) Voltage triangle.

Substituting Equation 12.41 into Equation 12.39 and canceling v_S gives

$$dB = 20 \log \frac{1}{\sqrt{2}} = -3.01 \text{ dB}$$

Now that you know about f_L, f_H, midband, and dBs, let us summarize what is going on in a circuit at f_L and f_H. Obviously, if $R_{ckt} = X_C$ then the voltage loss across R_{ckt} must be equal to the voltage loss across X_C as shown in Figure 12.11. Further, if you recall your work with phasors in your circuits class, you recognize that the phase angle for the circuit is 45° and the output voltage is reduced to $1/\sqrt{2}$ times its midband value. As we have just seen, that is equivalent to a drop of about -3 dB. At the cutoff frequency we often say that the output is "down by 3 dB." Finally, the power delivered to the load at f_L is half its midband value. In fact, the cutoff frequencies are often referred to as the "half power points" for this reason.

12.13 Bode Plot

Since the dB concept lets us express circuit gain on a logarithmic scale, it would be convenient for us to express the frequency on a logarithmic scale too. That is especially true when the flat and uninteresting midband region extends over many orders of magnitude. If you remember your circuits class, that is precisely what a Bode plot does. It is a graph of dB on the vertical axis vs log f on the horizontal axis. Figure 12.12 is an idealized Bode plot for the circuit of Figure 12.10.

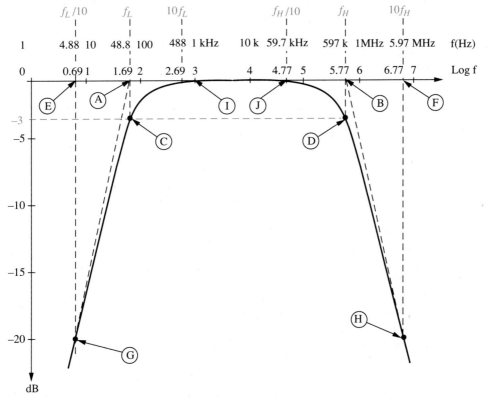

FIGURE 12.12 A Bode plot.

To construct a Bode plot of a circuit's frequency response, you must convert all voltage readings to dBs using Equation 12.39 and you must either use "semilog" graph paper, or else you must take the log of all frequencies.

An idealized Bode plot can be constructed from only two pieces of data. They are f_L and f_H. A good procedure is:

1. Locate both f_L and f_H on the horizontal axis (points A and B on Figure 12.12). Remember to find their logs if you are using linear graph paper. Next, locate points C and D by moving vertically down from points A and B a distance of -3 dB (the output is down by 3 dB at cutoff). These are the cutoff points and the Bode plot must pass through them.

2. Locate point E on the horizontal axis by going down in frequency from f_L by one "decade" (divide f_L by 10). Locate point F on the horizontal axis by going up in frequency from f_H by one decade (multiply f_H by 10). Next, locate points G and H by moving vertically down from points E and F a distance of -20 dB. (Equation 12.39 says that the slope of the graph, or the "roll-off" should be 20 dB per decade.) The Bode plot must pass through points E and F.

3. Locate point I on the horizontal axis by going up in frequency from f_L by one decade (multiply f_L by 10). Locate point J on the horizontal axis by going down in frequency from f_H by one decade (divide f_H by 10). The Bode plot joins the horizontal axis at these points which represent the end points of the midband.

This procedure is illustrated in Figure 12.12.

If you go to the lab, hook up the circuit of Figure 12.10, take frequency response data, and construct an experimental Bode plot, you will find that it compares reasonably well with an idealized Bode plot for the circuit.

12.14   Unbypassed Emitter Circuits

You will find that in many circuits only part of the emitter resistance is bypassed. This situation is illustrated by the circuit of Figure 12.13a. In this case, as we shall see shortly, A_v is much less than the fully bypassed circuit, R_{in} is much greater, and so is the complexity of the resulting equivalent circuit as shown in Figure 12.13b. The above changes are the result of the fact that, as we have mentioned on numerous occasions before, when you look from the base circuit across to the other side of the transistor, due to the fact that the currents are so much larger, the resistances look as if they are much larger—the Miller effect again.

First let us examine the effect on A_v of having an unbypassed R_E. We will take the same approach as we did in Section 12.8.

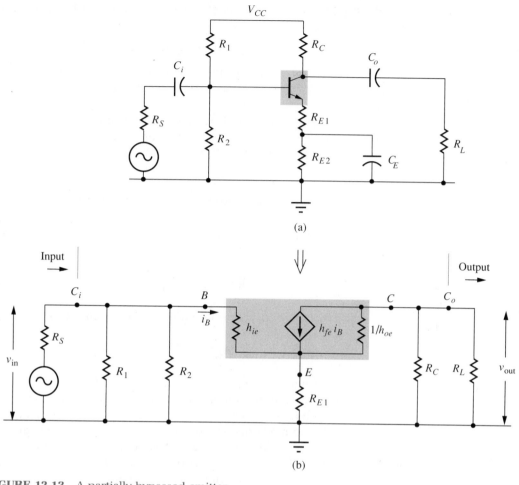

FIGURE 12.13 A partially bypassed emitter.
(a) Circuit diagram.
(b) Equivalent circuit.

From Section 12.8 we have

$$A_v \equiv \frac{v_{out}}{v_{in}} \qquad\qquad (12.17)$$

Looking first at the output circuit of Figure 12.13b, you will see that the presence of $1/h_{oe}$ complicates the situation a good deal. Although we could include it in our analysis, since $1/h_{oe}$ is very large for most modern transistors, ignoring it will introduce no significant error. Thus whenever we have to deal with unbypassed emitter circuits, we will ignore $1/h_{oe}$. We can then write

$$v_{\text{out}} \approx -h_{fe}i_B(R_C \| R_L) \tag{12.42}$$

You will notice that Equation 12.42 is essentially the same as Equation 12.18, except for $1/h_{oe}$. Again the negative sign simply indicates a reversal of phase.

Looking at the input circuit we see quite a different situation. Using the same approach as we used in the development of Equation 12.19, with reference to the input circuit of Figure 12.13b we can write

$$v_{\text{in}} = i_B h_{ie} + h_{fe}i_B R_{E1} \tag{12.43}$$

Notice that Equation 12.43 is quite different from Equation 12.19 in Section 12.7.

Substituting Equations 12.42 and 12.43 in Equation 12.17 and cancelling i_B from both numerator and denominator, we have

$$A_v = -\frac{h_{fe}(R_C \| R_L)}{h_{ie} + h_{fe}R_{E1}} \tag{12.44}$$

Notice that if $h_{ie} \ll h_{fe}R_{E1}$ then Equation 12.44 simplifies to

$$A_v \approx -\frac{R_C \| R_L}{R_{E1}} \tag{12.45}$$

Again the negative signs simply indicate a reversal of phase.

The approximation of Equation 12.45 eliminates the uncertainty of the transistor characteristics from the determination of amplifier gain. Distortion is also reduced since distortion is often the result of variation of transistor characteristics with either signal frequency or amplitude. Unfortunately, you will find that the approximation of Equation 12.45 cannot be used unless A_v is under about 10 for most transistors.

When a circuit is to be designed that is to have a given target voltage gain, a successful procedure is to go through the design procedure presented in Chapter 10 and then solve Equation 12.44 for the required value of R_{E1}. Then let R_{E2} make up the remainder of the R_E value required by the design procedure of Chapter 10.

As you can see from Figure 12.13, R_{out} is not changed appreciably by the introduction of R_{E1} due to the fact that $1/h_{oe}$ is very large. Thus the value of C_o changes very little. However, as we have already suggested, the input circuit is quite different.

Our approach for determining the value of R_{in} will be quite similar to that used in Section 12.3, except that this time as we look into the input, we will see h_{ie} in series with the unbypassed emitter resistor, R_{E1}. And do not forget that R_{E1} will look h_{fe} times bigger than it really is since we are looking from the base circuit to the emitter. Thus

$$R_{\text{in}} = R_1 \| R_2 \| (h_{ie} + h_{fe}R_{E1}) \tag{12.46}$$

Obviously, C_i will not need to be nearly as big as it was in Section 12.4 but, using Equation 12.46 to determine R_{in}, Equation 12.6 will still apply.

The equation required to calculate the value of C_E for the circuit of Figure 12.13 will be a bit more complicated than Equation 12.15 which was developed in Section 12.7 but the approach will be the same. Again, in order to visualize the circuit of which C_E is a part, think about how a charge might get from the positive plate of C_E around to the negative plate. This gives

$$R_{ckt} = R_{E2} \left\| \left(R_{E1} + \frac{h_{ie} + R_1 \| R_2 \| R_S}{h_{fe}} \right) \right. \tag{12.47}$$

Equation 12.15 now becomes

$$C_E = \frac{1}{2\pi f_L \left(R_{E2} \left\| \left(R_{E1} + \dfrac{h_{ie} + R_1 \| R_2 \| R_S}{h_{fe}} \right) \right. \right)} \tag{12.48}$$

EXAMPLE 12.12

Redesign the circuit of Example 12.7 so that it will have a voltage gain of 10.

SOLUTION:

The circuit of Example 12.7 is reproduced here in Figure 12.14a. According to Example 12.1, $h_{ie} = 651\ \Omega$ so, solving Equation 12.44 for R_{E1}, we have

$$R_{E1} = \frac{R_C \| R_L}{A_v} - \frac{h_{ie}}{h_{fe}} \tag{12.49}$$

$$= \frac{1.2\ \text{k}\Omega \| 1\ \text{k}\Omega}{10} - \frac{651\ \Omega}{135} = 49.7\ \Omega \qquad (\text{Use } 47\ \Omega)$$

In order to have a stable circuit, R_E must remain unchanged so

$$R_{E2} = R_E - R_{E1} \tag{12.50}$$

$$= 390\ \Omega - 47\ \Omega = 343\ \Omega \qquad (\text{Use } 330\ \Omega)$$

In both cases above the closest standard 10 percent resistor is specified. The new circuit is drawn in Figure 12.14b.

R_{in} now becomes

$$R_{in} = R_1 \| R_2 \| (h_{ie} + h_{fe}R_{E1}) \tag{12.46}$$

$$= 27\ \text{k}\Omega \| 6.8\ \text{k}\Omega \| (651\ \Omega + 135 * 47\ \Omega) = 3.06\ \text{k}\Omega$$

Notice that R_{in} is now considerably larger than the 581 Ω value obtained for the fully emitter bypassed circuit of Example 12.1.

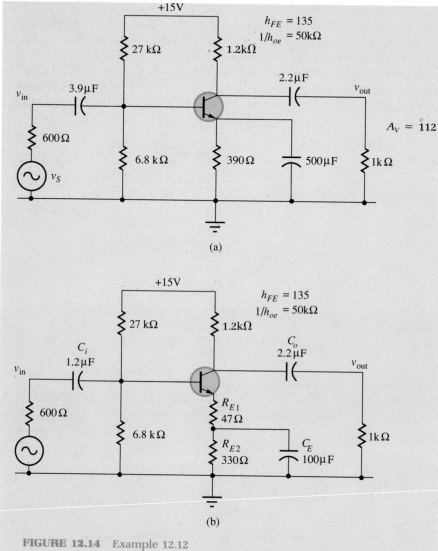

FIGURE 12.14   Example 12.12
(a) The circuit copied from Example 12.7.
(b) Revised circuit.

If we use the design target of $f_L = 35$ Hz as was the case in Example 12.2, then

$$C_i = \frac{1}{2\pi f_L(R_S + R_{in})} \tag{12.6}$$

$$= \frac{1}{2\pi * 35 \text{ Hz}(600\ \Omega + 3.06\ \text{k}\Omega)} = 1.24\ \mu\text{F} \qquad (\text{Use } 1.2\ \mu\text{F})$$

This compares to $C_i = 3.9\ \mu\text{F}$ calculated in Example 12.2 for the fully emitter bypassed circuit.

Looking at the equivalent circuit of Figure 12.13b we can adapt Equation 12.7 to express the output impedance for this circuit as

$$R_\text{out} = R_C \parallel (1/h_{oe} + R_{E1})$$
$$= 1.2\ \text{k}\Omega \parallel (50\ \text{k}\Omega + 47\ \Omega) = 1.17\ \text{k}\Omega \qquad (12.51)$$

To three figures this is the same result as was obtained in Example 12.3. This supports the claim made earlier that R_{E1} has no noticeable effect on the circuit output resistance. Thus the value of the output coupling capacitor C_o will not change either.

The value of C_E is given by

$$C_E = \cfrac{1}{2\pi f_L \left(R_{E2} \left\| \left(R_{E1} + \cfrac{h_{ie} + R_2 \parallel R_1 \parallel R_S}{h_{fe}} \right) \right.\right)} \qquad (12.48)$$

$$= \cfrac{1}{2\pi * 35\ \text{Hz} \left(330\ \Omega \left\| \left(47\ \Omega + \cfrac{651\ \Omega + 6.8\ \text{k}\Omega \parallel 27\ \text{k}\Omega \parallel 600\ \Omega}{135} \right) \right.\right)}$$

$$= 95.2\ \mu\text{F} \qquad\qquad (\text{Use } 100\ \mu\text{F})$$

The fact that R_{E1} is in series with the resistances seen in the base circuit causes C_E to be much lower than that calculated for the fully bypassed circuit of Example 12.4.

You will find that the unbypassed emitter circuit can be considered as sort of a transition circuit, having some of the characteristics of the common emitter circuit studied in this chapter and some of the characteristics of the emitter follower circuit to be studied in Chapter 13.

12.15 Circuit Design and Analysis Recipes

Chapters 8, 9, 10, and 12 contain all of the information needed to design or analyze most simple transistor amplifier stages. In fact, if you have really mastered the material presented in these chapters, you are well on your way to understanding electronics.

At this point, for your convenience, let us collect the important equations used and place them in two tables. Table 12.1 is valid for circuit

TABLE 12.1   Common Emitter Amplifier Design Procedure

STD. Choose nearest standard resistor value.

1. $R_C \stackrel{*}{=} R_L \parallel (-1/h_{oe})$ · · · STD. (12.12)

2. $R_E \stackrel{*}{=} \dfrac{R_{DC}}{4} = \dfrac{R_C}{3}$ · · · (STD. if step 10 is omitted)   (10.15)

3. $I_C \stackrel{*}{=} \dfrac{V_{CC}}{R_{DC} + R_{AC}} = \dfrac{V_{CC}}{R_C + R_E + R_C \parallel R_L}$ · · · (9.25)

4. $V_{BE} \approx 26 \text{ mV} * \ln\dfrac{I_C}{10 \text{ fA}}$ · · · (8.4)

5. $V_E = I_C R_E$ · · · (10.18)

6. $V_B = V_{BE} + V_E$ · · · (10.19)

7. $V_C = V_{CC} - I_C R_C$ · · · (10.5)

8. $I_B = \dfrac{I_C}{h_{FE}}$ · · · (9.7)

9. $h_{ie} = \dfrac{26 \text{ mV}}{I_B}$ · · · (11.3)

10. $R_{E1} = \dfrac{R_C \parallel R_L}{A_v} - \dfrac{h_{ie}}{h_{fe}}$ · · · STD. (12.49)

11. $R_{E2} = R_E - R_{E1}$ · · · STD. (12.50)

12. $R_B = \dfrac{S}{1 - S} h_{FE} R_E \stackrel{*}{\approx} \dfrac{1}{10} h_{FE} R_E$ · · · (10.16)

13. $V_{BB} = V_B + I_B R_B$ · · · (10.20)

14. $R_1 = R_B \dfrac{V_{CC}}{V_{BB}}$ · · · STD. (10.27)

15. $R_2 = R_1 \dfrac{V_{BB}}{V_{CC} - V_{BB}}$ · · · STD. (10.29)

16. $R_{in} = R_1 \parallel R_2 \parallel (h_{ie} + h_{fe} R_{E1})$ · · · (12.46)

17. $R_{out} = R_C \parallel (1/h_{oe} + R_{E1})$ · · · (12.51)

18. $C_i = \dfrac{1}{2\pi f_L (R_S + R_{in})}$ · · · (12.6)

19. $C_o = \dfrac{1}{2\pi f_L (R_{out} + R_L)}$ · · · (12.13)

20. $C_E = \dfrac{1}{2\pi f_L \left(R_{E2} \parallel \left(R_{E1} + \dfrac{h_{ie} + R_1 \parallel R_2 \parallel R_S}{h_{fe}} \right) \right)}$ · · · (12.48)

*Designer's formula only.

design, and Table 12.2 is valid for circuit analysis. Figure 12.15 contains the circuits and load lines relevant to both tables. You should not try to memorize these tables. Your time would be more profitably spent studying the electronics involved in their development.

TABLE 12.2   Common Emitter Amplifier Analysis Procedure

1. $V_{BB} = V_{CC}\dfrac{R_2}{R_1 + R_2}$ $\qquad$ (10.22)

2. $R_B = R_1 \parallel R_2$ $\qquad$ (10.23)

3. $S = \dfrac{R_B}{R_B + h_{FE}R_E}$ $\qquad$ (10.10)

4. $I_C = \dfrac{V_{BB} - V_{BE}}{\dfrac{R_B}{h_{FE}} + R_E} = $ | | | . mA $\qquad$ (10.12)

5. $V_{BE} = 26 \text{ mV} * \ln\dfrac{I_C}{10 \text{ fA}} = $ | | | V $\qquad$ (8.4)

6. $V_E = I_C R_E$ $\qquad$ (10.18)
7. $V_B = V_{BE} + V_E$ $\qquad$ (10.19)
8. $V_C = V_{CC} - I_C R_C$ $\qquad$ (10.5)

9. $I_B = \dfrac{I_C}{h_{FE}}$ $\qquad$ (9.7)

10. $h_{ie} = \dfrac{26 \text{ mV}}{I_B}$ $\qquad$ (11.3)

11. $R_{in} = R_1 \parallel R_2 \parallel (h_{ie} + h_{fe}R_{E1})$ $\qquad$ (12.46)
12. $R_{out} = R_C \parallel (1/h_{oe} + R_{E1})$ $\qquad$ (12.51)

13. $A_v = \dfrac{-h_{fe}(1/h_{oe} \parallel R_C \parallel R_L)}{h_{ie} + h_{fe}R_{E1}}$ $\qquad$ (12.44)

14. $A_i = A_v\dfrac{R_{in}}{R_L}$ $\qquad$ (12.25)

15. $v_{\text{outPmax}} = \begin{Bmatrix} V_C - V_E \\ I_C R_{AC} \end{Bmatrix} \begin{Bmatrix} \text{whichever} \\ \text{is} \\ \text{smaller} \end{Bmatrix}$

16. $f_L = \begin{Bmatrix} \dfrac{1}{2\pi\, C_i(R_S + R_{in})} \\[2ex] \dfrac{1}{2\pi\, C_o(R_{out} + R_L)} \\[2ex] \dfrac{1}{2\pi\, C_E\left(R_{E2} \parallel \left(R_{E1} + \dfrac{h_{ie} + R_1 \parallel R_2 \parallel R_S}{h_{fe}}\right)\right)} \end{Bmatrix} \begin{Bmatrix} \text{whichever} \\ \text{is} \\ \text{larger} \end{Bmatrix}$

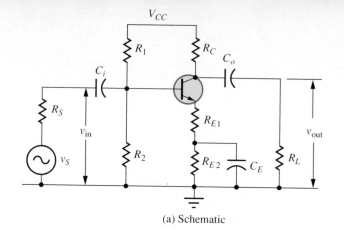

(a) Schematic

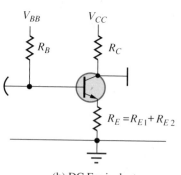

(b) DC Equivalent

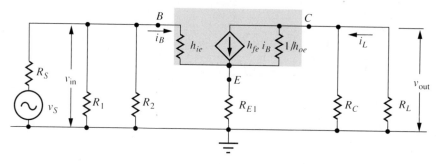

(c) AC Equivalent

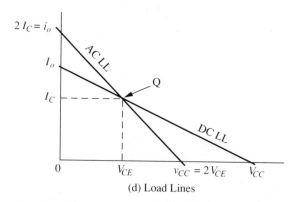

(d) Load Lines

FIGURE 12.15 Circuits used in the circuit design and the circuit analysis procedures.

EXAMPLE 12.13

Use the procedure of Table 12.1 to redesign the circuit of Example 12.7. Compare the gain of the resulting circuit with that of Example 12.7.

● See Appendix G for a SuperCalc solution.

SOLUTION:

From Example 12.7 we have

$$V_{CC} = 15 \text{ V} \qquad R_S = 600 \text{ }\Omega \qquad R_L = 1 \text{ k}\Omega$$
$$f_L = 35 \text{ Hz} \qquad h_{FE} = 135 \qquad 1/h_{oe} = 50 \text{ k}\Omega$$

Using the above values in the design procedure of Table 12.1, assuming that maximum gain is desired gives

1. $R_C \stackrel{*}{=} R_L \parallel (-1/h_{oe})$ STD. (12.12)

    $= 1 \text{ k}\Omega \parallel (-50 \text{ k}\Omega = 1.02 \text{ k}\Omega$ (Use 1 kΩ)

2. $R_E \stackrel{*}{=} \dfrac{R_{DC}}{4} = \dfrac{R_C}{3}$ (STD. if step 10 is omitted)   (10.15)

    $= \dfrac{1 \text{ k}\Omega}{3} = 333 \text{ }\Omega$ (Use 330 Ω)

Note: Since maximum gain is sought, all of R_E will be bypassed so R_{E1} and R_{E2} will not be used. Thus R_E is standardized.

3. $I_C \stackrel{*}{=} \dfrac{V_{CC}}{R_{DC} + R_{AC}} = \dfrac{V_{CC}}{R_C + R_E + R_C \parallel R_L}$ (9.25)

    $= \dfrac{15 \text{ V}}{1 \text{ k}\Omega + 330 \text{ }\Omega + 1 \text{ k}\Omega \parallel 1 \text{ k}\Omega} = 8.20 \text{ mA}$

4. $V_{BE} \approx 26 \text{ mV} * \ln \dfrac{I_C}{10 \text{ fA}}$ (8.4)

    $= 26 \text{ mV} * \ln \dfrac{8.20 \text{ mA}}{10 \text{ fA}} = 0.713 \text{ V}$

5. $V_E = I_C R_E$ (10.18)

    $= 8.20 \text{ mA} * 330 \text{ }\Omega = 2.70 \text{ V}$

6. $V_B = V_{BE} + V_E$ (10.19)

    $= 0.713 \text{ V} + 2.70 \text{ V} = 3.42 \text{ V}$

7. $V_C = V_{CC} - I_C R_C$ (10.5)

    $= 15 \text{ V} - 8.20 \text{ mA} * 1 \text{ k}\Omega = 6.80 \text{ V}$

8. $I_B = \dfrac{I_C}{h_{FE}}$ (9.7)

    $= \dfrac{8.20 \text{ mA}}{135} = 60.7 \text{ }\mu\text{A}$

9. $h_{ie} = \dfrac{26 \text{ mV}}{I_B}$ (11.3)

    $= \dfrac{26 \text{ mV}}{60.7 \text{ } \mu A} = 428 \text{ } \Omega$

10. $R_{E1} = 0 \text{ } \Omega$ (maximum gain is sought so all of R_E is bypassed)

11. $R_{E2} = R_E - R_{E1}$

    $= 330 \text{ } \Omega - 0 \text{ } \Omega = 330 \text{ } \Omega$

12. $R_B = \dfrac{S}{1-S} h_{FE} R_E \overset{*}{\approx} \dfrac{1}{10} h_{FE} R_E$ (10.16)

    $= \dfrac{1}{10} 135 * 330 \text{ } \Omega = 4.46 \text{ k}\Omega$

13. $V_{BB} = V_B + I_B R_B$ (10.20)

    $= 3.42 \text{ V} + 60.7 \text{ } \mu A * 4.46 \text{ k}\Omega = 3.69 \text{ V}$

14. $R_1 = R_B \dfrac{V_{CC}}{V_{BB}}$ STD. (12.27)

    $= 4.46 \text{ k}\Omega \dfrac{15 \text{ V}}{3.69 \text{ V}} = 18.1 \text{ k}\Omega$ (Use 18 kΩ)

15. $R_2 = R_1 \dfrac{V_{BB}}{V_{CC} - V_{BB}}$ STD. (10.29)

    $= 18 \text{ k}\Omega \dfrac{3.69 \text{ V}}{15 \text{ V} - 3.69 \text{ V}} = 5.87 \text{ k}\Omega$ (Use 5.6 kΩ)

16. $R_{in} = R_1 \parallel R_2 \parallel (h_{ie} + h_{fe} R_{E1})$ (12.46)

    $= 18 \text{ k}\Omega \parallel 5.6 \text{ k}\Omega \parallel (428 \text{ } \Omega + 135 * 0 \text{ } \Omega) = 389 \text{ } \Omega$

17. $R_{out} = R_C \parallel (1/h_{oe} + R_{E1})$ (12.51)

    $= 1 \text{ k}\Omega \parallel (50 \text{ k}\Omega + 0 \text{ } \Omega) = 980 \text{ } \Omega$

18. $C_i = \dfrac{1}{2\pi f_L (R_S + R_{in})}$ (12.6)

    $= \dfrac{1}{2\pi * 35 \text{ Hz } (600 \text{ } \Omega + 389 \text{ } \Omega)} = 4.60 \text{ } \mu F$ (Use 4.7 μF)

19. $C_o = \dfrac{1}{2\pi f_L (R_{out} + R_L)}$ (12.13)

    $= \dfrac{1}{2\pi * 35 \text{ Hz } (980 \text{ } \Omega + 1 \text{ k}\Omega)} = 2.3 \text{ } \mu F$ (Use 2.2 μF)

20. $C_E = \dfrac{1}{2\pi f_L \left(R_{E2} \left\| \left(R_{E1} + \dfrac{h_{ie} + R_1 \| R_2 \| R_S}{h_{fe}} \right) \right) \right)}$ (12.48)

$= \dfrac{1}{2\pi * 35 \text{ Hz} \left(330 \ \Omega \left\| \left(0 \ \Omega + \dfrac{428 \ \Omega + 18 \text{ k}\Omega \| 5.6 \text{ k}\Omega \| 600 \ \Omega}{135} \right) \right) \right)}$

$= 657 \ \mu\text{F}$ (Use 500 μF)

$A_v = -\dfrac{h_{fe}(1/h_{oe} \| R_C \| R_L)}{h_{ie} + h_{fe}R_{E1}}$ (12.20)

$= -\dfrac{135(50 \text{ k}\Omega \| 1 \text{ k}\Omega \| 1 \text{ k}\Omega)}{428 \ \Omega + 135 * 0 \ \Omega} = -156$

The value obtained in Example 12.7 is $A_v = -112$.

$A_i = A_v \dfrac{R_{\text{in}}}{R_L}$ (12.25)

$= -156\dfrac{389 \ \Omega}{1 \text{ k}\Omega} = -60.7$

The value obtained in Example 12.7 is $A_i = -65$.

$A_P = A_v A_i$ (12.29)

$= (-156)(-60.7) = 9470$

The value obtained in Example 12.7 is $A_P = 7280$

The improved gain values in Example 12.13 compared to Example 12.7 are due to better impedance matching.

12.16 Troubleshooting   ◄◄◄◄◄◄◄

It is possible for the insulation between the plates of a capacitor to break down and allow the capacitor to begin to conduct. The resulting leakage current can affect the Q point of the circuit. You should be alert to this possibility, especially when servicing older equipment. Another problem you will encounter is the loss of capacity with age. This may cause a loss of low frequency response or, according to Equation 12.44, a loss in signal gain. An easy way to test for this possibility is to "bridge" the suspected capacitor with a known good capacitor while monitoring the signal output.

12.17 Summary

In this chapter the results of the previous chapters were brought together and culminated in the circuit design recipe and the circuit analysis recipe presented in Section 12.15. If the approach presented is used thoughtfully, it will go a long way toward meeting your needs as a technologist. Consequently we will now turn our attention to the emitter follower circuit.

GLOSSARY

Bode plot: A frequency response diagram in which the gain or phase shift of a circuit is plotted as a function of frequency. (Section 12.13)

Decibel (dB):   A logarithmic unit used to express the ratio of one signal power level to another. (Section 12.12)

Gain: The ratio of the output of a circuit to its input. (Section 12.8)

Impedance (Z):   Total opposition to the flow of current, either AC or DC. The vector sum of the circuit resistance plus reactance. (Section 12.3)

PROBLEMS

Section 12.3

1. Given the circuit of Figure 12.16. If the DC collector current is $I_C = 4$ mA, find:
   (a)   The input impedance of the amplifier.
   (b)   The value of C_i if $f_L = 20$ Hz.

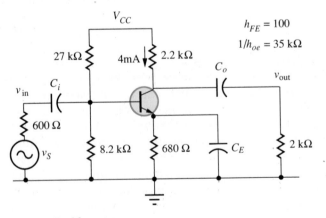

FIGURE 12.16   Problem 1.

Section 12.4

2. Given the circuit of Figure 12.17. If the DC collector voltage is $V_C = 5.25$ VDC, and the open circuit voltage of the signal source is $v_S = 25$ mVPP, find:

   (a)   The signal voltage v_{in} measured at the input to the amplifier.

   (b)   The value of the input coupling capacitor C_i required for a low frequency cutoff of $f_L = 20$ Hz. Specify the nearest standard value.

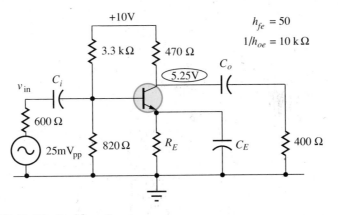

FIGURE 12.17   Problem 2.

3. When the circuit of Figure 12.16 is hooked up it is found that if v_S is set to $v_S = 10$ mVPP before the signal generator is connected, after it is connected, $v_{in} = 6$ mVPP. Based on this evidence, what must be the value of h_{ie}?

4. When the circuit of Figure 12.17 is hooked up using $C_i = 10$ μF, it is found that $f_L = 25$ Hz. Based on this evidence, assuming that C_E and C_o are large, what must be the value of h_{ie}?

Section 12.6

5. Given the circuit of Figure 12.16, find:

   (a)   The output impedance, R_{out} of the amplifier.

   (b)   The value of the output coupling capacitor C_o that will give $f_L = 20$ Hz. Specify the nearest standard value.

6. Given the circuit of Figure 12.17, find:

   (a)   The output impedance, R_{out} for the amplifier.

   (b)   The value of the output coupling capacitor C_o that will give $f_L = 20$ Hz. Specify the nearest standard value.

Section 12.7

7. Given the circuit of Figure 12.16, find the value of the emitter bypass capacitor C_E that will give $f_L = 20$ Hz. Specify the nearest standard value.

Section 12.8

8. Find the voltage gain, the current gain, and the power gain for the circuit of Figure 12.16.
9. Find the voltage delivered to the load by the circuit of Figure 12.17.
10. Find the voltage gain, the current gain, and the power gain for the circuit of Figure 12.18.

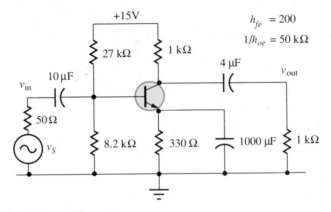

FIGURE 12.18   Problem 10.

Section 12.9

11. Find the low frequency cutoff, f_L, for the circuit of Figure 12.18.
12. Change the value of the capacitors in Figure 12.18 so that they will all have the same value of f_L as that calculated in Problem 11. Specify the nearest standard values.

Section 12.10

13. If the transistor of Figure 12.16 has capacitances of $C_{ib} = 25$ pF and $C_{ob} = 5$ pF, find the high frequency cutoff, f_H.
14. If the transistor of Figure 12.17 has capacitances of $C_{ib} = 25$ pF and $C_{ob} = 5$ pF, find the high frequency cutoff, f_H.

Section 12.12

15. A certain stereo requires a 100 mV signal into its 100 kΩ record player input and delivers 35 W to an 8 Ω loudspeaker. Find its voltage gain, its current gain, and its power gain. Express the power gain in dB.
16. The stereo in Problem 15 claims a channel separation of 25 dB. When it is delivering 35 W of signal to the left speaker, how many volts of that signal would you expect to see across the right speaker?

Section 12.13

17. Construct an idealized Bode plot for a "high fidelity" amplifier ($f_L = 20$ Hz, $f_H = 20,000$ Hz).
18. The frequency response data for a certain amplifier stage is contained in Table 12.3. Construct a Bode plot from these data and estimate f_L and f_H for the amplifier.

TABLE 12.3   Frequency Response

Data Index	F Hz	V_{out} V
1	10	.02
2	27	.08
3	86	.54
4	153	.9
5	233	1.1
6	402	1.2
7	680	1.36
8	1314	1.4
9	3027	1.4
10	6140	1.44
11	12080	1.46
12	20220	1.46
13	30410	1.46
14	47300	1.46
15	101300	1.44
16	211000	1.44
17	299200	1.4
18	398900	1.34
19	642900	1.24
20	1064300	1.04
21	2036600	.66
22	3010500	.47
23	3991500	.34

Section 12.14

19. Alter the circuit of Problem 1 so that it will have a voltage gain of approximately $A_v = 5$. (Find the new values for R_{E1}, R_{E2}, C_E, and C_i).
20. Alter the circuit of Problem 1 so that it will have a voltage gain of approximately $A_v = 5$. (Find the new values for R_{E1}, R_{E2}, C_E, and C_i).

Section 12.15

21. Using a 2N4401 transistor (specifications are in Appendix F), design and specify all components (use nearest standard values) for a fully

bypassed common emitter amplifier using voltage divider biasing. Plan for R_L = 1.5 kΩ, f_L = 30 Hz, S = 5%, and V_{CC} = 9 V. Assume that the signal source for the amplifier has an internal resistance of R_S = 100 Ω.

22. Redesign the circuit of Problem 21 so that it will have a voltage gain of approximately A_v = 5.

23. Do a complete analysis of the circuit of Figure 12.18. Determine the maximum open circuit peak-to-peak voltage the signal source v_S can have before clipping will be observed in the output.

24. Do a complete analysis of the circuit of Figure 12.19.

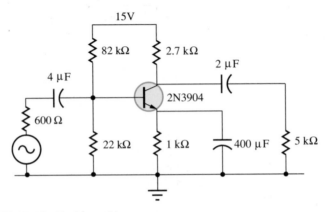

FIGURE 12.19   Problem 24.

Troubleshooting ▶   **Section 12.16**

25. If the emitter bypass capacitor in Figure 12.19 became an open circuit, determine the effect on:
    (a)   The Q point.
    (b)   The AC signal.

26. Repeat Problem 25 if the emitter bypass capacitor became a dead short.

27. Determine the effect on the Q point of the circuit in Figure 12.19 if the output coupling capacitor became a dead short.

Three terminal regulators mounted on a circuit
board. (Photograph by Joseph Nettis/Stock, Boston.)

Chapter 13
Emitter Follower Circuits

SPECIAL TERMS

Follower circuit
Darlington circuit
Series pass transistor

13.1   Introduction and Objectives

This chapter is a logical extension of Section 12.14. There you got a glimpse of what happens when the emitter is not connected to the AC ground. We will expand on that topic in this chapter by studying what is known as the "common collector," or the "emitter follower" circuit. Although a common collector h parameter model has been developed, since the data books usually supply the common emitter parameters, and since you are already familiar with the common emitter h parameter model, we will adapt it to the emitter follower circuit.

After having studied this chapter you should understand the more important characteristics of the emitter follower circuit and in the process you will have opportunity to further develop your ability to design and analyze transistor circuits. You should be able to:

- Adapt the design and analysis recipes we developed in Chapter 12 to the common emitter circuit and use them to design and analyze common emitter amplifier circuits.
- Draw the AC equivalent circuit for the emitter follower circuit using common emitter h parameters and use it to determine the circuit input and output impedances and voltage gain.
- Draw a Darlington circuit and its equivalent circuit using common emitter h parameters.
- Design and analyze circuits containing Darlington circuits.
- Design and analyze regulator circuits containing a series pass transistor.

13.2   Using Common Emitter Parameters

Figure 13.1a contains a typical common collector circuit which is also known by the descriptive name "emitter follower." Notice that there is no collector resistor. The collector is connected directly to the power supply. Of course, you remember that the power supply acts like an AC ground (see Section 11.9 if you are unsure of this point). That is where the name "Common Collector" comes from. The collector is "common."

Further, you will notice in Figure 13.1a that the signal output is from the emitter. Since the base to emitter voltage is nearly constant at $V_{BE} \approx$ 0.7 V, if the base voltage V_B increases, the emitter voltage V_E will increase too in order to keep V_{BE} nearly constant. On the other hand, if V_B decreases, V_E will also decrease. The emitter voltage follows the base voltage. Thus the name "Emitter Follower." The circuit in Figure 13.1a is sometimes called a *common collector circuit*, and it is sometimes called an *emitter follower circuit*. You should recognize it by either name; both are widely used. Its identifying characteristic is that the signal is taken from the emitter terminal of the transistor.

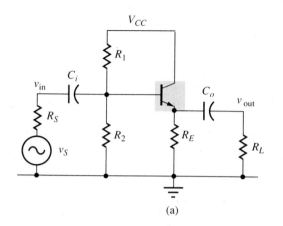

(a)

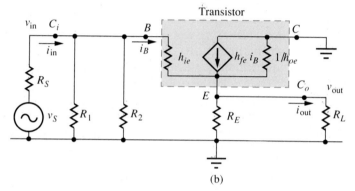

(b)

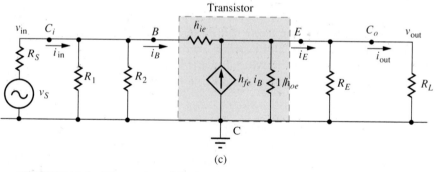

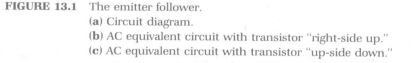

(c)

FIGURE 13.1   The emitter follower.
(a) Circuit diagram.
(b) AC equivalent circuit with transistor "right-side up."
(c) AC equivalent circuit with transistor "up-side down."

Some authors introduce common collector h parameters at this point (h_{ic}, h_{fc}, etc.). Since they are specifically intended to describe this circuit, their use should simplify the circuit description. However, the common emitter h parameters are almost as easy to use. Since you are already familiar with them, and since the manufacturers' data sheets list only them, we will use h_{ie}, h_{fe}, and $1/h_{oe}$ for our AC equivalent circuit for the emitter follower. Figure 13.1b is the most straightforward way to draw the AC equivalent circuit. The difficulty with Figure 13.1b is that it leaves the grounded collector inelegantly dangling in space. Figure 13.1c is a much neater way to draw the equivalent circuit. If you look carefully you will see that Figure 13.1c is the same circuit as Figure 13.1b, except that the transistor has been "turned upside down." After all, if the collector is at AC ground, why not turn it around so that it looks like it is grounded?

The circuit of Figure 13.1c emphasizes some very important characteristics of the emitter follower circuit that make it useful to the circuit designer. For one thing, it indicates that if the base current i_B is to flow in the direction indicated, then v_{out} must be less than v_{in}. The circuit must have a voltage gain of less than one. In fact, as you can see from Figure 13.1c, the output will always be less than the input by the amount $i_B h_{ie}$.

Second, as the load looks back into the circuit it will see not only R_E and $1/h_{oe}$, but it will also see h_{ie} and R_1, R_2, and R_S. Do not forget that the combination of h_{ie}, R_1, R_2, and R_S will be divided by h_{fe} because it is on the base side of the transistor. That will mean that the output resistance will be low.

Third, from Figure 13.1c you can see that since h_{ie} no longer goes directly to AC ground, the input resistance will be higher than it was for the common emitter circuit. This is not a lot different from what we encountered in Section 12.14.

Finally, since the transistor is turned upside down in Figure 13.1c, the current $h_{fe}i_B$ now flows downward toward ground rather than upward away from ground as it did in the common emitter circuit. That means that we no longer have phase inversion. The current flowing to the load is in phase with the current from the source. That is indicated in Figure 13.1c by the fact that both the current from the source and the current to the load are pictured as flowing toward the right. We will look more closely at these characteristics later in this chapter.

13.3 Circuit Design

The circuit design procedure introduced in Table 12.1 is easily adapted to the emitter follower circuit. Just remember that in this case, $R_C = 0\ \Omega$, $R_{E2} = 0\ \Omega$, and the output is being taken from the emitter. Obviously then the first couple of steps in Table 12.1 have to be modified. In fact, step 1 can be eliminated; $R_C = 0\ \Omega$. Since the output is being taken from the emitter, it seems reasonable that Step 2 should be

2. $R_E = R_L$ (13.1)

We will examine the above assumption later.

The basic idea behind Step 3 of the design procedure does not change. Since we still want the Q point in the center of the AC load line

$$I_C \stackrel{*}{=} \frac{V_{CC}}{R_{DC} + R_{AC}} \tag{9.25}$$

Looking at the circuit of Figure 13.1 we see that $R_{DC} = R_E$ and $R_{AC} = R_E \parallel R_L$. Step 3 of the design procedure then becomes

$$3.\ I_C \stackrel{*}{=} \frac{V_{CC}}{R_E + R_E \parallel R_L} \tag{13.2}$$

Steps 4 through 6 remain unchanged.

Step 7 can obviously be eliminated since $V_C = V_{CC}$.

Steps 8 and 9 remain unchanged.

Steps 10 and 11 are eliminated since there is only one emitter resistor and it is R_{E1} as far as the equations are concerned.

Steps 12 through 15 are unchanged.

Looking at the equivalent circuit of Figure 13.1c, Step 16 becomes

$$16.\ R_{in} = R_1 \parallel R_2 \parallel \left(h_{ie} + h_{fe}(1/h_{oe} \parallel R_E \parallel R_L) \right) \tag{13.3}$$

Again looking at the equivalent circuit of Figure 13.1c, Step 17 becomes

$$17.\ R_{out} = R_E \parallel 1/h_{oe} \parallel \frac{h_{ie} + R_2 \parallel R_1 \parallel R_S}{h_{fe}} \tag{13.4}$$

Steps 18 and 19 remain unchanged.

Finally, Step 20 is obviously eliminated since there is no bypass capacitor.

EXAMPLE 13.1

Design an emitter follower circuit to drive a 1 kΩ load using a 10 VDC power supply and a 2N4401 transistor. Assume $R_S = 600\ \Omega$ and $f_L = 20$ Hz.

SOLUTION:

From Appendix F, $h_{fe} = 135$, $1/h_{oe} = 1/30\ \mu S = 33.3$ kΩ

$$2.\ R_E = R_L \tag{13.1}$$
$$= 1\ \text{k}\Omega \qquad\qquad (\text{Use }1\,\text{k}\Omega)$$

3. $I_C = \dfrac{V_{CC}}{R_E + R_E \parallel R_L}$ (13.2)

   $= \dfrac{10 \text{ V}}{1 \text{ k}\Omega + 1 \text{ k}\Omega \parallel 1 \text{ k}\Omega} = 6.67 \text{ mA}$

4. $V_{BE} \approx 26 \text{ mV} * \ln\dfrac{I_C}{10 \text{ fA}}$ (8.4)

   $= 26 \text{ mV} * \ln\dfrac{6.67 \text{ mA}}{10 \text{ fA}} = 0.708 \text{ V}$

5. $V_E = I_C R_E$ (10.18)

   $= 6.67 \text{ mA} * 1 \text{ k}\Omega = 6.67 \text{ V}$

6. $V_B = V_{BE} + V_E$ (10.19)

   $= 0.708 \text{ V} + 6.67 \text{ V} = 7.37 \text{ V}$

7. $V_C = V_{CC} = 10 \text{ V}$

8. $I_B = \dfrac{I_C}{h_{FE}}$ (9.7)

   $= \dfrac{6.67 \text{ mA}}{135} = 49.5 \text{ }\mu\text{A}$

9. $h_{ie} = \dfrac{26 \text{ mV}}{I_B}$ (11.3)

   $= \dfrac{26 \text{ mV}}{49.4 \text{ }\mu\text{A}} = 527 \text{ }\Omega$

12. $R_B = \dfrac{S}{1 - S} h_{FE} R_E \overset{*}{\approx} \dfrac{1}{10} h_{FE} R_E$ (10.16)

   $= \dfrac{135 * 1 \text{ k}\Omega}{10} = 13.5 \text{ k}\Omega$

13. $V_{BB} = V_B + I_B R_B$ (10.17)

   $= 7.37 \text{ V} + 49.4 \text{ }\mu\text{A} * 13.5 \text{ k}\Omega = 8.04 \text{ V}$

14. $R_1 = R_B \dfrac{V_{CC}}{V_{BB}}$ STD. (10.27)

   $= 13.5 \text{ k}\Omega \dfrac{10 \text{ V}}{8.04 \text{ V}} = 16.8 \text{ k}\Omega$ (Use 18 kΩ)

15. $R_2 = R_1 \dfrac{V_{BB}}{V_{CC} - V_{BB}}$ STD. (10.29)

   $= 18 \text{ k}\Omega \dfrac{8.04 \text{ V}}{10 \text{ V} - 8.04 \text{ V}} = 73.9 \text{ k}\Omega$ (Use 68 kΩ)

16. $R_{in} = R_1 \parallel R_2 \parallel \left(h_{ie} + h_{fe}(1/h_{oe} \parallel R_E \parallel R_L) \right)$ (13.3)

   $= 18 \text{ k}\Omega \parallel 68 \text{ k}\Omega \parallel \left(527 \text{ }\Omega + 135(33.3 \text{ k}\Omega \parallel 1 \text{ k}\Omega \parallel 1 \text{ k}\Omega) \right)$

   $= 11.7 \text{ k}\Omega$

17. $R_{\text{out}} = R_E \parallel 1/h_{oe} \parallel \dfrac{h_{ie} + R_2 \parallel R_1 \parallel R_S}{h_{fe}}$ (13.4)

$= 1 \text{ k}\Omega \parallel 33.3 \text{ k}\Omega \parallel \dfrac{527 \ \Omega + 68 \text{ k}\Omega \parallel 18 \text{ k}\Omega \parallel 600 \ \Omega}{135}$

$= 8.10 \ \Omega$

18. $C_i = \dfrac{1}{2\pi f_L (R_S + R_{\text{in}})}$ (12.6)

$= \dfrac{1}{2\pi * 20 \text{ Hz}(600 \ \Omega + 11.7 \text{ k}\Omega)}$ (Use 0.68 µF)

$= 0.645 \ \mu F$

19. $C_o = \dfrac{1}{2\pi f_L (R_{\text{out}} + R_L)}$ (12.13)

$= \dfrac{1}{2\pi * 20 \text{ Hz}(8.10 \ \Omega + 1 \text{ k}\Omega)} = 7.89 \ \mu F$ (Use 8 µF)

Figure 13.2 contains the circuit.

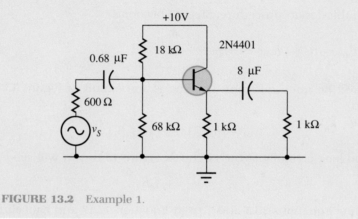

FIGURE 13.2 Example 1.

The preceding example illustrates two of the valuable characteristics of the emitter follower circuit that we mentioned in Section 13.2. Notice the value of R_{in}. The common emitter circuit typically has an input impedance around 1 kΩ. This circuit easily achieves an input impedance at least an order of magnitude greater than that. With a little effort we could push it another decimal place higher. There was a time when this was one of the most valuable characteristics of the emitter follower. Today the field effect transistor (FET) has largely taken over the task of providing high input impedance. We will study FETs in Part III of this book.

The second valuable characteristic of the emitter follower illustrated in Example 13.1 is its low output impedance. It is typically two orders of magnitude lower than the output impedance of a common emitter circuit. This makes the emitter follower valuable for driving low impedance loads such as the loud speaker of your stereo. In Chapter 15 you will learn about that application of the emitter follower circuit.

13.4   Circuit Analysis

As was the case with circuit design, the analysis of an existing emitter follower circuit involves rather obvious adaptations of the ideas presented in Chapter 12. All that is required is an understanding of Table 12.2 and the equivalent circuit of Figure 13.1c. Again, remember that $R_{DC} = R_E$ and $R_{AC} = R_E \| R_L$. Looking at Table 12.2, Steps 1 through 10 are unchanged with the obvious simplification of Step 8: $V_C = V_{CC}$.

From the design procedure, Step 11 becomes

$$11.\ R_{in} = R_1 \| R_2 \| \left(h_{ie} + h_{fe}(1/h_{oe} \| R_E \| R_L) \right) \tag{13.3}$$

Also from the design procedure, Step 12 becomes

$$12.\ R_{out} = R_E \| 1/h_{oe} \left\| \frac{h_{ie} + R_2 \| R_1 \| R_S}{h_{fe}} \right. \tag{13.4}$$

Step 13 will be quite different. Looking at the circuit of Figure 13.1c, we can write

$$v_{out} = h_{fe}i_B(1/h_{oe} \| R_E \| R_L) \tag{13.5}$$

and if you look carefully at the circuit of Figure 13.1c you will see that

$$v_{in} = v_{out} + i_B h_{ie} \tag{13.6}$$

Substituting Equations 13.5 and 13.6 in Equation 12.17 and canceling i_B as we did in Section 12.7, we have

$$13.\ A_v = \frac{h_{fe}(1/h_{oe} \| R_E \| R_L)}{\text{NUM.} + h_{ie}} \tag{13.7}$$

Notice that the letters NUM (NUMerator) are used to stand for the entire expression in the numerator. That will save us a lot of symbol writing, and it will also emphasize the fact that the denominator contains everything in the numerator plus it contains h_{ie}.

Equation 13.7 supports the statement made in Section 13.2 that the voltage gain must be less than one. In fact, as you can see from either Figure 13.1c or from Equation 13.7, except for h_{ie}, the voltage gain would be exactly one.

Steps 14, 15, and 16 remain unchanged with the obvious deletion of C_E in Step 16.

The estimation of the high frequency cutoff, f_H, will be very much different from what it was in Section 12.10 although our approach will be much the same. Figure 13.3 contains the equivalent circuit with C_{ib} and C_{ob} added. If you compare Figure 13.3 with Figure 12.9 you will see that C_{ib} and C_{ob} have essentially exchanged places. First considering the effect of C_{ob} alone we have

$$f_H = \frac{1}{2\pi\ C_{ob}(R_2 \parallel R_1 \parallel R_S \parallel (h_{ie} + 1/h_{oe} \parallel R_E \parallel R_L))} \tag{13.8}$$

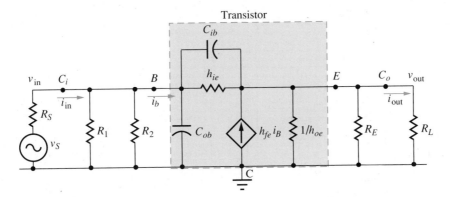

FIGURE 13.3   High frequency AC equivalent circuit.

Next considering the effect of C_{ib} alone, using the same arguments that we used in obtaining Equation 12.35, we have

$$f_H = \frac{1}{2\pi\ C_{ib}A_v(h_{ie} \parallel (R_2 \parallel R_1 \parallel R_S + 1/h_{oe} \parallel R_E \parallel R_L))} \tag{13.9}$$

Of course, the observed high frequency cutoff will be the smallest of the values obtained from Equations 13.8 and 13.9. In fact, it will probably be a little less than that, since wire and resistor capacitances are not considered in these equations.

EXAMPLE 13.2

Analyze the circuit of Figure 13.2 and compare with Example 13.1.

● (For a PSpice solution see Appendix H.)

SOLUTION:
Adapting Table 12.2 gives

1. $V_{BB} = V_{CC}\dfrac{R_2}{R_1 + R_2}$ 　　　　　　　　　　　　　(10.22)

$= 10\text{ V}\dfrac{68\text{ k}\Omega}{18\text{ k}\Omega + 68\text{ k}\Omega} = 7.91\text{ V}$

2. $R_B = R_1 \parallel R_2$ 　　　　　　　　　　　　　　　　　(10.23)

$= 18\text{ k}\Omega \parallel 68\text{ k}\Omega = 14.2\text{ k}\Omega$

3. $S = \dfrac{R_B}{R_B + h_{FE}R_E}$ 　　　　　　　　　　　　　(10.10)

$= \dfrac{14.2\text{ k}\Omega}{14.2\text{ k}\Omega + 135 * 1\text{ k}\Omega} = 0.0954$

$= 9.54\%$ 　　　　　　　　　　　(Design: 10%)

4. $I_C = \dfrac{V_{BB} - V_{BE}}{\dfrac{R_B}{h_{FE}} + R_E}$ 　　　　　　　　　　(10.12)

$= \dfrac{7.91\text{ V} - V_{BE}}{\dfrac{14.2\text{ k}\Omega}{135} + 1\text{ k}\Omega}$

	6.52	6.51	mA

5. $V_{BE} = 26\text{ mV} * \ln\dfrac{I_C}{10\text{ fA}} =$

	0.707	0.707	V	(8.4)

6. $V_E = I_C R_E$ 　　　　　　　　　　　　　　　　　　(10.18)

$= 6.51\text{ mA} * 1\text{ k}\Omega = 6.51\text{ V}$ 　　　(Design: 6.67 V)

7. $V_B = V_{BE} + V_E$ 　　　　　　　　　　　　　　　　(10.19)

$= 0.707\text{ V} + 6.51\text{ V} = 7.22\text{ V}$ 　　(Design: 7.37 V)

8. $V_C = V_{CC}$

$= 10\text{ V}$

9. $I_B = \dfrac{I_C}{h_{FE}}$ 　　　　　　　　　　　　　　　　　(9.7)

$= \dfrac{6.51\text{ mA}}{135} = 48.2\ \mu\text{A}$ 　　　　(Design: 49.5 μA)

10. $h_{ie} = \dfrac{26\text{ mV}}{I_B}$ 　　　　　　　　　　　　　　(11.3)

$= \dfrac{26\text{ mV}}{48.2\ \mu\text{A}} = 539\ \Omega$

11. $R_{in} = R_1 \parallel R_2 \parallel \left(h_{ie} + h_{fe}(1/h_{oe} \parallel R_E \parallel R_L) \right)$ 　　(13.3)

$= 18\text{ k}\Omega \parallel 68\text{ k}\Omega \parallel (539\ \Omega + 135(33.3\text{ k}\Omega \parallel 1\text{ k}\Omega \parallel 1\text{ k}\Omega))$

$= 11.7\text{ k}\Omega$

12. $R_{\text{out}} = R_E \parallel 1/h_{oe} \left\| \dfrac{h_{ie} + R_2 \parallel R_1 \parallel R_S}{h_{fe}} \right.$ (13.4)

$= 1 \text{ k}\Omega \parallel 33.3 \text{ k}\Omega \left\| \dfrac{539 \ \Omega + 68 \text{ k}\Omega \parallel 18 \text{ k}\Omega \parallel 600 \ \Omega}{135} \right.$

$= 8.19 \ \Omega$

13. $A_v = \dfrac{h_{fe}(1/h_{oe} \parallel R_E \parallel R_L)}{\text{NUM} + h_{ie}}$ (13.7)

$= \dfrac{135(33.3 \text{ k}\Omega \parallel 1\text{k}\Omega \parallel 1 \text{ k}\Omega)}{\text{NUM} + 539 \ \Omega} = 0.992$

14. $A_i = A_v \dfrac{R_{\text{in}}}{R_L}$ (12.25)

$= 0.992 \dfrac{11.7 \text{ k}\Omega}{1 \text{ k}\Omega} = 11.6$

15. $v_{\text{outPmax}} = \begin{Bmatrix} V_C - V_E \\[2mm] I_C R_{\text{AC}} \end{Bmatrix} \begin{Bmatrix} \text{whichever} \\ \text{is} \\ \text{smaller} \end{Bmatrix}$

$= \begin{cases} 10 \text{ V} - 6.51 \text{ V} = 3.49 \text{ V} \\[3mm] 6.51 \text{ mA}(1 \text{ k}\Omega \parallel 1 \text{ k}\Omega) = 3.26 \text{ V} \end{cases}$

The above result indicates that the Q point is not exactly in the center of the AC load line. Clipping will be observed nearly 1/4 volt sooner at the bottom than at the top of the waveform.

16. $f_L = \begin{Bmatrix} \dfrac{1}{2\pi \, C_i(R_S + R_{\text{in}})} \\[4mm] \dfrac{1}{2\pi \, C_o(R_{\text{out}} + R_L)} \end{Bmatrix} \begin{Bmatrix} \text{whichever} \\ \text{is} \\ \text{larger} \end{Bmatrix}$

$= \begin{cases} \dfrac{1}{2\pi * 0.68 \ \mu\text{F}(600 \ \Omega + 11.7 \text{ k}\Omega)} = 19 \text{ Hz} \\[4mm] \dfrac{1}{2\pi * 8 \ \mu\text{F}(8.19 \ \Omega + 1 \text{ k}\Omega)} = 19.7 \text{ Hz} \end{cases}$ (Design: 20 Hz)

If you went to the lab and hooked this circuit up (and you should!) you would find that it functions very close to the prediction of Example 13.2.

13.5   Circuit Gain

You discovered in Section 13.4 that the voltage gain of the emitter follower is always less than one. You will find later that this is a characteristic of all follower type circuits. You may have felt that your time was being wasted by studying circuits that "did not produce any gain." That feeling would be correct, except for the two important characteristics of the emitter follower that we pointed out in Section 13.3. They are its relatively high input impedance and its relatively low output impedance.

In connection with Example 13.2 you saw that the circuit's high input impedance and its low output impedance combined to produce a fairly high current gain in spite of the low voltage gain. Since the voltage gain is so near to unity, the circuit power gain is essentially the same as the current gain. Thus it is not uncommon to have a power gain of 15 to 20 dB for an emitter follower.

Probably its low output impedance is the most attractive characteristic of the emitter follower circuit.

13.6   The Darlington Circuit

If you go back and look at almost any of the transistor circuits we have talked about, you will recognize that those circuits would work better if the current gain h_{fe} of the transistor were higher. For example, if h_{fe} = 1000 in Example 13.1, the circuit input impedance increases from R_{in} = 11.7 kΩ to R_{in} = 83.4 kΩ, the output impedance decreases from R_{out} = 8.10 Ω to R_{out} = 4.48 Ω.

If you look at Section 8.5 where we talked about the construction of transistors, you will find that physical limitations make it difficult to construct transistors with h_{fe} much greater than about 300. However, the emitter follower idea can be used to hook two or more transistors together to produce circuits which can have a current gain as high as 100,000 or more. Figure 13.4a is one such circuit. It is known as a "Darlington circuit" in honor of the developer of the circuit.

You can construct a Darlington circuit using two identical transistors, or perhaps more commonly Q_2 can have a higher current rating and/or a higher power rating than Q_1. Notice that the Darlington circuit has only three terminals for external connections; the emitter terminal of Q_2, the base terminal of Q_1, and the collector terminal. As you might guess, in this day of microelectronics it is possible for manufacturers to make both transistors of a Darlington circuit on the same chip and put it in the same kind of package as is used for a single transistor. Such a package is known as a *Darlington transistor*.

It is impossible to distinguish a Darlington transistor from a conventional transistor just by looking at the package. However, as you can see

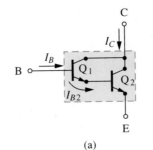

(a)

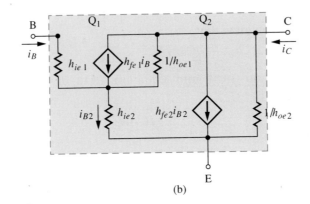

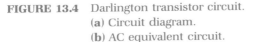

(b)

FIGURE 13.4 Darlington transistor circuit.
(a) Circuit diagram.
(b) AC equivalent circuit.

from Figure 13.4a there are two B-E junctions to bias if the Darlington is to be turned on; so one clue that you are dealing with a Darlington is that V_{BE} will be higher than you would otherwise expect. Another clue is the very high value of h_{FE}. Any time you see an h_{FE} of 1000 or more, you can be quite sure that you are dealing with a Darlington.

Darlington transistors are commercially available in both NPN and PNP types. Appendix F contains some typical data book listings for some Darlington transistors.

Figure 13.4b contains the h parameter equivalent circuit for a Darlington circuit. Due to the very high gain of the circuit, $1/h_{oe1}$ can usually be ignored. Then

$$i_{B2} \approx h_{fe1}i_B \tag{13.10}$$

Substituting Equation 13.10 for i_{B2} in Q_2 we have the overall current gain for the Darlington transistor

$$h_{fe} \doteq h_{fe1}h_{fe2} \tag{13.11}$$

Equation 13.11 tells us why the gain of a Darlington transistor is so much higher than that of a conventional transistor.

Figure 13.4b also shows us that h_{ie} will be quite large. The Miller effect tells us that h_{ie2} will be multiplied by h_{fe1}. This gives us

$$h_{ie} = h_{ie1} + h_{fe1}h_{ie2} \tag{13.12}$$

Substituting the approximation of Equation 11.3 into Equation 13.12 gives us

$$h_{ie} \approx \frac{26 \text{ mV}}{I_B} + h_{fe1}\frac{26 \text{ mV}}{I_{B2}} \tag{13.13}$$

But you should recognize that

$$I_{B2} \approx I_{C1} = h_{fe1}I_B \tag{13.14}$$

Substituting Equation 13.14 into Equation 13.13, h_{fe1} can be canceled leaving

$$h_{ie} \approx 2\frac{26 \text{ mV}}{I_B} \tag{13.15}$$

In order to solve for $1/h_{oe}$ for the Darlington, it is obvious from Figure 13.4b that if $1/h_{oe1}$ is ignored

$$1/h_{oe} = 1/h_{oe2} \tag{13.16}$$

The only other special consideration involved in working with a Darlington is the calculation of V_{BE}. Using the approximation we have used so often before, for Q_1 we have

$$V_{BE1} \approx 26 \text{ mV} * \ln \frac{I_{C1}}{10 \text{ fA}} \tag{8.4}$$

but since

$$I_{C1} = I_{B2} = \frac{I_C}{h_{FE2}} \tag{9.7}$$

we have

$$V_{BE1} = 26 \text{ mV} * \ln \frac{I_C}{h_{FE2} * 10 \text{ fA}} \tag{13.17}$$

For Q_2 we have

$$V_{BE2} = 26 \text{ mV} * \ln \frac{I_C}{10 \text{ fA}} \tag{13.18}$$

From Figure 13.4a it is obvious that

$$V_{BE} = V_{BE1} + V_{BE2} \tag{13.19}$$

Substituting Equations 13.17 and 13.18 into Equation 13.19 gives

$$V_{BE} = 26 \text{ mV}\left(\ln \frac{I_C}{h_{FE2} * 10 \text{ fA}} + \ln \frac{I_C}{10 \text{ fA}}\right) \qquad (13.20)$$

Equation 13.20 works well if you are dealing with a Darlington circuit made up of individual transistors so that h_{FE2} is known or can be estimated. If you are dealing with a Darlington transistor, the overall h_{FE} is usually given but not h_{FE2}. If we assume that $h_{FE1} = h_{FE2}$ then from Equation 13.11

$$h_{FE2} \approx \sqrt{h_{FE}} \qquad (13.21)$$

If you have the junction equation programmed into your calculator, and you should, you would probably like to have Equation 13.20 in a better form. If you are really good at using logarithms, you could show that substituting Equation 13.21 into Equation 13.20 and simplifying yields

$$V_{BE} = 2 * 26 \text{ mV} * \ln\left(\frac{I_C}{10 \text{ fA}}\right) - 13 \text{ mV} * \ln(h_{FE}) \qquad (13.22)$$

EXAMPLE 13.3

Determine appropriate values of R_1 and R_2 for the circuit of Figure 13.5. Specify the nearest standard resistor sizes.

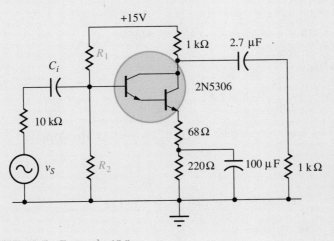

FIGURE 13.5   Example 13.3.

SOLUTION:

Referring to Appendix F, we find that a 2N5306 is a Darlington transistor with h_{FE} between 7,000 and 70,000. Thus the most likely value is

$h_{FE} = \sqrt{7,000 * 70,000} = 22,100$. Since no other parameters are listed, we will assume that $1/h_{oe}$ is large. Much of the design procedure of Table 12.1 can be adapted to this circuit. Beginning at Step 3 we have

$$3. \ I_C \stackrel{*}{=} \frac{V_{CC}}{R_{DC} + R_{AC}} = \frac{V_{CC}}{R_C + R_E + R_C \parallel R_L + R_{E1}} \quad (9.25)$$

$$= \frac{15 \text{ V}}{1 \text{ k}\Omega + 68 \ \Omega + 220 \ \Omega + 1 \text{ k}\Omega \parallel 1 \text{ k}\Omega + 68 \ \Omega}$$

$$= 8.08 \text{ mA}$$

Replacing Equation 8.4 with Equation 13.22 for the Darlington transistor, we have

$$4. \ V_{BE} = 2 * 26 \text{ mV} * \ln\frac{I_C}{10 \text{ fA}} - 13 \text{ mV} * \ln(h_{FE}) \quad (13.22)$$

$$= 2 * 26 \text{ mV} * \ln\frac{8.08 \text{ mA}}{10 \text{ fA}} - 13 \text{ mV} * \ln(22,100)$$

$$= 1.30 \text{ V}$$

$$5. \ V_E = I_C R_E \quad (10.18)$$

$$= 8.08 \text{ mA}(68 \ \Omega + 220 \ \Omega) = 2.33 \text{ V}$$

$$6. \ V_B = V_{BE} + V_E \quad (10.19)$$

$$= 1.30 \text{ V} + 2.33 \text{ V} = 3.62 \text{ V}$$

Although Step 7 is not really needed, we will solve for V_C so that if you hook up the circuit (you should!), you can check the Q point.

$$7. \ V_C = V_{CC} - I_C R_C \quad (10.5)$$

$$= 15 \text{ V} - 8.08 \text{ mA} * 1 \text{ k}\Omega = 6.92 \text{ V}$$

$$8. \ I_B = \frac{I_C}{h_{FE}} \quad (9.7)$$

$$= \frac{8.08 \text{ mA}}{22,100} = 366 \text{ nA}$$

Since they are not needed, Steps 9, 10, and 11 are skipped.

$$12. \ R_B = \frac{S}{1 - S} h_{FE} R_E \stackrel{*}{\approx} \frac{1}{10} h_{FE} R_E \quad (10.16)$$

$$= \frac{1}{10} 22,100(68 \ \Omega + 220 \ \Omega) = 636 \text{ k}\Omega$$

$$13. \ V_{BB} = V_B + I_B R_B \quad (10.17)$$

$$= 3.62 \text{ V} + 366 \text{ nA} * 636 \text{ k}\Omega = 3.86 \text{ V}$$

$$14. \ R_1 = R_B \frac{V_{CC}}{V_{BB}} \quad \text{STD.} \ (10.27)$$

$$= 636 \text{ k}\Omega \frac{15 \text{ V}}{3.86 \text{ V}} = 2.48 \text{ M}\Omega \qquad (\text{Use } 2.7 \text{ M}\Omega)$$

15. $R_2 = R_1 \dfrac{V_{BB}}{V_{CC} - V_{BB}}$ STD.(10.29)

$= 2.7 \text{ M}\Omega \dfrac{3.86 \text{ V}}{15 \text{ V} - 3.86 \text{ V}} = 934 \text{ k}\Omega$ (Use 1 MΩ)

You should notice that the use of the Darlington transistor does not change the design procedure a great deal, but it changes the numerical results a great deal.

EXAMPLE 13.4
Find the low frequency cutoff for the circuit of Figure 13.6.

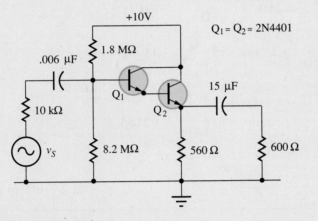

FIGURE 13.6   Example 13.4.

SOLUTION:
In the circuit of Figure 13.6 we have two 2N4401 transistors hooked up as a Darlington pair in an emitter follower circuit. From Appendix F, the important transistor characteristics are: $h_{FE} = 135$, and $1/h_{oe} = 1/30 \text{ } \mu S = 33.3 \text{ k}\Omega$. For the Darlington,

$$h_{FE} = h_{FE1} h_{FE2} \tag{13.11}$$
$$= 135 * 135 = 18{,}200$$

From the circuit analysis procedure of Table 12.2 we have

1. $V_{BB} = V_{CC} \dfrac{R_2}{R_1 + R_2}$ (10.22)

$= 10 \text{ V} \dfrac{8.2 \text{ M}\Omega}{1.8 \text{ M}\Omega + 8.2 \text{ M}\Omega} = 8.2 \text{ V}$

2. $R_B = R_1 \parallel R_2$ $\qquad$ (10.23)

$\qquad = 1.8 \text{ M}\Omega \parallel 8.2 \text{ M}\Omega = 1.48 \text{ M}\Omega$

3. $S = \dfrac{R_B}{R_B + h_{FE}R_E}$ $\qquad$ (10.10)

$\qquad = \dfrac{1.48 \text{ M}\Omega}{1.48 \text{ M}\Omega + 18{,}200 * 560 \ \Omega} = 0.126 = 12.6\%$

The circuit is not as stable as it should be, but hopefully the operation will be close to the following predictions.

4. $I_C = \dfrac{V_{BB} - V_{BE}}{\dfrac{R_B}{h_{FE}} + R_E}$ $\qquad$ (10.12)

$\qquad = \dfrac{8.20 \text{ V} - V_{BE}}{\dfrac{1.48 \text{ M}\Omega}{18{,}200} + 560 \ \Omega} = \dfrac{8.20 \text{ V} - V_{BE}}{641 \ \Omega}$

5. $V_{BE} = 2 * 26 \text{ mV} * \ln \dfrac{I_C}{10 \text{ fA}} - 13 \text{ mV} * \ln(h_{FE})$ $\qquad$ (13.22)

$\qquad = 2 * 26 \text{ mV} * \ln \dfrac{I_C}{10 \text{ fA}} - 13 \text{ mV} * \ln(18{,}200)$

$\qquad = 2 * 26 \text{ mV} * \ln \dfrac{I_C}{10 \text{ fA}} - 0.128 \text{ V}$

Iterating Steps 4 and 5 above gives

4. $I_C = \dfrac{8.20 \text{ V} - V_{BE}}{641 \ \Omega} =$	11.7	10.7 mA	(10.12)
5. $V_{BE} = 2 * 26 \text{ mV} * \ln \dfrac{I_C}{10 \text{ fA}} - 0.128 \text{ V} =$	1.32	1.31 V	

Skipping the Q point calculations

9. $I_B = \dfrac{I_C}{h_{FE}}$ $\qquad$ (9.7)

$\qquad = \dfrac{10.7 \text{ mA}}{18{,}200} = 590 \text{ nA}$

Since we are dealing with a Darlington circuit, Steps 10, 11, and 12 become

10. $h_{ie} = \dfrac{52 \text{ mV}}{I_B}$ $\qquad$ (13.15)

$\qquad = \dfrac{52 \text{ mV}}{590 \text{ nA}} = 88.1 \text{ k}\Omega$

11. $R_{in} = R_1 \parallel R_2 \parallel \left(h_{ie} + h_{fe}(1/h_{oe} \parallel R_E \parallel R_L) \right)$ (13.3)

$\qquad = 1.8 \text{ M}\Omega \parallel 8.2 \text{ M}\Omega \parallel (88.1 \text{ k}\Omega + 18{,}200(33.3 \text{ k}\Omega \parallel 560 \ \Omega \parallel 600 \ \Omega))$

$\qquad = 1.16 \text{ M}\Omega$

12. $R_{out} = R_E \parallel 1/h_{oe} \parallel \dfrac{h_{ie} + R_1 \parallel R_2 \parallel R_S}{h_{fe}}$ (13.4)

$\qquad = 560 \ \Omega \parallel 33.3 \text{ k}\Omega \parallel \dfrac{88.1 \text{ k}\Omega + 1.8 \text{ M}\Omega \parallel 8.2 \text{ M}\Omega \parallel 10 \text{ k}\Omega}{18{,}200} = 5.33 \ \Omega$

Skipping Steps 13, 14, and 15

16. $f_L = \left\{ \begin{array}{c} \dfrac{1}{2\pi \, C_i(R_S + R_{in})} \\[2mm] \dfrac{1}{2\pi \, C_o(R_{out} + R_L)} \end{array} \right\} \left\{ \begin{array}{c} \text{whichever} \\ \text{is} \\ \text{larger} \end{array} \right\}$

$\qquad = \dfrac{1}{2\pi * 0.006 \ \mu\text{F}(10 \text{ k}\Omega + 1.16 \text{ M}\Omega)} = 22.7 \text{ Hz}$

or

$\qquad = \dfrac{1}{2\pi * 15 \ \mu\text{F}(5.33 \ \Omega + 600 \ \Omega)} = 17.5 \text{ Hz}$

Since the input circuit cuts off at about 22.7 Hz, the low frequency cutoff will be no lower than $f_L = 22.7$ Hz.

Notice that the Darlington circuit makes two rather ordinary transistors deliver rather high quality performance.

13.7 Regulator Circuits Revisited

In Chapter 7 we talked about power supply regulator circuits. The circuit we used there is reproduced here as Figure 13.7a. That regulator could be called a "brute force regulator" because, as we said in Chapter 7, not only does R_S carry the current to the Zener diode, but it also carries the load current. As a result, R_S must carry a large current, and when the load is disconnected, all of that current must be absorbed by the Zener diode. Therefore, the power supply is always fully loaded. Whatever current the external load does not take, the Zener diode must take. This means high power dissipation, heat production, and inefficiency. In Chapter 7 we mentioned that in later chapters we would describe better regulator cir-

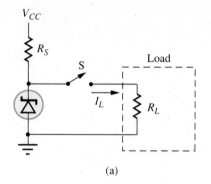

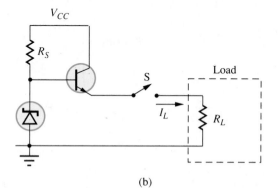

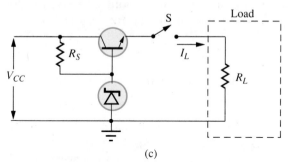

FIGURE 13.7   Regulator circuit.
(a) Zener regulator.
(b) Zener regulator with a series pass transistor added.
(c) Typical circuit diagram.

cuits. Now that you know about the emitter follower, we can talk about one such circuit.

Consider the circuit in Figure 13.7b. Obviously, the transistor is hooked up as an emitter follower. Consequently, the emitter voltage follows the base, so if the Zener regulates the base voltage, the emitter should be fairly well regulated too. Here R_S supplies only the Zener current and the transistor base current which, as you know, is quite small, and is given by

$$I_B = \frac{I_L}{h_{FE}} \tag{9.7}$$

The transistor supplies the load current. Now when the load is disconnected, the transistor shuts off minimizing the power dissipation. The transistor in this kind of circuit is often called a "series pass transistor" because it is in series with the load. The circuit in Figure 13.7c emphasizes this fact. You should see that Figure 13.7c and Figure 13.7b are, in fact, identical circuits. Because it is easier to draw, you will find Figure 13.7c more commonly used to picture circuits containing series pass transistors.

EXAMPLE 13.5
Repeat Example 7.4 using the circuit of Figure 13.7.

SOLUTION:
Example 7.4 called for the design of a regulator intended to operate a tape recorder ($V_L = 6$ VDC, $I_L = 100$ mADC) from an automobile electrical system ($V_{min} = 13$ VDC, $V_{max} = 15$ VDC). Figure 13.8 contains the relevant circuit.

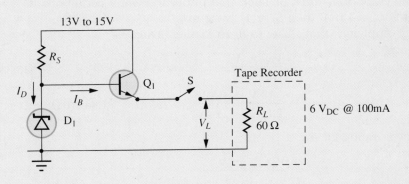

FIGURE 13.8 Example 13.5.

Selection of Q_1:

$$P_{max} = (V_{CCmax} - V_E)I_C$$
$$= (15 \text{ V} - 6 \text{ V})100 \text{ mA} = 900 \text{ mW}$$

Looking in Appendix F, the data for a 2N6715 are:

$$P_{DISS} = 2 \text{ W} \qquad \text{(Adequate to handle 900 mW)}$$
$$I_{Cmax} = 2 \text{ A} \qquad \text{(Adequate to handle 100 mA)}$$
$$h_{FE} = 100$$

To use a transistor with a 1 W rating would be inviting disaster.

Selection of D_1:

$$V_{BE} = 26 \text{ mV} * \ln \frac{I_C}{10 \text{ fA}} \qquad (8.4)$$

$$= 26 \text{ mV} * \ln \frac{100 \text{ mA}}{10 \text{ fA}} = 0.778 \text{ V}$$

$$V_B = V_{BE} + V_E \qquad (10.19)$$
$$= 0.778 \text{ V} + 6 \text{ V} = 6.78 \text{ V}$$

$$I_B = \frac{I_C}{h_{FE}} \qquad (9.7)$$

$$= \frac{100 \text{ mA}}{100} = 1 \text{ mA}$$

Looking at Appendix F, the data for a 1N5235 are:

$$V_D = 6.8 \text{ V} \qquad \text{(Close enough to 6.78 V)}$$
$$I_T = 20 \text{ mA} \qquad \text{(Much greater than 1 mA)}$$

Selection of R_S:

The value of R_S must be such that under the worst conditions ($V_{CC} = 13$ V; $I_B = 1$ mA) then $I_D = I_T = 20$ mA.

Applying Ohm's Law to R_S in Figure 13.8 gives

$$R_S = \frac{V_{CC} - V_B}{I_D + I_B} \qquad (13.23)$$

$$= \frac{13 \text{ V} - 6.8 \text{ V}}{20 \text{ mA} + 1 \text{ mA}} = 295 \text{ }\Omega \qquad \text{(Use } 270 \text{ }\Omega\text{)}$$

Power rating for R_S:

$$P_{RS} = \frac{(V_{CCmax} - V_B)^2}{R_S} \qquad (13.24)$$

$$= \frac{(15 \text{ V} - 6.8 \text{ V})^2}{270 \text{ }\Omega} = 0.249 \text{ W} \qquad \text{(Use } 1/2 \text{ W)}$$

If you compare the results of Example 13.7 with that of Example 7.4 you will see that the addition of a series pass transistor makes the demands on R_S and the Zener diode in a regulator much less severe. That does not necessarily cause the regulator to do a better job, however. It places the Zener one step removed from the load it is trying to control, and as you know V_{BE} is not a quantity on which we have a very good handle. Thus although this regulator is used in many practical applications, you will have to wait until we get to Chapter 21 to learn about really high quality regulators.

13.8 Troubleshooting ◄◄◄◄◄◄◄◄

Since emitter follower circuits are not encountered as often as is the common emitter circuit, it is possible for the unwary to be misled by them. For instance, in the common emitter circuit, if the collector voltage is about right, the circuit is probably working fairly well. In the emitter follower circuit, since the collector is connected to the power supply, its voltage will always be at V_{CC} whether the circuit is working normally or not. For the same reason, both V_E and V_B will look abnormally high and the signal voltage gain will look abnormally low. Before you begin troubleshooting, be certain that you know what is "normal" for the circuit with which you are working.

13.9 Summary

In this chapter we adapted the procedures of the common emitter circuits and the common emitter h parameters to the emitter follower circuit. We found that in general the procedures were much the same, but the ungrounded emitter caused the input impedance to be much higher, the output impedance to be much lower, and the voltage gain to be much lower, being slightly less than one. Finally, the Darlington circuit and the series pass transistor were seen to be applications of the emitter follower circuit.

Glossary

Darlington circuit: A circuit consisting of two transistors. The collectors are connected together in such a way that they act as a single transistor having a gain that is the product of the gain of the separate transistors. The collectors are connected together and the emitter of the input transistor is connected to the base of the output transistor. (Section 13.6)

Follower circuit: Any amplifier circuit in which the output follows the input. Such circuits typically have a gain of one or less and no phase inversion. (Section 13.2)

Series pass transistor: A transistor placed in series with a load, usually in the emitter follower configuration for the purpose of supplying current to the load. (Section 13.7)

PROBLEMS

Section 13.3

1. Determine the value of R_1 and R_2 for the circuit in Figure 13.9 if $V_{CC} = 9$ V.
2. Determine the value of R_1 and R_2 for the circuit in Figure 13.9 if $V_{CC} = 25$ V.

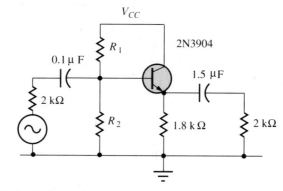

FIGURE 13.9   Problem 1.

Section 13.4

3. Determine the input impedance R_{in} and the output impedance R_{out} of the circuit in Figure 13.10.

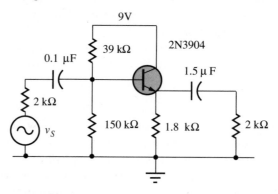

FIGURE 13.10   Problem 3.

4. Determine the Q point voltages for the circuit of Example 13.2 if $h_{FE} = 300$. Compare with the original values. Is the circuit still usable?

Section 13.5

5. Determine the power gain for the circuit in Example 13.2. Express your answer in dB.
6. Determine the voltage gain, current gain, and power gain for the circuit of Problem 3. Express the power gain in dB.

7. Determine the low frequency cutoff for the circuit of Figure 13.2.
8. Determine the high frequency cutoff for the circuit of Problem 3.

Section 13.6

9. Repeat Problem 21 in Chapter 12 using a 2N5306 Darlington transistor. Compare the resulting circuits.
10. Determine the value of C_i in Figure 13.5 that will give approximately the same value of f_L as is given by C_o.
11. Repeat Problem 1 if the 2N3904 transistor is replaced with a 2N5306 Darlington transistor.
12. Determine the input impedance R_{in} and the output impedance R_{out} of the circuit of Figure 13.11.

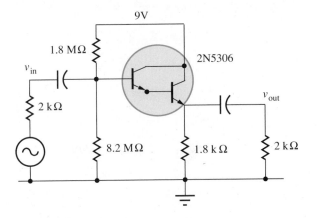

FIGURE 13.11   Problem 12.

Section 13.7

13. (a) Estimate the maximum voltage that will appear at the regulator output in Example 13.5 when switch S in Figure 13.8 is opened.
    (b) Repeat A above if a 560 Ω "bleeder resistor" is connected between the emitter of the transistor and ground.
14. Estimate the outcome if the output terminals of the regulator of Example 13.5 became shorted together.

Section 13.8   ◀ Troubleshooting

15. Suppose that the collector and emitter terminals of the 2N3904 in Figure 13.10 became shorted together. Determine:
    (a) The new Q point.
    (b) The power dissipated in the emitter resistor.
16. Repeat Problem 15 if, instead of the above fault, the 39 kΩ resistor became a dead short.
17. Repeat Problem 15 if, instead of the above fault, the 150 kΩ resistor became a dead short.

Class A amplifiers are very inefficient—and very
expensive. (Photograph by Gabe Palmer/The Stock
Market.)

Chapter 14
Class A Power Amplifiers

SPECIAL TERMS

Class A amplifier
Amplifier efficiency
Transformer coupled amplifier
Impedance matching
Emitter swamping

347

14.1 Introduction and Objectives

In the past we have briefly mentioned transistor power rating. In this chapter we will study the concept in greater detail. At the same time we will look at how power is delivered to external loads by transistor amplifying circuits. We will look briefly at transformer coupled circuits. Finally we will introduce the idea of circuit efficiency.

After studying this chapter you should begin to understand transistor power dissipation ratings, their meaning and their use. You should also have some understanding of the special considerations required for dealing with high power circuits and for dealing with transformer coupled circuits. You should be able to:

- Design and analyze simple Class A capacitively coupled power circuits.
- Design and analyze simple Class A transformer coupled power circuits.
- Determine the electrical efficiency of both of the above amplifier types.

14.2 Class A Operation

Historically, electronic circuits have been classified according to the way they respond to AC signals. A circuit is said to be operating Class A if it conducts current 100 percent of the time. A better definition in the case of a transistor circuit might be that in Class A operation, the transistor is never saturated and never cut off. In terms of the load line, the operation never attempts to go beyond either end of the AC load line. All of the transistor circuits we have considered so far in this book have been Class A circuits.

Two advantages of this class of operation are its simplicity and its linearity. Its biggest disadvantage is its low efficiency. To avoid saturation or cutoff, the load line must be long enough to encompass any expected signal amplitude. Thus for circuits designed to deliver relatively large amounts of power, high Q point voltages and currents are required, and that means high power dissipation.

14.3 Transistor Power Dissipation

Since $P = VI$, any time that voltage is lost across a device that has current flowing through it, power is dissipated by that device. Some devices such as electric motors have the ability to use at least some of the power delivered to them to do some other form of work. The power absorbed by a transistor is always converted to heat. The amount of power, P_{DISS} that a transistor must dissipate under quiescent conditions is given by

$$P_{\text{DISS}} = V_{CE}I_C \tag{14.1}$$

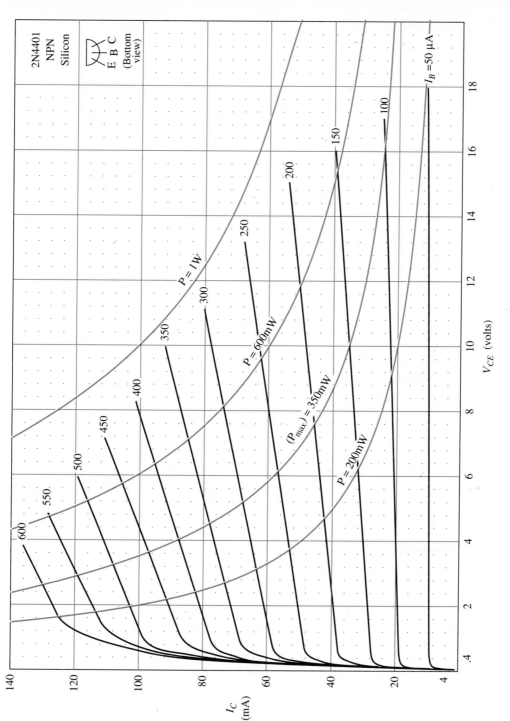

FIGURE 14.1 Characteristics for a 2N4401.

When power is applied to a transistor circuit, its temperature will rise. If its temperature is higher than its surroundings, it will then lose heat to those surroundings. The greater the temperature difference, the more rapidly it will lose heat. Thus its temperature will rise until heat is lost at the same rate as it is produced. If the transistor's temperature has to go too high in order to achieve that balance, the transistor can be destroyed. The rate at which a transistor can dissipate heat is dependent on such things as the type of case in which it is housed (metal will dissipate heat faster than plastic), the amount of ventilation provided to the transistor, and the amount of heat sinking used.

In Figure 14.1 several power dissipation curves have been drawn on a 2N4401 characteristic graph. Those power dissipation curves were obtained by graphing Equation 14.1 using the indicated values for P_{DISS}. If you are mathematically inclined you will recognize that the power dissipation curves are hyperbolas with asymptotes formed by the I_C and the V_{CE} axes.

EXAMPLE 14.1

Construct a 500 mW power dissipation curve for a transistor. Use $I_{Cmax} = 250$ mA and $V_{CEmax} = 20$ V.

SOLUTION:

Solving Equation 14.1 for I_C we have

$$I_C = \frac{P_{DISS}}{V_{CE}}$$

$$= \frac{500 \text{ mW}}{V_{CE}}$$

Substituting arbitrary values for V_{CE} over the stipulated range yields Table 14.1. Graphing these values yields the required graph shown in Figure 14.2.

TABLE 14.1   Power Dissipation

Data Index	VCE Volts	IC mA
1	2	250
2	3	167
3	4	125
4	5	100
5	6	83.3
6	7.5	66.7
7	10	50
8	15	33.3
9	20	25

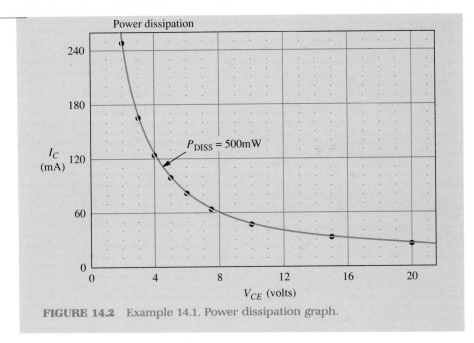

FIGURE 14.2   Example 14.1. Power dissipation graph.

It is obvious that if a transistor is to operate at a given value of P_{DISS}, then the transistor's Q point must be located on that particular power dissipation curve. Although the power dissipation curve is based on relatively long term thermal considerations, in class A operation it is not considered good design practice to permit the AC load line to ever go above the transistor manufacturer's specified maximum power dissipation curve.

In Section 9.6 when we first introduced the idea of an AC load line, we showed that in order to achieve midpoint bias, the horizontal axis intercept of the AC load line, v_{CC} must be twice V_{CE}, and the vertical axis intercept, i_O must be twice I_C. If you know some calculus you can prove that if the Q point is located on the maximum power dissipation curve, and if the AC load line meets the conditions for midpoint bias, then the AC load line will be tangent to the maximum power dissipation curve. One such proof is included in Appendix D.

EXAMPLE 14.2
Locate a Q point at any arbitrary point on the maximum power dissipation curve drawn in Example 14.1. Draw a tangent to the above curve at the above point. Compare the above Q point values of I_C and V_{CE} with the vertical and horizontal intercepts of the above tangent line.

SOLUTION:

In Figure 14.3 a Q point was arbitrarily placed at $I_C = 50$ mA and $V_{CE} = 10$ V. Using a straight edge, a line which visually appeared to be tangent to the maximum power curve was drawn. Reading from the graph, the vertical intercept of the tangent line is approximately $i_O = 100$ mA and the horizontal intercept is approximately $v_{CC} = 20$ V. These values appear to be approximately twice the chosen values for I_C and V_{CE} respectively. Thus the Q point appears to be at the center of the tangent line.

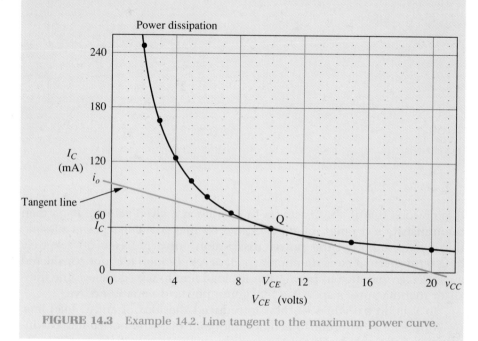

FIGURE 14.3 Example 14.2. Line tangent to the maximum power curve.

The rather surprising result of the above situation is that for a midpoint biased Class A circuit, the transistor dissipates the largest amount of heat at the Q point. When the signal causes I_C to increase, the attendant decrease in V_{CE} results in a decrease in power dissipation, and when the signal causes I_C to decrease, there is also a decrease in power dissipation.

14.4 Designing for Output Power

For the circuit design procedure of Chapter 12 we assumed that the load resistance R_L and the supply voltage V_{CC} were given. It seems reasonable that the load resistance might be given, but you may have wondered how

V_{CC} is determined in a practical situation. In most electronic devices there is at least one amplifier circuit that must deliver a specified maximum AC signal power to a given load. The output circuit connected to the loud-speaker of your stereo would be one example. The deflection circuits for your computer's video monitor or your TV set would be another example. In that case the designer is not free to choose V_{CC}. Its value is dependent on the required signal output power, P_{Lrms}, and the external load resistance, R_L, to which that power is to be delivered.

In order to calculate the value of V_{CC} required to deliver a given power to a given load, we will first calculate the required collector current, I_C. Referring to the circuit in Figure 14.4a, we will assume that all of the AC load line is to be used and that the Q point is at the center of the load line as shown in Figure 14.4c. From the power equation we have

$$P_{Lrms} = \frac{V_{Lrms}^2}{R_L}$$

or

$$V_{Lrms} = \sqrt{P_{Lrms}R_L} \qquad (14.2)$$

From the equivalent circuit of Figure 14.4b we can write

$$i_{Crms} = \frac{V_{Lrms}}{R_{AC}} = \frac{V_{Lrms}}{R_C \| R_L} \qquad (14.3)$$

Substituting Equation 14.2 into Equation 14.3 gives

$$i_{Crms} = \frac{\sqrt{P_{Lrms}R_L}}{R_{AC}} \qquad (14.4)$$

Converting rms to Peak we have

$$i_{CP} = \sqrt{2}\, i_{Crms} \qquad (14.5)$$

Substituting Equation 14.4 into Equation 14.5 gives

$$i_{CP} = \frac{\sqrt{2P_{Lrms}R_L}}{R_{AC}} \qquad (14.6)$$

If we assume midpoint bias and if we assume that all of the load line is to be used as shown in Figure 14.4c, from the figure you can see that

$$I_C = i_{CP} \qquad (14.7)$$

Substituting Equation 14.6 into Equation 14.7 gives

$$I_C = \frac{\sqrt{2P_{Lrms}R_L}}{R_{AC}} \qquad (14.8)$$

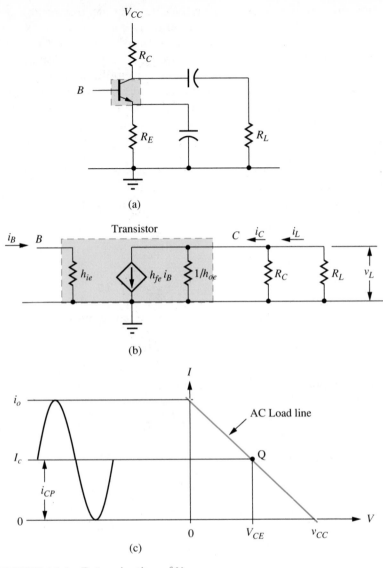

FIGURE 14.4   Determination of V_{CC}.
(a) Circuit diagram.
(b) AC equivalent circuit.
(c) Collector current waveform.

In order to determine the supply voltage, V_{CC} required to supply the above current, Equation 9.25 can be solved for V_{CC}. The result is

$$V_{CC} \overset{*}{=} I_C(R_{DC} + R_{AC})$$

(14.9)

Equations 14.8 and 14.9 must be used in the design procedure of Table 12.1 in place of Step 3 when the circuit is required to deliver a specified power to the load.

EXAMPLE 14.3

Design the circuit of Figure 14.5a given that it must deliver 25 mWrms to the 600 Ω load. Assume $h_{FE} = 100$, $1/h_{oe} = 20$ kΩ, and $f_L = 100$ Hz.

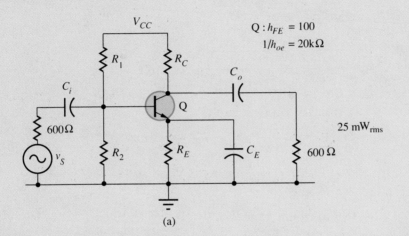

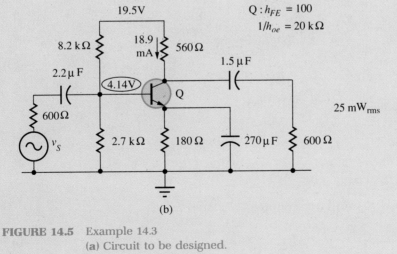

FIGURE 14.5   Example 14.3
(a) Circuit to be designed.
(b) Completed design.

SOLUTION:

Returning to the design procedure in Table 12.1 we have

1. $R_C \overset{*}{=} R_L \parallel (-1/h_{oe})$ STD.   (12.12)

    $= 600 \ \Omega \parallel (-20 \ \text{k}\Omega) = 619 \ \Omega$ (Use $560 \ \Omega$)

2. $R_E \overset{*}{=} \dfrac{R_{DC}}{4} = \dfrac{R_C}{3}$ (STD. if Step 10 is omitted)   (10.15)

    $= \dfrac{560 \ \Omega}{3} = 187 \ \Omega$ (Use $180 \ \Omega$)

All of R_E is bypassed according to Figure 14.5a so the nearest standard value is specified for R_E and Steps 10 and 11 will be omitted.

3. $I_C \overset{*}{=} \dfrac{\sqrt{2 P_{Lrms} R_L}}{R_{AC}}$ (14.8)

    $= \dfrac{\sqrt{2 * 25 \ \text{mWrms} * 600 \ \Omega}}{560 \ \Omega \parallel 600 \ \Omega} = 18.9 \ \text{mA}$

3.1 $V_{CC} = I_C(R_{DC} + R_{AC})$ (14.9)

    $= 18.9 \ \text{mA}(560 \ \Omega + 180 \ \Omega + 560 \ \Omega \parallel 600 \ \Omega) = 19.5 \ \text{V}$

4. $V_{BE} \approx 26 \ \text{mV} * \ln\dfrac{I_C}{10 \ \text{fA}}$ (8.4)

    $= 26 \ \text{mV} * \ln\dfrac{18.9 \ \text{mA}}{10 \ \text{fA}} = 0.735 \ \text{V}$

5. $V_E = I_C R_E$ (10.18)

    $= 18.9 \ \text{mA} * 180 \ \Omega = 3.40 \ \text{V}$

6. $V_B = V_{BE} + V_E$ (10.19)

    $= 0.735 \ \text{V} + 3.40 \ \text{V} = 4.14 \ \text{V}$

7. $V_C = V_{CC} - I_C R_C$ (10.5)

    $= 19.5 \ \text{V} - 18.9 \ \text{mA} * 560 \ \Omega = 8.88 \ \text{V}$

8. $I_B = \dfrac{I_C}{h_{FE}}$ (9.7)

    $= \dfrac{18.9 \ \text{mA}}{100} = 189 \ \mu\text{A}$

9. $h_{ie} = \dfrac{26 \ \text{mV}}{I_B}$ (11.3)

    $= \dfrac{26 \ \text{mV}}{189 \ \mu\text{A}} = 137 \ \Omega$

Steps 10 and 11 are omitted (see Step 2).

12. $R_B = \dfrac{S}{1 - S} h_{FE} R_E \overset{*}{\approx} \dfrac{1}{10} h_{FE} R_E$ (10.16)

    $= \dfrac{100 * 180 \ \Omega}{10} = 1.80 \ \text{k}\Omega$

13. $V_{BB} = V_B + I_B R_B$ (10.20)

       $= 4.14 \text{ V} + 189 \text{ } \mu\text{A} * 1.80 \text{ k}\Omega = 4.48 \text{ V}$

14. $R_1 = R_B \dfrac{V_{CC}}{V_{BB}}$ STD. (10.27)

      $= 1.80 \text{ k}\Omega \dfrac{19.5 \text{ V}}{4.48 \text{ V}} = 7.82 \text{ k}\Omega$ (Use 8.2 kΩ)

15. $R_2 = R_1 \dfrac{V_{BB}}{V_{CC} - V_{BB}}$ STD. (10.29)

      $= 8.2 \text{ k}\Omega \dfrac{4.48 \text{ V}}{19.5 \text{ V} - 4.48 \text{ V}} = 2.45 \text{ k}\Omega$ (Use 2.7 kΩ)

16. $R_{in} = R_1 \parallel R_2 \parallel (h_{ie} + h_{fe} R_{E1})$ (12.46)

      $= 8.2 \text{ k}\Omega \parallel 2.7 \text{ k}\Omega \parallel 137 \text{ } \Omega = 129 \text{ } \Omega$

17. $R_{out} = R_C \parallel (1/h_{oe} + R_{E1})$ (12.51)

      $= 560 \text{ } \Omega \parallel 20 \text{ k}\Omega = 545 \text{ } \Omega$

18. $C_i = \dfrac{1}{2\pi f_L (R_S + R_{in})}$ (12.6)

     $= \dfrac{1}{2\pi * 100 \text{ Hz}(600 \text{ } \Omega + 129 \text{ } \Omega)} = 2.18 \text{ } \mu\text{F}$ (Use 2.2 μF)

19. $C_o = \dfrac{1}{2\pi f_L (R_{out} + R_L)}$ (12.13)

     $= \dfrac{1}{2\pi * 100 \text{ Hz}(545 \text{ } \Omega + 600 \text{ } \Omega)} = 1.39 \text{ } \mu\text{F}$ (Use 1.5 μF)

20. $C_E = \dfrac{1}{2\pi f_L \left(R_{E2} \left\| \left(R_{E1} + \dfrac{h_{ie} + R_1 \parallel R_2 \parallel R_S}{h_{fe}} \right) \right. \right)}$ (12.48)

     $= \dfrac{1}{2\pi * 100 \text{ Hz} \left(180 \text{ } \Omega \left\| \dfrac{137 \text{ } \Omega + 8.2 \text{ k}\Omega \parallel 2.7 \text{ k}\Omega \parallel 600 \text{ } \Omega}{100} \right. \right)}$

     $= 274 \text{ } \mu\text{F}$ (Use 270 μF)

The results of Example 14.3 are shown on the circuit of Figure 14.5b.

14.5 Capacitor Coupled Circuit Efficiency

Circuit efficiency is defined to be the ratio of the rms signal power delivered to the external load, P_{Lrms} divided by the DC power delivered to the circuit by the power supply, P_{CC}. Or

$$\text{Eff} \equiv \frac{P_{L\text{rms}}}{P_{CC}} \tag{14.10}$$

Efficiency is a unitless term since it is a power divided by a power. It is usually expressed as a percent.

The efficiency of a capacitively coupled Class A amplifier has a theoretical maximum upper limit which we can use as a benchmark against which we can compare the performance of practical circuits. To calculate the theoretical maximum efficiency, assume an ideal transistor ($1/h_{oe} = \infty$) in the circuit of Figure 14.6a in which all biasing resistors, emitter resistors, etc. are ignored. Thus the power delivered by the power supply, P_{CC} will be the same as the power delivered to the collector circuit, P_C. Further, assume that the circuit is optimally designed using ideal components. Thus

$$P_{CC} = P_C; \qquad R_C = R_L; \qquad i_0 = 2I_C \tag{14.11}$$

Then, as you can see from Figure 14.6b, the maximum collector signal current, i_{CP}, is given by

$$i_{CP} = I_C \tag{14.7}$$

or, converting to the effective (rms) value

$$i_{C\text{rms}} = \frac{I_C}{\sqrt{2}} \tag{14.12}$$

Since in this case we are dealing with a theoretical situation, in which $R_C = R_L$, i_C will be divided equally between R_C and R_L. Thus

$$i_{L\text{rms}} = \frac{i_{C\text{rms}}}{2} = \frac{I_C}{2\sqrt{2}} \tag{14.13}$$

Substituting Equation 14.13 into the power equation gives

$$P_{L\text{rms}} = i_{L\text{rms}}{}^2 R_L = \left(\frac{I_C}{2\sqrt{2}}\right)^2 R_L$$

$$= \frac{I_C{}^2 R_L}{8} \tag{14.14}$$

Next an equation for P_{CC}, the DC power consumed by the circuit must be obtained in terms of I_C and R_L. To that end, remembering that the collector circuit is the only circuit to be considered, the power equation can be written

$$P_{CC} = P_C = V_{CC}I_C \tag{14.15}$$

Since an optimum design using ideal components is assumed, the circuit designer's equation can be applied. Solved for V_{CC} it can be written

$$V_{CC} = I_C(R_{DC} + R_{AC}) \tag{14.9}$$

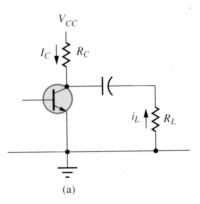

(a)

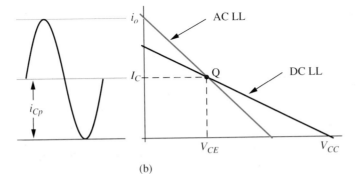

(b)

(c)

FIGURE 14.6 Maximum ideal circuit efficiency.
(a) Circuit diagram.
(b) Load lines.
(c) AC equivalent circuit.

Since R_C is the only resistor considered to be in the collector circuit

$$R_{DC} = R_C \qquad\qquad (14.16)$$

and since $R_C = R_L$

$$R_{AC} = R_C \parallel R_L = \frac{R_C}{2} \qquad\qquad (14.17)$$

Substituting Equations 14.16 and 14.17 into Equation 14.9 yields

$$V_{CC} = I_C\left(R_C + \frac{R_C}{2}\right) = \frac{3}{2}I_C R_C \tag{14.18}$$

Substituting Equation 14.18 into Equation 14.15 yields the needed power equation

$$P_{CC} = \frac{3}{2}I_C^2 R_C \tag{14.19}$$

Substituting Equations 14.14 and 14.19 into Equation 14.10 gives

$$\text{Eff} = \frac{\dfrac{I_C^2 R_L}{8}}{\dfrac{3}{2}I_C^2 R_C} = \frac{I_C^2 R_L}{8} * \frac{2}{3 I_C^2 R_C}$$

Finally, recognizing that $R_C = R_L$ and canceling like factors gives us the theoretical efficiency we have sought

$$\text{Eff} = \frac{1}{12} = 8.33\%$$

This result should be more than a little surprising to you. It says that under the best of conditions, and ignoring the power that must inevitably be lost in the base and the emitter biasing resistors, less than 10 percent of the power taken from the power supply actually shows up in the external signal. This just means that a Class A circuit cannot be used to deliver significant amounts of power without generating large amounts of heat.

Obviously the efficiency of real circuits is less, primarily due to the power consumed by the biasing components. For example, if we were dealing with a real circuit, Equation 14.15 would become

$$P_{CC} = V_{CC}(I_C + I_1) \tag{14.20}$$

where I_1 is the current flowing in R_1 of the base biasing circuit. The easiest way to obtain I_1 is to recognize that

$$I_1 = \frac{V_{CC} - V_B}{R_1} \tag{14.21}$$

EXAMPLE 14.4
Calculate the efficiency of the circuit in Figure 14.5b.

SOLUTION:
Since the power delivered to the load is given, all that is necessary is to calculate the power delivered to the circuit by the DC supply, put the result in Equation 14.10 and solve.

From Step 3 of the design procedure we have $I_C = 18.9$ mA. Step 6 tells us that $V_B = 4.14$ V. Equation 14.21 will then give us I_1, the DC current used by the base circuit.

$$I_1 = \frac{V_{CC} - V_B}{R_1} \qquad (14.21)$$

$$= \frac{19.5 \text{ V} - 4.14 \text{ V}}{8.2 \text{ k}\Omega} = 1.87 \text{ mA}$$

From Equation 14.20 we have the DC power delivered to the circuit, P_{CC}

$$P_{CC} = V_{CC}(I_C + I_1) \qquad (14.20)$$

$$= 19.5 \text{ V}(18.9 \text{ mA} + 1.87 \text{ mA}) = 405 \text{ mW}$$

Finally, the overall efficiency of the circuit is

$$\text{Eff} = \frac{P_{Lrms}}{P_{CC}} \qquad (14.10)$$

$$= \frac{25 \text{ mW}}{405 \text{ mW}} = 0.0617 = 6.17\%$$

In view of the theoretical upper limit of 8.33 percent, the above efficiency is not too bad. It would be most interesting for you to hook up this circuit in your own lab and determine the efficiency experimentally. If you do, you will find that due to saturation and cutoff problems, the maximum signal output will be slightly less than the desired 25 mW. Thus the experimentally determined efficiency will be down around 5 percent.

14.6   Transformer Coupled Circuit Efficiency

In addition to its low efficiency, the capacitor coupled common emitter circuit's relatively high output impedance is often a disadvantage. Both of these problems can be addressed to some extent by using transformer output coupling as shown in Figure 14.7a.

If you review what you learned about transformers and "impedance matching" in your circuits class, you will recognize that by correctly specifying the transformer turns ratio, the designer can cause the load resistance, R_L, to be transformed so that the transistor sees an AC load, R_{AC}, of any required value.

Since the DC collector resistance, R_C of the circuit in Figure 14.7a is just the DC resistance of the transformer primary winding, R_C is typically quite low. Thus the DC load line is quite steep. Since the slope of the AC load line is determined by the AC impedance as seen at the transformer primary winding, the AC load line will be much flatter. These relationships are shown in Figure 14.7b.

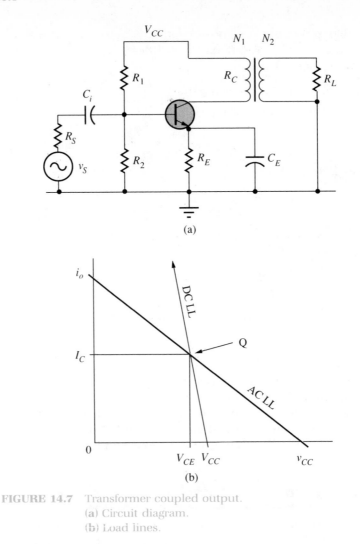

FIGURE 14.7   Transformer coupled output.
(a) Circuit diagram.
(b) Load lines.

You may at first wonder how v_{CC} can ever be greater than V_{CC}. Remember that the transformer primary is in fact an inductor, and as the current in it decreases there is an "inductive kick" which causes the voltage to go sufficiently high to dissipate the energy stored in the transformer's magnetic field. It is this inductive kick that causes v_{CC} to be greater than V_{CC}. In order to handle the inductive kick, the maximum voltage rating for the transistor must be at least twice V_{CE}.

Incidentally, the inductive kick also makes it dangerous to disconnect the loudspeaker of a transformer coupled stereo when the stereo is playing. When R_{AC} becomes "infinite," the AC load line becomes nearly horizontal making v_{CC} so high that the collector-emitter breakdown voltage, BV_{CEO} of the output transistor may be exceeded.

Using the approach of Section 14.5 it is possible to determine the theoretical maximum efficiency for the transformer coupled circuit. Again we will ignore all biasing components as shown in Figure 14.8a, and assume an optimum design using ideal components. In this case, that will include an ideal transformer. Thus, the DC load line will be a vertical line since ideal transformers are assumed to have no internal resistance and we are assuming that there is no emitter resistor. The AC load line will have a slope determined by R_{AC}. These relationships are shown in Figure 14.8b.

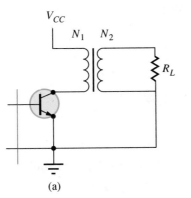

(a)

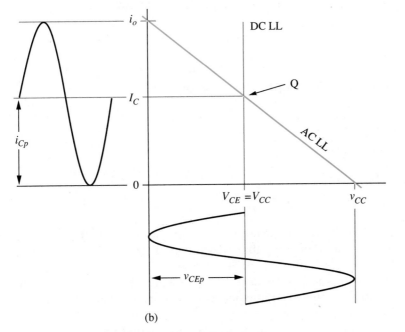

(b)

FIGURE 14.8 Ideal transformer coupled circuit.
(a) Circuit diagram.
(b) Load lines.

To obtain an expression for the AC power output to R_L we recognize that for ideal transformers, power output from the secondary, P_L equals power input to the primary. Thus you can show that

$$P_{Lrms} = v_{CErms}i_{Crms}$$

Or, expressed in terms of peak values

$$P_{Lrms} = \frac{v_{CEP}}{\sqrt{2}} \times \frac{i_{CP}}{\sqrt{2}} = \frac{v_{CEP}i_{CP}}{2} \qquad (14.22)$$

From Figure 14.8b it is evident that

$$V_{CE} = v_{CEP}$$

and

$$I_C = i_{CP}$$

Substituting into Equation 14.22 gives

$$P_{Lrms} = \frac{V_{CE}I_C}{2} \qquad (14.23)$$

Since the DC load line is vertical, you can see that

$$V_{CE} = V_{CC} \qquad (14.24)$$

Substituting Equation 14.24 into Equation 14.23 gives

$$P_{Lrms} = \frac{V_{CC}I_C}{2} \qquad (14.25)$$

Since we assume that the collector circuit is the only one involved, P_{CC}, the power delivered by the power supply is

$$P_{CC} = V_{CC}I_C \qquad (14.15)$$

Substituting Equation 14.25 and 14.15 into Equation 14.10 and canceling common factors we have

$$Eff = \frac{P_{Lrms}}{P_{CC}} = \frac{\dfrac{V_{CC}I_C}{2}}{V_{CC}I_C} = \frac{1}{2} = 50\% \qquad (14.26)$$

Compared to the capacitor coupled circuit, 50 percent efficiency may look fairly good, but remember that this means that ideally, under maximum signal conditions only half of the power from the DC power supply is delivered to the load in the form of a useful signal. The other half is wasted heating up the transistor. Further, since this is a Class A circuit, when no signal is present the transistor must absorb all of the power from the DC power supply. Thus the transistor's power dissipation rating, P_{DISS} is given by

$$P_{DISS} \geq 2P_{Lrms} \qquad (14.27)$$

Just as was the case with capacitor coupled circuits, the efficiency of real transformer coupled circuits will always be less than the ideal due primarily to the power consumed in the biasing circuits. The power consumed and the efficiency can be calculated in exactly the same way as it was done for capacitively coupled circuits using Equations 14.20, 14.21, and 14.10.

Although you will occasionally see transformer coupled circuits in the output of small TVs radios, etc., due to its low efficiency the circuit is rarely used where more than a watt or so of output power is required.

14.7 Transformer Coupled Circuit Design

As you have probably come to expect in circuit design, the designer of a Class A transformer coupled circuit must rely on judgment and past experience in optimizing a design. To give you some idea of the factors involved, refer again to the design procedure outlined in Table 12.1. Steps 1 through 3 will need to be changed.

The circuit we wish to design is that of Figure 14.7a. Let us assume that a given maximum rms power, P_{Lrms} is to be delivered to a given load, R_L such as the loudspeaker of your stereo. We will alter the design procedure of Table 14.1 in order to specify all of the components in Figure 14.7a.

Equation 14.27 gives us P_{DISS} which is one of the transistor specifications. We can then go to a transistor data book and pick a transistor that will meet our needs.

Since the transistor is transformer coupled to the load R_L, we are free to specify the transformer turns ratio such that the transistor will see the optimum value of AC impedance, R_{AC}. One method of obtaining a reasonable value for R_{AC} is as follows:

First we will determine the maximum value allowable, R_{ACmax}. To do that, recognize that v_{CC} on the AC load line must never exceed the collector-emitter breakdown voltage, BV_{CEO}, of the transistor we have chosen. We can then draw the AC load line from BV_{CEO} tangent to the P_{DISS} curve on the transistor characteristic graph. This has been done in Figure 14.9a.

As we showed in Section 14.3, the point of tangency is the Q point for midpoint bias. Thus we can write

$$P_{DISS} = V_{CE}I_C$$

or, substituting from Equation 14.27

$$P_{Lrms} = \frac{V_{CE}I_C}{2} \qquad (14.28)$$

From our work with load lines we know that for midpoint bias

$$V_{CEmax} = \frac{BV_{CEO}}{2} \qquad (14.29)$$

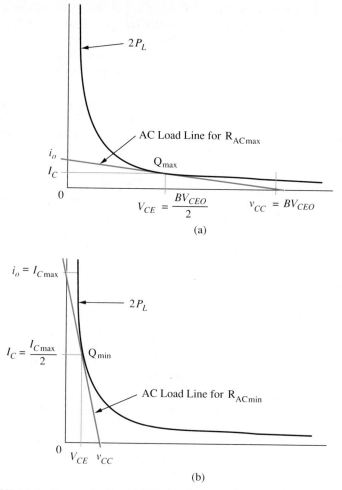

FIGURE 14.9 Determination of R_{AC} for the transformer coupled circuit.
(a) Maximum value of R_{AC}.
(b) Minimum value of R_{AC}.

Also from our work with AC load lines in Section 9.7, we know that for midpoint bias

$$R_{AC} = \frac{V_{CE}}{I_C} \qquad (9.24)$$

or

$$I_C = \frac{V_{CE}}{R_{AC}} \qquad (14.30)$$

Substituting Equations 14.30 and 14.29 into Equation 14.28 gives

$$P_{Lrms} = \frac{BV_{CEO}}{4} \frac{BV_{CEO}}{2R_{ACmax}} = \frac{BV_{CEO}^2}{8R_{ACmax}} \tag{14.31}$$

or

$$R_{ACmax} = \frac{BV_{CEO}^2}{8P_{Lrms}} \tag{14.32}$$

Equation 14.32 sets an upper limit for R_{AC}.

Similarly we can set a lower limit for R_{AC} by recognizing that the load line must never exceed the I_{Cmax} rating for the transistor. As before, if we have midpoint bias, the load line will be tangent to the power curve at the Q point and

$$I_C = \frac{I_{Cmax}}{2} \tag{14.33}$$

These relationships are shown in Figure 14.9b. In this case, we can write the power equation as

$$2P_{Lrms} = I_C^2 R_{ACmin} \tag{14.34}$$

or, substituting Equation 14.33 into Equation 14.34 and solving for R_{ACmin}

$$R_{ACmin} = \frac{2P_{Lrms}}{\left(\frac{I_{Cmax}}{2}\right)^2} = \frac{8P_{Lrms}}{I_{Cmax}^2} \tag{14.35}$$

Equations 14.32 and 14.35 set the upper and lower limits respectively for our choice of R_{AC}. Optimizing the selection of R_{AC} within these limits requires experience and good judgment on the part of the designer. Transistor, transformer, and power supply considerations all have to be weighed. A reasonable first choice for R_{AC} is obtained by averaging these two extreme values. Thus

$$R_{AC} \approx \frac{R_{ACmax} + R_{ACmin}}{2} \tag{14.36}$$

The next step is to go to the manufacturer's catalogs and select an impedance matching transformer that will transform R_L to the desired value of R_{AC}. In the event that a particular catalog provides transformer voltage ratios rather than impedance ratios, it may be necessary to use the impedance matching relationship

$$P_{pri} = P_{sec}$$

or

$$\frac{V_{pri}^2}{R_{pri}} = \frac{V_{sec}^2}{R_{sec}}$$

Thus the required transformer turns ratio, α is given by

$$\alpha \equiv \frac{V_{pri}}{V_{sec}} = \sqrt{\frac{R_{pri}}{R_{sec}}} = \sqrt{\frac{R_{AC}}{R_L}} \qquad (14.37)$$

The DC resistance of the transformer primary winding must be obtained from either the manufacturer's data or from direct ohmmeter measurement. This resistance is the DC collector resistance, R_C.

Next, using a procedure similar to that used in Section 14.4, the collector current, I_C must be calculated. We can write the power equation for the transformer primary as

$$P_{Crms} = (i_{Crms})^2 R_{AC} \qquad (14.38)$$

Substituting $P_{Lrms} = P_{Crms}$ and solving for i_{Crms} gives

$$i_{Crms} = \sqrt{\frac{P_{Lrms}}{R_{AC}}} \qquad (14.39)$$

Again, assuming midpoint bias and the use of the entire AC load line, we can write

$$I_C = i_{CP} = \sqrt{2}\, i_{Crms} \qquad (14.5)$$

Substituting Equation 14.39 into Equation 14.5 gives the circuit designer's equation for a transformer coupled Class A amplifier circuit

$$\bullet \qquad I_C \overset{*}{=} \sqrt{\frac{2P_{Lrms}}{R_{AC}}} \qquad (14.40)$$

The next step is to obtain a value for the emitter resistor R_E. Since the DC resistance of the transformer primary is typically quite low, using Step 2 of the design procedure in Table 12.1 to obtain the value for R_E will not lead to a stable circuit. However, since I_C is already known for this circuit, a reasonable value for R_E can be obtained as follows: If you return to Section 10.5 where the idea of "emitter swamping" was introduced, you will see that since we are assuming that we can specify what ever value of power supply voltage we wish, we can use

$$\bullet \qquad R_E \overset{*}{=} \frac{2\ \text{V}}{I_C} \qquad (14.41)$$

Finally, using the circuit designer's equation in the form presented in Section 14.4, V_{CC} is given by

$$\bullet \qquad V_{CC} \overset{*}{=} I_C(R_{DC} + R_{AC}) \qquad (14.9)$$

As we have done before, the "$\overset{*}{=}$" symbol indicates that the formulas developed in this section are designer's formulas based on the assumption of midpoint bias and use of the whole AC load line. Further, it is assumed that the only restriction placed on the designer is the value of the load resistance and the power that must be delivered to it. The use of these formulas when these assumptions are not met will yield erroneous results.

The design can now be completed by entering the design procedure of Table 12.1 at Step 4.

EXAMPLE 14.5

Design a Class A common emitter transformer coupled amplifier that will deliver 0.25 Wrms of audio power to a 4 Ω load with a low frequency cutoff of 100 Hz. Assume that the DC resistance of the primary winding of the transformer that you specify has a value of $R_C = 15 \ \Omega$.

SOLUTION:

Figure 14.10a contains the required circuit.

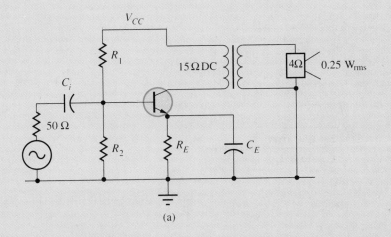

(a)

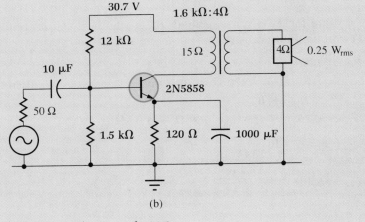

(b)

FIGURE 14.10 Example 14.5.

$$P_{DISS} = 2P_{Lrms} \tag{14.27}$$
$$= 2 * 250 \text{ mW} = 500 \text{ mW}$$

According to the data in Appendix F, a 2N5858 has a $P_{DISS} = 750$ mW so it should do the job with a comfortable safety margin. The rest of the required data are: $BV_{CEO} = 80$ V, $I_{Cmax} = 1$ A and $h_{FE} \approx 100$. Altering Table 12.1 to include the developments in this chapter, we have

$$1.1 \ R_{ACmax} \stackrel{*}{=} \frac{BV_{CEO}{}^2}{8P_{Lrms}} \tag{14.32}$$

$$= \frac{(80 \text{ V})^2}{8 * 250 \text{ mW}} = 3.20 \text{ k}\Omega$$

$$1.2 \ R_{ACmin} \stackrel{*}{=} \frac{8P_{Lrms}}{(I_{Cmax})^2} \tag{14.35}$$

$$= \frac{8 * 250 \text{ mW}}{(1 \text{ A})^2} = 2.00 \ \Omega$$

$$1.3 \ R_{AC} \stackrel{*}{=} \frac{R_{ACmax} + R_{ACmin}}{2} \tag{14.36}$$

$$= \frac{3.20 \text{ k}\Omega + 2.00 \ \Omega}{2} = 1.60 \text{ k}\Omega$$

$$2. \ I_C \stackrel{*}{=} \sqrt{\frac{2P_{Lrms}}{R_{AC}}} \tag{14.40}$$

$$= \sqrt{\frac{2 * 250 \text{ mW}}{1.60 \text{ k}\Omega}} = 17.7 \text{ mA}$$

$$3. \ R_E \stackrel{*}{=} \frac{2 \text{ V}}{I_C} \qquad \text{(STD.   if Step 10 is omitted)} \tag{14.41}$$

$$= \frac{2 \text{ V}}{17.7 \text{ mA}} = 113 \ \Omega \qquad \text{(Use 120 }\Omega\text{)}$$

Since a value for A_v is not specified, R_E is standardized and Steps 10 and 11 will be omitted.

$$3.1 \ P_{RE} = I_C{}^2 R_E$$
$$= (17.7 \text{ mA})^2 * 120 \ \Omega = 37.5 \text{ mW} \qquad \text{(Use 1/4 W)}$$

$$3.2 \ V_{CC} \stackrel{*}{=} I_C(R_{DC} + R_{AC}) \tag{14.9}$$
$$= (17.7 \text{ mA}) (15 \ \Omega + 120 \ \Omega + 1.6 \text{ k}\Omega) = 30.7 \text{ V}$$

$$4. \ V_{BE} \approx 26 \text{ mV} * \ln \frac{I_C}{10 \text{ fA}} \tag{8.4}$$

$$= 26 \text{ mV} * \ln \frac{17.7 \text{ mA}}{10 \text{ fA}} = 0.733 \text{ V}$$

5. $V_E = I_C R_E$ (10.18)

      $= 17.7 \text{ mA} * 120 \ \Omega = 2.12 \text{ V}$

This is close enough to the requirement of Equation 14.41.

6. $V_B = V_{BE} + V_E$ (10.19)

      $= 0.733 \text{ V} + 2.12 \text{ V} = 2.85 \text{ V}$

7. $V_C = V_{CC} - I_C R_C$ (10.5)

      $= 30.7 \text{ V} - 17.7 \text{ mA} * 15 \ \Omega = 30.4 \text{ V}$

8. $I_B = \dfrac{I_C}{h_{FE}}$ (9.7)

      $= \dfrac{17.7 \text{ mA}}{100} = 177 \ \mu\text{A}$

9. $h_{ie} = \dfrac{26 \text{ mV}}{I_B}$ (11.3)

      $= \dfrac{26 \text{ mV}}{177 \ \mu\text{A}} = 147 \ \Omega$

For maximum gain, Steps 10 and 11 of Table 12.1 are omitted.

12. $R_B \overset{*}{\approx} \dfrac{1}{10} h_{FE} R_E$ (10.16)

      $= \dfrac{100 * 120 \ \Omega}{10} = 1.2 \text{ k}\Omega$

13. $V_{BB} = V_B + I_B R_B$ (10.17)

      $= 2.85 \text{ V} + 177 \ \mu\text{A} * 1.2 \text{ k}\Omega = 3.07 \text{ V}$

14. $R_1 = R_B \dfrac{V_{CC}}{V_{BB}}$ STD.   (10.27)

      $= 1.2 \text{ k}\Omega \dfrac{30.7 \text{ V}}{3.07 \text{ V}} = 12.0 \text{ k}\Omega$ (Use 12 kΩ)

15. $R_2 = R_1 \dfrac{V_{BB}}{V_{CC} - V_{BB}}$ STD.   (10.29)

      $= 12 \text{ k}\Omega \dfrac{3.07 \text{ V}}{30.7 \text{ V} - 3.07 \text{ V}} = 1.33 \ \Omega$ (Use 1.5 kΩ)

16. $R_{\text{in}} = R_1 \parallel R_2 \parallel h_{ie}$ (12.46)

      $= 12 \text{ k}\Omega \parallel 1.5 \text{ k}\Omega \parallel 147 \ \Omega = 133 \text{ k}\Omega$

Since the circuit is transformer coupled, Step 17 is omitted.

18. $C_i = \dfrac{1}{2\pi f_L (R_S + R_{\text{in}})}$ (12.6)

      $= \dfrac{1}{2\pi * 100 \text{ Hz} (50 \ \Omega + 133 \ \Omega)} = 8.72 \ \mu\text{F}$ (Use 10 μF)

Since the circuit is transformer coupled, Step 19 is omitted.

20. $C_E = \dfrac{1}{2\pi f_L\left(R_E \left\| \dfrac{h_{ie} + R_1 \| R_2 \| R_S}{h_{fe}}\right.\right)}$ (12.48)

$= \dfrac{1}{2\pi * 100 \text{ Hz}\left(120\ \Omega \left\| \dfrac{147\ \Omega + 12\ \text{k}\Omega \| 1.5\ \text{k}\Omega \| 50\ \Omega}{100}\right.\right)}$

$= 829\ \mu\text{F}$

(Use 1000 μF)

You should notice that the value required for C_E in Example 14.5 above is quite large. This is due to the large currents required in high power circuits. Because it must be so large (and expensive!), it is often omitted in high power circuits. According to Section 12.13, its omission will increase the circuit input impedance and lower the circuit gain. Since the circuit's input impedance is quite low, raising it does not present any difficulty. The designer can often get the required gain less expensively in the voltage amplifier stages which precede the power amplifier in the circuit.

14.8 Circuit Analysis

The circuit analysis procedure presented in Table 12.2 of Section 12.15 can be used for the power circuits discussed in this chapter with some minor variations in the case of the transformer coupled circuit. One obvious variation is in Step 12. Looking at the equivalent circuit shown in Figure 14.11, if we assume that the transformer is ideal, we have

- $$R_{\text{out}} = 1/h_{oe}\left(\frac{N_2}{N_1}\right)^2$$ (14.42)

Similarly, as we have already seen, R_{AC} is R_L transformed by the transformer transformation ratio, or

- $$R_{AC} = R_L\left(\frac{N_1}{N_2}\right)^2$$ (14.43)

The transformer has a further effect on Step 13 of Table 12.2. It becomes

- $$A_v = \left(\frac{N_2}{N_1}\right)\frac{h_{fe}(1/h_{oe} \| R_{AC})}{h_{ie} + h_{fe}R_{E1}}$$ (14.44)

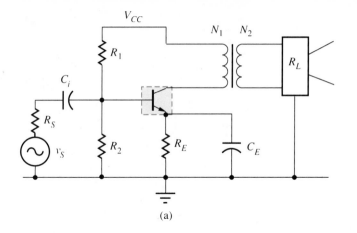

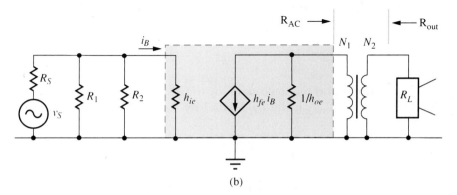

FIGURE 14.11 Transformer coupled equivalent circuit.
(a) Circuit diagram.
(b) Equivalent circuit.

Since we are dealing with power amplifiers, the following step should be included following Step 15

$$P_{Lrms} = \frac{\left(v_{CPmax}\dfrac{N_2}{N_1}\right)^2}{2R_L} \qquad (14.45)$$

●

EXAMPLE 14.6

Calculate the Q point voltages, and the efficiency of the circuit in Figure 14.12.

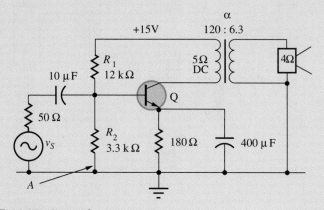

FIGURE 14.12 Example 14.6.

SOLUTION:

From Appendix F the required data for the 2N4401 transistor is: $h_{FE} \approx$ 135. The value of R_{AC} can be calculated by solving Equation 14.37 for $R_{pri} = R_{AC}$.

$$R_{AC} = R_L \left(\frac{N_1}{N_2} \right)^2 \tag{14.45}$$

$$= 4 \ \Omega \left(\frac{120 \ V}{6.3 \ V} \right)^2 = 1.45 \ k\Omega$$

Next, the necessary steps from Table 12.2 are performed

1. $V_{BB} = V_{CC} \dfrac{R_2}{R_1 + R_2}$ (10.22)

$$= 15 \ V \frac{3.3 \ k\Omega}{3.3 \ k\Omega + 12 \ k\Omega} = 3.24 \ V$$

2. $R_B = R_1 \parallel R_2$ (10.23)

$$= 12 \ k\Omega \parallel 3.3 \ k\Omega = 2.59 \ k\Omega$$

3. $S = \dfrac{R_B}{R_B + h_{FE} R_E}$ (10.10)

$$= \frac{2.59 \ k\Omega}{2.59 \ k\Omega + 135 * 180 \ \Omega} = 0.0963 = 9.63\%$$

Although Step 3 is not required, it is comforting to know that the circuit is reasonably stable.

4. $I_C = \dfrac{V_{BB} - V_{BE}}{\dfrac{R_B}{h_{FE}} + R_E}$ (10.12)

$= \dfrac{3.24\ V - V_{BE}}{\dfrac{2.59\ k\Omega}{135} + 180\ \Omega} = \dfrac{3.24\ V - V_{BE}}{199\ \Omega}$

4. $I_C = \dfrac{3.24\ V - V_{BE}}{199\ \Omega} =$	12.8	12.6	mA
5. $V_{BE} = 26\ mV * \ln\dfrac{I_C}{10\ fA} =$	0.725	0.725	V

(8.4)

6. $V_E = I_C R_E$ (10.18)
   $= 12.6\ mA * 180\ \Omega = 2.28\ V$
7. $V_B = V_{BE} + V_E$ (10.19)
   $= 0.725\ V + 2.28\ V = 3.00\ V$
8. $V_C = V_{CC} - I_C R_C$ (10.5)
   $= 15\ V - 12.6\ mA * 5\ \Omega = 14.9\ V$

Steps 6, 7, and 8 give the required Q point voltages.

In order to determine the circuit efficiency, the AC signal power delivered to the load and the DC power taken from the power supply must be obtained. To obtain the AC signal power to the load, we will recognize that most transformers are fairly efficient, so we can assume that the AC power delivered by the transistor to the primary will be the power delivered to the load. To that end, Step 15 of the circuit analysis procedure will give us the maximum unclipped signal amplitude at the collector, v_{Cmax} which is the signal amplitude to the transformer primary.

15. $v_{CPmax} = \begin{Bmatrix} V_C - V_E \\ I_C R_{AC} \end{Bmatrix} \begin{Bmatrix} \text{whichever} \\ \text{is} \\ \text{smaller} \end{Bmatrix}$

$= \begin{Bmatrix} 14.9\ V - 2.28\ V & = 12.7\ VP \\ 12.6\ mA * 1.45\ k\Omega & = 18.3\ VP \end{Bmatrix} = 12.7\ VP$

The Q point is not in the center of the AC load line since the above values of v_{CPmax} are not equal. In order to avoid clipping, the smallest value must be used. Thus

$$v_{CPmax} = 12.7\ VP$$

The AC signal power is then given by

$$P_{Lmax} = \frac{v_{Crms}^2}{R_{AC}} = \left(\frac{v_{CP}}{\sqrt{2}}\right)^2 \frac{1}{R_{AC}} = \frac{v_{CP}^2}{2R_{AC}}$$

$$P_{Lmax} = \frac{12.7 \text{ VP}^2}{2 * 1.45 \text{ k}\Omega} = 55.6 \text{ mWrms}$$

To solve for the DC power delivered by the power supply, P_{CC}, we will use the procedure first described in Section 14.5. The current delivered to the base circuit, I_1 is given by

$$I_1 = \frac{V_{CC} - V_B}{R_1} \tag{14.21}$$

$$= \frac{15 \text{ V} - 3.00 \text{ V}}{12 \text{ k}\Omega} = 1.00 \text{ mA}$$

Substituting in Equation 14.20 gives

$$P_{CC} = V_{CC}(I_C + I_1) \tag{14.20}$$
$$= 15 \text{ V}(12.6 \text{ mA} + 1 \text{ mA}) = 204 \text{ mW}$$

Finally the efficiency can be calculated

$$\text{Eff} \equiv \frac{P_{Lrms}}{P_{CC}} \tag{14.10}$$

$$= \frac{55.6 \text{ mW}}{204 \text{ mW}} = 0.273 = 27.3\%$$

As you can see, this circuit does not come close to the ideal maximum efficiency of 50 percent. The biggest reason is that the Q point is not in the center of the AC load line. You may wish to redesign the circuit and see if you can improve on it. To maximize your learning, you should go to the lab and hook up the circuit to see how close theory agrees with practice.

▶▶▶▶▶▶▶▶ 14.9 Troubleshooting

As we have said so often before, the first step in troubleshooting should be to check the Q point voltages. However, in the case of the transformer coupled circuit, the collector voltage is not the best indicator of circuit operation. As you saw in Example 14.6, the collector voltage for the transformer coupled circuit is usually very close to V_{CC}. The reason is that the resistance of the transformer primary winding is typically very low. Thus even though there may be current flowing, there is very little voltage loss across the transformer primary. As was the case with the emitter follower, the voltage lost across R_E is the best indicator of collector current.

In most power circuits resistances are relatively low. Thus currents can easily become destructively large. This makes it doubly important that you understand the circuit and know what the expected Q point values are before your attempt to troubleshoot the circuit. Further, due to the relatively high power levels involved, when problems develop in a device, the chances are usually quite good that the power amplifier stages will be involved.

EXAMPLE 14.7

Describe the effect on the circuit in Figure 14.12 if a cold solder joint developed at Point A, causing the 3.3 kΩ resistor, R_2, to become ungrounded.

SOLUTION:

The circuit becomes that of Figure 14.13. All of the current originally intended for the voltage divider now flows into the base. In terms of

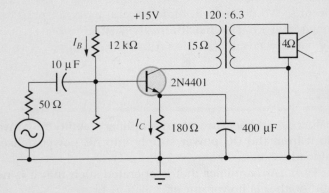

FIGURE 14.13 Example 14.7.

the circuit analysis procedure of Table 12.2, $V_{BB} = V_{CC} = 15$ V and $R_B = R_1 = 12$ kΩ. Entering the procedure at Step 4 gives

4. $I_C = 53.0$ mA	(12.6 mA normal)
5. $V_{BE} = 0.762$ V	(0.725 V normal)
6. $V_E = 9.53$ V	(2.28 V normal)
7. $V_B = 10.3$ V	(3.00 V normal)
8. $V_C = 14.7$ V	(14.9 V normal)

The power dissipated in the emitter resistor, P_{RE} is

$$P_{RE} = V_E I_C$$
$$= 9.53 \text{ V} * 53.0 \text{ mA} = 505 \text{ mW}$$

If R_E is a one-half watt resistor, it will get too hot to touch and may begin to smoke.

The power dissipated in the transistor, P_{DISS} is

$$P_{DISS} = V_{CE} I_C$$
$$= (14.7 \text{ V} - 9.53 \text{ V})(53.0 \text{ mA}) = 274 \text{ mW}$$

This is well below the rated maximum dissipation for the 2N4401 so the transistor may be warm to the touch, but it should not be damaged.

14.10   Summary

In this chapter we studied transistor power dissipation and dissipation ratings. We defined Class A operation and we showed how the design procedure of Table 12.1 can be altered to achieve a specified power output from both Class A capacitively coupled circuits and Class A transformer coupled circuits. We defined amplifier operating efficiency, how it may be determined, and what its maximum theoretical upper limit is for both capacitively coupled circuits and for transformer coupled circuits.

GLOSSARY

Amplifier efficiency: A measure of an amplifier's ability to convert power input to it from the DC power supply into AC power delivered to an external load. (Section 14.5)

Class A amplifier: An amplifier that is operated such that it is never saturated and neither is it ever cut off. (Section 14.2)

Emitter swamping: A biasing technique. See Section 10.5 for an explanation of the technique. (Section 14.7)

Impedance matching: The coupling of a power source to its load in such a way that there is maximum power transfer from the source to the load. (Section 14.6)

Transformer coupled amplifier: An amplifier that achieves DC isolation through the use of a transformer which couples the AC signal to the external load. (Section 14.6)

PROBLEMS

Section 14.3

1. Construct a 600 mW power dissipation curve for a transistor over a range of $I_{Cmax} = 1$ A to $V_{CEmax} = 45$ V.

2. If the transistor in Problem 1 is to have a collector current of $I_C = 25$ mA, what is the maximum allowable value of V_{CE} if the 600 mW power dissipation curve is not to be exceeded?

3. Find the power dissipated by the transistor in the circuit of Figure 14.5b.

4. Repeat Example 14.2 using Q point values of $V_{CE} = 5$ V and $I_C = 100$ mA.

Section 14.4

5. Find the value of I_C and V_{CE} at the point where an AC load line that is determined by $R_{AC} = 500$ Ω is tangent to the 600 mW power dissipation curve drawn in Problem 1.

6. Find the power dissipated in the transistor of Figure 14.5 at the Q point.

Section 14.5

7. For the Q point chosen in Example 14.2, find the instantaneous power dissipated in the transistor at the instant when it is operating on the given AC load line at the following values of V_{CE}:
   (a) $V_{CE} = 9$ V   (c) $V_{CE} = 11$ V
   (b) $V_{CE} = 10$ V (the Q point)
   At which of the above points is the transistor dissipation greatest?

8. Using a Class A capacitively coupled circuit similar to Figure 14.5a, determine the power supply voltage needed to deliver 500 mW of RMS power to a 100 Ω load.

9. The circuit of Figure 14.14 is used in a tape recorder. Use the circuit analysis procedure in Table 12.2 to determine

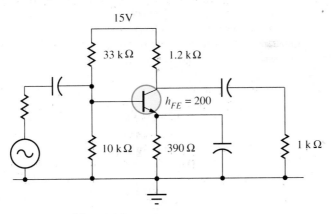

FIGURE 14.14   Problem 14.9.

(a)   The maximum peak amplitude of the output signal, v_{CPmax}.
(b)   The efficiency of the circuit.

Section 14.6

10. Under the ideal conditions described in Section 14.6, if the transistor in a transformer coupled Class A common emitter amplifier is to dissipate 500 mW and if $V_{CC} = 5$ V, find the required value of BV_{CEO} and R_{AC}.

11. For the circuit described in Problem 10, find the maximum power that can be delivered to an external load.

Section 14.7

12. Using the transistor characteristics in Appendix F and the method described in Section 14.7, find the optimum value of R_{AC} for a 2N3904 transistor if the circuit is to supply 100 mWrms to an external load.

13. Design a Class A common emitter transformer coupled amplifier that will deliver 500 mWrms of audio power to an 8 Ω load. Assume that the transformer you specify has a primary winding DC resistance of 10 Ω. Choose the transistor you will use from those listed in Appendix F.

Section 14.8

14. Calculate the efficiency of the circuit designed in Problem 13.

15. Reevaluate resistors R_1 and R_2 for the circuit in Example 14.6 in order to obtain a better Q point.

16. Calculate the efficiency of the circuit designed in Problem 15.

Troubleshooting ▶   ## Section 14.9

17. Predict the circuit behavior if R_1 of Figure 14.12 became an open circuit.

18. Predict the circuit behavior if the emitter bypass capacitor, C_E, of Figure 14.12 became a short circuit.

19. For the circuit of Figure 14.12, calculate the new Q point voltages if the transistor developed the following faults:
    (a)   The collector became an open circuit and the base became shorted to the emitter.
    (b)   The collector became shorted to the emitter and the base became an open circuit.

Class AB is usually used for high-power output.
(Photograph by Owen Franken/Stock, Boston.)

Chapter 15
Complementary Power Amplifiers

SPECIAL TERMS

Class B operation
Complementary circuit
Push-pull circuit
Crossover distortion
Class C operation
Class AB operation
Current mirror
Feedback
Totem pole circuit

15.1 Introduction and Objectives

In the last chapter you learned that Class A amplifiers are not very efficient circuits. That means that if they are to deliver a significant amount of AC power to a load, large power supplies, large transistors, and large heat sinks are required. In this chapter we will study the more efficient, and therefore more frequently used Class AB power amplifier circuit.

After having studied this chapter, you should: Know the definition of Class B, Class AB, and Class C circuits and why the Class B circuit is more efficient than the Class A circuit, you should understand the source of crossover distortion and how it can be minimized, and you should be able to:

- Identify a given circuit as Class A, Class AB, Class B, or Class C.
- Design, analyze, and troubleshoot a simple Class AB complementary push-pull power amplifier.
- Identify crossover distortion in a waveform, its cause and ways to minimize it.

15.2 Class B Operation

In Chapter 14 we defined Class A operation as operation such that the circuit conducts current all of the time. It is that requirement that causes the Class A circuit to have such low efficiency. In fact, since the power drawn from the DC supply, P_{CC} remains constant, the efficiency will approach zero as the input signal amplitude becomes small. To improve the efficiency it would be desirable for the DC supply current to be directly proportional to the signal amplitude. Then the smaller the signal, the cooler the circuit would run. Although your calculator might have some trouble calculating the efficiency when the signal amplitude became zero, the circuit would generate very little heat under that condition. That is exactly what Class B operation is intended to accomplish.

Class B operation is defined to be operation such that conduction occurs exactly half of the time.

If the diode is ideal, the circuit in Figure 15.1a will qualify as a Class B circuit. Figure 15.1b illustrates the fact that conduction will occur exactly half of the time.

One way to get Class B operation in a transistor circuit would be to operate an ideal transistor with no DC bias. Such a circuit is illustrated in Figure 15.2a. If there is no DC bias, the transistor will be cut off and the Q point will be at V_{CC} on the load line as illustrated in Figure 15.2b. The ideal transistor will conduct only when the signal is positive. When the signal is

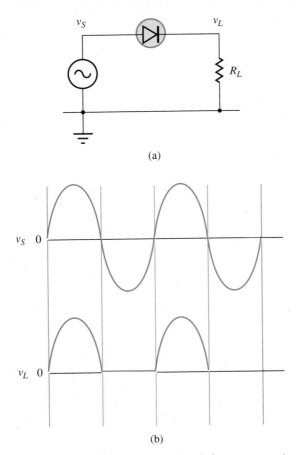

(a)

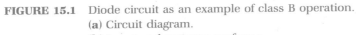

(b)

FIGURE 15.1 Diode circuit as an example of class B operation.
(a) Circuit diagram.
(b) Input and output waveforms.

zero or negative, the B-E junction will not be forward biased so no conduction will occur. This is illustrated in Figure 15.2b. Since, ideally the
collector current is half-wave rectified, you may recall from our study of
rectifiers in Section 4.5 that the average collector current is given by

$$I_{Cav} = \frac{i_{CP}}{\pi} \tag{4.8}$$

Thus the DC supply current is directly proportional to the amplitude of
the AC signal current and, according to our argument at the beginning of
this section, it should be a fairly efficient circuit.

The difficulty with the circuit is that it is not midpoint biased. Since
the Q point is all of the way to one end of the load line, fully half of the

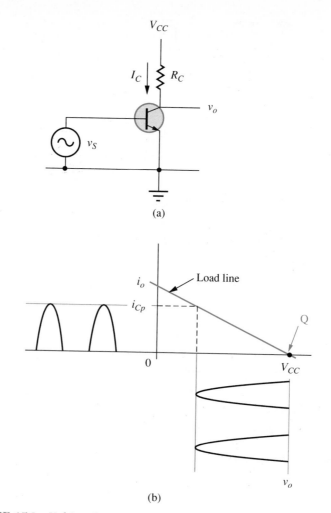

FIGURE 15.2   Unbiased transistor circuit as an example of class B operation.
(a) Circuit diagram.
(b) Load line.

signal waveform is clipped. That constitutes an intolerable amount of distortion. What is needed is to have the circuit cut off at the Q point for the sake of high efficiency and also to have the circuit midpoint biased for the sake of low distortion. Several circuits have been developed which attempt to achieve these apparently contradictory goals. Early attempts were not very sophisticated, often involving the use of one or more transformers. Given the cost, size, weight, and frequency characteristics of transformers, it is easy to see why other approaches have been sought. We will look rather carefully at one such circuit.

15.3 The Ideal Complementary Circuit

Perhaps the most popular output circuit is the complementary symmetry push-pull circuit. An idealized version of the circuit appears in Figure 15.3a.

There are three important features of the circuit in Figure 15.3a that you should notice. The first thing you should notice is that there are two resistors labeled "R_1." They are labeled the same because ideally they have the same resistance value. The circuit is symmetrical. As you will see shortly, this is necessary in order to provide midpoint bias. If the circuit of Figure 15.3a is midpoint biased

$$V_{B1} = V_{B2} = V_B = \frac{V_{CC}}{2} \tag{15.1}$$

The second important feature of the circuit of Figure 15.3a that you should notice is that it is, in fact, an emitter follower circuit. No matter which transistor you look at, the signal output is taken from the emitter. We spent some time in Section 13.2 pointing out that output from the emitter is one of the identifying characteristics of an emitter follower circuit. Q_2 takes the place of the emitter resistor for Q_1, and Q_1 takes the place of the emitter resistor for Q_2. Since the circuits we will be studying in this chapter are based on the emitter follower, if you have forgotten about them, you should take the time at this point to review Chapter 13.

The third important feature of the circuit of Figure 15.3a is that the transistors are NOT identical, they are complementary. That means that one, Q_1 is an NPN type while the other, Q_2 is a PNP type. Otherwise, their characteristics, h_{fe}, $1/h_{oe}$, etc. are ideally identical.

On a practical basis, it is difficult to meet the above ideal. A number of transistor types are intended to be complements. The 2N3904 and the 2N3906, the 2N4401 and the 2N4403, or the 2N3055 and the MJ2955 are good examples. Some complementary transistors are matched by the manufacturer and are sold together under a single part number. An example is the PTC1956 which consists of a PTC195 and a matching PTC196.

The easiest way to deal with a PNP transistor is to remember that all voltages and currents are reversed compared to the more familiar NPN circuit. The reversed current flow is indicated by the direction of the arrow on the emitter. The arrow indicates that for PNP transistors, current flows into the emitter terminal rather than out of it as it does in the NPN transistors we are used to. The reversed voltage polarity can be provided by either using a negative collector supply or, as in this case, by "turning the transistor upside down," grounding the collector and biasing the emitter positive. Then current will tend to flow in the direction indicated by the arrow on the emitter.

Since Q_2 is inverted, it may look a bit more normal to you if you turn the circuit upside down (or if you stand on your head!).

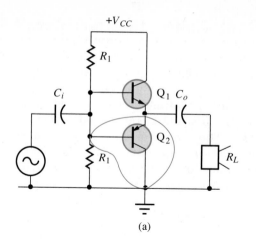

(a)

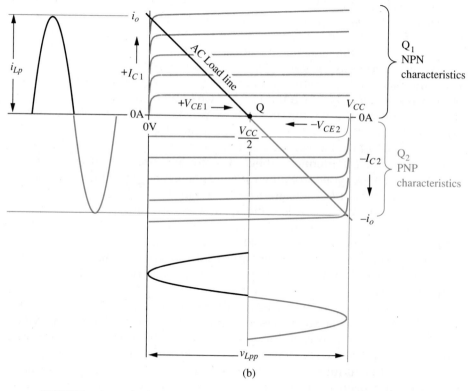

(b)

FIGURE 15.3 Ideal complementary circuit.
(a) Circuit diagram.
(b) Load line.

Since the circuit of Figure 15.3a is an emitter follower, then from Chapter 13 you know that since the emitter follows the base, ideally, if V_B is half of V_{CC}, then V_E should also be half of V_{CC}. Thus using identical resistors in the voltage divider biasing circuit meets our requirement for midpoint bias.

The word "ideally" used so liberally in this chapter means ignore a certain 0.7 V that may be bothering you. Since the bases are tied together and so are the emitters, neither base junction can be forward biased. Therefore, both transistors are shut off and $I_C = 0$ A. This meets our requirement for an efficient circuit.

Figure 15.3b shows the AC load line and the Q point for the ideal circuit of Figure 15.3a. You will notice that the characteristics for both transistors are represented. Since all voltages and currents for the PNP transistor are negative, the characteristics for it appear inverted, just as they would be displayed on a curve tracer. This allows us to picture the behavior of both transistors simultaneously using a single Q point and AC load line. Notice that the diagram indicates that at the Q point, $I_C = 0$ A. It also indicates that we have midpoint bias. Thus this circuit achieves both of the "apparently contradictory goals" we described in Section 15.2.

The name given to the circuit of Figure 15.3a is descriptive since, as we have mentioned, the transistors are complementary, and also the circuit is symmetric. The following explanation of the circuit's operation will show you why it is known as a *push-pull circuit*:

As shown in Figure 15.4a, when the signal is positive, Q_1 turns on and the circuit's operation moves from the Q point up the load line of Figure 15.3b. Current flows from V_{CC} through Q_1 and charges C_o. This pushes current down through R_L to ground. At the same time, the positive signal reverse biases Q_2. Thus none of the current flowing from V_{CC} can flow through Q_2 to ground. All of the current from the DC supply is used to charge C_o, and so to drive R_L.

As shown in Figure 15.4b, when the signal is negative, Q_1 is reverse biased and so it ceases to conduct. At the same time, the negative signal turns on Q_2 and the circuit's operation moves from the Q point down the load line of Figure 15.3b. Current flows from C_o through Q_2 to ground. As C_o discharges it pulls current up through R_L from ground. Thus the circuit "pushes" and "pulls" on alternate halves of the AC cycle. The resulting current waveforms are shown in Figure 15.4c.

The name "push-pull" is derived from the idea that on one half of the AC cycle one device "pushes" current into the load, and on the other half of the cycle another device "pulls" current from the load.

Notice that the power supply delivers current on only the positive half of the AC cycle and the transistors each conduct on alternate half cycles. Ideally, the load receives the full AC signal waveform. Further, all of the current from V_{CC} is used to drive R_L. Thus the circuit should be very efficient.

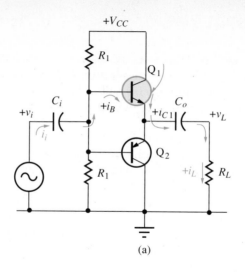

(a)

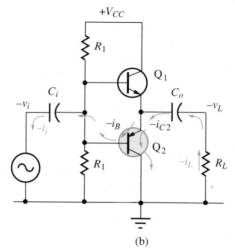

(b)

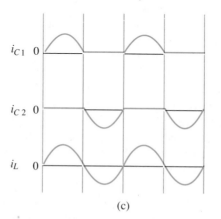

(c)

FIGURE 15.4 Ideal complementary circuit operation
(a) Current flow on the positive half of the input cycle. Q_1 conducts.
(b) Current flow on the negative half of the input cycle. Q_2 conducts.
(c) Current flow through each transistor, and the resulting current flow in R_L.

15.4 Ideal Circuit Efficiency

As we have done with the circuits in Chapter 14, we can calculate the maximum circuit efficiency for the complementary push-pull circuit under ideal conditions. As before, this will give us a bench mark against which to judge real circuits. We will again ignore circuit biasing and limitations imposed by nonideal components.

Recall from Section 14.5 that circuit efficiency is given by

$$\text{Eff} \equiv \frac{P_{Lrms}}{P_{CC}} \tag{14.10}$$

From Figure 15.3b you can see that the ideal maximum signal is

$$v_{LPP} = V_{CC}$$

or, converting to rms

$$v_{Lrms} = \frac{v_{LPP}}{2\sqrt{2}} = \frac{V_{CC}}{2\sqrt{2}} \tag{15.2}$$

From Figure 15.3b you can also see that ideally

$$i_{LP} = i_0$$

or, converting to rms

$$i_{Lrms} = \frac{i_{LP}}{\sqrt{2}} = \frac{i_0}{\sqrt{2}} \tag{15.3}$$

Substituting Equations 15.2 and 15.3 into the power equation gives

$$P_{Lrms} = v_{Lrms} i_{Lrms} \tag{15.4}$$
$$= \frac{V_{CC}}{2\sqrt{2}} * \frac{i_0}{\sqrt{2}}$$
$$= \frac{V_{CC} i_0}{4}$$

Turning our attention to the power from the DC supply, Figure 15.4c indicates that the current delivered to the circuit by the DC supply, which is i_{C1}, has a half-wave rectified waveform. Remember that in this analysis we are ignoring any current used in the biasing circuit. From Section 4.5 we know that the DC average of a half-wave rectified waveform is given by

$$I_{DC} = \frac{I_P}{\pi} \tag{4.8}$$

or, using the notation we have been using

$$I_{DC} = \frac{i_0}{\pi} \tag{15.5}$$

Substituting Equation 15.5 into the power equation gives

$$P_{CC} = V_{CC}I_{DC} = \frac{V_{CC}i_0}{\pi}$$ (15.6)

Substituting Equations 15.4 and 15.6 into Equation 14.10 gives

$$\text{Eff} = \frac{P}{P_{CC}} = \frac{\dfrac{V_{CC}i_0}{4}}{\dfrac{V_{CC}i_0}{\pi}} = \frac{\pi}{4} = 0.785 = 78.5\%$$

After some of the circuit efficiencies we observed in Chapter 14, this is not too bad.

Another advantage this circuit has for use as a power amplifier is that since it is an emitter follower type circuit, as you know from Chapter 13, it tends to have low output impedance. Both its efficiency and its low output impedance make the complementary symmetry push-pull circuit very attractive as the output stage of an amplifier.

15.5 Crossover Distortion

As you can see, the ideal complementary symmetry circuit is a very simple circuit. The difficulty is that ideal conditions seldom exist. Since we wish to operate the transistors at cutoff, the nonlinearity of the B-E junction is a serious problem.

In fact, when a signal is input to the circuit of Figure 15.4, if we assume that $V_E = V_{CC}/2$ then on the positive going input signal swing, V_B must reach something like $V_{CC}/2 + 0.6$ V before Q_1 will begin to turn on. In other words, the signal must reach $v_i = +0.6$ V before any signal is seen at the output. Similarly, on the negative signal swing, the signal must reach $v_i = -0.6$ V before Q_2 begins to turn on and a signal begins to be seen at R_L. Thus if you hooked up the circuit of Figure 15.4 and observed the waveforms, and you should, you would see something like Figure 15.5. The circuit has a "dead zone" when the signal crosses over the zero line. This is known as "crossover distortion." Crossover distortion is distortion that results from nonlinearity of circuit components near cutoff. It produces a rather characteristic "tinny" sound in a loud speaker which the high fidelity enthusiasts variously refer to as a "hard," or a "brittle" sound.

The cause of crossover distortion in the circuit of Figure 15.4 is that there is no base-emitter bias voltage, V_{BE}. Therefore the signal must supply the transistor turn on voltage.

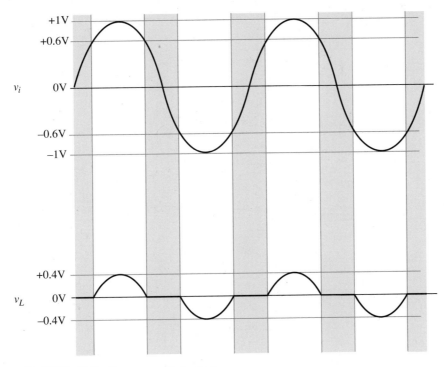

FIGURE 15.5 Crossover distortion.

EXAMPLE 15.1

If the input AC signal voltage, v_i, for the circuit of Figure 15.6 has a peak amplitude of 1 VP, calculate the peak amplitude of the output AC signal voltage, v_{LP}.

● (For a PSpice solution, see Appendix H.)

SOLUTION:

Since Q_2 can be assumed to be turned off when $v_{iP} = +1$ VP, at that instant in the analysis Q_2 can be ignored. Doing an AC KVL walk from v_i through the base of Q_1, through C_o, and R_L to ground then gives

$$v_{iP} - V_{BEP} - i_{LP}R_L = 0V$$

or

$$i_{LP} = \frac{v_{iP} - V_{BEP}}{R_L}$$

$$= \frac{1\ VP - V_{BEP}}{8\ \Omega} \tag{15.7}$$

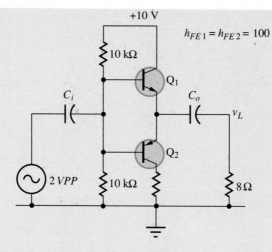

FIGURE 15.6   Example 15.1.

The familiar junction equation gives us

$$V_{BEP} = 26 \text{ mV} * \ln\left(\frac{I_{LP}}{10 \text{ fA}}\right)$$ (8.4)

Iterating Equations 15.7 and 8.4 as we have done so many times before gives

$I_{LP} = \dfrac{1 \text{ VP} - V_{BEP}}{8 \text{ }\Omega}$ =	37.5	30.9	31.5	mA
$V_{BEP} = 26 \text{ mV} * \ln\dfrac{I_{LP}}{10 \text{ fA}}$ =	0.753	0.748	0.748	V

Now we can write

$$v_{LP} = v_{iP} - V_{BEP}$$
$$= 1 \text{ VP} - 0.784 \text{ V} = 0.252 \text{ VP}$$

15.6   Class C Operation

If you look at Figure 15.5, you will notice that in fact the circuit is not operating Class B as we defined it in Section 15.2. Conduction is not half of the time, it is less. This type of operation is known as *Class C operation*. Class C operation is defined to be operation such that conduction occurs less than half of the time.

Almost all radio transmitters are operated Class C. The reason is that the distance the signal will radiate is strongly related to the instantaneous voltage of the signal. Since the lower voltage will not radiate very far anyhow, there is no point in turning the transmitter on except near the peaks of the RF waveform.

15.7 Class AB Operation

In fact, true Class B operation is difficult to achieve. As you probably have guessed by now, in electronics it is difficult to achieve anything EXACTLY. Operation is almost always either less than half of the time, or more than half of the time, rarely exactly half of the time. Operation that is between Class A (100 percent) and Class B (50 percent) is called *Class AB operation.* Class AB operation is defined to be operation such that conduction occurs less than 100 percent of the time, but more than 50 percent of the time. It is operation between Class A and Class B.

If a complementary circuit is operated Class AB, crossover distortion will be minimized (the purists would say it is never eliminated), but if the operation is close to Class B, the efficiency may still be acceptable. Here again we see the need for compromise in circuit design.

In order to draw a load line that represents, to some degree, the operation of a Class AB complementary symmetry circuit, the NPN characteristics and the PNP characteristics can be overlapped as shown in Figure 15.7. This representation shows that both transistors are slightly turned on and it also shows that we still have midpoint bias.

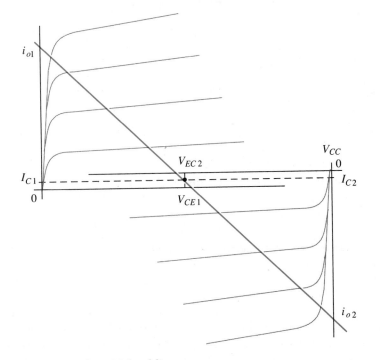

FIGURE 15.7 Class AB load line.

15.8   Diode Biasing

In order to have Class AB operation, V_{BE1} must be something near $+0.6$ V and V_{BE2} must be something near -0.6 V. In other words, V_{B1} must be something like 1.2 V higher voltage than V_{B2}.

Figure 15.8 shows a very neat way to obtain the required bias. Since both the diodes and the transistors are composed of P-N junctions, in principle their characteristics should match, giving us exactly the bias we need. This circuit arrangement is known as a "current mirror." In order to understand why, looking first at the diode circuit, we can write

$$V_{B1} - V_{B2} \approx 2 * 26 \text{ mV} * \ln \frac{I_D}{10 \text{ fA}} \qquad (15.8)$$

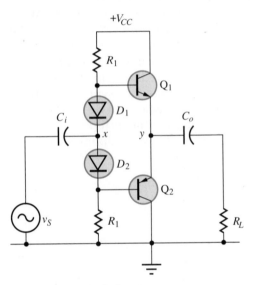

FIGURE 15.8   Unstabilized diode biasing.

Then looking at the transistor circuit, we can write

$$V_{B1} - V_{B2} \approx 2 * 26 \text{ mV} * \ln \frac{I_C}{10 \text{ fA}} \qquad (15.9)$$

From the above rather uncertain relationships we can make the approximation that

$$I_C \approx I_D \qquad (15.10)$$

Thus the collector current mirrors the diode current. In fact, in order to improve their mirroring properties, occasionally the transistors and the diodes that are intended for this type of circuit are matched and sold by the manufacturer under the same part number.

In practical devices using complementary circuits you will often find the diodes and the transistors mounted on the same heat sink. The reason is that as the transistor begins to heat up, V_{BE} begins to decrease tending to cause "thermal runaway" which we originally described in Section 3.12. In view of the resistances usually present in this type of circuit, thermal runaway can easily be catastrophic. However, if the biasing diode is in the same thermal environment as the transistor, as the transistor warms up and V_{BE} begins to decrease, the diode forward voltage also decreases. This reduces the bias and so tends to shut the transistor off inhibiting runaway.

EXAMPLE 15.2

Draw the load line and locate the Q point for the circuit in Figure 15.9a.

SOLUTION:

The current mirror idea tells us that $I_C \approx I_D$. Thus, to find I_C we will find I_D by doing a KVL analysis of the base biasing circuit. Since I_B is small compared to I_D, it can be ignored. Thus

$$V_{CC} - I_D R_1 - 2\,V_D - I_D R_1 = 0\ \text{V}$$

or

$$
\begin{aligned}
I_D &= \frac{V_{CC} - 2\,V_D}{2R_1} \\
&= \frac{9\ \text{V} - 2\,V_D}{2 * 1.2\ \text{k}\Omega} = \frac{4.5\ \text{V} - V_D}{1.2\ \text{k}\Omega}
\end{aligned}
\tag{15.11}
$$

Iterating Equation 15.11 and the junction equation gives

$I_D = \dfrac{4.50\ \text{V} - V_D}{1.2\ \text{k}\Omega} =$	3.17	3.18	3.18	mA
$V_D = 26\ \text{mV} * \ln \dfrac{I_D}{10\ \text{fA}} =$	0.689	0.689	0.689	V

Using the current mirror idea

$$I_{C1} = I_D \tag{15.10}$$
$$= 3.18\ \text{mA}$$

Due to the symmetry of the circuit of Figure 15.9a

$$V_E = \frac{V_{CC}}{2} = \frac{9\ \text{V}}{2} = 4.5\ \text{V}$$

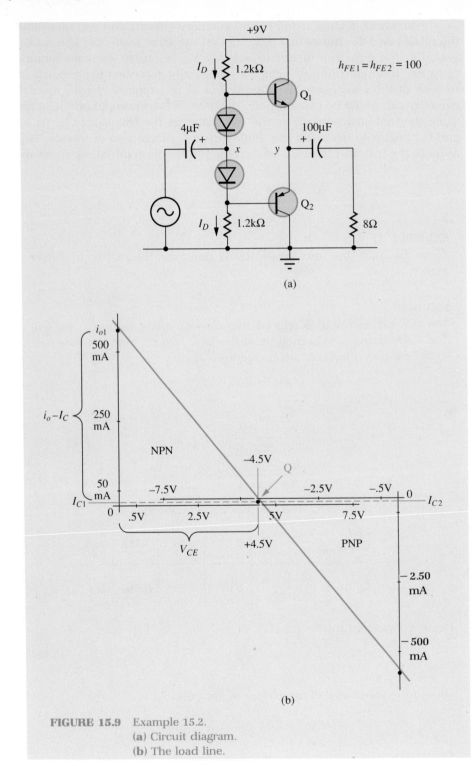

FIGURE 15.9 Example 15.2.
(a) Circuit diagram.
(b) The load line.

Thus

$$V_{CE1} = V_{C1} - V_E$$
$$= 9 \text{ V} - 4.5 \text{ V} = 4.5 \text{ V}$$

This locates the Q point for Q_1. A similar argument would locate the Q point for Q_2.

In order to draw the load line, we need to find i_0. This can be done by using the approach of Section 9.6 where we talked about the slope of the load line. From Figure 15.9b we can write

$$R_{AC} = \frac{V_{CE}}{i_0 - I_C}$$

If $1/h_{oe}$ is ignored, as we did in Chapter 13 for emitter follower circuits, from Figure 15.9a you can see that the only resistance in the circuit from an AC standpoint is R_L. Thus

$$i_{01} = \frac{V_{CE}}{R_{AC}} + I_C$$
$$= \frac{4.5 \text{ V}}{8 \text{ }\Omega} + 3.18 \text{ mA} = 566 \text{ mA}$$

Using these data and the idea of overlapping the characteristics for the NPN and the PNP transistors, the load line is drawn in Figure 15.9b.

15.9 Circuit Stabilization

As we mentioned in Section 15.8 when we introduced the current mirror idea, due to the use of the rather uncertain junction equation, diode biasing alone is not as precise as we would like it to be.

EXAMPLE 15.3
To find out what happens to the current mirror if the diodes and the transistors have different characteristics, assume that the junction equation for the diodes in Figure 15.10 is

$$V_D = 30 \text{ mV} * \ln \frac{I_D}{10 \text{ fA}} \tag{15.12}$$

and assume that the usual junction equation fits the transistors. Calculate and compare the diode current and the collector current.

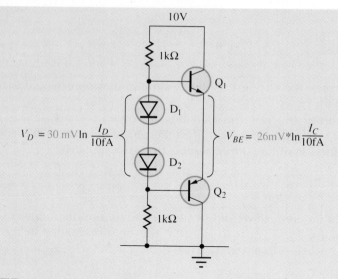

FIGURE 15.10 Example 15.3.

SOLUTION:

Ignoring the base current and doing a KVL walk from V_{CC} to ground through the diode circuit as we have done before, we can write the equation for the diode circuit

$$V_{CC} - 2\,I_D R_1 - 2\,V_D = 0\text{ V}$$

or

$$I_D = \frac{V_{CC} - 2\,V_D}{2\,R_1} \tag{15.11}$$

$$= \frac{10\text{ V} - 2\,V_D}{2 * 1\text{ k}\Omega} = \frac{5\text{ V} - V_D}{1\text{ k}\Omega} \tag{15.13}$$

Iterating Equations 15.13 and 15.12 gives

$I_D = \dfrac{5\text{ V} - V_D}{1\text{ k}\Omega} =$	4.30	4.20	4.20	mA
$V_D = 30\text{ mV} * \ln\dfrac{I_D}{10\text{ fA}} =$	0.804	0.803	0.803	V

Due to the symmetry of the circuit you can see that

$$V_{BE} = V_D = 0.803\text{ V}$$

Using the usual junction equation for the transistors we have

$$V_{BE} = 26\text{ mV} * \ln\frac{I_C}{10\text{ fA}} \tag{8.4}$$

Solving Equation 8.4 for I_C gives

$$I_C = 10 \text{ fA} * e^{V_{BE}/26 \text{ mV}}$$

$$= 10 \text{ fA} * e^{0.803 \text{ V}/26 \text{ mV}} = 258 \text{ mA} \qquad (15.14)$$

The diode current is not mirrored in the collector circuit at all. The collector current of 258 mA is over 50 times the diode current of 4.20 mA.

Although you will rarely see diodes whose characteristics vary as much as those in Example 15.3, you can see that the circuit is not very stable. In Section 10.5 you learned how to improve the stability of transistor circuits by adding an emitter resistor. We will take the same approach here. In order to preserve the symmetry of the circuit an emitter resistor is usually added to each transistor as shown in Figure 15.11.

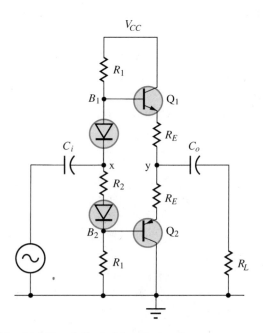

FIGURE 15.11 Stabilized diode biased circuit.

In Chapter 10 you learned that to have good stability R_E should be about one-third of R_C. Obviously that approximation will not work for this circuit. You also learned that good stability will result if there is about 2 V across R_E. That approach is not satisfactory either due to the fact that we wish to have very close to Class B operation. That means that if we have 2 V across R_E at the Q point, we will lose too much of the peak signal voltage across R_E. It is the designer's responsibility to decide on the best compromise between stability and efficiency. A reasonable minimum value is

$$R_E \approx \frac{R_L}{10} \tag{15.15}$$

Equation 15.15 will assure us that the effect of R_E on the signal will be small.

In order to enable the designer to control the difference in base voltages and thus to control the value of I_C, it is customary to include the resistor R_2 in the base biasing circuit as shown in Figure 15.11. You might think that two R_2 resistors should be used to preserve symmetry in the circuit. This is not usually done because the value of R_2 is usually small compared to other base circuit resistances. Thus its presence is not significant. In circuits where precise control is required, such as in high power stereo sets, R_2 is often an adjustable resistor so that service personnel can optimize the value of I_C. We will discuss this resistor more fully in Section 15.11.

EXAMPLE 15.4
Repeat Example 15.3 for the circuit of Figure 15.12.

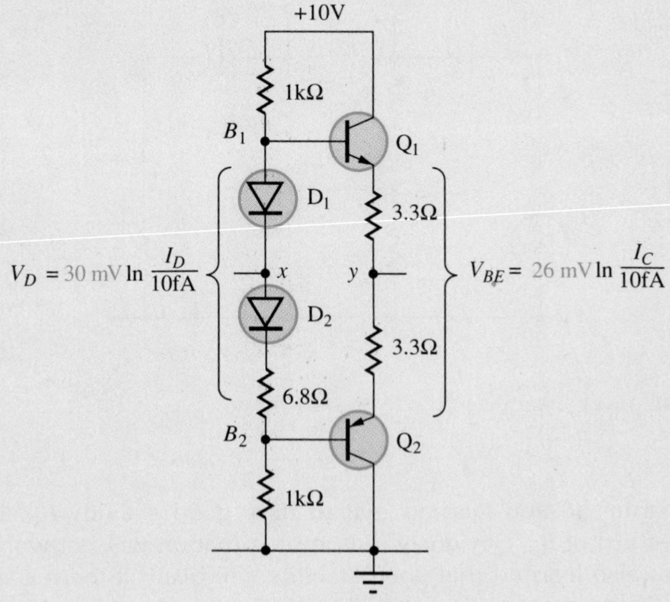

FIGURE 15.12   Example 15.4.

SOLUTION:

Repeating the KVL walk through the diode circuit gives

$$V_{CC} - 2\,V_D - I_D(2\,R_1 + R_2) = 0\ \text{V}$$

or

$$I_D = \frac{V_{CC} - 2\,V_D}{2\,R_1 + R_2}$$

$$= \frac{10\ \text{V} - 2\,V_D}{2 * 1\ \text{k}\Omega + 6.8\ \Omega} = \frac{5\ \text{V} - V_D}{1.0034\ \text{k}\Omega} \tag{15.16}$$

Iterating Equation 15.16 and the assumed diode equation, Equation 15.12 gives

$I_D = \dfrac{5\ \text{V} - V_D}{1.0034\ \text{k}\Omega}$	=	4.29	4.18	4.18	mA
$V_D = 30\ \text{mV} * \ln\dfrac{I_D}{10\ \text{fA}}$	=	0.804	0.803	0.803	V

Doing a KVL walk from B_1 to B_2 through the diodes gives

$$V_{B1,2} = V_{B1} - V_{B2} = 2\,V_D + I_D R_2$$
$$= 2 * 0.803\ \text{V} + 4.18\ \text{mA} * 6.8\ \Omega = 1.63\ \text{V}$$

Doing a KVL walk from B_1 to B_2 through the transistors gives

$$V_{B1,2} = 2\,V_{BE} + I_C * 2\,R_E \tag{15.17}$$

or

$$I_C = \frac{V_{B1,2} - 2\,V_{BE}}{2\,R_E}$$

$$= \frac{1.63\ \text{V} - 2\,V_{BE}}{2 * 3.3\ \Omega} = \frac{0.815\ \text{V} - V_{BE}}{3.3\ \Omega} \tag{15.18}$$

Iterating Equation 15.18 and the usual base-emitter junction equation, Equation 8.4 gives

I_C	34.8	19.4	24.0	22.4	22.9	22.7	22.8	mA
V_{BE}	0.715	0.736	0.741	0.739	0.740	0.740	0.740	V

The collector current of 22.8 mA is still over five times the diode current of 4.22 mA. Thus the circuit is still not sufficiently stable to handle the rather extreme variation in component characteristics postulated here, but as you can see, it is about 10 times better than the unstabilized circuit. It is also adequate to handle the variations you are likely to encounter in practice.

15.10   Signal Amplification

If you have tried to harmonize the description of how the complementary push-pull circuit works as given in Section 15.3, with the diode biasing circuit in Figure 15.8, or the more complete circuit in Figure 15.11, you might have had a little trouble. Figure 15.4 shows the positive part of the AC signal turning Q_1 on, but D_1 in Figure 15.11 will prevent the positive signal from reaching Q_1. Also on the negative part of the AC signal, D_2 will prevent the negative signal from reaching Q_2.

Of course you are right, no positive charges can ever flow from the input to the base of Q_1 in Figure 15.11. What actually happens is that as the positive signal causes the voltage at point X to go above its Q point value of $V_{CC}/2$, the forward bias on diode D_1 is reduced. Therefore, it conducts less forward current. This means that the current flowing from V_{CC} down through R_1 is reduced. Thus there is less voltage lost across R_1. In other words, the voltage at the base of Q_1 increases. Since the circuit is an emitter follower, the output voltage follows the base, producing the positive going output signal. A similar activity occurs with respect to Q_2 on the negative signal swing, causing V_{B2} and the output voltage to decrease. This produces the negative going output signal.

If you hook up the circuit of Example 15.3 or 15.4, and you should, you will discover that the presence of the biasing diodes prevents the circuit of Figure 15.11 from using the entire load line. To understand why, let us assume that at the peak of the positive input voltage swing, V_{XP}, the peak voltage at Point X in Figure 15.11 goes all the way up to V_{CC}. As shown in Figure 15.13, this stops all current from flowing through D_1. V_{B1} would

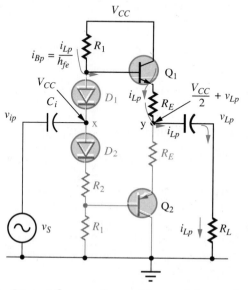

FIGURE 15.13   Current flow on the positive peak input signal swing.

then go all the way up to V_{CC} too except for the fact that the peak base current, i_{BP} must still flow from V_{CC} through R_1 to Q_1. Therefore R_1 forms a voltage divider with Q_1 and the emitter circuit. The only way to make V_{B1} go up close to V_{CC} is to make R_1 appear small compared to the resistance in the emitter circuit, $R_E + R_L$. However, if R_1 is made too small, too much current will be wasted in the biasing circuit. Again the circuit designer must decide on a compromise.

We encountered a similar situation with reference to the base biasing circuit for the common emitter circuit. There we found a reasonable compromise to be

$$R_B \approx \frac{h_{FE}R_E}{10} \tag{10.17}$$

We will take a similar tack here. Notice that in the complementary circuit, if the diodes and R_2 are ignored, the resistance that is analogous to R_B of the common emitter circuit is $R_1 \parallel R_1 = R_1/2$. Also, notice that in the complementary circuit, the AC resistance seen in the emitter circuit when either transistor conducts is $R_E + R_L$. Thus, for the complementary circuit Equation 10.17 becomes

$$R_1 \approx \frac{h_{FE}(R_E + R_L)}{5} \tag{15.19}$$

Notice that Equation 15.19 shows us that the higher the gain of the transistor, the larger R_1 can be, and so the smaller the current wasted in the biasing circuit. Thus a Darlington circuit, or the use of Darlington power transistors will improve the circuit efficiency.

Incidentally, you should notice that the above choice for R_1 destroys the current mirror idea. The current flowing in the base biasing diodes will be much higher than the collector current.

To determine the value of power supply voltage, V_{CC}, needed to deliver a given amount of power to a given load resistance, R_L, we will first find the peak current, i_{LP} required by the load. The power equation gives us

$$P_{Lrms} = i_{Lrms}{}^2 R_L = \frac{i_{LP}{}^2 R_L}{2}$$

or

$$i_{LP} = \sqrt{\frac{2\,P_{Lrms}}{R_L}} \tag{15.20}$$

Next, assume that the peak input AC signal voltage, v_{iP} is large enough so that on the positive peak of the AC signal, D_1 of Figure 15.11 is fully reverse biased. This means that it will not be conducting at all. Thus D_1, Q_2, and all of the circuitry relating to Q_2 can be ignored. The circuit will then look like Figure 15.13. Then the voltage at Point Y will have gone from its Q point value of $V_{CC}/2$ up to its maximum value of $V_{CC}/2 + v_{LP}$. Doing a KVL walk through the circuit from V_{CC} through R_1, Q_1, and R_E to point Y at that instant, we have

$$V_{CC} - \frac{i_{LP}R_1}{h_{fe}} - V_{BEP} - i_{LP}R_E = \frac{V_{CC}}{2} + v_{LP} \qquad (15.21)$$

Since only the AC portion of V_Y goes through C_o to R_L, applying Ohm's Law to the load gives

$$v_{LP} = i_{LP}R_L \qquad (15.22)$$

Substituting Equation 15.22 into Equation 15.21, combining terms, and solving for V_{CC} gives the required power supply equation

$$V_{CC} = 2\left(i_{LP}\left(\frac{R_1}{h_{fe}} + R_E + R_L \right) + V_{BEP} \right) \qquad (15.23)$$

Of course, we could have analyzed the circuit on the negative peak signal swing. Due to the symmetry of the circuit, the behavior would be comparable, and Equation 15.23 would have come out the same.

EXAMPLE 15.5

Determine appropriate values of R_1, R_E, and V_{CC} in order that the circuit of Figure 15.14 will deliver 0.5 W to the 10 ohm load.

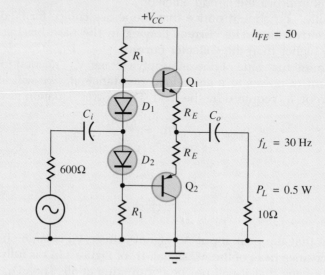

FIGURE 15.14   Example 15.5.

SOLUTION:
We will first calculate the appropriate resistor values:

1. $R_E = \dfrac{R_L}{10}$ \hfill (15.15)

   $= \dfrac{10\ \Omega}{10} = 1\ \Omega$ \hfill (Use $1\ \Omega$)

2. $R_1 = \dfrac{h_{FE}(R_E + R_L)}{5}$ \hfill (15.19)

   $= \dfrac{50(1\ \Omega + 10\ \Omega)}{5} = 110\ \Omega$ \hfill (Use $100\ \Omega$)

Next we will calculate the required value of i_{LP}:

3. $i_{LP} = \sqrt{\dfrac{2\ P_{Lrms}}{R_L}}$ \hfill (15.20)

   $= \sqrt{\dfrac{2 * 0.5\ W}{10\ \Omega}} = 316\ mA$

Next we will estimate the base to emitter voltage, V_{BEP} at the above peak collector current:

4. $V_{BEP} = 26\ mV * \ln \dfrac{i_{LP}}{10\ fA}$ \hfill (8.4)

   $= 26\ mV * \ln \dfrac{316\ mA}{10\ fA} = 0.808\ V$

Finally we have enough information to calculate V_{CC}:

5. $V_{CC} = 2\left(i_{LP}\left(\dfrac{R_1}{h_{fe}} + R_E + R_L\right) + V_{BEP} \right)$ \hfill (15.23)

   $= 2\left(316\ mA\left(\dfrac{100\ \Omega}{50} + 1\ \Omega + 10\ \Omega\right) + 0.808\ V \right)$

   $= 9.83\ V$

15.11 Completing the Design

Example 15.5 contains the first five steps in the design of the circuit of Figure 15.15a. We will complete the procedure in this section.

You will find the approach very similar to the development of Table 12.1 for the common emitter circuit.

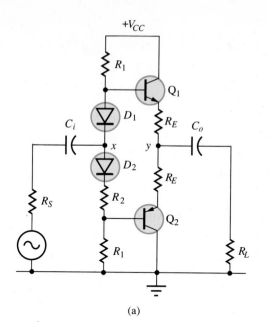

(a)

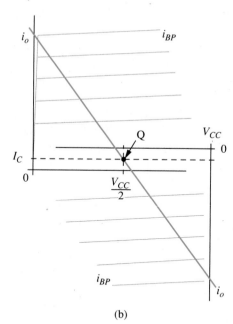

(b)

FIGURE 15.15   Complementary circuit design.
(a) Circuit diagram.
(b) Load line.

The procedure to be developed here assumes that the power that must be delivered to the load, the resistance of the load, the source resistance, and the transistor specifications are known. Since the transistor specifications cannot really be known until long after h_{FE} is used in the design procedure, some initial assumptions have to be made with reference to the transistors. Experience shows that an adequate transistor power dissipation rating is given by

$$P_{DISS} > \frac{P_{Lrms}}{2} \tag{15.24}$$

An adequate transistor voltage rating is given by

$$BV_{CEO} > 6\sqrt{P_{Lrms}R_L} \tag{15.25}$$

The above approximations are good enough for initial transistor selection.

After having completed the first five steps of the design procedure as outlined in Example 15.5, the remaining steps required for the design of the circuit of Figure 15.15a are:

6. I_C: As we mentioned in Section 15.7, the closer we get to Class A operation, the smaller will be the crossover distortion appearing in the output; but the closer we get to Class B operation, the greater the efficiency of the circuit. Again the designer must decide on a compromise. A reasonable minimum value for I_C might be

$$I_C \approx \frac{i_{LP}}{100} \tag{15.26}$$

7, 8. V_E: For midpoint bias, we wish to have $V_Y = \dfrac{V_{CC}}{2}$ in Figure 15.15a. Thus

$$V_{E1} = \frac{V_{CC}}{2} + I_C R_E \tag{15.27}$$

$$V_{E2} = \frac{V_{CC}}{2} - I_C R_E \tag{15.28}$$

9. V_{BE}: The junction equation gives the best estimate of V_{BE}.

$$V_{BE} = 26 \text{ mV} * \ln\frac{I_C}{10 \text{ fA}} \tag{8.4}$$

10, 11. V_B: Again referring to Figure 15.15a, it is obvious that

$$V_{B1} = V_{E1} + V_{BE} \tag{15.29}$$

$$V_{B2} = V_{E2} - V_{BE} \tag{15.30}$$

12. I_D: If we assume that the base currents are small compared to the current flowing in the base bias circuit, I_D, then from Figure 15.15 we have

$$I_D = \frac{V_{B2}}{R_1} \qquad \text{(15.31)}$$

13. V_D: The junction equation gives the best estimate of the voltage across the diodes.

$$V_D = 26 \text{ mV} * \ln\frac{I_D}{10 \text{ fA}} \qquad \text{(3.10)}$$

14. R_2: In the event that the diodes in Figure 15.15 do not give the required voltage difference between the transistor base terminals, R_2 makes up the difference. Doing a KVL walk through the biasing circuit from B_1 to B_2 gives

$$V_{B1} - 2\,V_D - I_D R_2 = V_{B2}$$

Solving for R_2 gives

$$R_2 = \frac{V_{B1} - 2\,V_D - V_{B2}}{I_D}$$

The above equation usually produces a negative value for R_2. This means that the diodes are producing more than enough voltage to provide the required bias. In that event, use only one diode. We then have

$$R_2 = \frac{V_{B1} - V_{B2} - NV_D}{I_D} \qquad \text{(15.32)}$$

where N represents the number of diodes used in the circuit.

For more precise control, use an adjustable resistor for R_2.

15. P_{DISS}: The power dissipated in each transistor must be checked to be certain that the transistors initially selected are appropriate. Obviously at the Q point the power dissipation is just $V_{CE}I_C = V_{E2}I_C$. We must add to that value the power dissipated when the signal is present. Thus

$$P_{\text{DISS}} = V_{E2}I_C + v_{CEav}i_{Cav}$$

From Section 15.2 we know that the transistors conduct on alternate halves of the AC waveform so the average values are the peak values divided by π. Therefore

$$P_{\text{DISS}} = V_{E2}I_C + \frac{v_{CEP}}{\pi} * \frac{i_{CP}}{\pi}$$

If we assume that the voltage swing uses all of the load line, which it nearly does with this design procedure, we can write

$$P_{\text{DISS}} = V_{E2}I_C + \frac{V_{CE}}{\pi} * \frac{i_{LP}}{\pi}$$

$$= V_{E2}\left(I_C + \frac{i_{LP}}{\pi^2}\right) \tag{15.33}$$

At this point the rating of the transistors originally selected must be checked to be certain they are appropriate. If they are not, another transistor must be selected and the design repeated.

16. h_{ie}: In Section 11.4 you learned that h_{ie} can be approximated by

$$h_{ie} \approx \frac{26 \text{ mV}}{I_B} \tag{11.3}$$

Of course, h_{ie} is never constant for any transistor because the transistor's operation varies up and down the load line about the Q point in response to the AC signal. In previous circuits, using the Q point value of the base current, I_B to calculate the average h_{ie} was reasonable since I_B went both above and below its Q point value in response to the AC signal. Thus its average value was I_B. The complementary circuit is a different situation. Here, as you can see in Figure 15.15b, the base current varies from cutoff up to

$$i_{BP} = \frac{i_0}{h_{fe}} = \frac{i_{LP}}{h_{fe}}$$

but it never goes below cutoff. In Section 15.3 we showed that its average value will then be

$$i_{Bav} = \frac{i_{LP}}{h_{fe}\pi}$$

Putting this idea into Equation 11.3 gives

$$h_{ie} = \frac{26 \text{ mV} * \pi h_{fe}}{i_{LP}} \tag{15.34}$$

17. R_{in}: When we stand at the input and look into the circuit of Figure 15.15a, the diodes and R_2 can be ignored since their effects are small. Further, remember that when one transistor is conducting, the other one is turned off and can be ignored. This is illustrated in Figure 15.16. Thus the

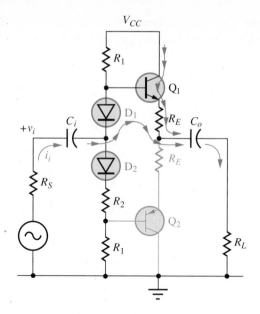

FIGURE 15.16   Input and output impedances of the complementary circuit.

input impedance is given by

$$R_{in} = R_1 \parallel R_1 \parallel \left(h_{ie} + h_{fe}(R_E + R_L) \right)$$

$$= \frac{R_1}{2} \parallel \left(h_{ie} + h_{fe}(R_E + R_L) \right) \tag{15.35}$$

18. R_{out}: Using the same arguments as above, the output impedance is given by

$$R_{out} = R_E + \frac{h_{ie} + \dfrac{R_1}{2} \parallel R_S}{h_{fe}} \tag{15.36}$$

19, 20. C_i and C_o: The coupling capacitor relationships are the same as the other circuits we have studied. They are

$$C_i = \frac{1}{2\pi f_L(R_S + R_{in})} \tag{12.6}$$

$$C_o = \frac{1}{2\pi f_L(R_{out} + R_L)} \tag{12.13}$$

This completes the design procedure. The steps in this design procedure are collected in Table 15.1.

TABLE 15.1   Complementary Bipolar Amplifier Design

* : Designer's formula only.

STD: Choose nearest standard resistor value.

Given: P_L R_L R_S f_L $\left\{ \begin{matrix} \text{Transistor} \\ \text{spec.} \\ \text{estimate} \end{matrix} \right\}$ $\left\{ \begin{matrix} P_{DISS} \geq \dfrac{1}{2}P_L \\ BV_{CEO} \geq 6\sqrt{P_L R_L} \end{matrix} \right\}$

1. $R_E \overset{*}{\approx} \dfrac{R_L}{10}$ \hfill (STD.) (15.15)

2. $R_1 \overset{*}{\approx} \dfrac{h_{FE}(R_E + R_L)}{5}$ \hfill (STD.) (15.19)

3. $i_{LP} = \sqrt{\dfrac{2\,P_{Lrms}}{R_L}}$ \hfill (15.20)

4. $V_{BEP} = 26\ \text{mV} * \ln\dfrac{i_{LP}}{10\ \text{fA}}$ \hfill (8.4)

5. $V_{CC} = 2\left(i_{LP}\left(\dfrac{R_1}{h_{fe}} + R_E + R_L\right) + V_{BEP} \right)$ \hfill (15.23)

6. $I_C \overset{*}{\approx} \dfrac{i_{LP}}{100}$ \hfill (15.26)

7. $V_{E1} = \dfrac{V_{CC}}{2} + I_C R_E$ \hfill (15.27)

8. $V_{E2} = \dfrac{V_{CC}}{2} - I_C R_E$ \hfill (15.28)

9. $V_{BE} = 26\ \text{mV} * \ln\dfrac{I_C}{10\ \text{fA}}$ \hfill (8.4)

10. $V_{B1} = V_{E1} + V_{BE}$ \hfill (15.29)

11. $V_{B2} = V_{E2} - V_{BE}$ \hfill (15.30)

12. $I_D = \dfrac{V_{B2}}{R_1}$ \hfill (15.31)

13. $V_D = 26\ \text{mV} * \ln\dfrac{I_D}{10\ \text{fA}}$ \hfill (3.10)

14. $R_2 = \dfrac{V_{B1} - V_{B2} - NV_D}{I_D}$ \hfill (STD.) (15.32)

15. $P_{DISS} = V_{E2}\left(I_C + \dfrac{i_{LP}}{\pi^2} \right)$ \hfill (Check trans. specs)   (15.33)

16. $h_{ie} \approx \dfrac{26\ \text{mV} * \pi h_{fe}}{i_{LP}}$ \hfill (15.34)

17. $R_{in} = \dfrac{R_1}{2} \,\|\, \left(h_{ie} + h_{fe}(R_E + R_L) \right)$ \hfill (15.35)

18. $R_{out} = R_E + \dfrac{h_{ie} + \dfrac{R_1}{2} \| R_S}{h_{fe}}$ \hfill (15.36)

19. $C_i = \dfrac{1}{2\pi f_L(R_S + R_{in})}$ \hfill (12.6)

20. $C_o = \dfrac{1}{2\pi f_L(R_{out} + R_L)}$ \hfill (12.13)

Design procedure for the circuit of Figure 15.15.

EXAMPLE 15.6

Design a complementary circuit that will supply 0.5 W of AC signal to a 10 Ω load. Assume a source impedance of 600 Ω and a cutoff frequency of 30 Hz. Select the transistors from those listed in Appendix F.

- (For a SuperCalc solution, see Appendix G.)

SOLUTION:

Given: $P_L = 0.5$ W $R_L = 10\ \Omega$ $R_S = 600\ \Omega$ $f_L = 30$ Hz.

For purposes of transistor selection, we have the estimates

$$P_{\text{DISS}} \geq \frac{P_L}{2} \tag{15.24}$$

$$= \frac{0.5\ \text{W}}{2} = 250\ \text{mW}$$

and

$$BV_{CEO} \geq 6\sqrt{P_L R_L} \tag{15.25}$$
$$= 6\sqrt{0.5\ \text{W} * 10\ \Omega} = 13.42\ \text{V}$$

Looking at the transistor data in Appendix F, either the 2N3904, 2N3906 transistors, or the 2N4401, 2N4403 transistors should be able to do the job. We will use the latter pair.

We will use $h_{fe} = 135$. Plugging numbers into Table 15.1 gives

1. $R_E \stackrel{*}{\approx} \dfrac{R_L}{10}$ STD. (15.15)

$= \dfrac{10\ \Omega}{10} = 1\ \Omega$ (Use 1 Ω)

2. $R_1 \stackrel{*}{\approx} \dfrac{h_{FE}(R_E + R_L)}{5}$ STD. (15.19)

$= \dfrac{135(1\ \Omega + 10\ \Omega)}{5} = 297\ \Omega$ (Use 270 Ω)

3. $i_{LP} = \sqrt{\dfrac{2\ P_{L\text{rms}}}{R_L}}$ (15.20)

$= \sqrt{\dfrac{2 * 0.5\ \text{W}}{10\ \Omega}} = 316\ \text{mA}$

4. $V_{BEP} = 26\ \text{mV} * \ln\dfrac{i_{LP}}{10\ \text{fA}}$ (8.4)

$= 26\ \text{mV} * \ln\dfrac{316\ \text{mA}}{10\ \text{fA}} = 0.808\ \text{V}$

5. $V_{CC} = 2\left(i_{LP}\left(\dfrac{R_1}{h_{fe}} + R_E + R_L\right) + V_{BEP}\right)$ (15.23)

$= 2\left(316 \text{ mA}\left(\dfrac{270 \text{ }\Omega}{135} + 1 \text{ }\Omega + 10 \text{ }\Omega\right) + 0.808 \text{ V}\right)$

$= 9.84 \text{ V}$

The above supply voltage is less than the BV_{CEO} rating for the transistors so the transistors have an adequate voltage rating.

6. $I_C \overset{*}{\approx} \dfrac{i_{LP}}{100}$ (15.26)

$= \dfrac{316 \text{ mA}}{100} = 3.16 \text{ mA}$

7. $V_{E1} = \dfrac{V_{CC}}{2} + I_C R_E$ (15.27)

$= \dfrac{9.84 \text{ V}}{2} + 3.16 \text{ mA} * 1 \text{ }\Omega = 4.922 \text{ V}$

8. $V_{E2} = \dfrac{V_{CC}}{2} - I_C R_E$ (15.28)

$= \dfrac{9.84 \text{ V}}{2} - 3.16 \text{ mA} * 1 \text{ }\Omega = 4.916 \text{ V}$

9. $V_{BE} = 26 \text{ mV} * \ln\dfrac{I_C}{10 \text{ fA}}$ (8.4)

$= 26 \text{ mV} * \ln\dfrac{3.16 \text{ mA}}{10 \text{ fA}} = 0.688 \text{ V}$

10. $V_{B1} = V_{E1} + V_{BE}$ (15.29)

$= 4.922 \text{ V} + 0.688 \text{ V} = 5.61 \text{ V}$

11. $V_{B2} = V_{E2} - V_{BE}$ (15.30)

$= 4.916 \text{ V} - 0.688 \text{ V} = 4.23 \text{ V}$

12. $I_D = \dfrac{V_{B2}}{R_1}$ (15.31)

$= \dfrac{4.23 \text{ V}}{270 \text{ }\Omega} = 15.7 \text{ mA}$

13. $V_D = 26 \text{ mV} * \ln\dfrac{I_D}{10 \text{ fA}}$ (3.10)

$= 26 \text{ mV} * \ln\dfrac{15.7 \text{ mA}}{10 \text{ fA}} = 0.730 \text{ V}$

14. $R_2 = \dfrac{V_{B1} - V_{B2} - NV_D}{I_D}$ (STD.) (15.32)

$= \dfrac{5.61 \text{ V} - 4.23 \text{ V} - 1 * 0.730 \text{ V}}{15.7 \text{ mA}} = 41.7 \text{ }\Omega$ (Use 39 Ω)

Notice that if two diodes are specified, Equation 15.32 gives a negative value for R_2.

Perhaps it would be better to use a 50 Ω pot for R_2 and specify that it be adjusted for the minimum setting that will remove observable crossover distortion in the output.

15. $$P_{\text{DISS}} = V_{E2}\left(I_C + \frac{i_{LP}}{\pi^2}\right) \qquad \text{(Check trans. specs)} \tag{15.33}$$

$$= 4.92 \text{ V}\left(3.16 \text{ mA} + \frac{316 \text{ mA}}{\pi^2}\right) = 173 \text{ mW}$$

The power estimate of Equation 15.24 is very generous so the transistors are more than adequate.

16. $$h_{ie} \approx \frac{26 \text{ mV} * \pi h_{fe}}{i_{LP}} \tag{15.34}$$

$$= \frac{26 \text{ mV} * \pi * 135}{316 \text{ mA}} = 34.9 \text{ Ω}$$

17. $$R_{\text{in}} = \frac{R_1}{2} \parallel \left(h_{ie} + h_{fe}(R_E + R_L)\right) \tag{15.35}$$

$$= \frac{270 \text{ Ω}}{2} \parallel (34.9 \text{ Ω} + 135(1 \text{ Ω} + 10 \text{ Ω})) = 124 \text{ Ω}$$

18. $$R_{\text{out}} = R_E + \frac{h_{ie} + \dfrac{R_1}{2} \parallel R_S}{h_{fe}} \tag{15.36}$$

$$= 1 \text{ Ω} + \frac{34.9 \text{ Ω} + \dfrac{270 \text{ Ω}}{2} \parallel 600 \text{ Ω}}{135} = 2.07 \text{ Ω}$$

19. $$C_i = \frac{1}{2\pi f_L(R_S + R_{\text{in}})} \tag{12.6}$$

$$= \frac{1}{2\pi * 30 \text{ Hz}(600 \text{ Ω} + 124 \text{ Ω})} = 7.33 \text{ μF} \qquad \text{(Use 8 μF)}$$

20. $$C_o = \frac{1}{2\pi f_L(R_{\text{out}} + R_L)} \tag{12.13}$$

$$= \frac{1}{2\pi * 30 \text{ Hz}(2.07 \text{ Ω} + 10 \text{ Ω})} = 439 \text{ μF} \qquad \text{(Use 400 μF)}$$

Notice that R_1 for Example 15.6 is different from R_1 for Example 15.5. The reason is the difference in h_{fe}. As we mentioned in Section 15.10, the higher value of R_1 will result in less current wasted in the biasing circuit and so the circuit will be more efficient.

You should take the time to hook up the circuit of Example 15.6. If you do, you should use something like a 50 Ω pot for R_2 and watch the output for crossover distortion as you vary R_2. Also you should check to see if the maximum unclipped output is close to 0.5 W. It should be.

15.12 Circuit Analysis

We will construct a circuit analysis procedure in much the same format as the design procedure of Section 15.11. Assume that you are given a circuit with all component values and the supply voltage known. The needed information is the Q point data, the power dissipated in each transistor, and the power delivered to the load. The basic complementary circuit is shown in Figure 15.15a which is reproduced here as Figure 15.17. You will find that most of the equations used here come from the design procedure.

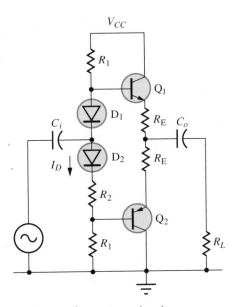

FIGURE 15.17 Basic complementary circuit.

1. i_{LP}: From Step 5 of the circuit design procedure we have

$$V_{CC} = 2\left(i_{LP}\left(\frac{R_1}{h_{fe}} + R_E + R_L\right) + V_{BEP}\right) \tag{15.23}$$

Since V_{CC} and the circuit resistors are now known, solving Equation 15.23 for i_{LP} gives

$$i_{LP} = \frac{\frac{V_{CC}}{2} - V_{BEP}}{\frac{R_1}{h_{fe}} + R_E + R_L} \qquad (15.37)$$

Equation 15.37 must be iterated with Equation 8.4 in Step 2.

2. V_{BEP}: From Step 4 of the circuit design procedure we have

$$V_{BEP} = 26 \text{ mV} * \ln \frac{i_{LP}}{10 \text{ fA}} \qquad (8.4)$$

3. P_{Lrms}: We have written the power equation before. It is

$$P_{Lrms} = (i_{Lrms})^2 R_L = \frac{(i_{LP})^2 R_L}{2} \qquad (15.38)$$

4. I_D: Assuming that the base currents are small enough to be ignored, doing a KVL walk through the base biasing circuit in Figure 15.17 gives

$$V_{CC} - I_D(2 R_1 + R_2) + NV_D = 0 \text{ V}$$

Where N stands for the number of biasing diodes used. In Figure 15.17, $N = 2$. Solving for I_D gives

$$I_D = \frac{V_{CC} - NV_D}{2 R_1 + R_2} \qquad (15.39)$$

As you have probably come to expect, Equation 15.39 must be iterated with Equation 3.10 from Step 5.

5. V_D:

$$V_D = 26 \text{ mV} * \ln \frac{I_D}{10 \text{ fA}} \qquad (3.10)$$

6. V_{B1}: Stopping the KVL walk through the base biasing circuit at B_1 and again ignoring base currents gives

$$V_{B1} = V_{CC} - I_D R_1 \qquad (15.40)$$

7. V_{B2}: Beginning the KVL walk through the base biasing circuit at B_2 and going from there to ground gives

$$V_{B2} = I_D R_1 \qquad (15.41)$$

8. I_C: Doing a KVL walk beginning at the base of Q_1 in Figure 15.17, going out of the emitter of Q_1, through the emitter resistors, into the emitter of Q_2, and terminating at the base of Q_2 we have

$$V_{B1} - V_{BE} - I_C * 2 R_E - V_{BE} = V_{B2}$$

Solving for I_C gives

$$I_C = \frac{V_{B1} - V_{B2} - 2 V_{BE}}{2 R_E}$$

or

$$I_C = \frac{\dfrac{V_{B1} - V_{B2}}{2} - V_{BE}}{R_E} \qquad (15.42)$$

Equation 15.42 must be iterated with Equation 8.4 of Step 9.

9. V_{BE}:

$$V_{BE} = 26 \text{ mV} * \ln \frac{I_C}{10 \text{ fA}} \qquad (8.4)$$

There is a serious problem with this iteration, however. We know that $\dfrac{V_{B1} - V_{B2}}{2}$ is going to be very small so that the transistors will be very nearly cut off. Under this condition the iteration procedure may not converge. In fact, if the program that you use to perform the iteration uses an initial value of $V_{BE} = 0.7$ V, Equation 15.42 will probably give you a negative value for the first approximation of I_C. One approach is to replace that negative value with some small positive value; 10 μA often works well. Go through Equations 8.4 and 15.42 once to see what the next approximation is for I_C. If it is less than your choice, replace your first choice for I_C with a value which is less than your first choice, but not nearly as small as the iteration calls for. On the other hand, if the equations give a next approximation for I_C that is greater than your first choice, replace your choice with a value which is greater than your choice, but not nearly as large as the iteration calls for. Repeat this manual intervention on each iteration until the result is close enough to meet your needs.

10. V_{E1}: Looking at the circuit of Figure 15.17, you can see that

$$V_{E1} = V_{B1} - V_{BE} \qquad (15.43)$$

11. V_{E2}: Again from the circuit of Figure 15.17

$$V_{E2} = V_{B2} + V_{BE} \qquad (15.44)$$

12. P_{DISS}: From the circuit design procedure of Section 15.11 we have

$$P_{\text{DISS}} = V_{E2}\left(I_C + \frac{i_{LP}}{\pi^2}\right) \qquad (15.33)$$

This completes the circuit analysis procedure. The equations from the above design procedure are collected in Table 15.2.

TABLE 15.2   Complementary Bipolar Amplifier Circuit Analysis

Given: V_{CC}, all circuit component values and transistor specs.

$$1.\ i_{LP} = \cfrac{\dfrac{V_{CC}}{2} - V_{BEP}}{\dfrac{R_1}{h_{fe}} + R_E + R_L} \left.\begin{array}{c} \\ \\ \end{array}\right\}\text{Iterate} \tag{15.37}$$

$$2.\ V_{BEP} = 26\ \text{mV} * \ln\frac{i_{LP}}{10\ \text{fA}} \tag{8.4}$$

$$3.\ P_{Lrms} = i_{Lrms}{}^2 R_L = \frac{i_{LP}{}^2 R_L}{2} \tag{15.38}$$

$$4.\ I_D = \frac{V_{CC} - NV_D}{2\ R_1 + R_2} \left.\begin{array}{c} \\ \\ \end{array}\right\}\text{Iterate} \tag{15.39}$$

$$5.\ V_D = 26\ \text{mV} * \ln\frac{I_D}{10\ \text{fA}} \tag{3.10}$$

$$6.\ V_{B1} = V_{CC} - I_D R_1 \tag{15.40}$$
$$7.\ V_{B2} = I_D R_1 \tag{15.41}$$

$$8.\ I_C = \cfrac{\dfrac{V_{B1} - V_{B2}}{2} - V_{BE}}{R_E} \left.\begin{array}{c} \\ \\ \end{array}\right\}\text{Iterate} \tag{15.42}$$

$$9.\ V_{BE} = 26\ \text{mV} * \ln\frac{I_C}{10\ \text{fA}} \tag{8.4}$$

$$10.\ V_{E1} = V_{B1} - V_{BE} \tag{15.43}$$
$$11.\ V_{E2} = V_{B2} + V_{BE} \tag{15.44}$$

$$12.\ P_{DISS} = V_{E2}\left(I_C + \frac{i_{LP}}{\pi^2}\right) \tag{15.33}$$

Analysis procedure for the circuit of Figure 15.15.

EXAMPLE 15.7

The circuit of Figure 15.18 is a simplified version of the audio output circuit used in a commercial TV set. Calculate the Q point information, the power dissipated in each transistor, and the power delivered to the load. h_{fe} for the transistors is 100.

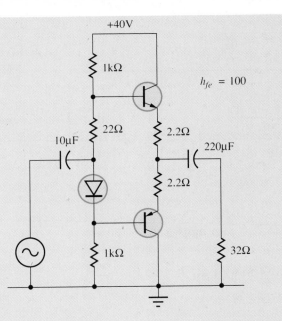

+40V

1kΩ

$h_{fe} = 100$

22Ω

2.2Ω

10μF

220μF

2.2Ω

1kΩ

32Ω

FIGURE 15.18   Example 15.7.

SOLUTION:

Inserting numbers in the equations of Table 15.2 gives

1. $i_{LP} = \dfrac{\dfrac{V_{CC}}{2} - V_{BEP}}{\dfrac{R_1}{h_{fe}} + R_E + R_L}$ (15.37)

$$= \frac{\dfrac{40\ V}{2} - V_{BEP}}{\dfrac{1\ k\Omega}{100} + 2.2\ \Omega + 32\ \Omega} = \frac{20\ V - V_{BEP}}{44.2\ \Omega}$$

2. $\quad i_{LP} = \dfrac{20\ V - V_{BEP}}{44.2\ \Omega} \quad =$	437	434	434	mA
$V_{BEP} = 26\ mV * \ln \dfrac{i_{LP}}{10\ fA} \quad =$	0.817	0.816	0.816	V

3. $P_{Lrms} = i_{Lrms}{}^2 R_L = \dfrac{i_{LP}{}^2 R_L}{2}$ (15.38)

$$= \frac{(434\ mA)^2 * 32\ \Omega}{2} = 3.01\ W$$

4. $I_D = \dfrac{V_{CC} - NV_D}{2\,R_1 + R_2}$

$= \dfrac{40\text{ V} - 1 * V_D}{2 * 1\text{ k}\Omega + 22\ \Omega} = \dfrac{40\text{ V} - V_D}{2.02\text{ k}\Omega}$ (15.39)

5.

$I_D = \dfrac{40\text{ V} - V_D}{2.02\text{ k}\Omega}$ =	19.4	19.4	19.4	mA
$V_D = 26\text{ mV} * \ln\dfrac{I_D}{10\text{ fA}}$ =	0.736	0.736	0.736	V

6. $V_{B1} = V_{CC} - I_D R_1$ (15.40)

$= 40\text{ V} - 19.4\text{ mA} * 1\text{ k}\Omega = 20.6\text{ V}$

7. $V_{B2} = I_D R_1$ (15.41)

$= 19.4\text{ mA} * 1\text{ k}\Omega = 19.4\text{ V}$

8. $I_C = \dfrac{\dfrac{V_{B1} - V_{B2}}{2} - V_{BE}}{R_E}$ (15.42)

$= \dfrac{\dfrac{20.6\text{ V} - 19.4\text{ V}}{2} - V_{BE}}{2.2\ \Omega} = \dfrac{0.581\text{ V} - V_{BE}}{2.2\ \Omega}$

You will note that more precision is being used in Step 8 than is being displayed. If you set your calculator to display a few more digits, $V_{B1} = 20.581\ldots$ V, and $V_{B2} = 19.419\ldots$ V. This is another illustration of the fact that you should always try to use all of the power of your calculator so that round off errors will be minimized.

9. $V_{BE} = 26\text{ mV} * \ln\dfrac{I_C}{10\text{ fA}}$ (8.4)

As we have come to expect, Steps 8 and 9 must be iterated but it will be much more difficult this time. The reason is that the "supply voltage" in this case is only 0.581 V according to Step 8. Thus it is not large compared to V_{BE}. In fact, if your first guess is $V_{BE} = 0.7$ V, Step 8 will give you a negative value for I_C and you are in serious trouble if you (or your program) thoughtlessly plug that value into Step 9. We will exercise our judgment to intervene in the iteration process as follows:

Beginning as usual with a first guess of $V_{BE} = 0.7$ V in Equation 15.42 gives $I_C = -53.9$ mA. This result, as we have pointed out, cannot be used in Equation 8.4. Since we know that I_C should be small (near

Class B operation), a reasonable first guess might be $I_C = 10\ \mu A$. Putting this value in Equation 8.4 gives $V_{BE} = 0.539$ V. Putting this value for V_{BE} back into Equation 15.42 gives $I_C = 19.4$ mA. Since this calculated result is bigger than our guess, the correct value is bigger than our 10 μA guess, but it is certainly not that much bigger, so I will guess $I_C = 100\ \mu A$.

Putting 100 μA into Equation 8.4 gives $V_{BE} = 0.599$ V. Equation 15.42 then gives $I_C = -7.84$ mA. Since this is less than my guess, the correct value is less than 100 μA. Thus the correct value must lie somewhere between the 10 μA first guess and the 100 μA second guess, so try $I_C = 50\ \mu A$.

Putting 50 μA in Equation 8.4 gives $V_{BE} = 0.581$ V and Equation 15.2 gives $I_C = 357\ \mu A$. Thus the correct value must be more than 50 μA but we established above that it is less than 100 μA, so try 60 μA. This gives $V_{BE} = 0.585$ V and $I_C = -1.80$ mA, which indicates a value less than 60 μA. Try 52 μA. The result is $V_{BE} = 0.582$ V and $I_C = -107$ μA so guess 51 μA. The result is $V_{BE} = 0.581$ V and $I_C = 123\ \mu A$. So guess 51.3 μA. As you can see, the iteration is converging (with some help!), on something like

$$I_C \approx 51.3\ \mu A \qquad V_{BE} \approx 0.581\ V$$

The result of this procedure is summarized in Table 15.3.

10. $V_{E1} = V_{B1} - B_{BE}$ (15.43)

$\qquad = 20.6\ V - 0.581\ V = 20.0\ V$

11. $V_{E2} = V_{B2} + V_{BE}$ (15.44)

$\qquad = 19.4\ V + 0.581\ V = 20.0\ V$

If you display a few more digits (about 3 more!) you will find that V_{E1} is slightly greater than V_{E2}, as you can see from Figure 15.18 that it should be.

12. $P_{\text{DISS}} = V_{E2}\left(I_C + \dfrac{i_{LP}}{\pi^2}\right)$ (15.33)

$$= 20.0\ V\left(51.3\ \mu A + \frac{434\ mA}{\pi^2}\right) = 881\ mW$$

Thus the results are:

$Q_1: V_C = 40\ V, \qquad V_B = 20.6\ V, \qquad V_E = 20.0\ V, \qquad I_C = 51.3\ \mu A$

$Q_2: V_C = 0\ V, \qquad V_B = 19.4\ V, \qquad V_E = 20.0\ V, \qquad I_C = 51.3\ \mu A$

$P_{\text{DISS}} = 881\ mW$

$P_L = 3.01\ W$

TABLE 15.3 Iteration With Intervention

$$I_C = \frac{0.581 \text{ V} - V_{BE}}{2.2 \text{ }\Omega} \qquad V_{BE} = 26 \text{ mV} * \ln \frac{I_C}{10 \text{ fA}}$$

	1	2	3	4	5	6	7	
I_C =	−53.9 mA	19.4 mA	−7.84 mA	357	−1.80 mA	−107	123	μA
Guess	10	100	50	60	52	51	51.3	μA
V_{BE} =	0.539	0.599	0.581	0.585	0.582	0.581	0.581	V

Rationale For Guess:

1. Must be a small positive value.   Guess 10 μA
2. Calculated value (19.4 mA) is larger than last
   guess so increase guess.   Guess 100 μA
3. Calculated value (−7.84 mA) is less than last
   guess so 10 μA < I_C < 100 μA   Guess 50 μA
4. Calculated value (357 μA) is larger than last
   guess so 50 μA < I_C < 100 μA   Guess 60 μA
5. Calculated value (−1.80 mA) is less than last
   guess so 50 μA < I_C < 60 μA   Guess 52 μA
6. Calculated value (−107 μA) is less than last
   guess so 50 μA < I_C < 52 μA   Guess 51 μA
7. Calculated value (123 μA) is larger than last
   guess so 51 μA < I_C < 52 μA, etc.   Guess 51.3 μA

Iteration of Steps 8 and 9 of Example 15.7. Note that the number actually used to calculate I_C is not 0.581 V. It is closer to 0.581434870 V.

▶▶▶▶▶▶▶▶ **15.13 Troubleshooting**

When trouble develops in electronic circuitry it most often occurs where voltage, current and power dissipation levels are the highest. That usually means, in the power supply or the output stages. Also, when trouble develops in the output circuitry, due to the small resistance values usually present, it is likely to have catastrophic results. This can include not only the destruction of expensive transistors, but circuit boards and power supplies as well, unless they are protected. Protection is the key. Before you attempt any kind of troubleshooting on a power circuit, you should know what sort of currents to expect and use fuses or current limiting circuitry that will prevent destructive current levels.

When excessively high collector current levels are found in a complementary circuit, you should look into the possibility of difficulty in the biasing circuit. It is important to notice that it is difficult to know how much current is flowing in the collector circuit of a complementary circuit by measuring collector circuit voltages. This is due to the fact that the resistances are so low. To determine the current, it is best to measure the voltage lost across the emitter resistors directly, rather than trying to calculate the voltage loss using $V_{E1} - V_{E2}$.

Again, the best troubleshooting procedure is to know what Q point voltages should be present, note what voltages are actually present, and then use careful logic to account for any discrepancies.

EXAMPLE 15.8

Given the circuit of Example 15.7, calculate the collector current and power dissipated in each transistor if the diode in the circuit of Figure 15.18 became an open circuit.

SOLUTION:

Doing a KVL walk from V_{CC} through R_1, into B_1, through R_{E1} and R_{E2}, out B_2, and through R_1 to ground as shown in Figure 15.19 gives

$$V_{CC} - R_1\frac{I_C}{h_{FE}} - V_{BE} - 2\,R_E I_C - V_{BE} - R_1\frac{I_C}{h_{FE}} = 0\text{ V}$$

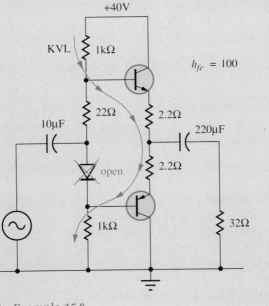

FIGURE 15.19   Example 15.8.

or

$$I_C = \frac{V_{CC} - 2\,V_{BE}}{\dfrac{2\,R_1}{h_{FE}} + 2\,R_E} = \frac{\dfrac{V_{CC}}{2} - V_{BE}}{\dfrac{R_1}{h_{FE}} + R_E} \tag{15.45}$$

$$= \frac{\dfrac{40\ \text{V}}{2} - V_{BE}}{\dfrac{1\ \text{k}\Omega}{100} + 2.2\ \Omega} = \frac{20\ \text{V} - V_{BE}}{12.2\ \Omega}$$

Iterating Equation 15.45 with the usual junction equation gives

$I_C = \dfrac{20\ \text{V} - V_{BE}}{12.2\ \Omega}$ =	1.58	1.57	1.57	A
$V_{BE} = 26\ \text{mV} * \ln \dfrac{I_C}{10\ \text{fA}}$ =	0.850	0.850	0.850	V

The power dissipated by each emitter resistor is

$$P_{RE} = I_C{}^2 R_E$$
$$= (1.57\ \text{A})^2 * 2.2\ \Omega = 5.42\ \text{W}$$

From the circuit of Figure 15.18 you can see that each transistor will dissipate half of the remaining power, or

$$P_{DISS} = \frac{V_{CC}I_C - 2\,P_{RE}}{2} = \frac{V_{CC}I_C}{2} - P_{RE}$$
$$= \frac{40\ \text{V} * 1.57\ \text{A}}{2} - 5.42\ \text{W} = 26.0\ \text{W} \tag{15.46}$$

With 5.42 Watts, going into emitter resistors that are probably rated at 0.5 W and 26 W going into transistors that are probably rated at less than 5 W, you can see that the circuit will be destroyed rather quickly.

If the circuit of Example 15.8 came to you for service, you would possibly find only one of the components "blown." Probably one of the transistors would be an open circuit. Obviously, to replace the defective transistor and hand it back to the customer would be the quickest way to disaster. One of the first things to check in such a situation is the difference in voltage between B_1 and B_2. If you read more than about 1.5 V there is probably trouble in the biasing circuit.

15.14 Circuit Variations

Due to the fact that the complementary circuit is an emitter follower circuit, the voltage gain will be slightly less than one. Thus the complementary circuit is almost always driven by some kind of voltage amplifier. Since V_{B2} is close to the kind of voltage you would expect at the collector of a common emitter voltage amplifier, direct coupling can be used as shown in Figure 15.20.

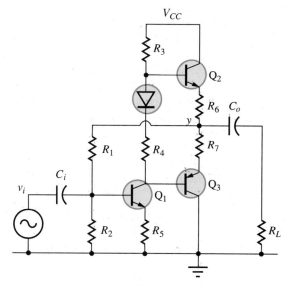

FIGURE 15.20   A direct coupled driver circuit.

An interesting feature of Figure 15.20 is that the base biasing resistors for Q_1 are not supplied directly from V_{CC}. The reason for taking the base bias from the output has to do with the phenomenon known as "feedback." Feedback can be defined as the return of output information back to the system.

It is feedback that causes a public address system to "howl" when an inattentive operator turns the volume up too loud. It is feedback that causes your automobile's cruise control to maintain a constant speed.

To understand the purpose of the feedback in the circuit of Figure 15.20, notice that if the voltage at Point Y began to increase for some reason, that would increase the base current to Q_1 and so cause a decrease in collector voltage. That means a decrease in base voltage at Q_2 and Q_3. Since the emitter follows the base in Q_2 and Q_3, that means the voltage at Point Y gets pulled back down. The feedback is in such a direction as to stabilize the circuit. Since it is "degenerative," the feedback in this circuit also tends to reduce the circuit gain slightly. We will look more closely at feedback later in this book.

To design the circuit of Figure 15.20, the design procedure of Table 15.1 and the procedure of Table 12.1 can be combined with some minor variations. Hopefully you will study such direct coupled circuits in depth in an advanced linear circuits course.

An important feature of the circuit designed in Example 15.6 is the size (and price!) of the output coupling capacitor, C_o. To eliminate it would save both space and money. Notice that both capacitors have been eliminated from the circuit of Figure 15.21. In the circuits we have been studying, we achieved midpoint bias by making $V_X = V_Y = V_{CC}/2$. It is the presence of this DC voltage that forced us to use coupling capacitors. Midpoint bias is achieved in the circuit of Figure 15.21 by using two power supplies. One supply has its negative terminal grounded and its positive terminal supplying $+V_{CC}$. The other supply has its positive terminal grounded and its negative terminal supplying $-V_{CC}$.

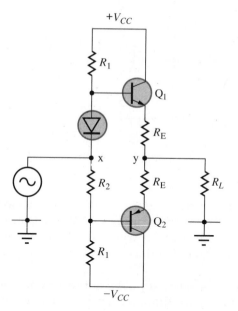

FIGURE 15.21   A circuit using a split power supply.

You can use the design procedure of Table 15.1 for the circuit of Figure 15.21. All that is needed is, after you finish the design, use half of the value for V_{CC} calculated in Step 5 of the design procedure for $+V_{CC}$, and use the other half for $-V_{CC}$. The midpoint between the dual supplies is then 0 VDC, and so no coupling capacitors are needed. In commercial devices that use this design some sophisticated circuitry is needed to assure that the midpoint does in fact stay exactly at 0 V. However, you can understand why it is done, and why dual power supplies are needed for such circuits. We will return to this idea in Chapter 19 when we study operational amplifiers.

One problem that the complementary circuit suffers from, particularly at higher output powers is the fact that it is very difficult for manufacturers to make transistors that are truly complements. Since one transistor is NPN and the other is PNP, it is hard to make their characteristics match. If they do not match, then you have a different amount of gain on the positive half of the AC cycle than on the negative half cycle. This is a distortion of the waveform, and must be avoided if possible.

A circuit which overcomes the above problem is the "totem pole" circuit shown in Figure 15.22. You will notice that the transistors are of the same type and in principle it should be possible to make them identical. As with the complementary circuit, the operation is Class AB with midpoint biasing.

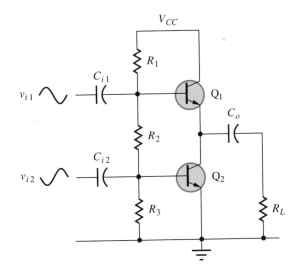

FIGURE 15.22 A totem pole output circuit.

The principle of operation of the totem pole circuit is the same as the complementary circuit. When Q_1 is turned on, Q_2 is turned off and so the output moves up toward V_{CC}. When Q_1 is turned off, Q_2 is turned on and so the output moves down toward ground.

Notice that two inputs are required for the totem pole circuit. Since the transistors are identical, in order to turn off Q_2 when Q_1 is turned on, the inputs must be out of phase. Some integrated circuits that are designed to drive totem pole outputs have two outputs that are out of phase for this reason. The totem pole output is commonly used in digital integrated circuits. You will study it in greater detail in more advanced courses.

15.15 Summary

In this chapter we defined Class B, Class C, and Class AB operation. We also described the complementary symmetry circuit and determined its ideal efficiency. We described crossover distortion, identified its cause, and developed circuitry intended to minimize it. Circuit analysis and design procedures which apply to the complementary symmetry circuit were developed. Finally some variations to the basic circuit were considered.

GLOSSARY

Class AB operation: Circuit operation such that conduction occurs less than 100 percent of the time, but more than half of the time. Operation between Class A and Class B. (Section 15.7)

Class B operation: Circuit operation such that conduction occurs exactly half of the time. (Section 15.2)

Class C operation: Circuit operation such that conduction occurs less than half of the time. (Section 15.6)

Complementary circuit: A series circuit consisting of both NPN and PNP transistors whose characteristics are as nearly matched as possible. (Section 15.3)

Crossover distortion: Distortion that results from nonlinearity of circuit components near cutoff. (Section 15.5)

Current mirror: A circuit which maintains the current in one branch equal to the current in another branch of the circuit. (Section 15.8)

Feedback: The return of output information back to the system which generated it. Section 15.14)

Push-pull circuit: A circuit in which the active devices conduct alternately, one "pushing" current into the external load and the other "pulling" current from the external load. (Section 15.3)

Totem pole circuit: A series circuit consisting of two or more active devices of the same type. (Section 15.14)

PROBLEMS

Section 15.2

1. Determine the average output voltage for the circuit of Figure 15.23 if the diode is assumed to be ideal.
2. Draw the collector current waveform for an ideal transistor that is being operated Class B.

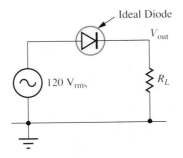

FIGURE 15.23   Problem 15.1.

Section 15.3

3. Determine the Q point value of V_{CE}, I_C, and I_B for the circuit of Figure 15.24 if the transistor is ideal and the operation is Class B.
4. From the list of transistors contained in Appendix F, select the transistor that is most nearly the complement of the following transistor:
   (a)   2N5810
   (b)   2N6727
5. Give a characteristic of the emitter follower circuit that is a real advantage for a power output amplifier.

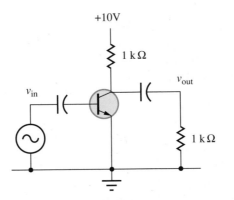

FIGURE 15.24   Problem 15.3.

Section 15.4

6. Determine the Q point values of V_E, V_B, and V_C for each of the transistors in Figure 15.25. Assume that the transistors are ideal.

7. What should be the voltage gain for the ideal circuit of Figure 15.25?

8. Draw the AC load line for the circuit of Figure 15.25 and locate the Q point. Assume that the transistors are ideal.

9. For the ideal circuit of Figure 15.25, calculate:
   (a)   The maximum unclipped rms power the circuit can deliver to the external load.
   (b)   The maximum power the circuit will take from the DC power supply. Do NOT ignore the biasing circuit.
   (c)   The efficiency of the circuit.

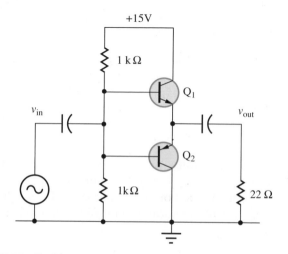

FIGURE 15.25   Problem 15.6.

Section 15.5

10. For the circuit of Figure 15.25, assume that the usual base-emitter junction equation applies. Draw a graph of v_{out} (vertical axis) vs v_{in} (horizontal axis) from $v_{out} = 0$ V to $v_{out} = 1$ V. *Hint:* Let v_{out} be the independent variable.

Section 15.6

11. If v_{in} in Figure 15.25 is a sine wave with a peak amplitude of $+1$ V, and if Q_1 begins to turn on at $+0.5$ V,
   (a)   How many degrees after $0+$ crossing does Q_1 begin to turn on?
   (b)   How many degrees after $0+$ crossing does Q_1 begin to turn off?
   (c)   For what percentage of the AC cycle does Q_1 conduct?
   (d)   What class of operation is this?

Section 15.8

12. For the circuit of Figure 15.26, calculate the required value of R_2 in order that:

    (a) $I_C = 0.1$ mA

    (b) $I_C = 1.0$ mA

    (c) $I_C = 10$ mA

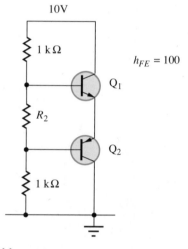

FIGURE 15.26   Problem 15.12.

13. For the circuit of Figure 15.27, calculate the required value of R_2 in order that:

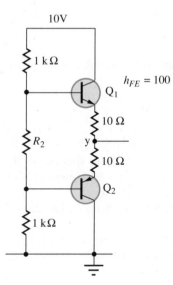

FIGURE 15.27   Problem 15.13.

(a) $I_C = 0.1$ mA
(b) $I_C = 1.0$ mA
(c) $I_C = 10$ mA

Section 15.9

14. Determine the value of the collector current, I_C for the circuit in Figure 15.28.
15. Draw the AC load line and locate the Q point for the circuit of Figure 15.28.

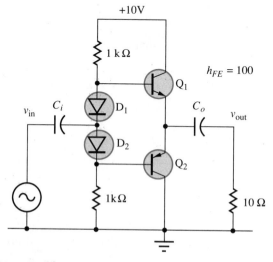

+10V

1 kΩ

$h_{FE} = 100$

Q_1

v_{in} C_i D_1 C_o v_{out}

D_2

Q_2

1kΩ

10 Ω

FIGURE 15.28   Problem 15.14.

Section 15.10

16. For the circuit of Figure 15.28, calculate the maximum unclipped output signal voltage.
17. Repeat Problem 16 if $h_{fe} = 1000$.

Section 15.11

18. Determine the required supply voltage in order that the circuit of Figure 15.28 will deliver 0.5 Wrms to the external load.
19. Using the procedure in Table 15.1, design a complementary amplifier capable of delivering 3 Wrms to an 8 Ω load. Assume $R_S = 600$ Ω and $f_L = 50$ Hz. Select the transistors to be used from those listed in Appendix F.
20. Complete the design of the circuit which was begun in Example 15.5. Compare the resulting circuit with that of Example 15.6 where higher gain transistors were used.

Section 15.12

21. Given the circuit in Figure 15.29, determine the maximum rms power that can be delivered to the 20 Ω load and for each transistor, find P_{DISS}, V_E, V_B, and I_C.

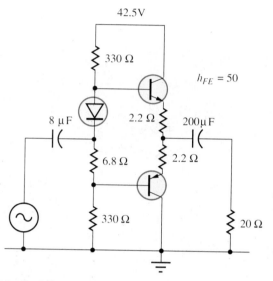

42.5V

330 Ω

$h_{FE} = 50$

8 μF

2.2 Ω

200μF

6.8 Ω

2.2 Ω

330 Ω

20 Ω

FIGURE 15.29   Problem 15.21.

Section 15.13

◄ Troubleshooting

22. Determine the effect if the diode in Figure 15.29 became:
    (a)   An open circuit.
    (b)   A short circuit.
23. Determine the effect if the collector terminal of the NPN transistor in Figure 15.29 became:
    (a)   An open circuit.
    (b)   Shorted to the emitter.
24. Determine the effect if the base terminal of the NPN transistor in Figure 15.29 became:
    (a)   An open circuit.
    (b)   Shorted to the collector.

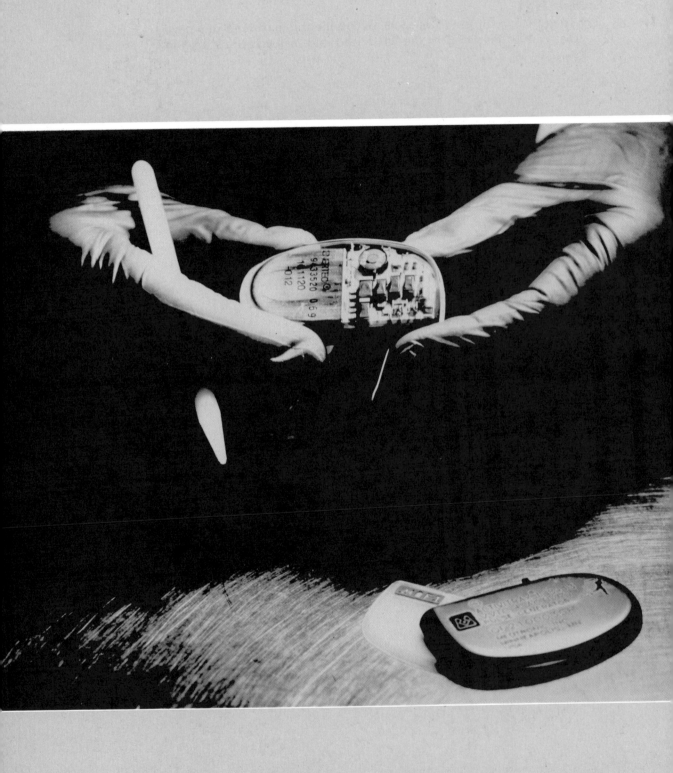

PART III
FIELD EFFECT
TRANSISTORS

A heart pacer that regulates the heartbeat of heart
attack victims. (Photograph by Ovak Arslanian.)

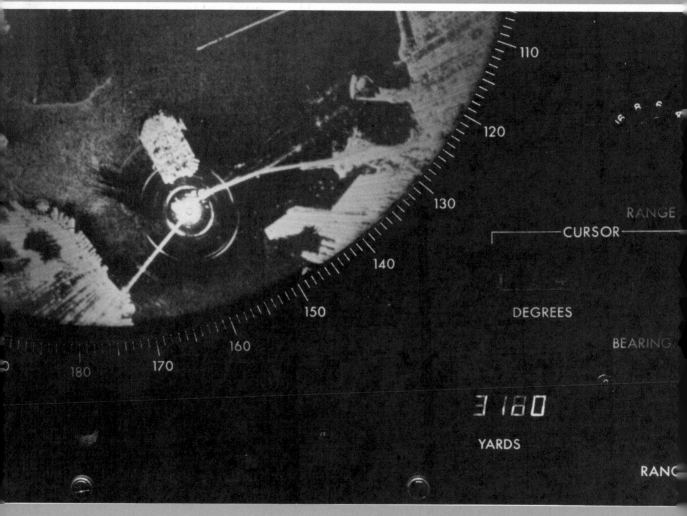

Electronics used by air traffic controllers.
(Photograph by Mark E. Gibson/The Stock Market.)

Chapter 16
The Junction field Effect
Transistor (JFET)

SPECIAL TERMS:

Channel
Source
Gate
Drain
Pinchoff ($V_{GS(OFF)}$)
Knee
Transfer characteristic
Transfer equation
Transconductance (g_{fs})

439

16.1   Introduction and Objectives

When we began our study of the P-N junction and bipolar transistors, we started with a very simple qualitative model which we used to describe in a rather elementary way the behavior of transistors. In this chapter we will attempt a similar approach with the JFET.

After studying this chapter you should understand the symbols and terminology used in connection with the JFET. You will begin to understand the common source circuit and the common source characteristic graph. Based on these understandings you should be able to:

- Recognize the symbol used for both N channel and P channel JFETs.
- Construct any two of the following from the third: 1. The JFET parameters, 2. The common source characteristic graph, and 3. The transfer characteristic.
- Construct the transconductance equivalent circuit for a given JFET.

16.2   Modeling the JFET

In Chapters 2 and 8 we tried to develop a qualitative model to describe the behavior of diodes and bipolar transistors respectively. We will take a similar approach here to describe the Junction Field Effect Transistor (JFET) and field effect devices in general. You may wish to look briefly at Chapters 2 and 8 to remind yourself of the approach we took there.

Let us begin our construction of a model for the JFET with a slab of N type semiconductor with terminals attached to each end of the slab as shown in Figure 16.1a. Our crystal will conduct current as shown in Figure 16.1b.

Because we have chosen to use an N type crystal, our device will be known as an N channel device. We could have chosen to talk about a P channel device just as well. If we had, the carriers would move in the opposite direction in the crystal. The situation here is analogous to the relationship between NPN and PNP transistors. Notice that since we are using an N channel, the carriers within the crystal are shown traveling in the opposite direction to the conventional current shown entering and leaving the device in Figure 16.1.

One terminal will be known as the "drain," D, perhaps because it drains carriers out of the crystal. The other terminal will be known as the "source," S, perhaps because it is the source of carriers moving into the crystal. You can see that the drain current, I_D, is equal to the source current, I_S, and that current is limited primarily by the number of carriers available in the channel.

To complete the device we need to add a control element. This will be accomplished by adding some P type semiconductor to the side of the crystal as shown in Figure 16.2a. If you look at the die pattern of a real JFET you will find that Figure 16.2a is a considerable simplification of reality, but the idea is the same.

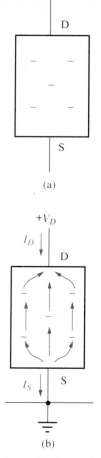

FIGURE 16.1 Conduction in an N channel.
(a) A slab of N type semiconductor.
(b) Negative carrier motion during conduction.

The terminal attached to the P type semiconductor is known as "the gate," G, probably because it controls the flow of carriers from the source across the crystal to the drain.

It is intended that the P-N junction formed by the gate and the channel of the JFET will always be reverse biased. From what you already know of junction behavior, you know that when the junction is reverse biased, a depletion region is formed at the junction as shown in Figure 16.2b.

You should immediately see two implications in the above statement. First, ideally there should be no gate current. Thus the input impedance to the gate should be infinite. This makes the JFET very attractive in circuits where high input impedance is needed. The input to your scope or your DMM would be two familiar examples.

Second, the existence of the depletion region tends to reduce the number of carriers in the channel and so to reduce the drain current, I_D.

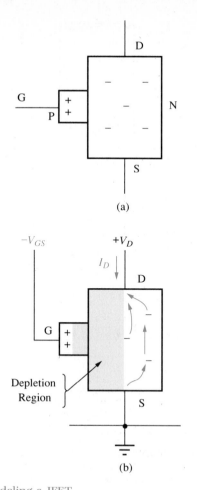

FIGURE 16.2 Modeling a JFET.
(a) Addition of the control terminal.
(b) Biasing the junction.

Of course, that is how the gate controls I_D. The greater the junction reverse bias, V_{GS}, the less I_D will be. In fact, if V_{GS} is made sufficiently negative, the depletion region will extend all the way across the channel. Since there are then no carriers in the channel, there can be no conduction and $I_D = 0$ A. This condition is known as "pinchoff."

We will identify the value of V_{GS} required to produce pinchoff as simply V_P. You will find that data books often use the more complicated term VGS(OFF). In the language of PSpice V_P is known as VTO.

The value of V_P is one of the two parameters commonly used to describe the behavior of a JFET. The other parameter is the "saturation current," I_{DSS}. From your previous understanding of semiconductors you know that the maximum current that such devices will carry is limited by the number of carriers present. Thus I_{DSS} is the maximum drain current that the device will conduct when $V_{GS} = 0$ V. We will have much more to say about both V_P and I_{DSS} later.

16.3 **Appearance and Symbols**

Since JFETs are packaged in the same type of case as any of a host of three-terminal devices, they are impossible to identify except by looking up the device part number in a data book. The standard N channel JFET symbol used in circuit diagrams is shown in Figure 16.3a beside the model

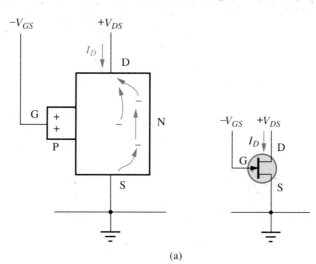

(a)

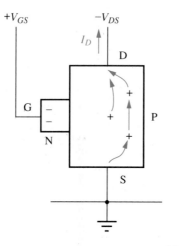
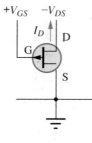

(b)

FIGURE 16.3 Symbols for a JFET.
(a) The N channel JFET.
(b) The P channel JFET.

which we developed in Section 16.2. The standard P channel JFET symbol used in circuit diagrams is shown in Figure 16.3b beside its modeled equivalent. Note that the arrow on the gate terminal indicates whether a given JFET is an N channel or a P channel device. Also notice the polarity of the bias voltage required for the two JFET types. This difference is required in order that the junction will be reverse biased in each case.

EXAMPLE 16.1

(a)   What should be the value of V_{GS} in the circuit of Figure 16.4?

(b)   What would be the value of V_{GS} if someone mistakenly used a +2 V power supply rather than the indicated −2 V supply?

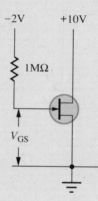

−2V   +10V

1MΩ

V_{GS}

FIGURE 16.4   Example 16.1.

●   (For a PSpice solution, see Appendix H.)

SOLUTION:

(a)   Since the gate-source junction is reverse biased, there should be no current flowing through the 1 MΩ resistor so no voltage should be lost across it. Thus $V_{GS} = -2$ V.

(b)   Reversing the gate power supply forward biases the gate to source junction producing a significant gate current I_G. The device no longer acts like an FET. The problem now becomes the familiar forward biased junction problem which we dealt with so often in connection with diodes and bipolar transistors. Applying Ohm's Law to the gate circuit of Figure 16.4 gives us

$$I_G = \frac{2 \text{ V} - V_{GS}}{1 \text{ M}\Omega}$$

Iterating the above equation with the junction equation gives

$$I_G = \frac{2\text{ V} - V_{GS}}{1\text{ M}\Omega} \quad = \quad \begin{array}{ccc} 1.30 & 1.51 & 1.51 \end{array} \quad \mu\text{A}$$

$$V_{GS} = 26\text{ mV} * \ln\frac{I_G}{10\text{ fA}} = \quad \begin{array}{ccc} 0.486 & 0.490 & 0.490 \end{array} \quad \text{V}$$

16.4 Ohmmeter Tests

Since the JFET contains a P-N junction, it is possible to use an ohmmeter to do some rather simple device-testing. The logic is much the same as that applied to bipolar transistors in Section 8.9.

Using an ohmmeter only it is impossible to distinguish between the source and the drain terminals since carriers can flow in either direction in the channel. In fact, a JFET will function with its source and drain terminals exchanged, although its characteristics will usually be altered somewhat.

In keeping with the arrow on the gate terminal, an N channel JFET will show conduction from the gate terminal to either the source or the drain terminal, but not in the reverse direction. A P channel JFET will show conduction either from the source to the gate, or from the drain to the gate terminal, but not in the reverse direction.

Although it is not designed to operate that way, if the junction of a JFET is forward biased so that it conducts current, no permanent damage will be done to the device provided the current is not so large that the device becomes overheated.

EXAMPLE 16.2
A given device is known to be a JFET but the leads are not identified, and it is not known whether it is a P channel or an N channel device. Outline a procedure for obtaining the above information using only an ohmmeter.

SOLUTION:
An ohmmeter is placed in its diode test mode. By process of elimination two leads are found that will conduct current no matter which way the ohmmeter is connected. These leads have to be connected to the channel. One lead must be the drain and the other lead must be

the source. With the negative lead of the ohmmeter connected to either of the above leads, the positive ohmmeter lead is moved to the third terminal of this device (the gate). If this configuration shows conduction, the gate is P type material and the device is an N channel JFET. This hypothesis is supported if there is no conduction when the leads are reversed. If no conduction is seen when the positive lead is connected to the gate terminal, the gate is N type material and the device is a P channel JFET. This hypothesis is supported if there is conduction when the leads are reversed.

16.5 Common Source Parameters and Characteristics

A common source characteristic graph for the JFET can be constructed in much the same way as the common emitter characteristic graph was drawn for bipolar transistors in Section 8.10. The variables to be plotted are the drain current I_D on the vertical axis versus the drain to source voltage V_{DS} on the horizontal axis. In this case, since the FET is a voltage sensitive device, the control parameter will be the gate to source voltage, V_{GS}.

A circuit that can be used to construct the common source characteristic graph is shown in Figure 16.5. To begin constructing the common source characteristic graph, we will first set the gate to source voltage at $V_{GS} = 0$ V. As we slowly increase V_{DS}, we find that I_D increases very rapidly as shown in the part of Figure 16.6 labeled "A."

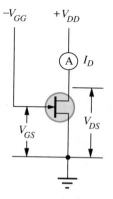

FIGURE 16.5 Circuit for plotting JFET common source characteristic graph.

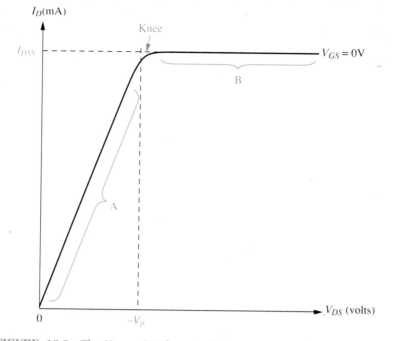

FIGURE 16.6 The $V_{GS} = 0V$ characteristic curve.

As we continue to increase V_{DS}, we find that eventually I_D ceases to increase. It then remains essentially constant at the "saturation current," I_{DSS} regardless of any further increase in V_{DS}. We have passed the "knee" of the curve in Figure 16.6 and we have reached the part of the curve labeled "B."

According to our model of Section 16.2, I_{DSS} is the current at which all of the carriers in the channel are being used to carry current. Since there are no more carriers, there can be no further increase in current.

A careful study of the electric field existing within the crystal would reveal that the knee of the drain characteristic curve should be located at the point where

$$V_{DSknee} = V_{GS} - V_P \qquad (16.1)$$

For the special case where $V_{GS} = 0$ V, $V_{DSknee} = -V_P$ as indicated on Figure 16.6. Thus if $V_P = -6$ V, the knee of the $V_{GS} = 0$ V curve should be located at about $V_{DS} = +6$ V.

You can see that the pinchoff voltage, V_P, which we first mentioned in Section 16.2 is an important device parameter. I_{DSS} is the other important device parameter which we will use along with V_P to describe the behavior of the JFET. Both I_{DSS} and V_P are listed on most device data sheets.

As we have mentioned, PSpice uses V_P which it calls VTO, but it does not use I_{DSS} directly. Instead it uses something it calls "BETA." We will find out why later. β is given by

$$\beta = \frac{I_{DSS}}{V_P^2} \qquad (16.2)$$

In order to obtain the rest of the curves that make up the characteristic graph as shown in Figure 16.7, the gate to source voltage V_{GS} is set at the fixed negative value of V_{GS1} and I_D vs V_{DS} data is again taken and plotted. This plot is labeled "V_{GS1}" in Figure 16.7. V_{GS} is then incremented and the procedure repeated until the complete family of curves is generated. The procedure is similar to that used in Section 8.10 with bipolar transistors. When the gate voltage is made so negative that no current flows, the JFET is pinched off and then $V_{GS} = V_P$ as we first mentioned in Section 16.2.

A procedure for obtaining the characteristic graph from PSpice is included in Appendix H. Notice the close correlation between the laboratory data in Figure 16.8a and the PSpice simulation in Figure 16.8b.

It is important to notice that there are definite upper and lower limits within which the JFET must operate. Those limits are where $V_{GS} = 0$ V and $V_{GS} = V_P$. In the case of the bipolar transistor no upper limit existed, unless the maximum power rating of the device was taken as the limit.

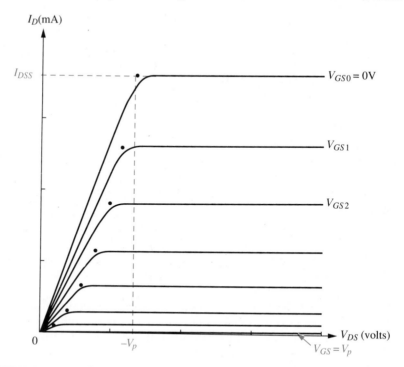

FIGURE 16.7   The JFET common source characteristic graph.

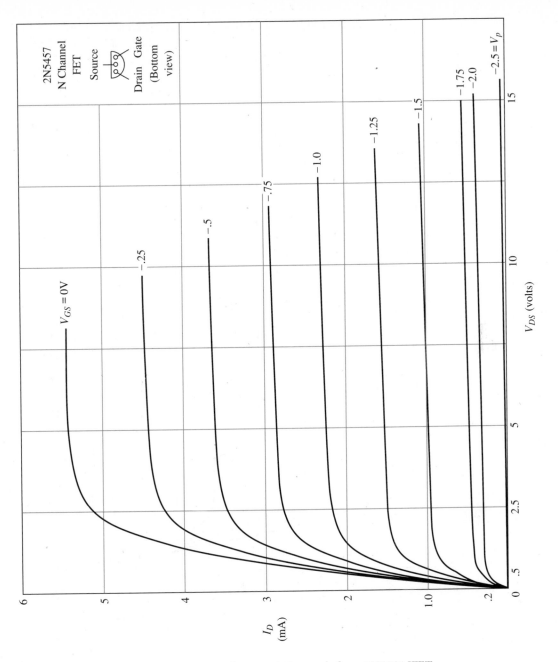

FIGURE 16.8 Common source characteristic graph for a 2N5457 JFET.
(a) Output from an x-y plotter connected to the circuit of Figure 16.5.

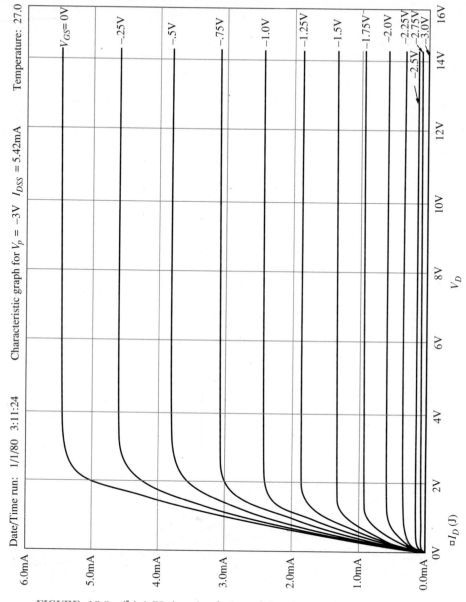

FIGURE 16.8 (b) A PSpice simulation of the circuit of Figure 16.5.

EXAMPLE 16.3

Given the characteristics for the particular 2N5457 JFET shown in Figure 16.8a, find V_P and I_{DSS}. Compare these values with the typical values listed on the manufacturer's data sheet in Appendix F.

SOLUTION:

According to the graph of Figure 16.8a, when $V_{GS} = -2.5$ V, the drain current is nearly pinched off. So the data book listing in Appendix F of $V_{GS(OFF)} = -3$ V does not seem to be unreasonable.

Again, reading from the graph, when $V_{GS} = 0$ V, I_D is slightly over 5.4 mA. So $I_{DSS} \approx 5.4$ mA. The typical value listed in Appendix F is $I_{DSS} = 5$ mA.

Notice that beyond the knee of the characteristic curves the JFET acts like a fairly good constant current source. JFETs are often used for this purpose.

EXAMPLE 16.4

A certain inexpensive power supply contains the circuit shown in Figure 16.9. Assume that the JFET has the characteristics shown in Figure 16.8a. Over what range of load resistance, R_L, will the circuit act as a constant current source?

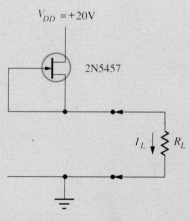

FIGURE 16.9 Example 16.4.

SOLUTION:

Since the gate terminal is connected to the source, $V_{GS} = 0$ V. Thus $I_L = 5.42$ mA provided V_{DS} is always above the knee of the characteristic curve. Obviously when $R_L = 0\ \Omega$ then $V_{DS} = V_{DD}$ and $I_L = 5.42$ mA. Reading from Figure 16.8a the minimum allowable value for V_{DS} is a bit over $-V_P$. To be safe we will use $V_{DSmin} \approx 5$ V. Thus the maximum value for R_L is given by:

$$R_{Lmax} = \frac{V_{DD} - V_{DSmin}}{I_{DSS}}$$

$$= \frac{20\ \text{V} - 5\ \text{V}}{5.42\ \text{mA}} = 2.77\ \text{k}\Omega$$

Thus R_L can be any value below about 2.7 kΩ.

16.6 The Transfer Characteristic

As we have seen, as long as the drain to source voltage, V_{DS} is above its knee value, the drain current I_D is almost completely independent of V_{DS}.

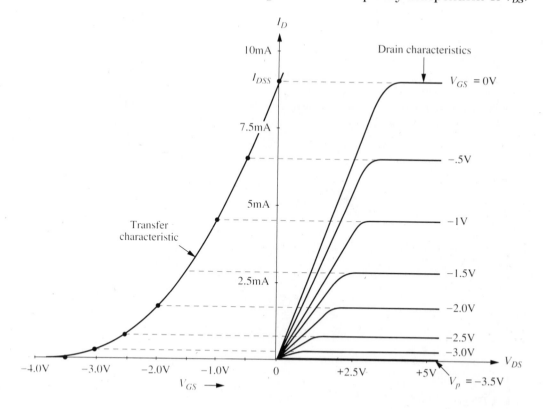

FIGURE 16.10 Transfer characteristic coupled to the JFET common source characteristics.

I_D is then dependent solely on the gate to source voltage, V_{GS}. Thus, another way to picture the behavior of the FET is to simply graph I_D vs V_{GS}. This graph is known as "the transfer characteristic." In Figure 16.10 a typical transfer characteristic is shown beside the common source characteristics from which it was plotted.

You should always remember that the transfer characteristic is only valid for drain to source voltages above the knee of the drain characteristic curves.

EXAMPLE 16.5

Construct a transfer characteristic graph from the drain characteristics of the 2N5457 presented in Figure 16.8.

SOLUTION:

Reading from Figure 16.8a, we see that the curve for $V_{GS} = 0$ V levels off at about $I_D = 5.42$ mA. These data along with the data for the other

TABLE 16.1   Transfer Characteristic

Data Index	V_{GS} (Volts)	I_D (A)
1	0	5.42
2	− .25	4.5
3	− .5	3.65
4	− .75	2.9
5	− 1	2.27
6	− 1.25	1.55
7	− 1.5	1.02
8	− 1.75	.5
9	− 2	.35
10	− 2.5	0

drain characteristic curves are recorded in Table 16.1. Using these data the transfer characteristic of Figure 16.11 is drawn.

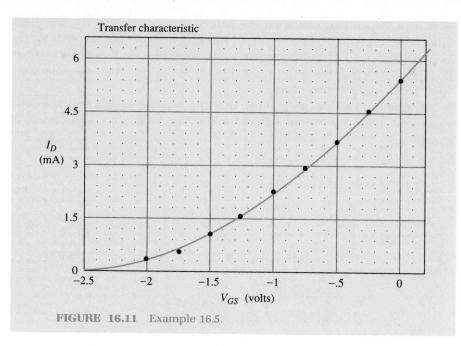

FIGURE 16.11   Example 16.5.

16.7 The Transfer Equation

If you look at the shape of the transfer characteristic of either Figure 16.10 or Figure 16.11, you will see that they look very much like parabolas. You may remember from your algebra class that the formula used to represent a parabola of this sort is

$$y = a(x - h)^2 + k$$

If the coordinates of the vertex of the parabola, (h, k) are taken to be $(V_P, 0)$ then, using our notation, we have

$$I_D = a(V_{GS} - V_P)^2 \qquad (16.3)$$

In order to evaluate the coefficient "a," we know that if $V_{GS} = 0$ V then $I_D = I_{DSS}$. Substituting these values in Equation 16.3 and solving for "a" gives

$$a = \frac{I_{DSS}}{(-V_P)^2} = \frac{I_{DSS}}{V_P^2} = \beta \qquad (16.2)$$

This is the BETA used by PSpice. Substituting Equation 16.2 into Equation 16.3, we have what is known as "the transfer equation."

$$I_D = \frac{I_{DSS}}{(-V_P)^2}(V_{GS} - V_P)^2 = I_{DSS}\left(1 - \frac{V_{GS}}{V_P}\right)^2 \qquad (16.4)$$

You will find that the transfer equation is also referred to as "the Shockley equation," in honor of the developer of the idea.

Solving for V_{GS} gives what is perhaps a more useful form of the transfer equation.

$$V_{GS} = V_P\left(1 - \sqrt{\frac{I_D}{I_{DSS}}}\right) \tag{16.5}$$

You should become well acquainted with this important equation. It allows us to compress all of the information in Figure 16.11 and much of the information in Figure 16.8a into a single equation that your calculator can easily handle.

EXAMPLE 16.6

Graph the transfer equation, Equation 16.4, using the values read from the graph in Example 16.3 for I_{DSS} and V_P. Compare the result with the graph drawn in Example 16.5.

SOLUTION:

$$I_D = I_{DSS}\left(1 - \frac{V_{GS}}{V_P}\right)^2 \tag{16.4}$$

$$= 5.42\ \text{mA}\left(1 - \frac{V_{GS}}{-2.5\ \text{V}}\right)^2 \tag{16.6}$$

The first two columns of Table 16.2 are the data from Table 16.1, the

TABLE 16.2   Transfer Characteristic Data

V_{GS} (Volts)	I_D (mA) Table 16.1	I_D (mA) Equation 16.6
0	5.42	5.42
−.250	4.50	4.39
−.500	3.65	3.47
−.750	2.90	2.66
−1.00	2.27	1.95
−1.25	1.55	1.36
−1.50	1.02	.867
−1.75	.500	.488
−2.00	.350	.217
−2.50	0	0

third column was obtained by substituting numbers from column one into Equation 16.6. Figure 16.12 is a graph of the two sets of data.

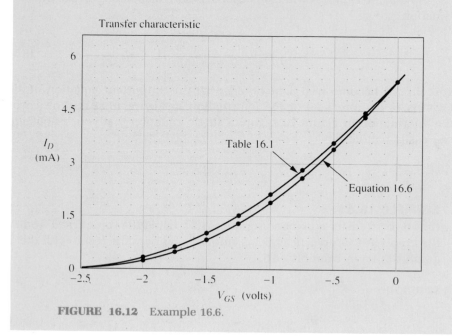

FIGURE 16.12 Example 16.6.

Example 16.6 indicates that if I_{DSS} and V_P are accurately known, Equation 16.4 or Equation 16.5 does an acceptable job of describing JFET behavior. Since manufacturer's data sheets usually list both I_{DSS} and V_P, we will use Equation 16.4 or Equation 16.5 extensively in working with JFET circuits.

16.8 Transconductance Equivalent Circuit

You may remember that in order to be able to apply our knowledge of circuit theory to circuits containing BJTs we had to develop a model of the device that contained only sources and resistors. We must now do the same thing with FETs. The JFET circuit is compared with the BJT circuit in Figure 16.13.

Based on everything we have said so far, you will recognize that the input section of the FET equivalent circuit must be quite different from the input section of the BJT equivalent circuit with which you are familiar. The BJT requires an input current I_B, and so we show an input resistance, h_{ie} in Figure 16.13b. On the other hand, the FET requires an input voltage, V_{GS}. You remember that since the gate is reverse biased, it does not con-

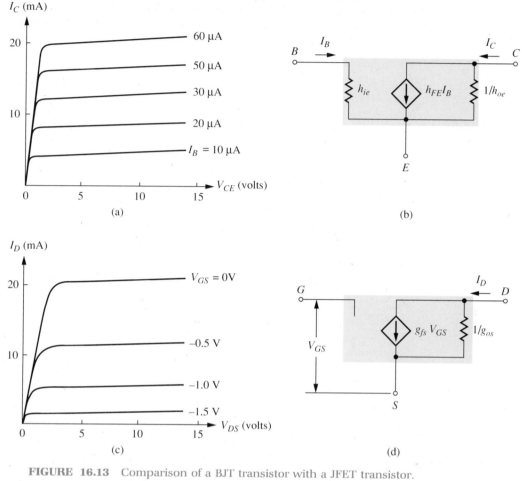

FIGURE 16.13 Comparison of a BJT transistor with a JFET transistor.
(a) Common emitter characteristic graph for a BJT.
(b) Equivalent circuit for a BJT.
(c) Common source characteristic graph for a JFET.
(d) Equivalent circuit for a JFET.

duct. To represent this fact, in the equivalent circuit of Figure 16.13d the gate terminal of the FET is not connected to anything. It is left dangling in space.

Comparing their characteristics, you should recognize that from an output standpoint, the two devices look quite similar. For the BJT, since the characteristics were nearly horizontal, we used a current source, $h_{fe}i_B$ and a resistor, $1/h_{oe}$ in parallel to represent the collector section. Just so, we will use a current source, $g_{fs}v_{GS}$ and a resistor, $1/g_{os}$ in parallel to represent the drain section of the FET. This leads us to the g parameter equivalent for an FET as shown in Figure 16.13d.

The resistor $1/g_{os}$ is completely analogous to $1/h_{oe}$ of the BJT. Its value is determined by the slope (1/slope, really!) of the characteristic curves. Using the language of calculus, g_{os} is usually defined as

$$g_{os} \equiv \frac{dI_D}{dV_{DS}} \tag{16.7}$$

or

$$1/g_{os} = \frac{dV_{DS}}{dI_D} \tag{16.8}$$

As you might expect, $1/g_{os}$ is typically very large and can often be ignored. The output conductance, g_{os}, or y_{os}, as it is called by some manufacturers, is one of the parameters that usually appears on data sheets. If you look in Appendix F, you will see that for a 2N5457, which is a commonly used JFET, $g_{os} = 10\ \mu S$ (or 10 μmhos) which gives $1/g_{os} = 100\ k\Omega$.

EXAMPLE 16.7

Using the characteristics for the 2N5457 in Figure 16.8a, determine the value of $1/g_{os}$ when $I_D = 3$ mA.

SOLUTION:

Figure 16.8a is reproduced here as Figure 16.14 with a line drawn approximately parallel to the characteristic curve that is nearest to $I_D = 3$ mA. Then $1/g_{os}$ is given by

$$1/g_{os} = \frac{dV_{DS}}{dI_D} \tag{16.8}$$

$$= \frac{V_{DS2} - V_{DS1}}{I_{D2} - I_{D1}} \tag{16.9}$$

$$= \frac{20\ V - 0\ V}{3\ mA - 2.9\ mA} = 200\ k\Omega$$

In terms of my ability to estimate the slope of the characteristic, this result agrees well with manufacturer's data.

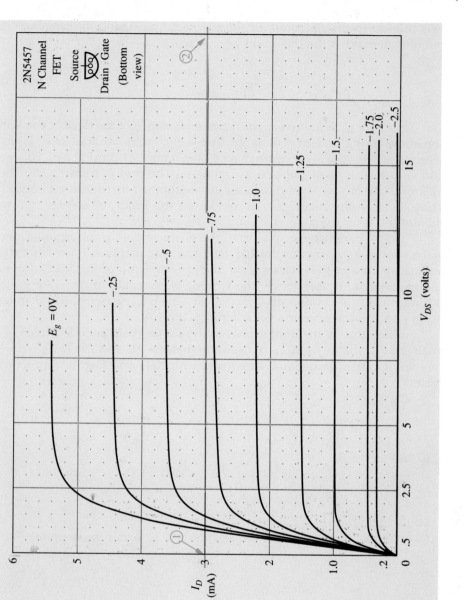

FIGURE 16.14 Example 16.7.

From Figure 16.13d, you can see that if we ignore $1/g_{os}$, as we will usually be able to do, we have

$$i_D = g_{fs}v_{GS} \qquad\qquad (16.10)$$

which leads to the definition of the parameter g_{fs}. Using the language of calculus to express the definition we can write

$$g_{fs} \equiv \frac{dI_D}{dV_{GS}}$$

(16.11)

The parameter g_{fs} is known as the *transconductance* of the device. You will sometimes see g_{fs} referred to as g_m. The name "transconductance" is descriptive since: (1) g_{fs} is a conductance. It is a current divided by a voltage and so it has units of Siemens (or mho). (2) The prefix "trans-" gives the idea of going across from one place to another, as in "transcontinental." Thus transconductance "goes across" from the input AC voltage, v_{GS}, to the output AC current, i_D.

The FET parameter g_{fs} relates to the characteristic graph in much the same way as h_{fe} relates to the BJT characteristic graph. It represents the vertical spacing of the characteristic lines. In the case of the JFET, due to the nonlinearity of the device, the spacing is not uniform. Thus for g_{fs} to be meaningful, the value of I_D or V_{GS} at which g_{fs} was measured must always be given whenever a value for g_{fs} is given.

Further, due to the nonlinearity of the device there is no DC value g_{FS} analogous to the h_{FE} of BJTs. To assume that the AC and the DC parameters are approximately equal as we did with BJTs will lead to erroneous results.

EXAMPLE 16.8
Determine the value of g_{fs} for the 2N5457 FET of Figure 16.8a at $I_D = 3$ mA.

SOLUTION:
Figure 16.8a is reproduced here as Figure 16.15. From the definition

$$g_{fs} \equiv \frac{dI_D}{dV_{GS}}$$

(16.11)

or

$$g_{fs} = \frac{I_{D2} - I_{D1}}{V_{GS2} - V_{GS1}}$$

(16.12)

Since we are looking for the vertical spacing of the characteristic curves at the point $I_D = 3$ mA, we will use the nearest plotted curves

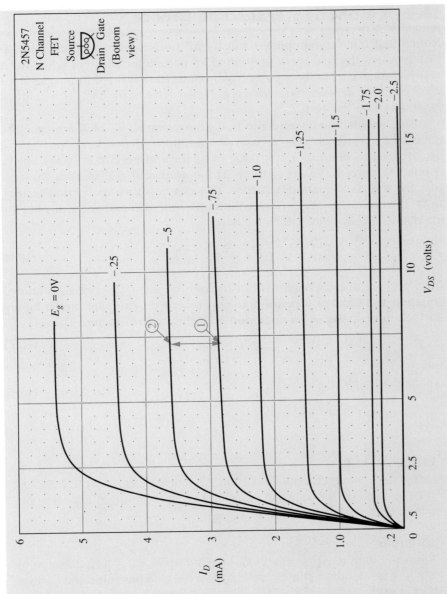

FIGURE 16.15 Example 16.8.

that bracket the current of interest. Thus we will go from $V_{GS1} = -0.75$ V to $V_{GS2} = -0.5$ V. Reading values from Figure 16.15, we have

$$g_{fs} = \frac{3.65 \text{ mA} - 2.9 \text{ mA}}{-0.5 \text{ V} - (-0.75 \text{ V})} = 3.00 \text{ mS}$$

16.9   Calculating Transconductance

Notice that our very simplified, but useful transconductance model of an FET requires that we know only one FET parameter and that is the transconductance, g_{fs}. Maybe we need two parameters if the output resistance, $1/g_{os}$ must be considered. However, as we have said, $1/g_{os}$ is usually so large it can be ignored.

Data sheets usually give the transconductance, g_{fs}, at some specified value of I_D. However, due to its very nonlinear dependence on I_D, it is usually best to determine the value of g_{fs} for a given circuit based on the Q point value of I_D. In order to avoid the graphical approach of Example 16.8, the transfer equation, Equation 16.4 can be substituted into Equation 16.11, the definition of g_{fs}. Then with a little calculus and some algebra (for the details see Appendix D.9), we have

$$g_{fs} = \frac{2}{-V_P}\sqrt{I_{DSS}I_D} \qquad (16.13)$$

Since almost all data book listings include V_P and I_{DSS}, once I_D is known, Equation 16.13 makes it possible to construct a g parameter model for any JFET circuit as long as its operation is in the "active" region. That is

$$I_D < I_{DSS} \qquad (16.14)$$

and

$$V_{DS} > V_{GS} - V_P \qquad (16.15)$$

EXAMPLE 16.9

Using the manufacturer's data from Appendix F, for a typical 2N5457 JFET determine the value of g_{fs} at $I_D = 3$ mA. Compare the value obtained with that of Example 16.8.

SOLUTION:

From the data in Appendix F, $V_P = -3$ V and $I_{DSS} = 5$ mA. Assuming that V_{DS} is adequate, the required $I_D = 3$ mA will be within the "active" region and our equations should apply. Thus the typical value of g_{fs} at $I_D = 3$ mA is given by

$$g_{fs} = \frac{2}{-V_P}\sqrt{I_{DSS}I_D} \qquad (16.13)$$

$$= \frac{2}{-(-3\text{ V})}\sqrt{5\text{ mA} * 3\text{ mA}} = 2.58\text{ mS}$$

In Example 16.8 $g_{fs} = 3.00$ mS for a discrepancy of about 20 percent.

Given the kind of variability we have come to expect for transistors, since the graph in Figure 16.15 was drawn from laboratory data obtained from a particular 2N5457 the above discrepancy should not be surprising.

16.10   Troubleshooting   ◀ ◀ ◀ ◀ ◀ ◀ ◀ ◀

Most of the troubleshooting procedures that work for BJTs also work for JFETs. You do need to remember that the input circuit for the JFET is usually a high impedance circuit so that test instrument loading can become a factor. You also need to remember that the gate junction must always be reverse biased. The device will cease to amplify if the gate junction becomes forward biased.

16.11   Summary

This chapter introduced the JFET, its appearance, and a simplified qualitative model accounting for its behavior. The parameters common source characteristics, and the transfer characteristic which give a more quantitative description of JFET behavior were also presented along with the transconductance equivalent circuit which allows us to use standard circuit analysis procedures on JFET circuits.

GLOSSARY

Channel: That portion of the crystal structure of an FET in which the drain current I_D flows, and to which the drain and the source terminals are connected. (Section 16.2)

Drain: That terminal of an FET which "drains" carriers out of the channel into the external circuit. (Section 16.2)

Gate: That terminal of an FET which controls the flow of drain current through the channel. (Section 16.2)

Knee: The place on a graph where there is a very marked change in slope. (Section 16.5)

Pinchoff $V_{GS(OFF)}$, or V_P: The value of gate to source voltage required to reduce the drain current I_D to zero. (Section 16.2)

Source: That terminal of an FET which injects carriers into the channel from the external circuit. (Section 16.2)

Transconductance: g_{fs} or g_m. An FET parameter which describes the change in drain current that will result from a given change in gate to source voltage. (Section 8)

Transfer characteristic: A graph which expresses the relationship between the drain current, I_D, and the gate to source voltage, V_{GS}. (Section 16.6)

Transfer equation: The transfer characteristic stated in the form of an equation. Also known as the Shockley equation. (Section 16.7)

PROBLEMS

Section 16.3

1. Draw the standard symbol for the following devices and label the S, G, and D terminals.
   (a)   A P channel JFET.
   (b)   An N channel JFET.
2. On the diagrams drawn in Problem 1, indicate the proper polarity of the voltage at each of the other terminals measured with respect to the source terminal. Also indicate the proper direction of current flow at each of the terminals.

Section 16.4

3. If the negative lead of an ohmmeter is connected to the source terminal of a P channel JFET, indicate whether or not the ohmmeter will show conduction when the positive lead is connected to:
   (a)   The drain terminal.
   (b)   The gate terminal.
4. Repeat Problem 3 for an N channel JFET.

Section 16.5

5. From the single characteristic curve of Figure 16.16, estimate the value of β.

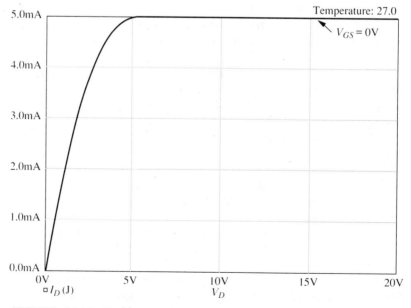

FIGURE 16.16   Problem 16.5.

6. Draw the $V_{GS} = 0$ V characteristic curve for an N channel JFET if $I_{DSS} = 10$ mA and $V_P = -3$ V.
7. Draw the $V_{GS} = 0$ V characteristic curve for a typical 2N5459 JFET using the data provided in Appendix F.
8. Over what range of load resistance, R_L, will the circuit of Figure 16.17 act like a fairly good constant current source?

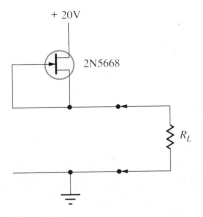

FIGURE 16.17   Problem 16.8.

Section 16.7

9. Write the transfer equation for the JFET of Figure 16.16.
10. Write the transfer equation for a typical 2N5670 JFET.
11. Complete the set of common source characteristic curves for the JFET in Problem 6 by incrementing V_{GS} in one volt steps from $V_{GS} = 0$ V to pinchoff.
12. Draw a complete set of common source characteristic curves for a typical 2N5670 JFET by incrementing V_{GS} in one volt steps from $V_{GS} = 0$ V to pinchoff.

Section 16.8

13. For the circuit of Figure 16.18, find the value of V_{GS} required to limit the current to $I_D = 1.5$ mA for a typical 2N5457 JFET.
14. Repeat Problem 13 if the drain resistor, R_D, became shorted ($R_D = 0\ \Omega$).
15. Repeat Problem 13 if $R_D = 10$ kΩ.

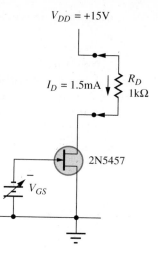

FIGURE 16.18 Problem 16.13.

Section 16.9

16. Use the definition of transconductance to determine the transconductance of the JFET whose characteristics are drawn in Figure 16.19

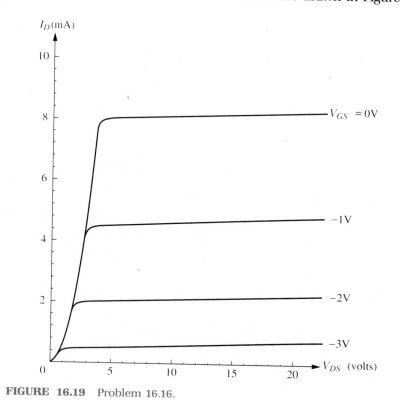

FIGURE 16.19 Problem 16.16.

when

(a) $I_D = 1$ mA.

(b) $I_D = 6$ mA.

17. Use the definition of transconductance and the characteristics you drew in Problem 11 to calculate the transconductance of the JFET at $I_D = 3$ mA.

18. Use Equation 16.13 to calculate the transconductances in Problem 16.

19. Construct a g parameter equivalent circuit of Problem 16.

20. Using data from Appendix F, construct a g parameter equivalent circuit for a 2N5670 JFET at $I_D = 5$ mA.

21. Repeat Problem 20 for a 2N5459 JFET at $I_D = 5$ mA.

Section 16.10   ◀ Troubleshooting

22. Describe the behavior of the circuit in Figure 16.17 if the drain of the 2N5668 became:

(a) An open circuit.

(b) Shorted to the source.

23. Describe the behavior of the circuit in Figure 16.17 if a 2N4445 transistor was substituted for the 2N5668.

24. Describe the behavior of the circuit in Figure 16.17 if a 2N5460 transistor was substituted for the 2N5668.

Automated assembly of electronic components.
(Courtesy of Staley Continental, Inc.)

Chapter 17
Common Source JFET Circuits

SPECIAL TERMS

Active region
Geometric average
Automatic gain control (AGC)
Gate leak resistor R_G
Bias line

17.1   Introduction and Objectives

From Chapter 16 you should have gotten some idea of what a JFET is and how it compares to a BJT. You should also have some understanding of the general features of the JFET common source characteristic graph, the transfer characteristic, the transfer equation, and the g parameter equivalent circuit.

When you finish this chapter, you should have a better understanding of the above ideas and how they are used in functional circuits containing JFETs. You should be able to:

- Design and analyze simple JFET amplifier circuits using a recipe approach similar to that presented in Chapter 12 for the BJT.

17.2   Load Line and Its Slope

The basic common source circuit is shown in Figure 17.1a. Figure 17.1b contains the circuit's g parameter equivalent circuit. At first glance the circuit does not seem to be very different from the common emitter circuits we drew for BJTs. In fact, as we found in Chapter 16, the drain circuit of the JFET and the collector circuit of the BJT function in essentially the same way. That means that we should be able to draw load lines much as we did for the BJT. However, there are some important differences.

For one thing, there is virtually no upper limit on I_C for the BJT. As we mentioned in Chapter 16, the JFET has an upper limit as well as a lower limit on the drain current. The "active region" for the JFET is between $I_D = 0$ A and $I_D = I_{DSS}$ as shown in Figure 17.2.

Another restriction on the active region which must be considered is the fact that the knee of the characteristic curves is not nearly as sharp and does not occur nearly as close to zero volts as it did with BJTs. In fact, as we mentioned in Chapter 16, the knee of the $V_{GS} = 0$ V curve occurs at $V_{DS} = -V_P$. This fact is also shown in Figure 17.2.

We will use the above restrictions combined with ideas from our work with BJTs to construct the AC load line for the JFET circuit.

In Section 12.5 we presented an argument for selecting the collector resistor. The result was

$$R_C = R_L \, \| \, (-1/h_{oe}) \qquad (12.12)$$

The same argument can be applied directly to the JFET circuit. Thus

$$R_D \stackrel{*}{=} R_L \, \| \, \left(\frac{-1}{g_{os}} \right) \qquad (17.1)$$

The asterisk above the equal sign in Equation 17.1 is intended to warn you that, like Equation 12.12, Equation 17.1 is a circuit designer's formula and

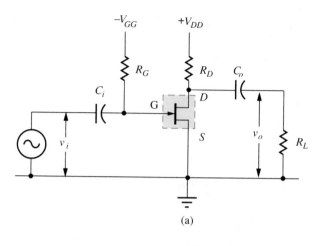

(a)

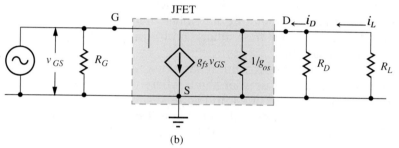

(b)

FIGURE 17.1 Basic common source amplifier circuit.
(a) Circuit diagram.
(b) The g parameter equivalent circuit.

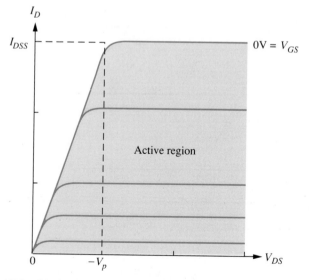

FIGURE 17.2 Limitations on the active region.

should not be used for any other purpose than the specification of the drain resistor.

Since R_D is a real resistor that you would have to pick from a box if you were to hook up the circuit, you should specify the nearest standard resistor value.

As you should recall from our study of BJTs, we defined R_{AC} to be the resistance the device "sees" as it "looks out at the circuit." For the BJT we had

$$R_{AC} = R_C \| R_L \qquad (9.17)$$

By looking at the equivalent circuit of Figure 17.1b you should recognize that for the JFET

$$R_{AC} = R_D \| R_L \qquad (17.2)$$

Also transferring from what you know about BJTs, $-1/R_{AC}$ gives the slope of the AC load line. You should remember that the negative reciprocal is required due to the fact that voltage and current axes "should" be exchanged on our graphs.

For a given value of R_{AC}, the restrictions on the active region which we mentioned above tell us a lot about where the AC load line should be placed. For example, load line A in Figure 17.3 probably makes little sense because the usable portion of the load line is much shorter than it could be. On the other hand, load line C is probably not a good choice either since the higher voltage required may increase the cost of the power supply and it will require the circuit to dissipate more heat. Usually a load

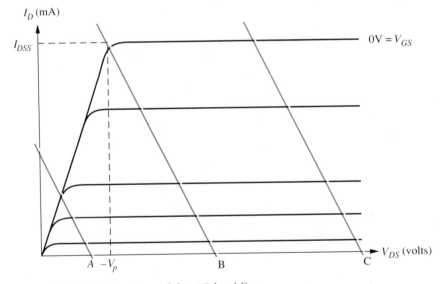

FIGURE 17.3 Location of the AC load line.

line near the location of line B is the best choice. Thus it is reasonable to require that the AC load line should go through the knee of the $V_{GS} = 0$ V curve. Thus the AC load line should go through the point whose coordinates are $(-V_P, I_{DSS})$.

Since we now know the coordinates of one point on the AC load line, and since we already know its slope, we have fixed the location of the load line. If you need help understanding this, you should reread Section 9.6 on the development of Equation 9.18. For the JFET AC load line we can then write

$$v_{DD} = -V_P + I_{DSS}R_{AC} \qquad (17.3)$$

We can now draw the AC load line through the horizontal intercept of v_{DD} and the point $(-V_P, I_{DSS})$.

EXAMPLE 17.1

Draw the AC load line for the circuit of Figure 17.4a.

SOLUTION:

The data book listing in Appendix F, contains the following data for the 2N5103: $V_P = -4.0$ V $I_{DSS} = 8$ mA $g_{os} = 100$ μS. Since the AC load line should go through the point $(-V_P, I_{DSS})$, it should go through the point whose coordinates are $(+4.0$ V, 8 mA). This point is located on the graph in Figure 17.4b.

To determine the slope of the AC load line we have

$$1/g_{os} = \frac{1}{100 \ \mu A} = 10 \ \text{k}\Omega$$

and

$$R_D = R_L \parallel (-1/g_{os}) \qquad (17.1)$$
$$= 2.2 \ \text{k}\Omega \parallel (-10 \ \text{k}\Omega) = 2.82 \ \text{k}\Omega \qquad (\text{Use } 2.7 \ \text{k}\Omega)$$

Since R_D is a real resistor, the nearest standard size is specified. The slope of the load line is then determined by

$$R_{AC} = R_D \parallel R_L \qquad (17.2)$$
$$= 2.7 \ \text{k}\Omega \parallel 2.2 \ \text{k}\Omega = 1.21 \ \text{k}\Omega$$

Then

$$v_{DD} = -V_P + I_{DSS}R_{AC} \qquad (17.3)$$
$$= -(-4 \ \text{V}) + 8 \ \text{mA} * 1.21 \ \text{k}\Omega = 13.7 \ \text{V}$$

This point is also located on Figure 17.4b and the AC load line is drawn.

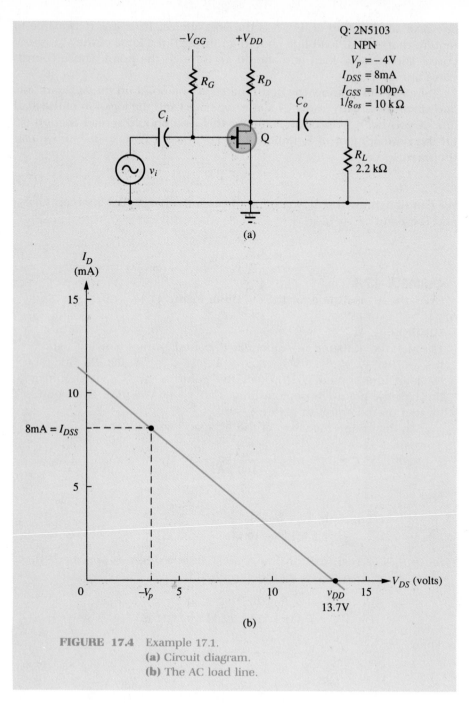

FIGURE 17.4   Example 17.1.
(a) Circuit diagram.
(b) The AC load line.

17.3 The Q Point Current I_D

The idea of an operating point is also directly transferable from our work with the BJT beginning with Section 9.3. Thus we will talk about the Q point voltages and current. We will also talk about the AC signal as it oscillates up and down the AC load line around the Q point. The obvious conclusion is that the Q point should be at the center of the AC load line just as it was in Section 9.7.

  If you really understand how JFETs work, you should immediately see at least one problem with the preceding statement. Remember that the upper limit of the active region for the JFET is the $V_{GS} = 0$ V characteristic line. Thus the usable portion of the AC load line effectively begins at $(-V_P, I_{DSS})$ and ends at the horizontal intercept of v_{DD} as shown in Figure 17.5 which was copied from Example 17.1.

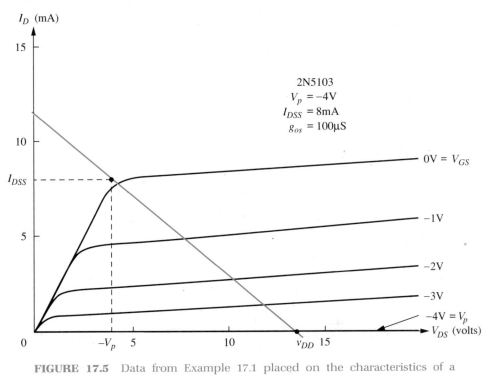

FIGURE 17.5 Data from Example 17.1 placed on the characteristics of a 2N5103 JFET.

  Based on the above argument, it sounds attractive to locate the Q point such that

$$I_D = \frac{I_{DSS}}{2} \qquad (17.4)$$

and

$$V_{GS} = \frac{V_P}{2} \tag{17.5}$$

There is a problem however. The junction equation says

$$I_D = I_{DSS}\left(1 - \frac{V_{GS}}{V_P}\right)^2 \tag{16.4}$$

Substituting Equation 17.5 into the transfer equation, Equation 16.4 results in

$$I_D = I_{DSS}\left(1 - \frac{V_P}{2V_P}\right)^2 = \frac{I_{DSS}}{4} \tag{17.6}$$

Obviously Equations 17.4 and 17.6 cannot both be true at the same time. The reason for the inconsistency is the inherent nonlinearity of the FET. Some kind of compromise must be sought. Given the notorious variability of the parameters we are dealing with, it would probably be close enough to just take a simple arithmetic average of Equation 17.4 and 17.6. That would give

$$I_D = \frac{\dfrac{I_{DSS}}{2} + \dfrac{I_{DSS}}{4}}{2} = 0.375 I_{DSS}$$

However, for second degree relationships of this sort it is more technically correct to use a geometric average. Therefore, since it involves pressing only one more button on your calculator, for a common source JFET small signal amplifier, we will use

$$I_D = \frac{I_{DSS}}{\sqrt{2*4}} \overset{*}{=} \frac{I_{DSS}}{\sqrt{8}} \tag{17.7}$$

Again the asterisk indicates a designer's formula. It is intended to give guidance in choosing an appropriate value for I_D and is not valid for any other purpose such as circuit analysis.

17.4 The Q Point Voltages V_{GS} and V_{DS}

Once the desired value for I_D is known, in order to determine the Q point voltages, V_{GS} and V_{DS}, we could construct the common source characteristic graph for the JFET, draw the AC load line, locate I_D, and read from the graph the required voltages. In Chapter 9 we pointed out that such a graphical solution is not very efficient. Once the value of I_D has been decided upon, it is usually easier to obtain V_{GS} from the transfer equation in the form of Equation 16.5

$$V_{GS} = V_P\left(1 - \sqrt{\frac{I_D}{I_{DSS}}}\right) \tag{16.5}$$

In order to derive an equation for V_{DS}, we will use the mathematician's definition of the slope of a line combined with what you already know of JFET circuit behavior. Your algebra teacher would say that the slope of a line is defined to be

$$m \equiv \frac{y_2 - y_1}{x_2 - x_1}$$

As we have mentioned so often before, since we in electronics draw our graphs with the axes exchanged, the AC circuit resistance, R_{AC}, is related to the slope of the AC load line by

$$R_{AC} = \frac{-1}{m}$$

Thus, for the graph of Figure 17.6 we can write

$$R_{AC} = -\left(\frac{V_2 - V_1}{I_2 - I_1}\right) = -\left(\frac{V_{DS} - (-V_P)}{I_D - I_{DSS}}\right)$$

or

$$R_{AC} = \frac{V_{DS} + V_P}{I_{DSS} - I_D} \qquad (17.8)$$

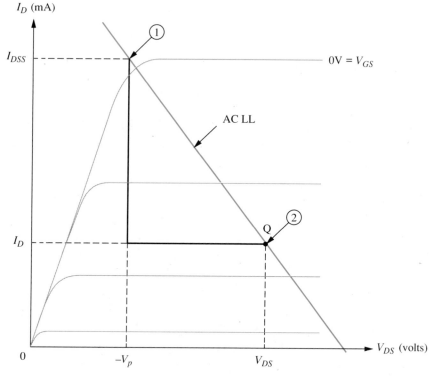

FIGURE 17.6   Determination of V_{DS}.

Solving Equation 17.8 for V_{DS} gives us the required equation:

$$\bullet \qquad V_{DS} = (I_{DSS} - I_D)R_{AC} - V_P \qquad (17.9)$$

When you use Equation 17.9, you need to remember that for N channel JFETs, V_P is a negative quantity so when it is subtracted, you actually add its absolute value.

EXAMPLE 17.2

Locate the Q point voltages and currents for the circuit of Example 17.1.

● (For a SuperCalc solution, see Appendix G.)

SOLUTION:

The circuit with the information from Example 17.1 is shown in Figure 17.7.

$$I_D = \frac{I_{DSS}}{\sqrt{8}} \qquad (17.7)$$

$$= \frac{8 \text{ mA}}{\sqrt{8}}$$

$$= 2.83 \text{ mA}$$

The transfer function then gives the gate to source voltage

$$V_{GS} = V_P\left(1 - \sqrt{\frac{I_D}{I_{DSS}}}\right) \qquad (16.5)$$

$$= -4 \text{ V}\left(1 - \sqrt{\frac{2.83 \text{ mA}}{8 \text{ mA}}}\right) = -1.62 \text{ V}$$

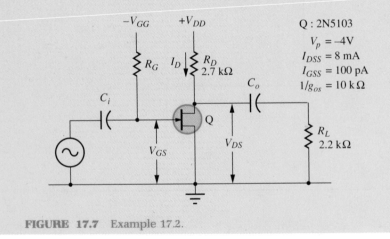

FIGURE 17.7 Example 17.2.

Equation 17.9 gives the drain to source voltage

$$V_{DS} = (I_{DSS} - I_D)R_{AC} - V_P \qquad (17.9)$$
$$= (8 \text{ mA} - 2.83 \text{ mA})1.21 \text{ k}\Omega - (-4 \text{ V})$$
$$= 10.3 \text{ V}$$

17.5 Drain Supply Voltage V_{DD}

In order to determine the required value of V_{DD}, refer to the simple grounded source circuit of Figure 17.1 which is reproduced here as Figure 17.8a. Doing a KVL walk from the drain supply, V_{DD} down to ground gives

$$V_{DD} - I_D R_D - V_{DS} = 0 \text{ V} \qquad (17.10)$$

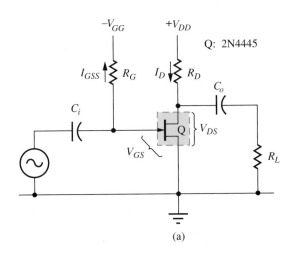

(a)

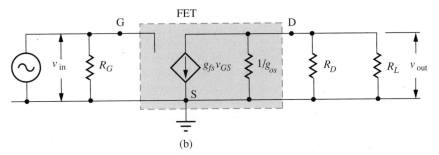

(b)

FIGURE 17.8   Basic circuit used for circuit analysis.
(a) Circuit diagram.
(b) The g parameter equivalent circuit.

Thus for the simple two-supply circuit the required drain power supply voltage is given by

$$V_{DD} = V_{DS} + I_D R_D \qquad (17.11)$$

After the value of V_{DD} is known, the drain to source breakdown voltage, BV_{DSS} specified by the manufacturer of the device you are planning to use should be checked to be certain that it exceeds V_{DD}. This is good design practice even though all of V_{DD} may never actually appear across the JFET.

17.6 Gate Supply Voltage V_{GG}

Since the gate junction of a correctly biased JFET is always reverse biased, ideally there will never be any current flowing through resistor R_G in Figure 17.8a. Thus there should never be any voltage lost across R_G and

$$V_{GG} = V_{GS} \qquad (17.12)$$

17.7 Resistor R_G

From a theoretical standpoint there is no way to calculate the value of R_G. The reason is that as we have said, theoretically $I_G = 0$ A. Thus, as far as the DC circuit is concerned, it should not make any difference whether R_G is big or small or even whether it exists at all!

From an AC standpoint, the value of R_G is more important. In fact, looking at the g parameter equivalent circuit of Figure 17.8b, we can say

$$R_{in} = R_G \qquad (17.13)$$

Thus the larger we make R_G the less load the circuit will place on the source. This would suggest that the larger we make R_G the better.

There is, however, some limitation on the value of R_G from a practical standpoint. You will notice that the data sheets often list a value for reverse gate leakage, I_{GSS}. If we assume that the voltage that is lost across R_G due to I_{GSS} should be small compared to V_P then we can say that the upper limit for R_G is

$$R_G \overset{*}{=} \frac{V_P}{10 * I_{GSS}} \qquad (17.14)$$

Since both V_P and I_{GSS} are negative quantities for an N channel JFET, R_G will be positive as it obviously must be. Also, since R_G is a real resistor you should specify a standard value for it. Usually the value given for R_G by Equation 17.14 is rather large. Practical considerations such as circuit component leakage currents and even the availability of resistors typically restrict R_G to under 100 MΩ.

EXAMPLE 17.3

Given the circuit of Figure 17.9, specify R_G, R_D, V_{GG}, and V_{DD}.

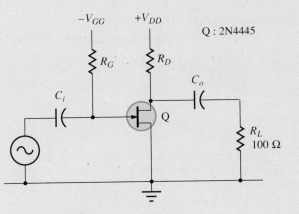

FIGURE 17.9   Example 17.3.

● (For a SuperCalc solution, see Appendix G.)

SOLUTION:

From the data book listing in Appendix F for the 2N4445:

$V_P = -10$ V; $I_{DSS} = 150$ mA; $I_{GSS} = -3$ nA.

Since no value is listed for g_{os} it will be assumed to be negligible

so $1/g_{os} = \dfrac{1}{0 \text{ S}} = \infty \ \Omega$

$$R_D = R_L \| (-1/g_{os}) \qquad\qquad (17.1)$$
$$= 100 \ \Omega \| \infty \ \Omega = 100 \ \Omega \quad (\text{Use } 100 \ \Omega)$$

$$R_{AC} = R_D \| R_L \qquad\qquad (17.2)$$
$$= 100 \ \Omega \| 100 \ \Omega = 50 \ \Omega$$

$$I_D = \frac{I_{DSS}}{\sqrt{8}} \qquad\qquad (17.7)$$

$$I_D = \frac{150 \text{ mA}}{\sqrt{8}} = 53.0 \text{ mA}$$

The junction equation gives

$$V_{GS} = V_P\left(1 - \sqrt{\frac{I_D}{I_{DSS}}}\right) \qquad\qquad (16.5)$$

$$= -10 \text{ V}\left(1 - \sqrt{\frac{53.0 \text{ mA}}{150 \text{ mA}}}\right) = -4.05 \text{ V}$$

The drain to source voltage is given by

$$V_{DS} = (I_{DSS} - I_D)R_{AC} - V_P \tag{17.9}$$
$$= (150 \text{ mA} - 53 \text{ mA}) * 50 \ \Omega - (-10 \text{ V}) = 14.8 \text{ V}$$

$$V_{DD} = V_{DS} + I_D R_D \tag{17.11}$$
$$= 14.8 \text{ V} + 53.0 \text{ mA} * 100 \ \Omega = 20.2 \text{ V}$$

The breakdown voltage, BV_{DSS} is listed as 25 V so the JFET is barely adequate.

$$V_{GG} = V_{GS} \tag{17.12}$$
$$= -4.05 \text{ V}$$

$$R_G \overset{*}{=} \frac{V_P}{10 * I_{GSS}} \tag{17.14}$$
$$= \frac{-10 \text{ V}}{10 * (-3 \text{ nA})} = 333 \text{ M}\Omega$$

$$(\text{Use } 10 \text{ M}\Omega)$$

17.8 Single Supply Bias

You will remember that for simplicity we began our study of BJT circuits using two power supplies, one of which we later eliminated. Likewise we have come to the place where it is time to eliminate one of the power supplies we have been using for JFET circuits. All that was needed in the case of the BJT was to reduce the collector supply voltage to the level needed by the base circuit. A voltage divider was the easy solution.

At first the situation looks more complicated in the case of the JFET since V_{GS} needs to be negative. However, when we say that V_{GS} has to be negative, all we are really saying is that the gate has to be at a lower voltage than the source. Remember that in Chapter 1 when we first talked about the use of subscripts we defined V_{GS} to mean

$$V_{GS} \equiv V_G - V_S \tag{17.15}$$

Thus, in Example 17.3, in order to achieve a -4.05 V gate to source voltage, we grounded the source terminal of the JFET and used a negative 4.05 V gate supply. This gives

$$V_{GS} \equiv V_G - V_S = (-4.05 \text{ V}) - 0 \text{ V} = -4.05 \text{ V}$$

Another way, which is just as acceptable to the JFET, would have been to ground the gate terminal and place the source terminal of the JFET at $+4.05$ V. That gives

$$V_{GS} = V_G - V_S = 0 \text{ V} - (+4.05 \text{ V}) = -4.05 \text{ V}$$

We will take the latter approach here.

Consider the circuit of Figure 17.10. We have a little notation problem in that there are now two source resistors in the circuit. Whenever there is a chance for confusion, we will distinguish them as the signal source resistor R_{SS}, and the JFET source resistor R_{SQ}.

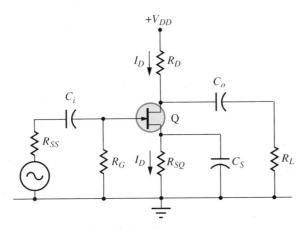

FIGURE 17.10   Source biased circuit.

As you already know, there should not be any DC current in R_G except for the leakage current I_{GSS}. Incidentally R_G is often called the "gate leak resistor" for this reason. Since I_{GSS} is negligible, there should be practically no voltage loss across R_G. Thus, for the circuit of Figure 17.10

$$V_G = 0 \text{ V} \tag{17.16}$$

and

$$\begin{aligned}V_{GS} &= V_G - V_S \\ &= 0 \text{ V} - V_S = -V_S\end{aligned} \tag{17.15}$$

or

$$V_S = -V_{GS} \tag{17.17}$$

Obviously the drain current I_D will flow through the source resistor R_{SQ} that we have now placed between the source terminal of the JFET and ground. Thus there will be a voltage loss across R_{SQ} and the source voltage V_S will be given by

$$V_S = I_D R_{SQ} \tag{17.18}$$

Equation 17.18 shows us that in order to obtain a given Q point, the value of R_{SQ} is given by

$$R_{SQ} = \frac{V_S}{I_D} \tag{17.19}$$

If the Q point is to be unaffected by the addition of R_{SQ} then the drain voltage V_D must be increased by V_S so that

●
$$V_D = V_{DS} + V_S \qquad (17.20)$$

and the supply voltage must also be increased by V_S so that Equation 17.11 becomes

●
$$V_{DD} = V_D + I_D R_D \qquad (17.21)$$

Notice that the circuit of Figure 17.10 is intrinsically stable. If the drain current starts to increase, that will increase the voltage lost across R_S which means more negative bias, V_{GS} which will in turn tend to bring I_D back down.

EXAMPLE 17.4
Redesign the circuit of Example 17.3 to operate from a single power supply.

● (For a SuperCalc solution, see Appendix G.)

SOLUTION:
The circuit is in Figure 17.11. From the data in Appendix F we have: $BV_{DSS} = 25$ V; $V_P = -10$ V; $I_{DSS} = 150$ mA; $I_{GSS} = -3$ nA. From Example 17.3 we have: $R_D = 100\ \Omega$; $R_{AC} = 50\ \Omega$; $I_D = 53.0$ mA; $V_{GS} = -4.05$ V; $V_{DS} = 14.8$ V.; $R_G = 10$ MΩ. In fact, the only information from Example 17.3 which does not apply is the value of V_{DD}, and, of course, the value of V_{GG}. We then have

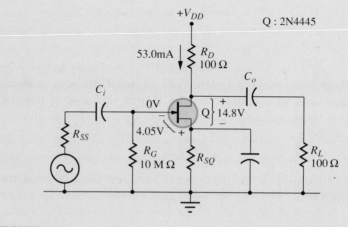

FIGURE 17.11 Example 17.4.

$$V_S = -V_{GS} \tag{17.17}$$
$$= -(-4.05 \text{ V}) = 4.05 \text{ V}$$

$$V_D = V_{DS} + V_S \tag{17.20}$$
$$= 14.8 \text{ V} + 4.05 \text{ V} = 18.9 \text{ V}$$

$$V_{DD} = V_D + I_D R_D \tag{17.21}$$
$$= 18.9 \text{ V} + 53.0 \text{ mA} * 100 \ \Omega = 24.2 \text{ V}$$

$$R_{SQ} = \frac{V_S}{I_D} \tag{17.19}$$

$$= \frac{4.05 \text{ V}}{53.0 \text{ mA}} = 76.4 \ \Omega \qquad (\text{Use } 82 \ \Omega)$$

17.9 The AC Circuit

All of our effort so far has been directed toward obtaining an acceptable load line and Q point. That is, we have been considering mostly DC conditions. Let us now assume that we have a reasonable Q point for the circuit of Figure 17.12a so that the g parameter equivalent circuit of Figure 17.12b applies and let us consider the capacitor size requirements of the circuit and the amount of gain the circuit will deliver.

First, looking at the input circuit of Figure 17.12b, as we pointed out in Section 17.7, the circuit input resistance is given by

$$R_{\text{in}} = R_G \tag{17.13}$$

Returning to the discussion on the low frequency cutoff, f_L, first presented in Section 12.3, we showed that

$$C = \frac{1}{2\pi f_L R_{\text{ckt}}} \tag{12.4}$$

You should recognize that the equation for determining C_i for this circuit is exactly the same as it was for BJT circuits

$$C_i = \frac{1}{2\pi f_L (R_{SS} + R_{\text{in}})} \tag{12.6}$$

If you have any questions about the logic behind Equation 12.6 you should reread Section 12.3.

As you contemplate the source bypass capacitor, C_S, if you returned to Chapter 12 to review the source of the previous equation, you may be bracing for "a formidable amount of calculator button-pushing," as we put it in Section 12.7. However, this time you are in for a pleasant surprise.

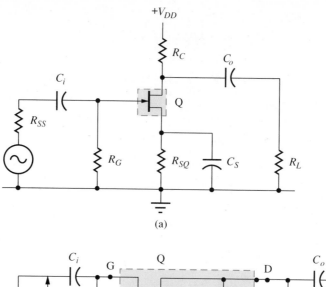

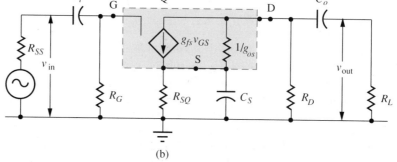

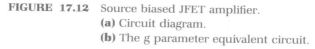

(b)

FIGURE 17.12   Source biased JFET amplifier.
(a) Circuit diagram.
(b) The g parameter equivalent circuit.

If you examine the h parameter equivalent circuit of Figure 12.5 for the BJT, you notice that the emitter bypass capacitor "sees" the entire input circuit in series with the relatively small resistor h_{ie}. That is where most of the button-pushing came from. In the circuit of Figure 17.12b you will notice that C_S cannot "see" the input circuit at all. As we said in Section 16.8, the gate terminal of the JFET is "dangling in space"; that is, the JFET equivalent of the h_{ie} term of the BJT is an infinite resistance. Further, C_S cannot see the output circuit very well either because it must "look" through $1/g_{os}$ which is a very large resistor. Thus R_{ckt} for the source bypass capacitor is essentially just R_{SQ}. Thus

$$C_S = \frac{1}{2\pi f_L R_{SQ}}$$

(17.22)

Turning our attention to the output circuit, as we have mentioned so often before, the circuit is virtually identical to that of the BJT circuit. Looking into the circuit at the output coupling capacitor, C_o, if we assume that C_S is a perfect conductor, we see

$$R_{\text{out}} = R_D \parallel 1/g_{os} \tag{17.23}$$

which is analogous to Equation 12.7. Then, comparing the circuit of Figure 17.12b to that of Figure 12.6b, we can write

$$C_o = \frac{1}{2\pi f_L(R_{\text{out}} + R_L)} \tag{12.13}$$

Turning our attention to the AC signal gain of the amplifier, you should remember that in Section 12.8 we defined gain to be

$$A \equiv \frac{\text{output}}{\text{input}} \tag{12.16}$$

or

$$A_v = \frac{v_{\text{out}}}{v_{\text{in}}} \tag{12.17}$$

Obviously for the circuit of Figure 17.12b,

$$v_{\text{in}} = v_{GS} \tag{17.24}$$

Also from Figure 17.12b you can see that, assuming that C_S acts like a dead short, the current from the current source $g_{fs}v_{GS}$ is delivered to the parallel combination of $1/g_{os}$, R_D, and R_L. Then we can say

$$v_{\text{out}} = -g_{fs}v_{GS}(1/g_{os} \parallel R_D \parallel R_L) \tag{17.25}$$

The minus sign in Equation 17.25 indicates phase inversion just as it did for the BJT.

Substituting Equations 17.25 and 17.24 into Equation 12.17 and canceling v_{GS} we have

$$A_v = -g_{fs}(1/g_{os} \parallel R_D \parallel R_L) \tag{17.26}$$

The current gain for the circuit of Figure 17.12 is also similar in concept to that for the BJT. Applying Ohm's Law to Figure 17.12b gives

$$i_{\text{out}} = \frac{v_{\text{out}}}{R_L} \tag{17.27}$$

and

$$i_{\text{in}} = \frac{v_{\text{in}}}{R_G} \tag{17.28}$$

Substituting Equations 17.27 and 17.28 in Equation 12.21 gives

$$A_i = \frac{i_{out}}{i_{in}}$$

$$= \frac{v_{out}}{R_L} * \frac{R_G}{v_{in}} \qquad (12.21)$$

or, substituting Equation 12.17 into the above equation, we have

$$A_i = A_v \frac{R_G}{R_L} \qquad (17.29)$$

EXAMPLE 17.5

Determine the capacitor values and the dB gain for the circuit of Example 17.4 if the internal resistance of the signal source is $R_{SS} = 600\ \Omega$ and if the low frequency cutoff is $f_L = 30$ Hz.

SOLUTION:

The circuit is redrawn with all of the known information added in Figure 17.13. Then

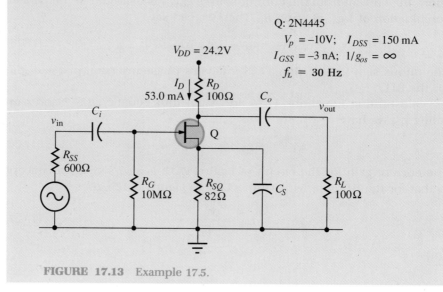

FIGURE 17.13 Example 17.5.

$$C_i = \frac{1}{2\pi f_L (R_{SS} + R_{in})} \qquad \text{(12.6)}$$

$$= \frac{1}{2\pi * 30 \text{ Hz}(600 \text{ }\Omega + 10 \text{ M}\Omega)}$$

$$= 530 \text{ pF} \qquad \text{(Use 560 pF)}$$

$$C_S = \frac{1}{2\pi f_L R_{SQ}} \qquad \text{(17.22)}$$

$$= \frac{1}{2\pi * 30 \text{ Hz} * 82 \text{ }\Omega}$$

$$= 64.7 \text{ }\mu\text{F} \qquad \text{(Use 75 }\mu\text{F)}$$

Since no value is listed in Appendix F for g_{os} for the 2N4445 we will assume that $1/g_{os}$ is large enough to be ignored. Then

$$R_{out} = R_D \parallel 1/g_{os} \qquad \text{(17.23)}$$

$$= 100 \text{ }\Omega \parallel \infty \text{ }\Omega = 100 \text{ }\Omega$$

$$C_o = \frac{1}{2\pi f_L (R_{out} + R_L)} \qquad \text{(12.13)}$$

$$= \frac{1}{2\pi 30 \text{ Hz} * (100 \text{ }\Omega + 100 \text{ }\Omega)} = 26.5 \text{ }\mu\text{F} \qquad \text{(Use 27 }\mu\text{F)}$$

To calculate the gain, g_{fs} is needed. From Section 16.9:

$$g_{fs} = \frac{2}{-V_P}\sqrt{I_{DSS}I_D} \qquad \text{(16.13)}$$

$$= \frac{2}{-(-10 \text{ V})}\sqrt{150 \text{ mA} * 53.0 \text{ mA}} = 17.8 \text{ mS}$$

$$A_v = -g_{fs}(1/g_{os} \parallel R_D \parallel R_L) \qquad \text{(17.26)}$$

$$= -17.8 \text{ mS}(\infty \text{ }\Omega \parallel 100 \text{ }\Omega \parallel 100 \text{ }\Omega)$$

$$= -0.892$$

$$A_i = A_v\frac{R_G}{R_L} \qquad \text{(17.29)}$$

$$= -0.892\frac{10 \text{ M}\Omega}{100 \text{ }\Omega}$$

$$= -89,200$$

From Section 12.8

$$A_P = A_v A_i \tag{12.29}$$
$$= (-0.892)(-89{,}200)$$
$$= 79{,}500$$

From Section 12.12

$$\text{dB} \equiv 10 * \log(A_P) \tag{12.37}$$
$$= 10 * \log(79{,}500)$$
$$= 49.0 \text{ dB}$$

There are a number of features about Example 17.5 that should surprise you and that you should pay attention to. First, looking at the capacitors we have specified, notice that the size of C_i in the above example is very small by BJT standards. The reason is the very high input impedance of the JFET circuit. Since the JFET has such high input impedance, the impedance of the capacitor can be quite high and still not be significant.

Second, the value of C_S is smaller than C_E of the BJT circuit also. In this case the reason is that as we have said, the source does not see the input circuit.

Next, looking at the gains obtained, notice that the voltage gain is nearly two decimal places smaller than you would get from a BJT circuit. Although somewhat larger voltage gains can be realized by using a very flat load line (a very much larger value of V_{DD}), if you need voltage gain and can put up with its low input resistance, use a BJT. On the other hand, notice the large current gain of the JFET. This is also due to the high input resistance. Thus a power gain upwards of 50 dB is not uncommon.

17.10   Circuit Design

Everything that we intend to present for the design of simple common source circuits has already been covered. All that remains is to collect the results in an easy to use table. That has been done in Table 17.1 with the relevant diagrams in Figure 17.14. Notice that the equations marked with an asterisk over the equal sign are design formulas only and should not be used for any other purpose.

TABLE 17.1   JFET Amplifier Design

* : Designer's formula only.
STD: Choose the nearest standard component value.

GIVEN: V_P I_{DSS} I_{GSS} $1/g_{os}$ R_{SS} R_L f_L

1. $R_D \overset{*}{=} R_L \parallel (-1/g_{os})$ (STD.) (17.1)

2. $R_{AC} = R_D \parallel R_L$ (17.2)

3. $I_D \overset{*}{=} \dfrac{I_{DSS}}{\sqrt{8}}$ (17.7)

4. $V_{GS} = V_P\left(1 - \sqrt{\dfrac{I_D}{I_{DSS}}}\right)$ (16.5)

5. $V_S = V_G - V_{GS} = -V_{GS}$ (17.17)

6. $V_{DS} = (I_{DSS} - I_D)R_{AC} - V_P$ (17.9)

7. $V_D = V_{DS} + V_S$ (17.20)

8. $V_{DD} = V_D + I_D R_D$ (17.21)

9. $R_G \overset{*}{\approx} \dfrac{-V_P}{10 * I_{GSS}}$ (STD.) (17.14)

10. $R_{SQ} = \dfrac{V_S}{I_D}$ (STD.) (17.19)

11. $R_{in} = R_G$ (17.13)

12. $R_{out} = R_D \parallel 1/g_{os}$ (17.23)

13. $C_i = \dfrac{1}{2\pi f_L(R_{SS} + R_{in})}$ (STD.) (12.6)

14. $C_S = \dfrac{1}{2\pi f_L R_{SQ}}$ (STD.) (17.22)

15. $C_O = \dfrac{1}{2\pi f_L(R_{out} + R_L)}$ (STD.) (12.13)

16. $g_{fs} = \dfrac{2}{-V_P}\sqrt{I_{DSS}I_D}$ (16.13)

17. $A_v = -g_{fs}(1/g_{os} \parallel R_D \parallel R_L)$ (17.26)

18. $A_i = A_v\dfrac{R_{in}}{R_L}$ (17.29)

Design procedure for the circuit of Figure 17.14

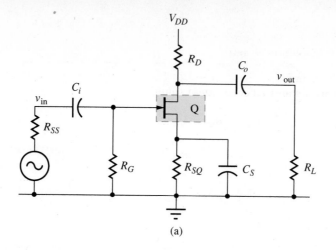

(a)

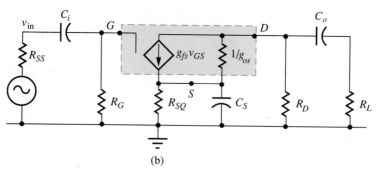

(b)

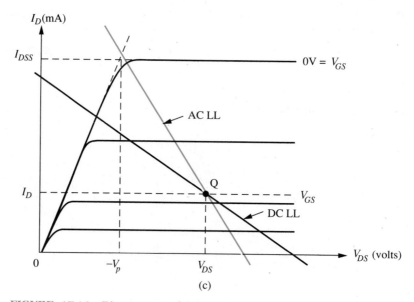

(c)

FIGURE 17.14 Diagrams used in connection with Table 17.1.
(a) Circuit diagram.
(b) The g parameter equivalent circuit.
(c) Graph of common source characteristics.

EXAMPLE 17.6

Using a 2N5457 JFET, design an amplifier that will drive a 1 kΩ load down to a low frequency cutoff of 30 Hz. Assume a signal source resistance of 100 kΩ.

●   (See Appendix G for a SuperCalc solution.)

SOLUTION:

The circuit to be designed is shown in Figure 17.15 from the data for the 2N5457 in Appendix F:

$$I_{DSS} = 5 \text{ mA}; \qquad I_{GSS} = -1 \text{ nA}; \qquad V_P = -3 \text{ V};$$
$$g_{os} = 10 \text{ μS}. \qquad \therefore \quad 1/g_{os} = 100 \text{ kΩ}$$

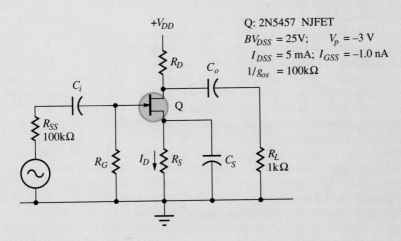

FIGURE 17.15   Example 17.6.

Plugging these data into the JFET amplifier design procedure

1. $R_D \overset{*}{=} R_L \| (-1/g_{os})$ (STD.)   (17.1)

   $= 1 \text{ kΩ} \| (-100 \text{ kΩ}) = 1.01 \text{ kΩ}$ (Use 1 kΩ)

2. $R_{AC} = R_D \| R_L$ (17.2)

   $= 1 \text{ kΩ} \| 1 \text{ kΩ} = 500 \text{ Ω}$

3. $I_D \overset{*}{=} \dfrac{I_{DSS}}{\sqrt{8}}$ (17.7)

   $= \dfrac{5 \text{ mA}}{\sqrt{8}} = 1.77 \text{ mA}$

4. $V_{GS} = V_P \left(1 - \sqrt{\dfrac{I_D}{I_{DSS}}} \right)$ (16.5)

$= -3 \text{ V} \left(1 - \sqrt{\dfrac{1.77 \text{ mA}}{5 \text{ mA}}} \right) = -1.22 \text{ V}$

5. $V_{DS} = (I_{DSS} - I_D)R_{AC} - V_P$ (17.9)

$= (5 \text{ mA} - 1.77 \text{ mA})500 \text{ }\Omega - (-3 \text{ V}) = 4.62 \text{ V}$

6. $V_S = V_G - V_{GS} = -V_{GS}$ (17.17)

$= 0 \text{ V} - (-1.22 \text{ V}) = 1.22 \text{ V}$

7. $V_D = V_{DS} + V_S$ (17.20)

$= 4.62 \text{ V} + 1.22 \text{ V} = 5.83 \text{ V}$

8. $V_{DD} = V_D + I_D R_D$ (17.21)

$= 5.83 \text{ V} + 1.77 \text{ mA} * 1 \text{ k}\Omega = 7.6 \text{ V}$

9. $R_G \stackrel{*}{\approx} \dfrac{-V_P}{10 * I_{GSS}}$ (STD.)   (17.14)

$= \dfrac{-(-3 \text{ V})}{10(-1 \text{ nA})} = 300 \text{ M}\Omega$ (Use 10 MΩ)

10. $R_{SQ} = \dfrac{V_S}{I_D}$ (STD.)   (17.19)

$= \dfrac{1.22 \text{ V}}{1.77 \text{ mA}} = 688 \text{ }\Omega$ (Use 680 Ω)

11. $R_{in} = R_G$ (17.13)

$= 10 \text{ M}\Omega$

12. $R_{out} = R_D \parallel 1/g_{os}$ (17.23)

$= 1 \text{ k}\Omega \parallel 100 \text{ k}\Omega = 990 \text{ }\Omega$

13. $C_i = \dfrac{1}{2\pi f_L (R_{SS} + R_{in})}$ (STD.)   (12.6)

$= \dfrac{1}{2\pi * 30 \text{ Hz}(100 \text{ k}\Omega + 10 \text{ M}\Omega)} = 525 \text{ pF}$ (Use 560 pF)

14. $C_S = \dfrac{1}{2\pi f_L R_{SQ}}$ (STD.)   (17.22)

$= \dfrac{1}{2\pi * 30 \text{ Hz} * 680 \text{ }\Omega} = 7.8 \text{ }\mu\text{F}$ (Use 8 μF)

15. $C_O = \dfrac{1}{2\pi f_L (R_{OUT} + R_L)}$ (STD.)   (12.13)

$= \dfrac{1}{2\pi * 30 \text{ Hz}(990 \text{ }\Omega + 1 \text{ k}\Omega)} = 2.67 \text{ }\mu\text{F}$ (Use 2.7 μF)

16. $g_{fs} = \dfrac{2}{-V_P}\sqrt{I_{DSS}I_D}$ \hfill (16.13)

$\phantom{16. g_{fs}} = \dfrac{2}{-(-3\text{ V})}\sqrt{5\text{ mA} * 1.77\text{ mA}} = 1.98\text{ mS}$

18. $A_v = g_{fs}(1/g_{os} \parallel R_D \parallel R_L)$ \hfill (17.26)

$ = 1.98\text{ mS}(100\text{ k}\Omega \parallel 1\text{ k}\Omega \parallel 1\text{ k}\Omega) = 0.986$

19. $A_i = A_v\dfrac{R_{in}}{R_L}$ \hfill (17.29)

$ = -0.986\,\dfrac{10\text{ M}\Omega}{1\text{ k}\Omega} = -9860$

20. $A_P = A_v A_i$ \hfill (12.29)

$ = (-0.986) * (-9860) = 9720$

21. $dB = 10 * \log{(A_P)}$ \hfill (12.37)

$ = 10 * \log(9720) = 39.9\text{ dB}$

Although most of the preceding result is not a lot different from what you are used to with BJT transistors, you should notice from Step 13 how small the input coupling capacitor is. This is the result of the fact that an FET is a voltage-sensitive device.

You should also notice how low the voltage gain is. This is also typical of JFET circuits. They have high power gain due to their high input impedance.

17.11 Circuit Analysis

As we have said in connection with BJTs, in order to understand the behavior of an existing circuit, the place to start is to determine the Q point. Once I_D, V_D, V_G, and V_S are known, then the circuit's response to applied signals can be determined. Ideally, circuit analysis begins with a fully labeled schematic diagram such as that shown in Figure 17.16. The value of all supply voltages, all resistor values, and the part number for the JFET are known. A data book or a manufacturer's data sheet will provide values for V_P, I_{DSS}, and $1/g_{os}$. Using this information, the Q point can be determined graphically as follows:

Substitute values in the transfer equation in the form

$$I_D = I_{DSS}\left(1 - \frac{V_{GS}}{V_P}\right)^2 \hfill (16.4)$$

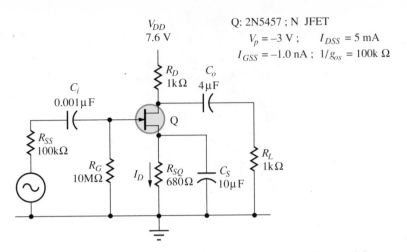

FIGURE 17.16   A capacitor coupled common source JFET amplifier.

and plot the transfer characteristic as we did in Section 16.7. The result is a graph like that of Figure 17.17. The variables in the transfer equation are V_{GS} and I_D. As we have found with almost every other device we have studied, in order to evaluate V_{GS} and I_D, the circuit must provide a second relationship between the same two variables. Once values are obtained for V_{GS} and I_D, the other Q point values can be easily calculated.

Doing a KVL walk through the circuit of Figure 17.16 from the gate terminal through the JFET to its source terminal and on through R_{SQ} to ground gives

$$V_G - V_{GS} - I_D R_{SQ} = 0 \text{ V} \tag{17.30}$$

Assuming as we have in the past that $V_G = 0$ V, and solving Equation 17.30 for I_D gives

$$I_D = -\frac{V_{GS}}{R_{SQ}} \tag{17.31}$$

Equation 17.31 is the equation of a straight line. This line is sometimes called the "self bias line," or more simply, the bias line. Since we have assumed that $V_G = 0$ V, the bias line passes through the origin. Otherwise, it would intersect the horizontal axis at V_G and Equation 17.31 would become

$$I_D = \frac{V_G - V_{GS}}{R_{SQ}} \tag{17.32}$$

Since we are interested in a graphical solution, the bias line of Equation 17.31 can be easily plotted on the transfer characteristic of Figure 17.17. All that is needed is to use Equation 17.31 to locate two points and then draw

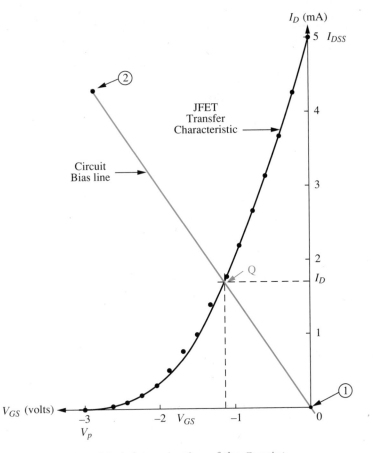

FIGURE 17.17 Graphical determination of the Q point.

a straight line through them. Of course, any two points will do, but one point that is easy to locate is the intercept. As we have said, for Equation 17.31 the intercept is the origin. This point is labeled point number one on Figure 17.17.

For the second point, in order to get the points as far apart as possible, we will use $V_{GS} = V_P$. Substituting this value for V_{GS} into Equation 17.31 locates the point labeled point number two on Figure 17.17. The bias line is drawn through points 1 and 2. The point where the bias line intersects the transfer characteristic is the Q point. This determines V_{GS} and I_D.

EXAMPLE 17.7

Determine the actual drain current for the circuit designed in Example 17.6 and compare with the design value.

● (For a SuperCalc solution, see Appendix G.)

SOLUTION:

The circuit and the JFET parameters are shown in Figure 17.16. For the 2N5457 transistor used, $V_P = -3$ V and $I_{DSS} = 5$ mA. Substituting values in the transfer equation gives

$$I_D = I_{DSS}\left(1 - \frac{V_{GS}}{V_P}\right)^2 \tag{16.4}$$

or

$$I_D = 5 \text{ mA}\left(1 - \frac{V_{GS}}{-3 \text{ V}}\right)^2 \tag{17.33}$$

The transfer characteristic in Figure 17.17 results from plotting Equation 17.33.

In order to obtain the bias line, the value of R_{SQ} from the circuit of Figure 17.18 is inserted in Equation 17.31. The result is

$$I_D = -\frac{V_{GS}}{R_{SQ}} \tag{17.31}$$

$$= -\frac{V_{GS}}{680 \ \Omega} \tag{17.34}$$

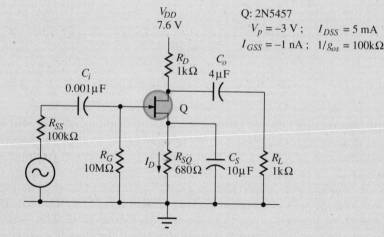

FIGURE 17.18 Circuit for Example 17.8.

A plot of Equation 17.34 gives the bias line shown in Figure 17.17. The bias line and the transfer characteristic intersect at the Q point. Reading from the graph gives something like $I_D = 1.75$ mA which agrees well with the design value of $I_D = 1.77$ mA from Example 17.6.

Since we have equations for both the transfer function and the bias line, we do not need to waste all the time to draw the above graphs. Equation 17.31 and the transfer function are two equations in the same two unknowns so we should be able to solve for I_D and V_{GS}. The difficulty is that since the transfer function is a second degree equation, we will have to solve a quadratic equation.

You might think that it would be easier to use iteration. The trouble with iteration in this case is that the equations do not converge very rapidly so you would have to use some intervention much like what we did in Section 15.12 in Steps 8 and 9 of the circuit analysis procedure for the complementary circuit.

Our approach will be to solve the quadratic equation. You will have to remember that

if
$$ax^2 + bx + c = 0 \tag{17.35}$$

then
$$x = \frac{-b \pm \sqrt{b^2 - 4ac}}{2a} \tag{17.36}$$

Fortunately, we will be working with numerical values so your calculator can handle most of the heavy work. Also, you may remember that the reason for the $\pm$ sign in Equation 17.36 is that a parabola has two branches. Since the transfer function consists of only the positive branch, we only use the positive part.

Our approach will be:

1. Reduce the bias line equation, Equation 17.31 and the transfer equation, Equation 16.4 to a single equation.
2. Express the resulting equation in the form of Equation 17.35.
3. Evaluate coefficients a, b, and c.
4. Insert the values for a, b, and c in Equation 17.36 and solve for the more positive branch only.

The two equations we need to solve are

$$I_D = -\frac{V_{GS}}{R_{SQ}} \tag{17.31}$$

and

$$I_D = I_{DSS}\left(1 - \frac{V_{GS}}{V_P}\right)^2 \tag{16.4}$$

Subtracting Equation 17.31 from Equation 16.4 and putting the result in quadratic form gives

$$I_{DSS}\left(1 - \frac{V_{GS}}{V_P}\right)^2 + \frac{V_{GS}}{R_{SQ}} = 0$$

$$\left(1 - \frac{V_{GS}}{V_P}\right)^2 + \frac{V_{GS}}{I_{DSS}R_{SQ}} = 0 \tag{17.37}$$

$$1 - \frac{2V_{GS}}{V_P} + \frac{V_{GS}^2}{V_P^2} + \frac{V_{GS}}{I_{DSS}R_{SQ}} = 0 \tag{17.38}$$

or

$$\left(\frac{1}{V_P^2}\right)V_{GS}^2 + \left(\frac{1}{I_{DSS}R_{SQ}} - \frac{2}{V_P}\right)V_{GS} + 1 = 0$$

$$\text{(a)} \quad x^2 \quad + \qquad \text{(b)} \qquad x \quad + c = 0 \tag{17.39}$$

Thus the coefficients for the quadratic formula are

$$a = \frac{1}{V_P^2} \tag{17.40}$$

$$b = \frac{1}{I_{DSS}R_{SQ}} - \frac{2}{V_P} \tag{17.41}$$

$$c = 1 \tag{17.42}$$

In the event that the DC voltage on the gate is not zero, then Equation 17.32 must be used in the above derivation instead of Equation 17.31. In that event, the only change is that instead of Equation 17.42 above we have

$$c = 1 - \frac{V_G}{I_{DSS}R_{SQ}} \tag{17.43}$$

We can now obtain numerical values for coefficients a, b, and c. Their values can then be substituted in the quadratic formula, Equation 17.36. We then have

$$V_{GS} = \frac{-b + \sqrt{b^2 - 4ac}}{2a} \tag{17.44}$$

This value of V_{GS} can then be substituted into Equation 17.31 or Equation 17.32 which will then give the value of I_D. Normal circuit analysis procedures will then give the remaining circuit values.

EXAMPLE 17.8
Repeat Example 17.7 using calculation rather than graphical analysis.

● (For a PSpice solution, see Appendix H.)

SOLUTION:
Figure 17.18 contains the fully labeled circuit including the JFET parameters. The values for the coefficients a, b, and c in the quadratic formula are:

$$a = \frac{1}{V_P^{\,2}} \tag{17.40}$$

$$= \frac{1}{(-3\ \text{V})^2} = 0.111$$

$$b = \frac{1}{I_{DSS}R_{SQ}} - \frac{2}{V_P} \tag{17.41}$$

$$= \frac{1}{5\ \text{mA} * 680\ \Omega} - \frac{2}{-3\ \text{V}} = 0.961$$

$$c = 1 \tag{17.42}$$

Substituting the above coefficients into the quadratic formula, Equation 17.44 gives

$$V_{GS} = \frac{-b + \sqrt{b^2 - 4ac}}{2a} \tag{17.44}$$

$$= \frac{-0.961 + \sqrt{(0.961)^2 - 4 * 0.111 * 1}}{2 * 0.111} = -1.21\ \text{V}$$

$$I_D = -\frac{V_{GS}}{R_{SQ}} \tag{17.31}$$

$$= -\frac{-1.21\ \text{V}}{680\ \Omega}$$

$$= 1.78\ \text{mA}$$

Compared to the graphical solution of $I_D = 1.75$ mA and the design value of $I_D = 1.77$ mA.

The above result agrees with the graphical solution of Example 17.7 to within my ability to read the graph of Figure 17.17. The reason it does not agree exactly with the design is that a standard resistor value was specified for R_S.

Once I_D is known, KVL will give us Q point values for V_S and V_D: A KVL walk from the source down to ground gives

$$V_S - I_D R_{SQ} = 0\ \text{V}$$

or

$$V_S = I_D R_{SQ} \tag{17.18}$$

A KVL walk from the supply down to the drain terminal gives the drain voltage.

$$V_D = V_{DD} - I_D R_D \qquad (17.45)$$

17.12   AC Circuit Analysis

After the Q point has been located, the AC analysis of a JFET circuit can be performed. To that end, it is advisable to have at least a rough idea of how the Q point relates to the drain characteristic curves. A sketch such as is shown in Figure 17.19 is often adequate. If the Q point is too close to one of the borders of the active region as defined in Figure 17.2, you should be aware that our circuit analysis equations may not apply and clipping may occur in the output signal. The Q points labeled A, B, and C in Figure 17.19 should certainly be suspect.

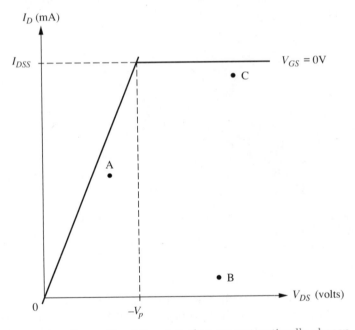

FIGURE 17.19   Examples of Q points that are not optimally chosen.

The g parameter equivalent circuit makes the task of circuit analysis quite straightforward. The circuit is shown in Figure 17.14, and is reproduced here as Figure 17.20. Most of the required relationships such as input resistance, output resistance, and circuit gain have already been developed.

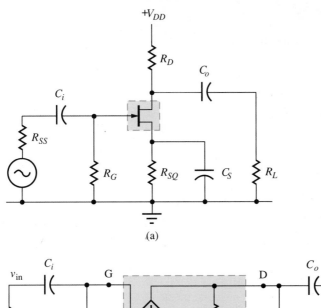

(a)

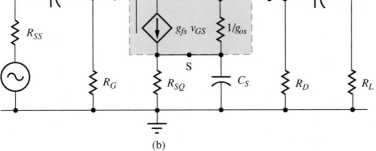

(b)

FIGURE 17.20 Common source circuit used for the determination of capacitor values.
(a) Circuit diagram.
(b) The g parameter equivalent circuit.

The determination of the low frequency cutoff, f_L, for the circuit is much the same as that for the BJT as discussed in Section 12.9 and again in Section 17.9. Transferring what we said in Chapter 12 to the JFET circuit of Figure 17.20, f_L will be determined by the capacitor circuit that cuts off at the highest frequency.

The circuit analysis ideas from Section 17.11 and Section 17.12 are collected together in the circuit analysis recipe of Table 17.2 with the associated diagrams of Figure 17.21. As is true of any such recipe, Table 17.2 should not be used blindly. Be certain you understand the principles on which the steps in the recipe are based; know when they apply, and when they do not.

TABLE 17.2   JFET Amplifier Analysis

GIVEN: The circuit and the JFET specs

DC Analysis:

1. Solve for V_G given that $I_G = 0$ A

2. $a = \dfrac{1}{V_P{}^2}$ (17.40)

3. $b = \dfrac{1}{I_{DSS}R_{SQ}} - \dfrac{2}{V_P}$ (17.41)

4. $c = 1 - \dfrac{V_G}{I_{DSS}R_{SQ}}$ (17.43)

5. $V_{GS} = \dfrac{-b + \sqrt{b^2 - 4ac}}{2a}$ (17.44)

6. $I_D = \dfrac{V_G - V_{GS}}{R_{SQ}}$ (17.32)

7. $V_D = V_{DD} - I_D R_D$ (17.45)
8. $V_S = I_D R_{SQ}$ (17.18)
9. $V_{DS} = V_D - V_S$ (locate Q point)   (17.20)

AC Analysis:

10. $R_{in} = R_G$ (17.13)
11. $R_{out} = R_D \parallel 1/g_{os}$ (17.23)

12. $f_L = \begin{cases} \dfrac{1}{2\pi C_i(R_{SS} + R_{in})} \\[2mm] \dfrac{1}{2\pi C_S R_{SQ}} \\[2mm] \dfrac{1}{2\pi C_o(R_{out} + R_L)} \end{cases}$ $\left.\begin{array}{l} \text{whichever} \\ \text{is} \\ \text{larger} \end{array}\right.$

(12.31)

(17.22)

(12.32)

13. $g_{fs} = \dfrac{2}{-V_P}\sqrt{I_{DSS}I_D}$ (16.13)

14. $v_{out} = -g_{fs}v_{GS}(1/g_{os} \parallel R_D \parallel R_L)$ (17.25)
15. $v_{in} = v_{GS}$ (17.24)

16. $A_v \equiv \dfrac{v_{out}}{v_{in}}$ (12.17)

17. $A_i = A_v\dfrac{R_{in}}{R_L}$ (17.29)

18. $dB_A = 10 * \log{(A_vA_i)}$ (12.37)

Analysis procedure for the circuit of Figure 17.21

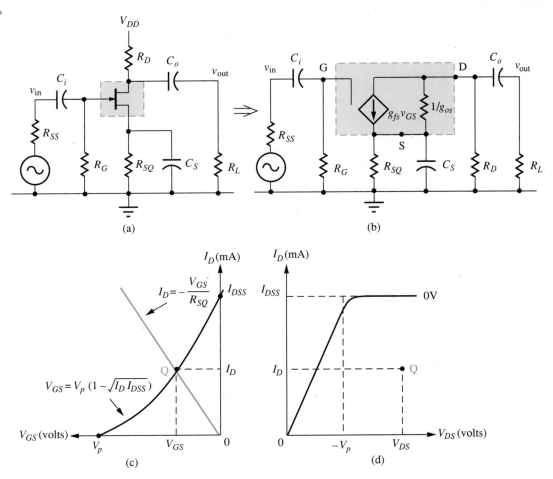

FIGURE 17.21   Diagrams used in connection with Table 17.2.
(a) Circuit diagram.
(b) The g parameter equivalent circuit.
(c) Transfer characteristic.
(d) Common source characteristics.

EXAMPLE 17.9

The circuit of Figure 17.22 was adapted from a piece of patient monitoring equipment used in a hospital. What should be the circuit's Q point voltages, voltage gain, and low frequency cutoff?

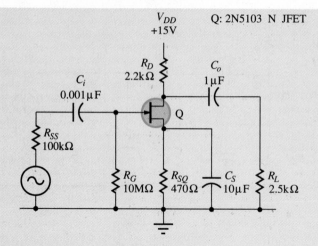

FIGURE 17.22   Circuit for Example 17.9.

● (For a PSpice solution, see Appendix H.)

SOLUTION:

JFET Amplifier Analysis

GIVEN: The circuit of Figure 17.22

2N5103: $V_P = -4$ V $I_{DSS} = 8$ mA $I_{GSS} = -100$ pA

$g_{os} = 100$ μS $= 1/10$ kΩ.

DC ANALYSIS:

1. $V_G = I_G R_G$

   $= 0$ V

2. $a = \dfrac{1}{V_P^2}$ (17.40)

   $= \dfrac{1}{(-4 \text{ V})^2} = 0.0625$

3. $b = \dfrac{1}{I_{DSS} R_{SQ}} - \dfrac{2}{V_P}$ (17.41)

   $= \dfrac{1}{8 \text{ mA} * 470 \text{ Ω}} - \dfrac{2}{-4 \text{ V}} = 0.766$

4. $c = 1 - \dfrac{V_G}{I_{DSS} R_{SQ}}$ (17.43)

   $= 1 - 0 = 1$

5. $V_{GS} = \dfrac{-b + \sqrt{b^2 - 4ac}}{2a}$ (17.44)

$= \dfrac{-0.766 + \sqrt{(0.766)^2 - 4 * 0.0625 * 1}}{2 * 0.0625} = -1.49 \text{ V}$

6. $I_D = \dfrac{V_G - V_{GS}}{R_{SQ}}$ (17.32)

$= \dfrac{0 \text{ V} - (-1.49 \text{ V})}{470 \ \Omega} = 3.16 \text{ mA}$

7. $V_D = V_{DD} - I_D R_D$ (17.45)

$= 15 \text{ V} - 3.16 \text{ mA} * 2.2 \text{ k}\Omega = 8.05 \text{ V}$

8. $V_S = I_D R_{SQ}$ (17.18)

$= 3.16 \text{ mA} * 470 \ \Omega = 1.49 \text{ V}$

9. $V_{DS} = V_D - V_S$ (17.20)

$= 8.05 \text{ V} - 1.49 \text{ V} = 6.56 \text{ V}$

The above Q point is located on the sketch in Figure 17.23. It looks like a reasonable Q point.

AC ANALYSIS:

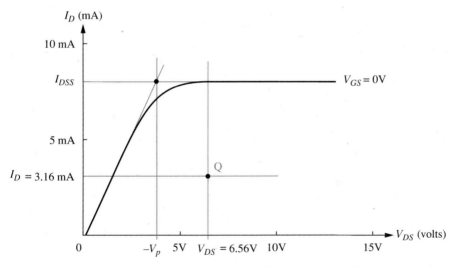

FIGURE 17.23   Example 17.9. Diagram for determining Q point.

10. $R_{in} = R_G = 10 \text{ M}\Omega$ (17.13)

11. $R_{out} = R_D \parallel 1/g_{os} = 2.2 \text{ k}\Omega \parallel 10 \text{ k}\Omega = 1.8 \text{ k}\Omega$ (17.23)

12. $f_L = \begin{cases} \dfrac{1}{2\pi C_i(R_{SS} + R_{in})} & (12.31) \\[4pt] = \dfrac{1}{2\pi * 0.001\ \mu F(100\ k\Omega + 10\ M\Omega)} = 15.8\ Hz \\[6pt] \dfrac{1}{2\pi C_S R_{SQ}} = \dfrac{1}{2\pi * 10\ \mu F * 470\ \Omega} = 33.9\ Hz & (17.22) \\[6pt] \dfrac{1}{2\pi C_o(R_{out} + R_L)} & (12.32) \\[4pt] = \dfrac{1}{2\pi * 1\ \mu F(1.8\ k\Omega + 2.5\ k\Omega)} = 37.0\ Hz = f_L \end{cases}$

13. $g_{fs} = \dfrac{2}{-V_P}\sqrt{I_{DSS} I_D}$ \hfill (16.13)

$= \dfrac{2}{-(-4\ V)}\sqrt{8\ mA * 3.16\ mA} = 2.51\ mS$

14. $v_{out} = -g_{fs} v_{GS}(1/g_{os} \parallel R_D \parallel R_L)$ \hfill (17.25)

$= -2.51\ mS * v_{GS}(10\ k\Omega \parallel 2.2\ k\Omega \parallel 2.5\ k\Omega)$

$= -2.63 * v_{GS}$

15. $v_{in} = v_{GS}$ \hfill (17.24)

16. $A_v \equiv \dfrac{v_{out}}{v_{in}}$ \hfill (12.16)

$= \dfrac{2.63 * v_{GS}}{v_{GS}} = 2.63$

17. $A_i = A_v \dfrac{R_{in}}{R_L}$ \hfill (17.29)

$= 2.63\dfrac{10\ M\Omega}{2.5\ k\Omega} = 10{,}500$

18. $dB = 10 * \log(A_v A_i)$ \hfill (12.37)

$= 10 * \log(2.63 * 10{,}500) = 44.4\ dB$

The Q point voltages are: $V_D = 8.05\ V$ \qquad $V_G = 0\ V$ \qquad $V_S = 1.49\ V$
The voltage gain is: $A_V = 2.63$
The cutoff frequency is: $f_L = 37\ Hz$

17.13 JFET Nonlinearity

On numerous occasions we have pointed out the fact that the drain characteristics for a JFET are not uniformly spaced. You may have been wondering how this nonlinearity affects an AC output waveform. As you can see from Figure 17.24, the result is distortion of the waveform. Notice that for the theoretical curves drawn, with the Q point at $V_{GS} = -2\ V$ and a

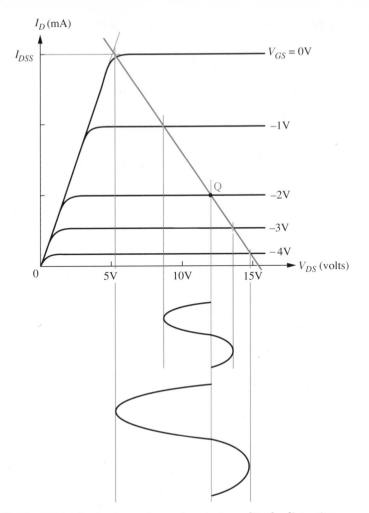

FIGURE 17.24 Output waveform showing amplitude distortion.

nice symmetrical 2*VPP* input extending up to -1 V and down to -3 V, the output is very unsymmetrical. From the Q point of about $V_{DS} = 12$ V the output extends down to about 8.5 V. That is a negative peak of about -3.5 V. It only goes up to about 14 V which gives a positive peak of only 2 V. This is an intolerable amount of distortion in many circuits.

As you can see from Figure 17.24, the larger the signal, the more noticeable the distortion becomes. Thus the simple circuits we have been studying are not very useful for amplifying large signals. This sort of distortion can be controlled through the use of feedback. The use of feedback for this purpose was first mentioned in Section 15.14. We will take up that topic again in a later chapter.

17.14 Automatic Gain Control (AGC)

The nonlinearity noted in the previous section can be used to advantage. An examination of the spacing of the drain characteristic curves or an examination of Equation 16.3 indicates that if the Q point is located at Q_1 in Figure 17.25, the gain will be higher than if the Q point is located at Q_2. Thus, if we are dealing with a very small signal that will use only a small fraction of the whole load line, we can vary the gain by moving the Q point up or down the load line.

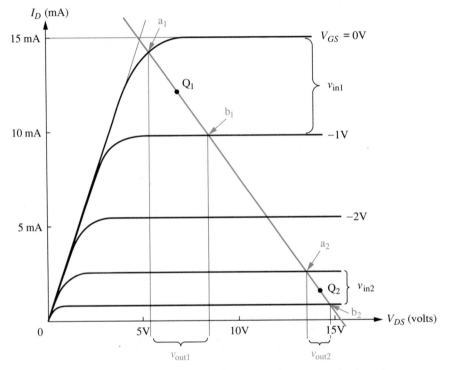

FIGURE 17.25   The dependence of circuit gain on Q point location.

EXAMPLE 17.10
Determine the gain at Q_1 and at Q_2 for the circuit represented by the load line in Figure 17.25.

SOLUTION:
Q_1: If we assume an input signal of $v_{GS} = 1 \ V_{PP}$ so that the operation will move up and down the load line from Point "a_1" to Point "b_1" on

Figure 17.25 then the output will go from about $a_1 = 5$ V to about $b_1 = 8.5$ V. Thus $v_{out1} \approx 3.5$ VPP (with phase reversal), and

$$A_{V1} = \frac{v_{out1}}{v_{in1}} = \frac{3.5 \ VPP}{1 \ VPP} = 3.5 \qquad (12.16)$$

Q_2: Again if $v_{in2} = 1$ VPP around Q_2 the output will go from about $a_2 = 13.5$ V to $b_2 = 15$ V or $v_{out2} \approx 1.5$ VPP. Thus

$$A_{V2} = \frac{v_{out2}}{v_{in2}} = \frac{1.5 \ VPP}{1 \ VPP} = 1.5 \qquad (12.16)$$

Consider the circuit shown in Figure 17.26. This circuit has a form of feedback but the signal is rectified by diode D_F and filtered by capacitor C_F before it is fed back to the input. Due to the orientation of D_F whenever the output is negative, C_F will charge to that negative peak (less about 0.6 V) which we will call V_F. Due to our choice of resistors R_1 and R_2, the gate bias voltage, V_{GS} will be close to $(10/11)\ V_F$. Therefore, when the signal is small, V_{GS} is small and the operation is in the vicinity of Q_1 in Figure 17.25. The result is we have maximum gain when the signal is small. When the signal is large, V_{GS} is large and the operation is in the vicinity of Q_2. We then have minimum gain when the signal is large—automatic gain control. Many radio receivers and "user friendly" tape recorders have this feature.

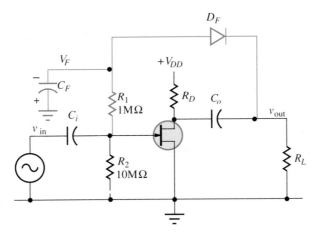

FIGURE 17.26 A simple AGC circuit.

▶▶▶▶▶▶▶▶   17.15   Troubleshooting

It is always desirable to have the device parameters available for the circuit you are troubleshooting. That is doubly true in the case of JFET circuits since they must always operate between the saturation line and pinchoff. Thus these parameters set the limits of normal operation.

Since the gate circuit is a high impedance circuit, you should be aware that any leakage in the gate coupling capacitor can adversely affect the Q point. Further, that fault is not easily seen by monitoring the gate terminal due to test instrument loading. Rather, monitor the drain as you bridge the gate leak resistor with a low value resistor such as 10 kΩ. If a marked change in V_D is observed, leakage may be occurring.

17.16   Summary

In this chapter you should have found that the basic ideas of load lines and Q point which applied to BJTs applies with some modification to the JFET. We made the necessary modifications to the circuit design and analysis procedures presented in Chapter 12 in order to adapt them to the JFET. Finally we examined the intrinsic nonlinearity of the JFET.

GLOSSARY

Active region: That area of the common source characteristic graph where the transfer equation applies. (Section 17.2)

Automatic gain control (AGC):   Circuitry that automatically adjusts the gain of an amplifier in order to maintain a relatively constant output amplitude. (Section 17.14)

Bias line: The self bias line is a line drawn on a device transfer characteristic which relates the bias voltage to the circuit current. (Section 17.11)

Gate leak resistor (R_G):   A resistor connected between the gate terminal of an FET and the gate supply. (Section 17.7)

Geometric average: A method of averaging consisting of finding the nth root of the product of n numbers. (Section 17.3)

PROBLEMS

Section 17.2

1. Make a rough sketch of I_D vs V_{DS} when $V_{GS} = 0$ V for a 2N5391 JFET. Identify the active region. For transistor parameters see the data book listing in Appendix F.

2. Repeat Problem 1 for the MPF102 JFET whose drain characteristics are shown in Figure 17.27.

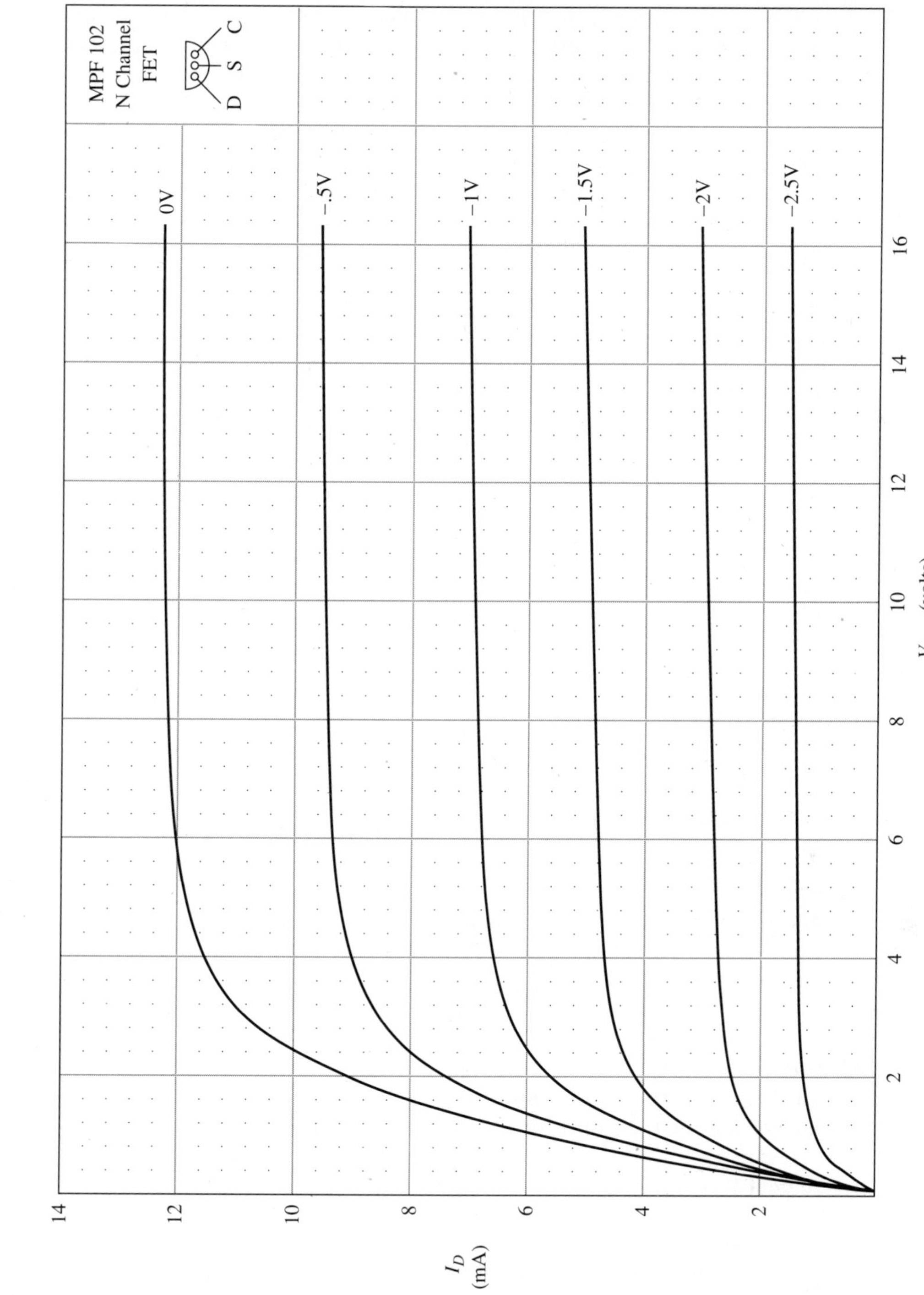

FIGURE 17.27 Problem 17.2.

3. A 2N5457 JFET common source amplifier is used to drive a BJT amplifier whose input resistance is 2 kΩ. Specify the resistance of the drain resistor that should be used. Use data from the data book listing in Appendix F for the 2N5457.

4. Repeat Problem 3 but use the data for the 2N5670 from the data book listing in Appendix F.

5. Determine the value of R_{AC} for the JFET circuit of Problem 3.

6. Specify R_D and determine R_{AC} for the MPF102 JFET whose characteristics are in Figure 17.27 if the circuit is to drive a 3 kΩ load. Assume $V_{GS} = -1.5\ V_{DC}$.

7. Draw the AC load line for Problem 5.

8. Draw the AC load line for Problem 6.

9. Specify the drain resistor and draw the AC load line for the circuit of Figure 17.28.

10. Repeat Problem 9 replacing the 2N5670 JFET with a 2N5668.

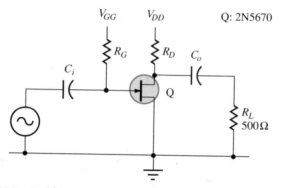

FIGURE 17.28   Problem 17.9.

Section 17.3

11. According to the data book listing in Appendix F, what would be the best Q point current for the following JFETs
   (a)   2N5457
   (b)   2N5459

Section 17.4

12. Find the gate to source voltage V_{GS} for the JFETs in Problem 11.

13. Locate the best Q point for the circuit of Problem 9 and determine I_D, V_{DS}, and V_{GS}.

14. Repeat Problem 13 with reference to Problem 10.

Section 17.5

15. Find the value of power supply voltage, V_{DD} required for the circuit of Figure 17.29. For the above value of V_{DD}, find the drain current, I_D if the drain to source voltage V_{DS} is increased to $V_{DS} = 10$ V.

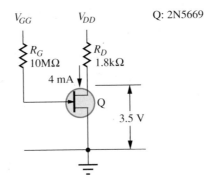

FIGURE 17.29   Problem 17.15.

Section 17.6

16. Given the circuit of Figure 17.29, find the value of the gate supply voltage, V_{GG}
    (a)   Ignoring any gate leakage current, I_{GSS}.
    (b)   Taking into account I_{GSS}.

Section 17.7

17. What "should" be the value of the gate resistor, R_G, in Figure 17.29?
18. Given the circuit of Figure 17.30, calculate appropriate values for V_{GG}, R_G, V_{DD}, and R_D.
19. Repeat Problem 18 if the JFET is a 2N5670 and the load resistor is $R_L = 500\ \Omega$.

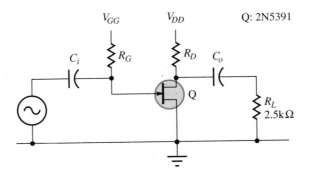

FIGURE 17.30   Problem 17.18.

Section 17.8

20. For the circuit of Figure 17.31, calculate V_{DD}, R_D, R_{SQ}, and R_G.

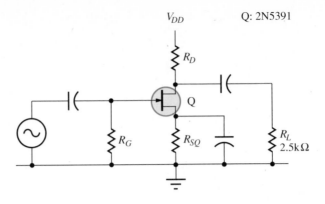

FIGURE 17.31   Problem 17.20.

Section 17.9

21. Assuming a lower cutoff frequency of $f_L = 20$ Hz for the circuit of Figure 17.32, calculate the value of all of the capacitors in the circuit.

22. Formulas for determining the value of resistances R_{in} and R_{out} seen looking into the circuit of Figure 17.12 were derived in Section 17.9 under the assumption that the source bypass capacitor C_S was large enough so that resistor R_{SQ} could be ignored. Derive formulas for R_{in} and R_{out} assuming that there is no bypass capacitor C_S in the circuit.

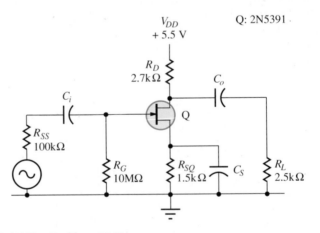

FIGURE 17.32   Problem 17.21.

23. Equation 17.26 was derived in Section 17.9 based on the assumption that "C_S acts like a dead short." Derive an equation for A_V that assumes that there is no C_S so R_{SQ} must be taken into account. Assume that $1/g_{os}$ is large enough to be ignored.

24. Determine the voltage gain for the circuit of Figure 17.7 used in Example 17.2 in Section 17.4.

Section 17.10

25. Using the design procedure of Table 17.1, design the circuit of Figure 17.33 and specify all of the components for it.

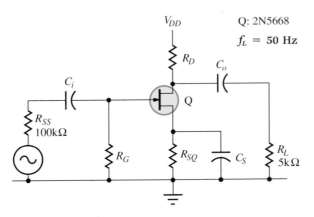

FIGURE 17.33 Problem 17.25.

Section 17.11

26. The first amplifier stage of a certain radio is shown in Figure 17.34. Determine the Q point value of I_D and V_{GS} graphically and then determine V_D, V_G, and V_S.

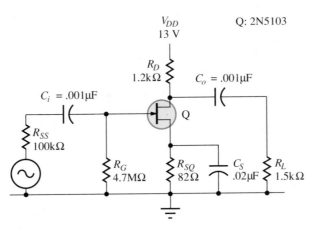

FIGURE 17.34 Problem 17.26.

27. Repeat Problem 26 for the circuit shown in Figure 17.35.
28. Repeat Problem 26 using a nongraphical solution.
29. Repeat Problem 27 using a nongraphical solution.

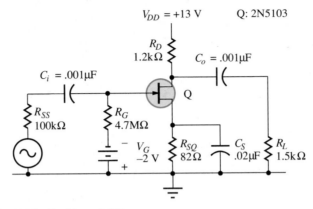

FIGURE 17.35   Problem 17.27.

Section 17.12

30. Determine the low frequency cutoff for the circuit of Figure 17.34 and find its dB gain at midband.
31. Repeat Problem 30 for the circuit of Figure 17.35.

Troubleshooting ▶   Section 17.15

32. For the circuit of Figure 17.34, determine the effect on the Q point if resistor R_G became:
    (a) An open circuit.
    (b) A dead short.
33. For the circuit of Figure 17.34, determine the effect on the Q point if capacitor C_S became:
    (a) An open circuit.
    (b) A dead short.
34. For the circuit of Figure 17.34, determine the effect on the gain of the circuit if capacitor C_C became an open circuit.

An application of MOSFET technology. (Courtesy of
Channel Master.)

Chapter 18
MOSFETs and MOSFET Circuits

SPECIAL TERMS
MOSFET
Punch through
Depletion mode
Enhancement mode
Substrate
Current sink

18.1   Introduction and Objectives

At this point you should have a reasonably good idea how JFETs behave. Since we will base much of our work with MOSFETs on your understanding of JFETs, if your memory of our approach in Chapter 16 is getting a little stale, you should review the early sections of that chapter at this point. We will build on that knowledge as we study MOSFETs in this chapter.

After having studied this chapter you should have an introductory understanding of how MOSFETs function and how they differ from the JFETs with which you are already familiar. At the conclusion of this chapter you should be able to:

- Show how depletion MOSFETs differ from enhancement MOSFETs.
- Design and analyze simple common source amplifier circuits containing a MOSFET.
- Show how MOSFETs are used in push-pull power amplifier circuits.

18.2   Depletion MOSFET Construction and Symbols

We will describe in detail only the N channel device, the NMOSFET. As you might guess, everything we say about the NMOSFET can be transferred to the PMOSFET by exchanging doping types, voltage polarities, and current direction. As we mentioned in connection with the other devices we have studied, due to the greater mobility of the negative carriers, the N channel device has better characteristics. It is used much more frequently than is the P channel device.

We will begin our construction of a MOSFET with a block of lightly doped P type semiconductor which we will call the "substrate." Into the substrate we will diffuse an N type channel. As with the JFET, the source and drain terminals will be connected to opposite ends of the channel. Next we will cover the surface of the channel with silicon dioxide, (SiO_2). Glass is made of SiO_2. It is a very good insulator. Finally we will deposit a metal conductor on top of the SiO_2 to which we will connect the gate terminal. The name of the device is derived from the fact that it is constructed of a **M**etal (the gate) and an **O**xide (the SiO_2 insulator) deposited on a **S**emiconductor (the channel) and it is an **FET**. Thus MOSFET. PSpice simplifies it by leaving off the last part. Thus it talks about NMOS and PMOS transistors. We will adopt that abbreviation on occasion.

Figure 18.1 contains a diagram which may at least conceptually resemble the real device. You will notice that four terminals are shown connected to the MOSFET. The source, gate, and drain terminals just as in a JFET, plus a terminal connected to the substrate. The substrate is usually

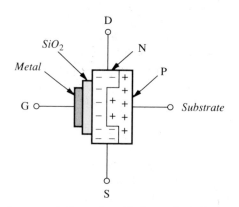

FIGURE 18.1 Representation of an N channel depletion MOSFET.

connected to the source. This connection is often made internally so that only the source, gate, and drain terminals are available to the circuit designer. However, some MOSFET packages have four terminals so that the substrate terminal is available. In practice, when such MOSFETs are used, the designer often just connects the substrate terminal to the source terminal. You will find out later that the turn-on voltage of an enhancement MOSFET can be varied by varying the bias on the substrate.

Figure 18.2 shows the circuit symbols used for the common forms of the depletion MOSFET.

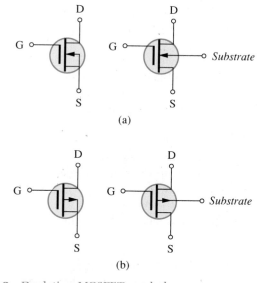

FIGURE 18.2 Depletion MOSFET symbols.
(a) N channel.
(b) P channel.

18.3   Sensitivity to Electrostatic Damage

As you know, on a dry day it is possible to generate a painfully high level of static electrical charge by shuffling your feet on the carpet, or even by just running a comb through your hair. The voltage involved can easily exceed many thousands of volts, although the charge involved is very small.

If such a charge is brought into contact with any of the electronic devices we have studied so far, the chances for damage are small. If the polarity happens to be such as to forward bias any junction encountered, the junction simply conducts and the small charge involved is harmlessly dissipated. Even if the polarity happens to be such as to reverse bias the junction, it will simply go into its Zener region and again conduct until the charge is dissipated. It will then go back to its normal operation, none the worse for the experience.

It is important to notice that this is NOT the case with MOS type devices. Due to the SiO_2 insulator, MOS devices will not conduct. Thus static charge can easily build up to destructively high voltage levels. If their voltage rating is exceeded, MOS devices experience a phenomenon known as "punch through." A hole is actually punched through the SiO_2 insulation layer causing it to become a conductor. The damage is permanent and the device is then useless.

MOS devices can easily be destroyed by such simple acts as removing the new device from the container in which it was purchased. To guard against such damage, MOS devices are usually packed with their leads connected together with conductive foam, or a small wire clip. This protection should be kept in place until the device is safely installed in the circuit in which it is to be used.

Many of the newer MOS devices contain some internal circuitry which functions like a pair of back-to-back diodes placed between the gate and the source. When the voltage goes above the rating of the device, the diode goes into Zener conduction providing a measure of overvoltage protection for the device. This protection may degrade the input resistance slightly, but it adds a great deal to the durability of the device. Even so it is good practice to use a static protected work station when working with MOS devices. Further, you should try to develop the habit of holding transistors or chips in such a way that your fingers always touch all of the terminals of the device. This will reduce the risk of destructively high voltage differences existing between the terminals of the device.

18.4   Depletion MOSFET Behavior

In the "depletion mode" MOSFETs behave very much like the JFETs with which you are familiar. In the case of the NMOS transistor diagrammed in

Figure 18.1, placing a negative voltage on the gate terminal has the effect of repelling the negative carriers in the channel. This effectively "depletes" the carriers in the channel and so reduces the drain current. The drain characteristics and the transfer characteristic will look the same as that for a JFET.

When a positive voltage is placed on the gate terminal of an NMOS-FET, it will behave quite differently from a JFET. A positive voltage at the gate of an N channel JFET simply forward biases the junction tending to cause current to flow in the gate. The device then acts more like a rectifier than an amplifier.

The insulating layer of SiO_2 between the gate and the channel of the MOSFET prevents gate current regardless of polarity of the gate. Thus we are not limited to only those voltages which reverse bias the gate.

As you should expect, when the gate terminal of the transistor of Figure 18.1 is made positive with respect to the substrate, more negative carriers are attracted into the channel, thus enhancing its ability to carry current. This is the "enhancement mode" of operation. Depletion MOS-FETs can typically operate in either the depletion or the enhancement mode. For this reason they are often called "depletion-enhancement" MOSFETs. At the risk of some ambiguity we will shorten the name a bit.

18.5 Depletion MOSFET Characteristics

From what we have said so far, you should not be surprised to learn that MOSFET behavior can be described using the same type of drain characteristics, transfer characteristic, and transfer equation as we used to describe the JFET.

I_{DSS} and V_P have nearly the same meaning for the MOSFET as for the JFET. Thus the transfer function still applies so

$$I_D = I_{DSS}\left(1 - \frac{V_{GS}}{V_P}\right)^2 \qquad (16.4)$$

and

$$V_{GS} = V_P\left(1 - \sqrt{\frac{I_D}{I_{DSS}}}\right) \qquad (16.5)$$

The PSpice model for a MOSFET fits most real devices remarkably well if the default model parameters PSpice calls "VTO" and "KP" are changed to fit the device being used. VTO and KP relate to values customarily given in the data sheets as follows:

$$VTO = V_P \tag{18.1}$$

and

$$KP = 2\frac{I_{DSS}}{V_P^2} \tag{18.2}$$

MOSFET characteristics are different from JFET characteristics in that the $V_{GS} = 0$ V curve is no longer the topmost curve of the set. However, I_{DSS} is still defined to be the drain current when $V_{GS} = 0$ V. A typical set of curves is shown in Figure 18.3. You should notice that operation below the

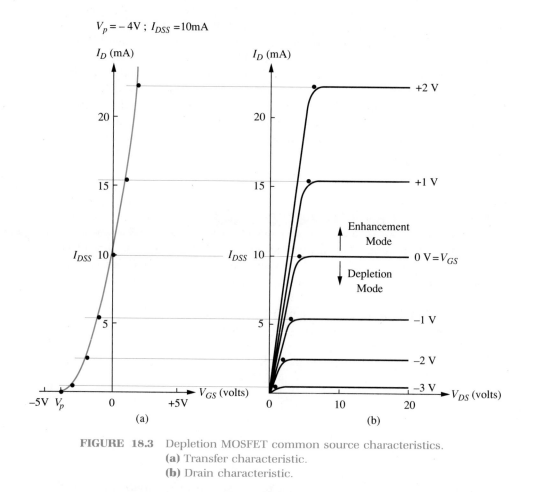

FIGURE 18.3 Depletion MOSFET common source characteristics.
(a) Transfer characteristic.
(b) Drain characteristic.

$V_{GS} = 0$ V line is depletion mode operation and operation above the $V_{GS} = 0$ V line is enhancement mode operation.

The knee of the drain characteristic curves is determined in the same way as it was for the JFET

$$V_{DSknee} = V_{GS} - V_P \qquad (16.1)$$

EXAMPLE 18.1

Construct a set of drain characteristics for a 3SK121Y depletion MOSFET.

- (For a PSpice solution see Appendix H.)

SOLUTION:

The data listed in Appendix F for the 3BK121Y MOSFET includes $I_{Dmax} = 50$ mA; $V_P = -4$ V; $I_{DSS} = 35$ mA. Thus the transfer equation is

$$I_D = I_{DSS}\left(1 - \frac{V_{GS}}{V_P}\right)^2 \qquad (16.4)$$

$$= 35 \text{ mA}\left(1 - \frac{V_{GS}}{-4 \text{ V}}\right)^2$$

With the knee of the curves at

$$V_{DSknee} = V_{GS} - V_P \qquad (16.1)$$

$$= V_{GS} - (-4 \text{ V})$$

Data from the above equations is collected in Table 18.1 and plotted in Figure 18.4 using 1 V increments for V_{GS}.

TABLE 18.1   3SK121Y MOSFET

$I_{DMAX} = 50$ mA; $V_P = -4$V; $I_{DSS} = 35$ mA

V_{GS} (V)	I_D (mA)	V_{DSknee} (V)
−4	0	—
−3	2.19	+1
−2	8.75	+2
−1	19.7	+3
0	35.0	+4
+1	54.7	+5

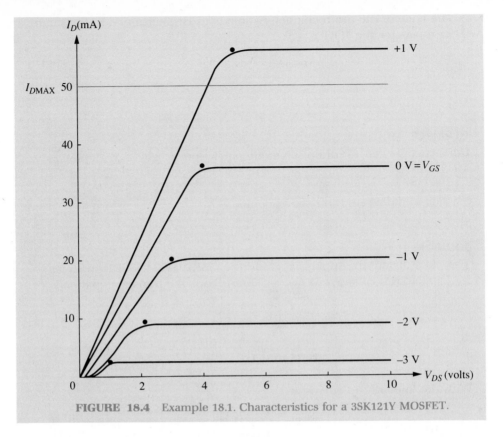

FIGURE 18.4   Example 18.1. Characteristics for a 3SK121Y MOSFET.

In order for you to see how close the model we have been developing approaches the behavior of actual devices, you should obtain several samples of the transistor used in Example 18.1 and, using care to avoid static damage, hook up the circuit of Figure 18.5. Then construct the graph of

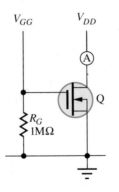

FIGURE 18.5   Circuit for constructing common source characteristics.

Figure 18.4 from experimental data. A given transistor may not agree too well, but the average of several transistors should come close.

18.6 Depletion MOSFET Circuit Design

Since MOSFET and JFET characteristics are similar in many ways, we will develop a common source amplifier design procedure for MOSFETs that is based on the procedure for JFETs in Table 17.1. The biggest difference between them has to do with the location of the Q point.

Due to the fact that depletion MOSFETs can operate in both the depletion and the enhancement modes, in principle, at least, there is no upper limit to the active region of the characteristic curves. For reasons of circuit simplicity, it is common for the circuit designer to bias MOSFETs to operate at

$$V_{GS} \overset{*}{=} 0 \text{ V} \tag{18.3}$$

Since the designer sets the bias voltage at zero volts, the source biasing resistor used in the JFET circuit can be eliminated. The typical small signal common source amplifier then becomes the circuit shown in Figure 18.6.

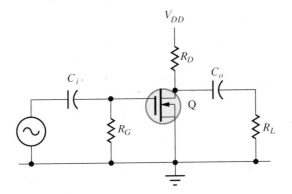

FIGURE 18.6 A typical capacitor coupled depletion MOSFET amplifier stage.

We can then develop the design procedure for the circuit of Figure 18.6 as follows:

Since the ideas are the same, Steps 1 and 2 of the JFET design procedure are unchanged, except that g_{os} is almost always negligible for MOSFETs.

If the designer chooses $V_{GS} = 0$ V then, either by definition, or by substituting in the transfer equation, I_D must be

$$I_D = I_{DSS}\left(1 - \frac{V_{GS}}{V_P}\right)^2 \tag{16.4}$$

$$= I_{DSS}\left(1 - \frac{0\text{ V}}{V_P}\right)^2 = I_{DSS} \tag{18.4}$$

Placement of the Q point at $V_{GS} = 0$ V limits the negative peak swing of the AC input signal, v_{in}, to V_P (a negative quantity) as shown in Figure 18.7. You might be tempted to think that to achieve midpoint bias, the circuit should be designed so that the load line is long enough to allow the positive peak of the AC input signal to go up to a positive value of $-V_P$. Due

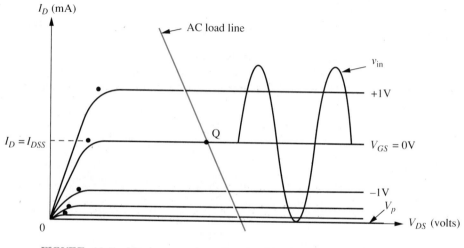

FIGURE 18.7   Maximum value of v_{in} is given by V_P.

to the nonlinear nature of the MOSFET, as shown in Figure 18.8, it is more common to plan for the AC load line to go through the knee of the curve given by

$$V_{GSknee} \overset{*}{\approx} -\frac{V_P}{2} \tag{18.5}$$

It is important to notice that as usual the asterisk over the equal sign in Equation 18.5 indicates that the equation is a designer's rule of thumb and should not be used for any other purpose. It is subject to change in light of the designer's experience and the application at hand.

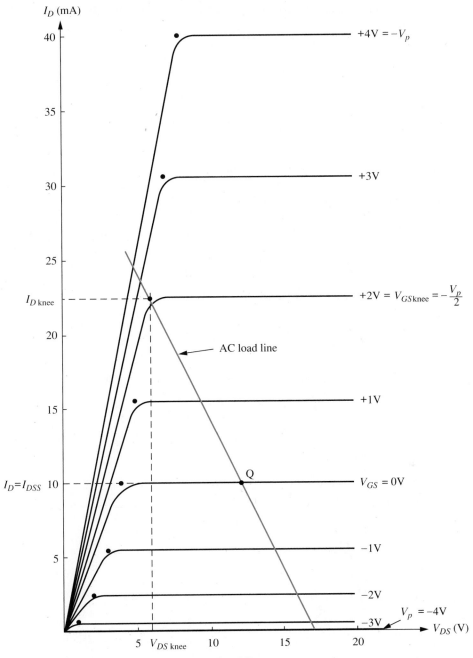

FIGURE   18.8   A typical MOSFET load line.

In Section 16.3 we showed that at the knee of the characteristic defined by Equation 18.5

$$V_{DSknee} = V_{GSknee} - V_P \tag{16.1}$$

The transfer equation will give the drain current at the knee of the curve defined by Equation 18.5

$$I_{Dknee} = I_{DSS}\left(1 - \frac{V_{GSknee}}{V_P}\right)^2 \tag{16.4}$$

Of course, the above equation could be simplified if Equation 18.5 were substituted into it, but, since a different value may be chosen for V_{GSknee}, we will not alter Equation 16.4.

Equations 16.1 and 16.4 give us the coordinates of one point on the AC load line (the knee of the curve) and its slope ($-1/$slope, remember?) is given by

$$R_{AC} = R_D \parallel R_L \tag{17.2}$$

Thus we can draw the AC load line as shown in Figure 18.9.

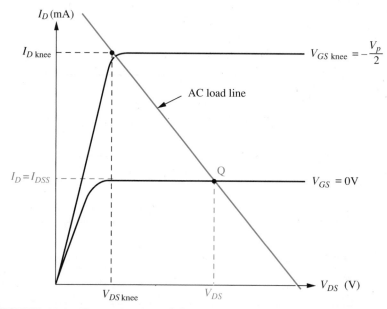

FIGURE 18.9 Determination of the Q point for a depletion MOSFET.

Using an approach much like that used in the development of Equation 17.9 in Section 17.4, we have

$$V_{DS} = (I_{Dknee} - I_D)R_{AC} + V_{DSknee} \qquad \text{(18.6)}$$

From this point onward the design procedure is the same as that for the JFET in Table 17.1 with a few rather obvious changes in notation and omissions such as the elimination of the source resistor and bypass capacitor. The preceding equations are collected in the MOSFET Amplifier Design procedure in Table 18.2 with the associated circuit diagram in Figure 18.10.

TABLE 18.2 MOSFET Amplifier Design

* : Designer's formula only.

STD.: Choose the nearest standard component value.

GIVEN: V_P I_{DSS} I_{GSS} R_{SS} R_L f_L

1. $R_D = R_L \,\|\, (-1/g_{os}) \approx R_L$ (STD.) (17.1)
2. $R_{AC} = R_D \,\|\, R_L$ (17.2)
3. $V_{GS} \stackrel{*}{=} 0 \text{ V}$ (18.3)
4. $I_D = I_{DSS}\left(1 - \dfrac{V_{GS}}{V_P}\right)^2 \stackrel{*}{=} I_{DSS}$ (18.4)
5. $V_{GSknee} \stackrel{*}{\approx} -\dfrac{V_P}{2}$ (18.5)
6. $V_{DSknee} = V_{GSknee} - V_P$ (16.1)
7. $I_{Dknee} = I_{DSS}\left(1 - \dfrac{V_{GSknee}}{V_P}\right)^2$ (16.4)
8. $V_{DS} = (I_{Dknee} - I_D)R_{AC} + V_{DSknee}$ (18.6)
9. $V_{DD} = V_{DS} + I_D R_D$ (18.7)
10. $R_G \stackrel{*}{\approx} \dfrac{V_P}{10 * I_{GSS}}$ (STD.) (17.14)
11. $R_{in} = R_G$ (17.13)
12. $R_{out} = R_D \,\|\, 1/g_{os} \approx R_D$ (17.23)
13. $C_i = \dfrac{1}{2\pi f_L (R_{SS} + R_{in})}$ (STD.) (12.6)
14. $C_o = \dfrac{1}{2\pi f_L (R_{out} + R_L)}$ (STD.) (12.13)
15. $g_{fs} = \dfrac{2}{-V_P}\sqrt{I_{DSS} I_D} \stackrel{*}{=} \dfrac{2}{V_P} I_{DSS}$ (16.13)
16. $A_v = -g_{fs}(R_D \,\|\, R_L)$ (17.26)
17. $A_i = A_v \dfrac{R_{in}}{R_L}$ (17.29)
18. $dB_A = 10 * \log(A_v A_i)$ (12.37)

Design procedure for the circuit of Figure 18.10

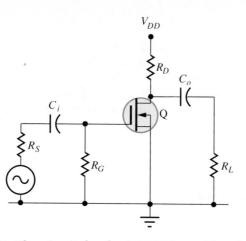

FIGURE 18.10 The circuit for the MOSFET amplifier design procedure of Table 18.2.

EXAMPLE 18.2

Design an amplifier that will drive a 950 Ω load down to 20 Hz from a 200 kΩ source using an MFE3001 transistor.

● (For a SuperCalc solution, see Appendix G.)

SOLUTION:

The circuit is shown in Figure 18.11.

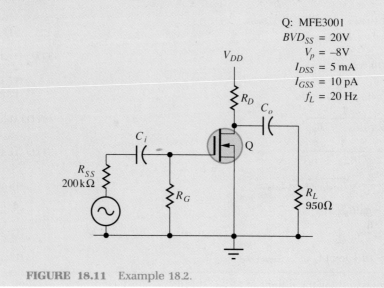

Q: MFE3001
$BVD_{SS} = 20V$
$V_p = -8V$
$I_{DSS} = 5$ mA
$I_{GSS} = 10$ pA
$f_L = 20$ Hz

FIGURE 18.11 Example 18.2.

MOSFET Amplifier Design

GIVEN:

MFE3001; $BV_{DSS} = 20$ V, $V_P = -8$ V, $I_{DSS} = 5$ mA, $I_{GSS} = 10$ pA, $R_{SS} = 200$ kΩ, $R_L = 950$ Ω, $F_L = 20$ Hz

1. $R_D = R_L$ (17.1)

     $= 950$ Ω   (Use 1 kΩ)

2. $R_{AC} = R_D \parallel R_L$ (17.2)

     $= 1$ kΩ $\parallel$ 950 Ω $= 487$ Ω

3. $V_{GS} = 0$ V   (18.3)

4. $I_D = I_{DSS}$ (18.4)

     $= 5$ mA

5. $V_{GSknee} = -\dfrac{V_P}{2}$ (18.5)

     $= -\dfrac{-8 \text{ V}}{2} = +4$ V

6. $V_{DSknee} = V_{GSknee} - V_P$ (16.1)

     $= 4$ V $- (-8$ V$) = 12$ V

7. $I_{Dknee} = I_{DSS}\left(1 - \dfrac{V_{GSknee}}{V_P}\right)^2$ (16.4)

     $= 5$ mA$\left(1 - \dfrac{4.0 \text{ V}}{-8 \text{ V}}\right)^2 = 11.3$ mA

8. $V_{DS} = (I_{Dknee} - I_D)R_{AC} + V_{DSknee}$ (18.6)

     $= (11.3$ mA $- 5$ mA$)487$ Ω $+ 12$ V $= 15.0$ V

9. $V_{DD} = V_{DS} + I_D R_D$ (18.7)

     $= 15.0$ V $+ 5$ mA $* 1$ kΩ $= 20$ V

The transistor's breakdown voltage rating, BV_{DSS}, is barely adequate.

10. $R_G \approx \dfrac{V_P}{10 * I_{GSS}}$ (STD.) (17.14)

     $= \dfrac{-8 \text{ V}}{10(-10 \text{ pA})} = 80$ GΩ   (Use 10 MΩ)

11. $R_{in} = R_G$ (17.13)

     $= 10$ MΩ

12. $R_{out} = R_D$ (17.23)

     $= 1$ kΩ

13. $C_i = \dfrac{1}{2\pi f_L(R_S + R_{in})}$ (STD.) (12.6)

     $= \dfrac{1}{2\pi * 20 \text{ Hz}(200 \text{ kΩ} + 10 \text{ MΩ})} = 780$ pF   (Use 820 pF)

14. $C_o = \dfrac{1}{2\pi f_L (R_{out} + R_L)}$ (STD.)(12.13)

$= \dfrac{1}{2\pi * 20 \text{ Hz}(1 \text{ k}\Omega + 950 \text{ }\Omega)} = 4.08 \text{ }\mu\text{F}$ (Use 4 µF)

15. $g_{fs} \overset{*}{=} \dfrac{2}{-V_P} I_{DSS}$ (16.13)

$= \dfrac{2}{-(-8 \text{ V})} 5 \text{ mA} = 1.25 \text{ mS}$

16. $A_v = -g_{fs}(R_D \parallel R_L)$ (17.26)

$= -1.25 \text{ mS}(1 \text{ k}\Omega \parallel 950 \text{ }\Omega) = 0.609$

17. $A_i = A_v \dfrac{R_{in}}{R_L}$ (17.29)

$= -0.609 \dfrac{10 \text{ M}\Omega}{950 \text{ }\Omega} = -6410$

18. $\text{dB}_A = 10 * \log (A_v A_i)$ (12.37)

$= 10 * \log (0.609 * 6410) = 35.9 \text{ dB}$

18.7 Depletion MOSFET Circuit Analysis

The analysis of the simple zero bias circuit is very straightforward. You already know that if $V_{GS} = 0$ V then

$$I_D = I_{DSS} \qquad (18.4)$$

Once I_D is known, a KVL analysis of the circuit will give the other voltages of interest. It is always advisable to sketch the drain characteristics and locate the Q point just to verify that it is in the active region and that the anticipated signal will not produce clipping.

If it is desired, the AC load line can be drawn through the Q point and v_{DD}. Notice that a lower case "v" is used indicating AC. We defined v_{CC} for BJTs in Section 9.6. Using the notation of FETs, v_{DD} is given by

$$v_{DD} = V_{DS} + I_D R_{AC} \qquad (9.18)$$

EXAMPLE 18.3
Find the dB gain of the circuit of Figure 18.12.

● (For a PSpice solution, see Appendix H.)

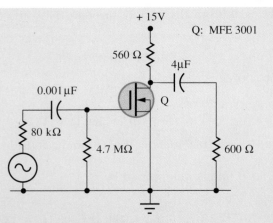

FIGURE 18.12 Circuit for Example 18.3.

SOLUTION:

MFE3001 data from Appendix F: $V_P = -8$ V; $I_{DSS} = 5$ mA. The above data are shown on the graph in Figure 18.13. From the circuit of Figure 18.12, $V_{GS} = 0$ V and so

$$I_D = I_{DSS} \tag{18.3}$$
$$= 5 \text{ mA}$$

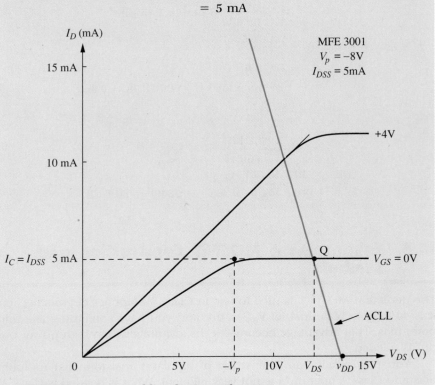

FIGURE 18.13 Load line for Example 18.3.

Doing a KVL walk from the supply through R_D to the drain terminal gives

$$V_{DD} - I_D R_D = V_D \qquad (18.8)$$

or, since the source is grounded,

$$V_{DS} = V_{DD} - I_D R_D \qquad (18.9)$$
$$= 15 \text{ V} - 5 \text{ mA} * 560 \ \Omega = 12.2 \text{ V}$$

Equations 18.3 and 18.9 define the Q point. It is located on Figure 18.13.

The AC load line is determined by

$$v_{DD} = V_{DS} + I_D R_{AC} \qquad (9.18)$$
$$= 12.2 \text{ V} + 5 \text{ mA} * (560 \ \Omega \parallel 600 \ \Omega) = 13.6 \text{ V}$$

The AC load line is drawn on Figure 18.13. As long as the signal amplitude is less than about 3 VP or 6 VPP, clipping should not occur.

For the MFE3001 operating at the above located Q point

$$g_{fs} = \frac{2}{-V_P} I_{DSS} \qquad (16.13)$$

$$= \frac{2}{-(-8 \text{ V})} \, 5 \text{ mA} = 1.25 \text{ mS}$$

The gain is given by

$$A_V = -g_{fs} R_D \parallel R_L \qquad (17.26)$$
$$= -1.25 \text{ mS} * (560 \ \Omega \parallel 600 \ \Omega) = -0.362$$

$$A_i = A_V \frac{R_{in}}{R_L} \qquad (17.29)$$

$$= -0.362 \frac{4.7 \text{ M}\Omega}{600 \ \Omega} = -2840$$

$$dB_A = 10 * \log (A_V A_i) \qquad (12.37)$$
$$= 10 * \log (-0.362 * -2840) = 30.1 \text{ dB}$$

18.8 Enhancement MOSFET Construction and Symbols

The depletion MOSFET is nice to use in Class A amplifier circuits because it is so easy to bias. Just let $V_{GS} = 0$ V and you have something like midpoint bias. That advantage becomes a disadvantage when high power output is needed.

You may recall from our study of bipolar transistors that midpoint bias and Class A operation is not very efficient. What is needed is Class AB operation with near cutoff biasing. Since zero gate voltage does not mean zero drain current, the depletion MOSFET does not work well in Class AB operation. The enhancement MOSFET is better suited to meet that need.

You can think of an enhancement MOSFET as being built exactly the same as the depletion device we talked about in Section 18.2 with the exception that we will leave out the channel that joined the source and the drain in Figure 18.1. Thus, conceptually at least, an enhancement MOSFET looks like Figure 18.14. The standard circuit symbols used for the enhancement MOSFET are shown in Figure 18.15. The line that represents

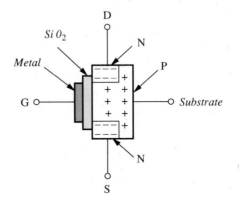

FIGURE 18.14 Representation of an N channel enhancement MOSFET.

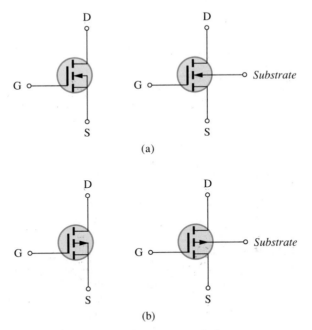

(a)

(b)

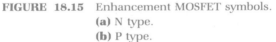

FIGURE 18.15 Enhancement MOSFET symbols.
    (a) N type.
    (b) P type.

the channel in a depletion MOSFET is replaced by a broken line to emphasize the fact that there is no channel connecting the source and the drain in the enhancement device. Obviously, both N type and P type devices are made and what we say about the N type, you can easily transfer to the P type.

18.9 Enhancement MOSFET Behavior

In order to understand how the enhancement MOSFET works, consider the circuit shown in Figure 18.16. It is a good circuit for drawing the device characteristic curves, by the way. To begin, if $V_{GS} = 0$ V, since there is no channel connecting the drain and the source, there will be no current flow. To apply a negative voltage to the gate does not help in the case of the N type device. It gives you less than no channel at all (if that is possible).

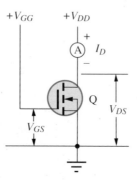

FIGURE 18.16 Circuit used to draw drain characteristics.

On the other hand, if the gate is made sufficiently positive with respect to the substrate, then you might imagine a layer of negative charges being attracted to the surface of the substrate as shown in Figure 18.17a. We have created a channel and current begins to flow. The higher the positive voltage placed on the gate, the deeper the channel and so we have more drain current as shown in Figure 18.17b. Several different shapes have been used in efforts to increase the area of the conducting surface and so to increase the current handling ability of the device. One of the most successful designs is what is known as the "VMOS design" in which a V-shaped grove is etched into the substrate into which the SiO_2 insulator and the metal gate structure are deposited. This approach allows the construction of very high power MOSFET devices.

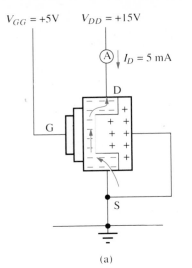

$V_{GG} = +5V$ $V_{DD} = +15V$

(A) $I_D = 5$ mA

(a)

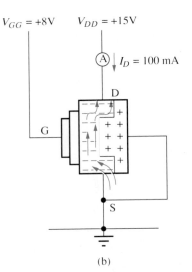

$V_{GG} = +8V$ $V_{DD} = +15V$

(A) $I_D = 100$ mA

(b)

FIGURE 18.17   Representation of enhancement MOSFET conduction.
(a) Small V_{GS} gives a shallow channel and so a small value of I_D.
(b) Large V_{GS} produces a deeper channel and so a larger value of I_D.

18.10 Enhancement MOSFET Characteristics

As you might guess from its name and what has been said so far, the enhancement MOSFET is a field effect device and so its transfer characteristic is the more positive branch of a parabola very similar to what you are used to for the depletion device. The only difference is that the parabola is shifted horizontally so that rather than the graph of Figure 18.3a, we now have something like Figure 18.18a. The vertex of the parabola is still at $V_{GS} = V_P$ and $I_D = 0$ A but V_P is now a positive value. Since the transistor begins to conduct when V_{GS} exceeds V_P, sometimes you will find V_P referred to as the "turn on voltage" and labeled V_{TO}. That is the notation used by PSpice. We will use V_P, because that is what most manufacturers use.

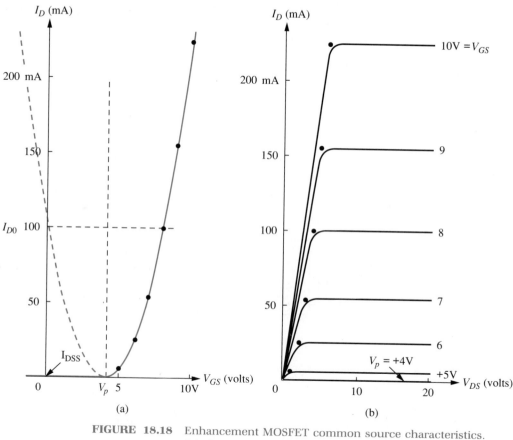

FIGURE 18.18 Enhancement MOSFET common source characteristics.
(a) The transfer characteristic.
(b) Drain characteristics.

In Section 16.5 we defined I_{DSS} to be the drain current when $V_{GS} = 0$ V. According to the ideal model we have presented for the enhancement MOSFET, when $V_{GS} = 0$ V, there is no channel and so we should have $I_{DSS} = 0$ A. In fact, no current will flow until V_{GS} exceeds V_P. Thus I_{DSS} is not even on the parabola of the transfer characteristic in Figure 18.18. In fact, I_{DSS} has a whole new meaning. It is a measure of the amount of leakage current that flows before the device is turned on. If you look at the device data sheets you will find that I_{DSS} is typically in the microamp range and can usually be ignored.

The drain characteristics for the enhancement device shown in Figure 18.18b are no different from those of the depletion device of Figure 18.3b except for the shift in V_{GS}. Therefore almost all of the ideas and equations we have developed for the depleted device can be adapted to the enhancement device with the exception of statements involving I_{DSS}.

You may have noticed that two pieces of information are required in order to describe FET behavior. For JFETs manufacturers typically provide V_P and I_{DSS}. Since I_{DSS} is now essentially irrelevant, the second piece of information that the manufacturers usually supply is the transconductance, g_{fs} measured at a specified value of I_D. Since g_{fs} is not constant, but depends on the value of I_D at which it was measured, we will refer to the values listed in the data sheets as "g_{fsx} measured at I_{Dx}."

In order that we can continue to use the transfer equation, you will notice that the more negative branch of the parabola that is the transfer characteristic in Figure 18.18a has a vertical intercept that we will call "I_{D0}." The "zero" in the subscript will serve to remind us that it is an intercept. Of course, the transistor can never operate at I_{D0}, or at any other point along the more negative branch of the parabola in Figure 18.18a for that matter. However, from a mathematical standpoint, I_{D0} can be used in place of I_{DSS} in the JFET equations.

In order to obtain a value for I_{D0} from the manufacturer's data, we will solve Equation 16.13 for I_{DSS}. We get

$$I_{D0} = I_{DSS} = \frac{\left(\dfrac{V_P g_{fsx}}{2}\right)^2}{I_{Dx}} \tag{18.10}$$

Equation 16.4, the transfer equation then becomes

$$I_D = I_{D0}\left(1 - \frac{V_{GS}}{V_P}\right)^2 \tag{18.11}$$

When we solve the transfer equation for V_{GS}, as we did in Equation 16.5, we have to be careful to remember that we are dealing with a parabola as shown in Figure 18.18a. Therefore, for a given value of I_D, the mathematician would give us two values of V_{GS}. Thus a more mathematically correct version of Equation 16.5 is

$$V_{GS} = V_P\left(1 - \pm\sqrt{\frac{I_D}{I_{DSS}}}\right) \tag{18.12}$$

Since V_P is positive for the enhancement MOSFET, and since we are interested in only the more positive value of V_{GS}, we will choose the negative value for the square root in Equation 18.12. Since it is subtracted, from one in Equation 18.12, that will give us the more positive value that we seek. Then, using I_{D0} in place of I_{DSS}, we have the enhancement MOSFET version of the transfer equation solved for V_{GS}

$$V_{GS} = V_P\left(1 + \sqrt{\frac{I_D}{I_{D0}}}\right) \tag{18.13}$$

EXAMPLE 18.4
Construct a transfer characteristic graph for a VN0106N9 MOSFET.

SOLUTION:
From Appendix F the data for a VN0106N9 transistor includes: $V_P = 2.4$ V; $g_{fs} = 400$ mS at $I_D = 500$ mA.

$$I_{D0} = \frac{\left(\dfrac{V_P g_{fsx}}{2}\right)^2}{I_{Dx}} \tag{18.10}$$

$$= \frac{\left(\dfrac{2.4\text{ V} * 400\text{ mS}}{2}\right)^2}{500\text{ mA}} = 461\text{ mA}$$

The transfer equation is then given by

$$I_D = I_{D0}\left(1 - \frac{V_{GS}}{V_P}\right)^2 \tag{18.11}$$

$$= 461\text{ mA}\left(1 - \frac{V_{GS}}{2.4\text{ V}}\right)^2$$

Figure 18.19 is a plot of the above equation from $V_{GS} = 2.4$ V to $V_{GS} = 5$ V which is the required transfer characteristic.

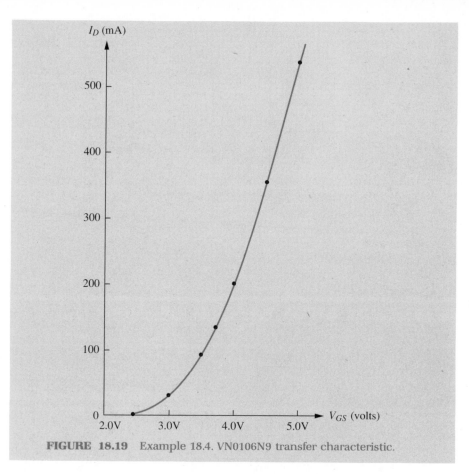

FIGURE 18.19   Example 18.4. VN0106N9 transfer characteristic.

18.11   Enhancement MOSFET Circuits

It is possible to bias enhancement MOSFETs in much the same way as we biased bipolar transistors in Chapter 10 provided you remember that there is no gate current. Obviously that precludes fixed current bias because there is no gate current to fix. All that is needed to bias most MOSFET circuits is to use Equation 18.10 to obtain I_{D0}, put it in the transfer equation, Equation 18.11 or 18.13, and use it along with your knowledge of circuits to obtain a satisfactory circuit.

EXAMPLE 18.5
Given the circuit of Figure 18.20, find the value of resistors R_D and R_G that will give $V_{DS} = 5$ V.

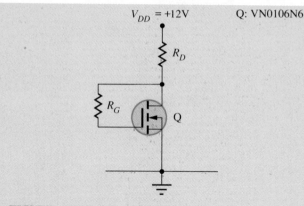

FIGURE 18.20 Example 18.5.

SOLUTION:

From the data in Appendix F for the VN0106N6: $V_P = 2.4$ V; $g_{fs} = 250$ mS (typical), at $I_D = 0.5$ A; $I_{GSS} = 100$ nA

$$I_{D0} = \frac{\left(\dfrac{V_P g_{fsx}}{2}\right)^2}{I_{Dx}} \tag{18.10}$$

$$= \frac{\left(\dfrac{2.4 \text{ V} * 250 \text{ mS}}{2}\right)^2}{0.5 \text{ A}} = 180 \text{ mA}$$

The transfer equation is then

$$I_D = I_{D0}\left(1 - \frac{V_{GS}}{V_P}\right)^2 \tag{18.11}$$

$$= 180 \text{ mA}\left(1 - \frac{V_{GS}}{2.4 \text{ V}}\right)^2$$

From the circuit of Figure 18.20, if $I_G = 0$ A then $V_{GS} = V_{DS}$. Thus $V_{GS} = 5$ V. Substituting in the transfer equation gives

$$I_D = 180 \text{ mA}\left(1 - \frac{V_{GS}}{2.4 \text{ V}}\right)^2$$

$$= 180 \text{ mA}\left(1 - \frac{5 \text{ V}}{2.4 \text{ V}}\right)^2 = 211 \text{ mA}$$

From the circuit of Figure 18.20

$$R_D = \frac{V_{DD} - V_{DS}}{I_D}$$

$$= \frac{12\text{ V} - 5\text{ V}}{211\text{ mA}} = 33.1\ \Omega \qquad \text{(Use 33 }\Omega\text{)}$$

The power rating for R_D is

$$P_{RD} = I_D{}^2 R_D$$

$$= (211\text{ mA})^2 * 33\ \Omega = 1.47\text{ W} \qquad \text{(Use 2 W)}$$

As we did in connection with the development of Equation 17.14, we will assume that the voltage lost across R_G should be small (1/10) compared to V_{GS}. Then

$$R_G = \frac{V_{GS}}{10 I_{GSS}}$$

$$= \frac{5\text{ V}}{10 * 100\text{ nA}} = 5.0\text{ M}\Omega \qquad \text{(Use 4.7 M}\Omega\text{)} \qquad (18.14)$$

Although you will find enhancement MOSFETs used in Class A circuits such as that of Example 18.5, as we mentioned at the beginning of Section 18.8, they are more commonly used in "CMOS" circuits. Most small computers would not exist if it were not for CMOS circuits. The acronym CMOS refers to the fact that complementary MOSFETs are used in the circuit. You may remember that we first introduced the complementary circuit in connection with Class B, or more accurately, Class AB bipolar circuits in Chapter 15.

In Chapter 15, we pointed out that the term "complementary" referred to the fact that both an NPN and a PNP transistor were included in the circuit. Thus it should come as no surprise to you that a CMOS circuit contains both an N channel MOSFET and a P channel MOSFET as shown in the simplified circuit of Figure 18.21.

Most of what we said in Chapter 15 applies to the circuit of Figure 18.21: The transistors should be complements. That means that they should have identical characteristics with Q_1 being an N channel device and Q_2 being P channel. We will have to make some allowances for the fact that it is very difficult to make V_P come out exactly the same for both transistors.

The resistors labeled R_1 are identical so that midpoint bias is achieved. The circuit is a source follower type with the drain of Q_1 connected to $+V_{DD}$ and the drain of Q_2 connected to ground, or perhaps more commonly, connected to $-V_{DD}$, thereby eliminating the need for capaci-

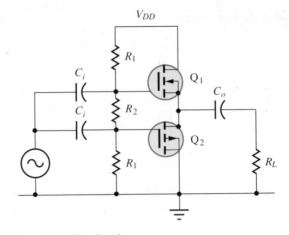

FIGURE 18.21 A CMOS circuit.

and allowing DC operation. Resistor R_2 is chosen such that the voltage loss across it is sufficient to provide the required value of V_{GS} for each transistor. Since V_{Gs} may be fairly large, the value of resistor R_2 may be quite significant compared to R_1. Thus, in order to preserve symmetry for the AC signal, if capacitor coupling is used it is not uncommon to use two identical input coupling capacitors as shown in Figure 18.21.

18.12 CMOS Power Amplifier Design

As we have said, all that is needed in order to design or analyze simple CMOS circuits is an application of the ideas we have developed for MOSFETs to the principles developed for BJT complementary circuits in Chapter 15.

Before we begin to develop a design procedure, a couple of words of caution are in order. First, remember that we are talking about very simple circuits. What we do here should be looked on as only a first step toward understanding the vastly more complex circuits you are likely to encounter later on. But it is a first step.

Second, if you hook up the circuits we are about to talk about, and you should, remember that : (1.) MOSFETs are static sensitive devices so exercise the cautions we talked about in Section 18.3. (2.) You are dealing with a high power circuit. That means very low (zero?) resistance. If both transistors ever get turned on at the same time, the possibility of generating some rather expensive smoke is very real.

Assume that it is our task to design the CMOS power amplifier shown in Figure 18.22. The amplifier is to deliver a specified RMS signal power,

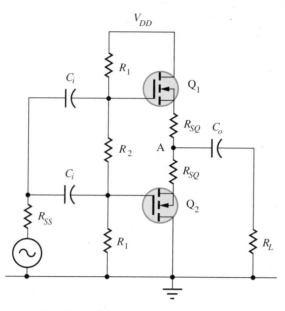

FIGURE 18.22   Circuit used for design of CMOS circuit.

P_{Lrms} to a specified load resistance, R_L. Using the ideas from Table 15.1 we have the preliminary estimate of transistor requirements

$$P_{DISS} \geq \frac{P_{Lrms}}{2} \qquad (15.24)$$

$$BV_{DSS} \geq 6\sqrt{P_{Lrms}R_L} \qquad (15.25)$$

Based upon the above estimated transistor requirements we can go to the data books and select an appropriate pair of MOSFETs. Remember that typically the higher the ratings you specify, the more the finished product will cost, but higher ratings will also result in more reliable operation and a longer life.

Once the transistors have been specified, a data book will provide values for V_P, g_{fsx}, I_{Dx}, and I_{GSS}. We can then begin our design procedure:

$$1. \ I_{D0} = \frac{\left(\dfrac{V_P g_{fsx}}{2}\right)^2}{I_{Dx}} \qquad (18.10)$$

Due to the fact that MOSFETs have a negative temperature coefficient, the need for source resistors is not as great as it was in Chapter 15 and they are often omitted. However, to provide for sample-to-sample variations among transistors, they should be included. Adapting Step 1 of Table 15.1 to our MOSFET notation we can say

2. $R_{SQ} \overset{*}{\approx} \dfrac{R_L}{10}$ 　　　　　　　　　　　　　　　(STD.) (15.15)

You should again be warned that Equation 15.15 and all other equations marked with the asterisk are rules of thumb, designer's formulas that are only valid if you are the designer and are free to specify the circuit component values.

In order to locate the AC load line, refer to Figure 18.23 which is analogous to Figure 15.15b for the BJT circuit. You will notice that the charac-

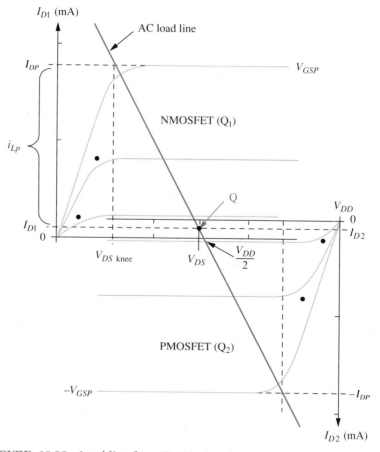

FIGURE 18.23　Load line for a CMOS circuit.

teristics for the PMOSFET (Q_2) are inverted just as they were for the PNP BJT in Chapter 15. From Section 15.10 we know that the peak drain current required to deliver the required power to the load is

3. $I_{DP} \approx i_{LP} = \sqrt{\dfrac{2P_{Lrms}}{R_L}}$ \hfill (15.20)

From the transfer equation for MOSFETs we have

4. $V_{GSP} = V_P\left(1 + \sqrt{\dfrac{I_{DP}}{I_{D0}}}\right)$ \hfill (18.13)

The idea that the maximum signal amplitude should not extend beyond the knee of the V_{GSP} characteristic curve comes from Section 18.6. In Section 16.3 we learned that the knee of the V_{GSP} characteristic is given by

5. $V_{DSknee} = V_{GSP} - V_P$ \hfill (16.1)

Extending the logic we used in Section 18.6 for the development of Equation 18.7 to the complementary circuit whose characteristics are shown in Figure 18.23 gives

6. $V_{DD} = 2[I_{DP}(R_{SQ} + R_L) + V_{DSknee}]$ \hfill (18.15)

In order that the operation will be slightly Class AB, as we did in Chapter 15, at the Q point we will have

7. $I_D \overset{*}{\approx} \dfrac{I_{DP}}{100}$ \hfill (15.26)

8. $V_{GS} = V_P\left(1 + \sqrt{\dfrac{I_D}{I_{D0}}}\right)$ \hfill (18.13)

From Figure 18.22, if the voltage at the point marked Point A is to be half the supply voltage, which it must be for midpoint bias, then

9. $V_{S1} = \dfrac{V_{DD}}{2} + I_D R_{SQ}$ \hfill (18.16)

10. $V_{S2} = \dfrac{V_{DD}}{2} - I_D R_{SQ}$ \hfill (18.17)

Again, from an examination of Figure 18.22

11. $V_{G1} = V_{S1} + V_{GS}$ \hfill (18.18)

12. $V_{G2} = V_{S2} - V_{GS}$ \hfill (18.19)

In order to determine reasonable values for the gate biasing resistors, we will recognize that in order for good voltage divider action, the current through R_1 must be at least 10 times I_{GSS}. Thus going from the gate of Q_2 to ground we have

13. $R_1 \overset{*}{\approx} \dfrac{V_{G2}}{10 I_{GSS}}$

(STD.) (18.20)

Applying the voltage divider idea to the gate circuit, we can say

$$\frac{R_2}{V_{G1} - V_{G2}} = \frac{R_1}{V_{G2}}$$

(18.21)

or

14. $R_2 = R_1 \dfrac{V_{G1} - V_{G2}}{V_{G2}}$

(STD.) (18.22)

Looking at the input, since the use of two input capacitors in Figure 18.22 effectively bypasses resistor R_2, and remembering that the MOSFETs will not draw any current, we have

$$R_{in} = R_1 \parallel R_1$$

or

15. $R_{in} = \dfrac{R_1}{2}$

(18.23)

Looking at the output, assuming as we always have that $1/g_{os} = \infty \ \Omega$ and remembering that only one transistor conducts at a time,

16. $R_{out} = R_{SQ}$

(18.24)

The coupling capacitors can now be determined using the same approach as we have used before. However, since we are using two input capacitors, the total capacitance will be split equally between them, thus

17. $C_i = \left(\dfrac{1}{2}\right) \dfrac{1}{2\pi f_L (R_{SS} + R_{in})}$

(STD.) (18.25)

18. $C_o = \dfrac{1}{2\pi f_L (R_{out} + R_L)}$

(STD.) (12.13)

This completes the design of a simple CMOSFET power amplifier. The preceding design steps are collected together in Table 18.3.

TABLE 18.3 CMOSFET Amplifier Design

* : Designer's formula only.
STD.: Choose nearest standard component value.

GIVEN: P_{Lrms} R_L R_{SS} f_L $\left\{\begin{array}{l}\text{Transistor} \\ \text{spec.} \\ \text{estimate:}\end{array}\right\}$ $\left\{\begin{array}{l} P_{DISS} \geq \frac{1}{2}P_L \\ BV_{DSS} \geq 6\sqrt{P_{Lrms}R_L} \end{array}\right\}$

1. $I_{D0} = \dfrac{\left(\dfrac{V_P g_{fsx}}{2}\right)^2}{I_{Dx}}$ (18.10)

2. $R_{SQ} \overset{*}{\approx} \dfrac{R_L}{10}$ (STD.) (15.15)

3. $I_{DP} \approx i_{LP} = \sqrt{\dfrac{2P_{Lrms}}{R_L}}$ (15.20)

4. $V_{GSP} = V_P\left(1 + \sqrt{\dfrac{I_{DP}}{I_{D0}}}\right)$ (18.13)

5. $V_{DSknee} = V_{GSP} - V_P$ (16.1)

6. $V_{DD} = 2[I_{DP}(R_{SQ} + R_L) + V_{DSknee}]$ (18.15)

7. $I_D \overset{*}{\approx} \dfrac{I_{DP}}{100}$ (15.26)

8. $V_{GS} = V_P\left(1 + \sqrt{\dfrac{I_D}{I_{D0}}}\right)$ (18.13)

9. $V_{S1} = \dfrac{V_{DD}}{2} + I_D R_{SQ}$ (18.16)

10. $V_{S2} = \dfrac{V_{DD}}{2} - I_D R_{SQ}$ (18.17)

11. $V_{G1} = V_{S1} + V_{GS}$ (18.18)

12. $V_{G2} = V_{S2} - V_{GS}$ (18.19)

13. $R_1 \overset{*}{\approx} \dfrac{V_{G2}}{10 I_{GSS}}$ (STD.) (18.20)

14. $R_2 = R_1\dfrac{V_{G1} - V_{G2}}{V_{G2}}$ (STD.) (18.22)

15. $R_{in} = \dfrac{R_1}{2}$ (18.23)

16. $R_{out} = R_{SQ}$ (18.24)

17. $C_i = \left(\dfrac{1}{2}\right)\dfrac{1}{2\pi f_L(R_{SS} + R_{in})}$ (STD.) (18.25)

18. $C_o = \dfrac{1}{2\pi f_L(R_{out} + R_L)}$ (STD.) (12.13)

Design procedure for the circuit of Figure 18.22.

EXAMPLE 18.6

Redesign the power amplifier of Example 15.6 using MOSFETs. Select the transistors from those listed in Appendix F.

● (For a SuperCalc solution, see Appendix G.)

SOLUTION:

From Example 15.6: $P_L = 0.5$ W $R_L = 10$ Ω $R_{SS} = 600$ Ω
$f_L = 30$ Hz. The circuit to be designed is shown in Figure 18.24. Substituting in the procedure of Table 18.3 we have

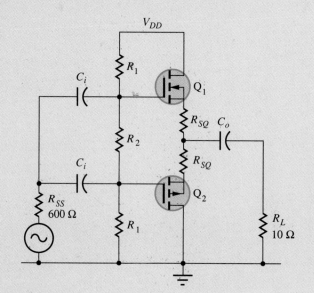

FIGURE 18.24 Example 18.6.

CMOSFET Amplifier Design

GIVEN: P_{Lrms} R_L R_{SS} f_L $\left\{ \begin{array}{c} \text{Transistor} \\ \text{spec.} \\ \text{estimate:} \end{array} \right\}$ $\left\{ \begin{array}{c} P_{DISS} \geq \dfrac{1}{2} P_{Lrms} \\ BV_{DSS} \geq 6\sqrt{P_{Lrms}R_L} \end{array} \right\}$

GIVEN: $P_{Lrms} = 0.5$ W $R_L = 10$ Ω $R_{SS} = 600$ Ω $f_L = 30$ Hz

TRANSISTOR SELECTION:

$$P_{DISS} \geq \frac{P_{Lrms}}{2} \tag{15.24}$$

$$= \frac{0.5 \text{ W}}{2} = 0.25 \text{ W}$$

$$BV_{DSS} \geq 6\sqrt{P_{Lrms}R_L} \qquad (15.25)$$
$$= 6\sqrt{0.5 \text{ W} * 10 \text{ }\Omega} = 13.4 \text{ V}$$

The transistors from Appendix F that are supposed to be matched and that come closest to the required power rating are:

Q_1: VN0106N6: $V_P = 2.4$ V, $I_{GSS} = 100$ nA, $g_{fs} = 250$ mS @ 0.5 A

Q_2: VP0106N6: $V_P = 3.5$ V, $I_{GSS} = 100$ nA, $g_{fs} = 200$ mS @ 0.5 A

1. $I_{D0} = \dfrac{\left(\dfrac{V_P g_{fsx}}{2}\right)^2}{I_{Dx}}$ $\qquad\qquad$ (18.10)

$$= \dfrac{\left(\dfrac{3.5 \text{ V} * 200 \text{ mS}}{2}\right)^2}{0.5 \text{ A}} = 245 \text{ mA}$$

2. $R_{SQ} \overset{*}{\approx} \dfrac{R_L}{10}$ $\qquad\qquad$ (STD.) (15.15)

$$= \dfrac{10 \text{ }\Omega}{10} = 1 \text{ }\Omega \qquad\qquad \text{(Use 1 }\Omega)$$

3. $I_{DP} \approx i_{LP} = \sqrt{\dfrac{2P_{Lrms}}{R_L}}$ $\qquad\qquad$ (15.20)

$$= \sqrt{\dfrac{2 * 0.5 \text{ W}}{10 \text{ }\Omega}} = 316 \text{ mAP}$$

4. $V_{GSP} = V_P\left(1 + \sqrt{\dfrac{I_{DP}}{I_{D0}}}\right)$ $\qquad\qquad$ (18.13)

$$= 3.5 \text{ V}\left(1 + \sqrt{\dfrac{316 \text{ mAP}}{245 \text{ mA}}}\right) = 7.48 \text{ VP}$$

5. $V_{DSknee} = V_{GSP} - V_P$ $\qquad\qquad$ (16.1)

$$= 7.48 \text{ V} - 3.5 \text{ V} = 3.98 \text{ V}$$

6. $V_{DD} = 2[I_{DP}(R_{SQ} + R_L) + V_{DSknee}]$ $\qquad\qquad$ (18.15)

$$= 2[316 \text{ mA}(1 \text{ }\Omega + 10 \text{ }\Omega) + 3.98 \text{ V}] = 14.9 \text{ V}$$

7. $I_D \overset{*}{\approx} \dfrac{I_{DP}}{100}$ $\qquad\qquad$ (15.26)

$$= \dfrac{316 \text{ mA}}{100} = 3.16 \text{ mA}$$

8. $V_{GS} = V_P\left(1 + \sqrt{\dfrac{I_D}{I_{D0}}}\right)$ (18.13)

$\quad = 3.5\text{ V}\left(1 + \sqrt{\dfrac{3.16\text{ mA}}{245\text{ mA}}}\right) = 3.90\text{ V}$

9. $V_{S1} = \dfrac{V_{DD}}{2} + I_D R_{SQ}$ (18.16)

$\quad = \dfrac{14.9\text{ V}}{2} + 3.16\text{ mA} * 1\ \Omega = 7.46\text{ V}$

10. $V_{S2} = \dfrac{V_{DD}}{2} - I_D R_{SQ}$ (18.17)

$\quad = \dfrac{14.9\text{ V}}{2} - 3.16\text{ mA} * 1\ \Omega = 7.45\text{ V}$

11. $V_{G1} = V_{S1} + V_{GS}$ (18.18)
$\quad = 7.46\text{ V} + 3.90\text{ V} = 11.4\text{ V}$

12. $V_{G2} = V_{S2} - V_{GS}$ (18.19)
$\quad = 7.45\text{ V} - 3.90\text{ V} = 3.55\text{ V}$

13. $R_1 \stackrel{*}{\approx} \dfrac{V_{G2}}{10 I_{GSS}}$ (STD.) (18.20)

$\quad = \dfrac{3.55\text{ V}}{10 * 100\text{ nA}} = 3.55\text{ M}\Omega$ (Use 3.3 MΩ)

14. $R_Z = R_1 \dfrac{V_{G1} - V_{G2}}{V_{G2}}$ (STD.) (18.22)

$\quad = 3.3\text{ M}\Omega \dfrac{11.4\text{ V} - 3.55\text{ V}}{3.55\text{ V}} = 7.29\text{ M}\Omega$ (Use 6.8 MΩ)

15. $R_{\text{in}} = \dfrac{R_1}{2}$ (18.23)

$\quad = \dfrac{3.3\text{ M}\Omega}{2} = 1.65\text{ M}\Omega$

16. $R_{\text{out}} = R_{SQ}$ (18.24)
$\quad = 1\ \Omega$

17. $C_i = \left(\dfrac{1}{2}\right)\dfrac{1}{2\pi f_L (R_{SS} + R_{\text{in}})}$ (STD.) (18.25)

$\quad = \left(\dfrac{1}{2}\right)\dfrac{1}{2\pi * 30\text{ Hz}(600\ \Omega + 1.65\text{ M}\Omega}$

$\quad = 1.61\text{ nF} = 0.0061\ \mu\text{F}$ (Use 0.0016μF)

$$18.\ C_o = \frac{1}{2\pi f_L(R_{out} + R_L)} \qquad \text{(STD.) (12.13)}$$

$$= \frac{1}{2\pi * 30\ \text{Hz}(1\ \Omega + 10\ \Omega)} = 482\ \mu\text{F} \qquad \text{(Use 500 }\mu\text{F)}$$

You should notice from the above example that the input impedance of CMOS power amplifiers is very high and the output impedance is very low, making them very attractive for high power circuits.

18.13 Digital CMOS

The CMOS circuits we have discussed so far are source follower type circuits. They work well for reasonably linear circuits. Where switching is all that is required, as is the case in the ones and zeros world of digital circuits, the common source CMOS circuit of Figure 18.25 is used extensively. In fact it may be safe to say that small computers as we know them would not exist if it were not for that circuit.

In order to understand how the circuit of Figure 18.25 works, look first at transistor Q_2. As you can see, it is an N channel MOSFET in a common source circuit. Thus it will begin to conduct when the gate voltage, V_G, is more positive than the source by the amount V_P, the turn on voltage. Similarly, Q_1 is a P channel MOSFET in a common source circuit. You may have to turn Figure 18.25 upside down in order to see that Q_1 is, in fact,

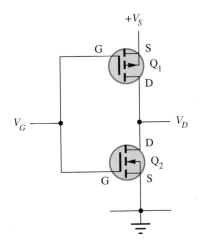

FIGURE 18.25 A CMOS inverter.

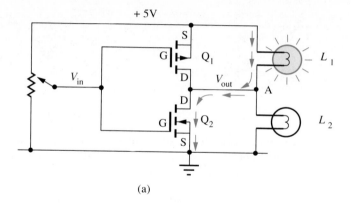

(a)

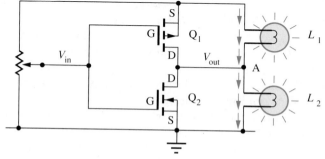

(b)

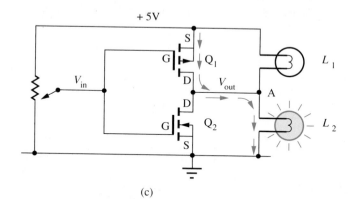

(c)

FIGURE 18.26 Operation of a CMOS inverter.
(a) Input high, output low.
(b) Ambiguous.
(c) Input low, output high.

in a common source circuit. It considers the positive supply terminal as the common terminal and it considers ground as its negative power supply. Thus it sees no gate bias voltage when $V_G = +V_{DD}$. It will begin to conduct when V_G drops below V_S by the amount V_P.

To understand the circuit behavior, assume that $V_S = +5$ V and $V_P = 3$ V. Now, if we start with $V_{in} = +5$ V as shown in Figure 18.26a, we find that as we have just said, Q_1 sees no gate bias and so it is turned off. However, Q_2 sees a high forward bias and so it will conduct pulling the output down close to 0 V. The circuit then acts as a current sink. As a result, it will ground Point A in Figure 18.26a and current will flow as shown. Lamp L_1 will burn brightly and L_2 will not be lit.

When the gate voltage in Figure 18.26 drops below $V_G = 3$ V, then Q_2 turns off. However, Q_1 does not turn on until $V_G = 2$ V since we have assumed a supply voltage of $V_{DD} = 5$ V and a turn on voltage of $V_P = 3$ V. Thus between $V_G = 3$ V and $V_G = 2$ V, we have the ambiguous case of Figure 18.26b where neither transistor is conducting. The lamps are then in series across the supply. Both lamps will glow dimly.

When the gate voltage drops below $V_G = 2$ V, Q_1 turns on as shown in Figure 18.26c. The circuit then becomes a current source and Lamp L_2 burns brightly.

From the above description you can see that the circuit is an inverter. When the input is high, the output is low. When the input is low, the output is high. You can also see one of the strong advantages of CMOS digital circuits: Since both transistors should never be on at the same time, and since the gates are well insulated, the device wastes very little power.

18.14   Troubleshooting   ◄◄◄◄◄◄◄◄◄

Problems in electronic circuits most frequently occur where the power is the highest. That means either in the power supply or in the output circuit. Thus these should be the first places for the troubleshooter to look.

As we mentioned in connection with BJT power circuits, a great deal of care should be used in troubleshooting these circuits. The resistances are small (zero?!) and power MOSFETs are expensive. Many servicers plug the device under test into a variable transformer and turn it up slowly as they monitor the circuit being tested. At least you should use a fuse in the circuit. When complementary circuits develop problems, the fault is often in the driving circuit. If you simply replace the output transistors without looking for the underlying cause of the failure, you will probably be replacing them again (at your own expense!).

18.15 Summary

In this chapter we introduced the MOSFET. We showed how they are different from JFETs, and how the depletion type MOSFET is different from the enhancement type. We showed how the circuit design and analysis procedures developed for JFETs must be altered for each type of MOSFET and how their unique characteristics can be used to advantage in circuits.

GLOSSARY

Current sink: Any device that will accept current from an external source and conduct it to ground. (Section 18.13)

Depletion mode: The mode of MOSFET operation in which the gate to source voltage is always of such polarity as to inhibit or reduce the drain current. (Section 18.4)

Enhancement mode: The mode of MOSFET operation in which the gate to source voltage is such as to increase the drain current. (Section 18.4)

MOSFET: Metal Oxide Semiconductor Field Effect Transistor. A transistor that is formed by a metal conductor separated by an oxide insulator from the semiconductor channel whose conduction it controls. (Section 18.2)

Punch through: A form of MOSFET failure that may occur when the gate voltage goes so high that the gate insulation is "punched through" so that it becomes a conductor. (Section 18.3)

Substrate: The body of a MOSFET semiconductor into which the channel is formed. (Section 18.9)

PROBLEMS

Section 18.2

1. Draw a diagram similar to Figure 18.1 for a P channel MOSFET. Beside your diagram draw the standard schematic symbol for the device. Identify all of the terminals.
2. For the circuits of Figure 18.27, indicate whether the polarity of the power supply (PS) should be positive (+) or negative (−).

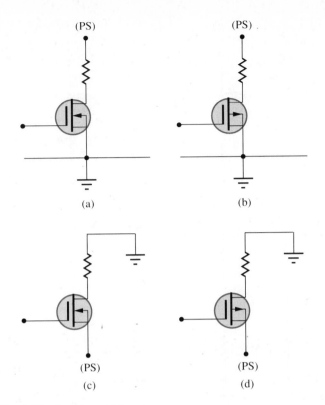

FIGURE 18.27   Problem 18.2.

Section 18.3

3. Draw a circuit that will give some overvoltage protection to an NMOS-FET without excessively degrading its performance.

Section 18.4

4. Identify each of the circuits in Figure 18.28 as either depletion mode or enhancement mode operation.

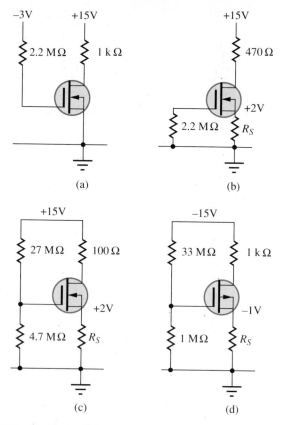

FIGURE 18.28   Problem 18.4.

Section 18.6

5. If the transistor in Figure 18.28b is an MFE130 MOSFET (See Appendix F for data), determine the nearest standard value for R_S.
6. Repeat Problem 5 for the circuit of Figure 18.28c.

Section 18.7

7. If the transistor in Figure 18.28a is an MFE130, determine the value of V_D.
8. Repeat Problem 7 for the circuit of Figure 18.28b.
9. Repeat Problem 7 for the circuit of Figure 18.28c.
10. Construct a transfer characteristic graph for a 3SK121Y MOSFET using the data from Appendix F.
11. Construct a set of drain characteristic curves from the data you generated in Problem 10.

Section 18.10

12. Construct a set of drain characteristic curves for an MFE3001 MOSFET using the data from Appendix F.
13. Repeat the design problem of Example 18.2 using a 2SK238 transistor.
14. For the circuit of Example 18.2, sketch the AC load line and estimate the value of the maximum unclipped output signal voltage.
15. Repeat Problem 14 with reference to the circuit of Problem 13.

Section 18.11

16. Using the data from Appendix F, write the transfer equation for a VN0106N6 transistor.
17. Construct a transfer characteristic for the transistor of Problem 16.

Section 18.12

18. Using the procedure of Table 18.3 design a circuit that will deliver 15 W to an 8 Ω load. Assume that the source has an output resistance of 600 Ω and that the cutoff frequency is to be 20 Hz.
19. Estimate the value of the input signal voltage required to drive the circuit of Problem 18.

Section 18.14

◀ Troubleshooting

20. For the circuit of Example 18.6, estimate the approximate drain current for each transistor and the effect on the other circuit components if resistor R_2 became:
    (a)   A short circuit.
    (b)   An open circuit.
21. For the circuit of Example 18.6, estimate the effect on the Q point and on the other circuit components if the output became shorted to ground ($R_L = 0\ \Omega$) with:
    (a)   No signal applied.
    (b)   The maximum signal applied.

PART IV
LINEAR INTEGRATED CIRCUITS

Integrated circuits are the heart of modern
computers. (Photograph by John Huet.)

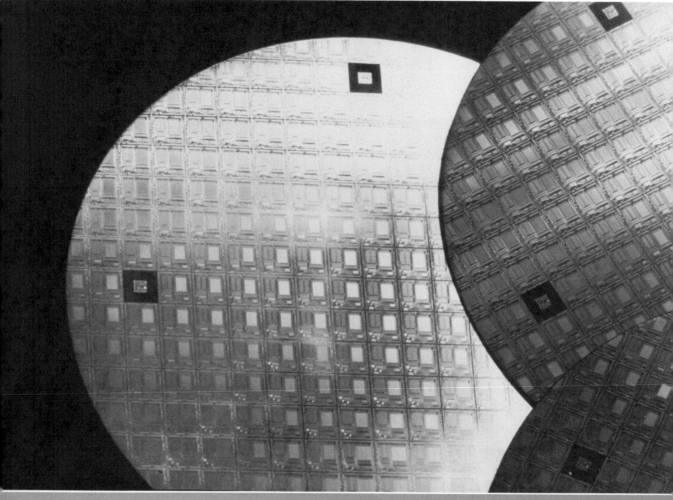

Inspection of a semiconductor wafer. (Courtesy of
National Semiconductor.)

Chapter 19
Introduction to Op Amps

OUTLINE

SPECIAL TERMS

Op amp

Inverting input

Noninverting input

Comparator

Current mirror

Diff amp

Rail

Bandwidth

Unity gain bandwidth

Rise time

Slew rate

19.1   Introduction and Objectives

In this chapter we will introduce the idea of an operational amplifier (op amp) and define an ideal op amp. We will also look briefly at the circuitry used in real op amps and how close they approach the ideal. Since most op amps in use today contain BJTs, in this chapter we will be discussing them and their equivalent circuit. Thus you may wish to refresh your memory by skimming over parts of Chapters 8 through 15.

After having studied this chapter you should begin to understand what an ideal op amp is. You should have a very limited idea of how the internal circuitry of real op amps work, and you should learn some of the limitations of real op amps. You should then be able to:

- List the characteristics of an ideal op amp.
- Use the above characteristics in the analysis and design of simple comparator and level detector circuits containing op amps.
- Solve simple gain bandwidth problems with reference to op amps.
- Do a limited amount of circuit analysis on the internal circuit of an op amp.

19.2   Ideal Op Amps

An "ideal operational amplifier" is defined to be a DC coupled amplifier that has two identifying characteristics. They are:

1. The input resistance is infinite.

$$R_{in} = \infty \ \Omega \tag{19.1}$$

2. The voltage gain is negative infinity.

$$A_V = -\infty \tag{19.2}$$

You can probably see some immediate use for an amplifier with infinite input resistance. As one example, a scope or a DMM with $R_{in} = \infty \ \Omega$ would never load any circuit to which it was connected. You can probably think of other examples such as your stereo or your tape deck where $R_{in} = \infty \ \Omega$ would be an advantage.

The infinite gain requirement is a bit more difficult to understand. For example, can you imagine how a scope with infinite gain would behave? The smallest input would cause the trace to go completely off the screen. After all, what is infinity times any small number? Such a scope would be useless.

On the other hand, when the spot on the above scope, or the spot that paints the picture on your TV set, has completed a trace across the screen, you would like it to return to the left side of the screen as quickly as possible so that it could get started on another trace. You would like the retrace amplifier to have infinite gain.

As we get into working with op amps you will find many reasons why it is desirable for real op amps to approach the ideal of Equations 19.1 and 19.2 as closely as possible.

Early op amps were single input devices. The input was an "inverting input" since the output waveform was inverted with respect to the input. That is why the gain in Equation 19.2 is $-\infty$ rather than $+\infty$. A positive input voltage will then produce a negative output and a negative input will produce a positive output.

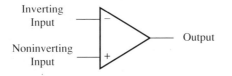

FIGURE 19.1   Symbol representing an op amp.

Most modern op amps have "differential" inputs. That means that they have both an "inverting," or a "$-$" input, and they have a "noninverting" or a "$+$" input. The circuit symbol used to represent the device is shown in Figure 19.1.

19.3 Real Op Amps

You probably realize that Equations 19.1 and 19.2 are not realistic equations. In this world nothing is infinite. Numbers have been attached to the dimensions of the smallest subatomic particle and to the distance to the most distant star. Although you might question how accurately those distances are known, neither of them is infinite. Nothing in our experience is infinite. Obviously then there is no such thing as an ideal op amp. In order to have some understanding of the limitations of real op amps, we will look briefly at the internal circuitry of the 741 which is the most commonly used op amp.

Hopefully the first thing you thought about when you saw Equation 19.1 was an FET input. Many modern op amps use FET inputs. However, we will limit our brief look inside op amps to the more commonly used 741 which is the industry standard. You will have the opportunity to study the internal circuitry of several op amp types if you take an advanced linear devices course later on.

The 741 comes in several versions such as 741, 741A, 741C, 741E, etc. Their specifications vary somewhat. The most popular, probably because it is the least expensive, is the 741C (commercial), which typically sells for less than a dollar.

A simplified version of the 741's internal circuitry is shown in Figure 19.2. A more complete circuit is shown on the manufacturer's data sheet in Appendix F.

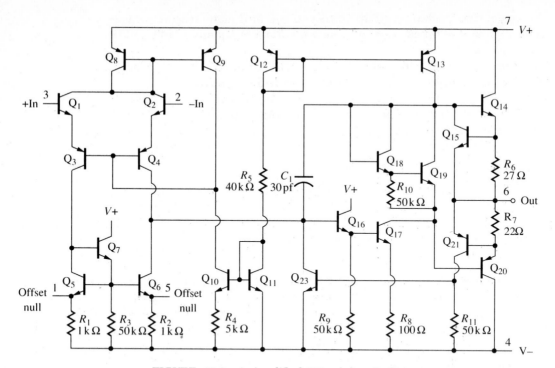

FIGURE 19.2   A simplified 741 schematic diagram.

Lest you be overwhelmed by the complexity of that "simplified" circuit in Figure 19.2, look briefly at the output circuit. You should recognize in transistors Q_{14} and Q_{20} the complementary push-pull circuit we studied in Chapter 15. Notice that a "split" power supply is required if the output is to be symmetrical about zero. You will find that your knowledge of transistor behavior will make other circuits within the 741 understandable as well. We will return to a more thorough study of the output circuit a little later.

In our analysis of the 741 circuit we will use three assumptions. They are:

1. The junction equation applies

$$V_{BE} = 26 \text{ mV} * \ln\left(\frac{I_C}{10 \text{ fA}}\right) \tag{8.4}$$

2. The transistors in the circuit are all identical.
3. For each transistor, the base current is so much smaller than the collector current that it can be ignored.

Probably the easiest place to begin an analysis of the 741 circuit is with the voltage reference circuit consisting of transistors Q_{11} and Q_{12}.

19.4 IC Considerations

Notice that the 741 is an integrated circuit (IC), containing dozens of components housed in an incredibly small space, but there is only one capacitor among those components, and it is not a coupling capacitor. Thus all of the amplifier stages are direct coupled. There simply is not room to do otherwise. This design has the great advantage that the circuit has no low frequency cutoff. It will amplify DC signals. It has the disadvantage that two power supplies are required if the output is to have both positive and negative values. Also the designer of the IC must be careful that the overall Q point produces no net DC offset between the input and the output. This requires the use of some level shifting circuits that may seem strange to you.

Another thing which tends to make op amp circuits look strange is the fact that manufacturers find that some components are easier to form on a chip than are others. As we have mentioned, manufacturers will go to almost any length to avoid having to use capacitors. Likewise, since transistors must be constructed anyway, it is easier to make additional transistors than to construct diodes. If you look at the 741 schematic on the manufacturer's data sheet in Appendix F, or the simplified version of the 741 schematic in Figure 19.2, you will see several transistors whose collector terminals are connected directly to their base terminals. Transistors Q_8 and Q_{12} are two examples. They are really just diodes. You should be able to locate several other examples of transistors used as diodes in the 741.

19.5 Current Mirrors

A circuit development that contributed a great deal to the development of the modern op amp is the current mirror circuit. We first mentioned the current mirror concept in connection with diode biasing of complementary push-pull output circuits in Section 15.8.

If you remember the characteristics of current sources, you can see why op amp designers like to use them. For one thing, an ideal current source has infinite internal resistance. This at least points in the direction indicated by Equation 19.1. Similarly, if you remember our early encounters with transistors, high resistance means flat load lines and flat load lines point in the direction indicated by Equation 19.2.

In Figure 19.2, transistors Q_{12} and Q_{13} are one example of a current mirror circuit. The relevant portion of Figure 19.2 is reproduced here as Figure 19.3. We will identify other current mirrors as we analyze other parts of the 741 circuit.

In order to analyze the behavior of the above current mirror notice that, as explained in Section 19.4, Q_{12} is really just a diode. It can be con-

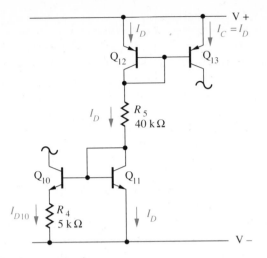

FIGURE 19.3 A current mirror.

sidered to be a voltage reference source whose voltage is determined by its collector current I_{C12}. According to the junction equation

$$V_{BE12} = 26 \text{ mV} * \ln\left(\frac{I_{C12}}{10 \text{ fA}}\right) \qquad (8.4)$$

From the circuit it is obvious that $V_{BE13} = V_{BE12}$. Thus, since we assume that the transistors have identical turn-on characteristics, $I_{C13} = I_{C12}$. That is, I_{C13} "mirrors" I_{C12} independently of what is connected to the collector of Q_{13}.

Again referring to Figure 19.3, a similar argument can be made with respect to transistors Q_{11} and Q_{10}—almost. Transistor Q_{11} is really a voltage reference diode just as Q_{12} is, but that voltage is across the series combination of the B-E junction of Q_{10} and resistor R_4. Thus I_{C10} is much less than I_{C11}, but its value is still determined by the current flowing in Q_{11}.

You should notice that Q_8 and Q_9 in Figure 19.2 also constitute a current mirror circuit with the collector current in Q_8 determining the current in Q_9.

EXAMPLE 19.1

If ± 15 V power supplies are used for the circuit of Figure 19.2, estimate the input circuit bias current, I_{C8} flowing in Q_8 and the output circuit bias current, I_{C13} flowing in Q_{13}.

SOLUTION:

The relevant circuit is shown in Figure 19.4. As usual, we assume that all base currents are negligibly small compared to collector currents, and that the characteristics of all transistors are identical.

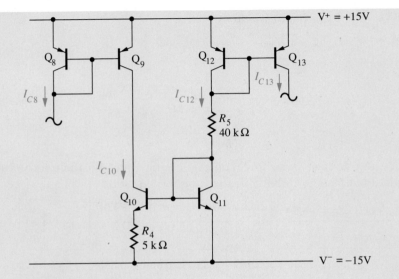

FIGURE 19.4   Example 19.1.

A KVL walk from V^+ through Q_{12}, R_5, and Q_{11} to V^- in Figure 19.4 gives

$$V^+ - V_{BE12} - I_{C12}R_5 - V_{BE11} = V^-$$

or

$$I_{C12} = \frac{15\ V - (-15\ V) - 2\ V_{BE}}{40\ k\Omega} \tag{19.3}$$

$$= \frac{15\ V - V_{BE}}{20\ k\Omega}$$

Iterating Equation 19.3 with the junction equation gives

$I_{C12} = \dfrac{15\ V - V_{BE}}{20\ k\Omega}$	=	715	718	718	μA
$V_{BE12} = 26\ mV * \ln\dfrac{I_{C12}}{10\ fA} =$		0.650	0.650	0.650	V

Thus

$$I_{C13} = I_{C12} = 718\ \mu A$$
$$V_{BE11} = V_{BE12} = 0.650\ V$$
$$V_{B10} = V_{B11} = V_{BE11} + V^-$$
$$= 0.650\ V + (-15\ V) = -14.4\ V$$

A KVL walk from B_{10} to V^- gives

$$V_{B10} - V_{BE10} - I_{C10}R_4 = V^-$$

or

$$I_{C10} = \frac{-14.4 \text{ V} - (-15 \text{ V}) - V_{BE10}}{5 \text{ k}\Omega} \tag{19.4}$$

$$= \frac{0.650 \text{ V} - V_{BE10}}{5 \text{ k}\Omega}$$

Since the voltage is so low, some guessing must be included when Equation 19.4 is iterated with the junction equation.

I_{C10} =	-10.0	22.2	20.1	18.6	19.0	18.9	μA
Guess =	$+10$	15					μA
V_{BE10} =	0.539	0.549	0.557	0.555	0.555	0.555	V

In the first step in the above iteration, my calculator calculated

$$I_{C10} = \frac{0.650 \text{ V} - 0.7 \text{ V}}{5 \text{ k}\Omega} = -10.0 \text{ μA}$$

Since I know that I_{C10} cannot be negative, but will be some small positive value, I inserted $I_{C10} = 10$ μA. The calculator then solved the junction equation to give $V_{BE10} = 0.539$ V. It then solved Equation 19.4 to obtain $I_{C10} = 22.2$ μA. I then compromised with the calculator by inserting $I_{C10} = 15$ μA. It then calculated $V_{BE10} = 0.549$ V and $I_{C10} = 20.1$ μA. At that point I decided that the equations would converge without further help so I quit guessing!

Since Q_9 is in series with Q_{10}

$$I_{C9} = I_{C10} = 18.9 \text{ μA}$$

Since Q_8 and Q_9 form a current mirror

$$I_{C8} = I_{C9} = 18.9 \text{ μA}$$

You may object that in above example the behavior of Q_8 and Q_9 is described incorrectly since Q_8 is really the voltage reference diode and so I_{C9} actually mirrors I_{C8}. Of course, you are correct. This means that in Figure 19.4 both I_{C11} and I_{C8} are trying to dictate the value of I_{C10}. What happens if they are different? The solution to the problem lies in the fact that, as you can see in Figure 19.2, the collectors of both Q_9 and Q_{10} are connected to the bases of Q_3 and Q_4. This connection is labeled "I_{sense}" in Figure 19.5. If, for example I_{C8} should happen to be too low, thus forcing I_{C9} to be less

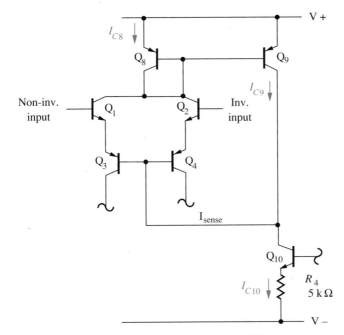

FIGURE 19.5 Circuit for determining input amplifier bias current.

than I_{C10} then the voltage at the sense terminal will drop. This will cause Q_1, Q_2, Q_3, and Q_4 all to conduct more, thus increasing I_{C8} until it equals the value required by I_{C10}.

19.6 The Differential Amplifier

The input portion of the 741 circuit is shown in Figure 19.6. You will notice that there are two inputs to the circuit. The "inverting input" goes to the base of transistor Q_2 and the "noninverting input" goes to the base of transistor Q_1. These are the inputs to the op amp from the outside world. The inverting input is connected to pin 2 of the 741 and the noninverting input is connected to pin 3.

Their names indicate the effect the inputs have on the op amp output. If the voltage at pin 2 is higher than the voltage at pin 3 then the effect is inverted so the op amp output goes negative. If V_2 is less than V_3, again the effect is inverted and the output goes positive. On the other hand, if viewed from the perspective of the noninverting input, the output is in phase with the input.

If $V_2 = V_3$ then, regardless of whether V_2 and V_3 are positive, negative or zero, the output should be zero volts. That is why this amplifier is known as a "differential amplifier," or a "diff amp" as it is often called.

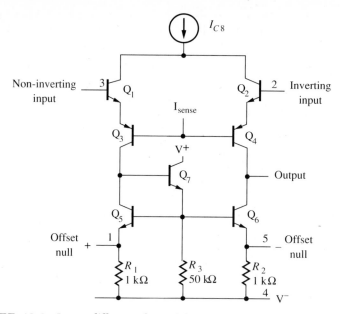

FIGURE 19.6 Input differential amplifier.

Ideally it amplifies only the difference between the input voltages. How close a specific amplifier approaches this ideal is indicated by its "common mode rejection ratio," CMRR, which is defined to be the ratio of its differential gain divided by its common mode gain. We will have more to say about CMRR in Section 21.8.

In Section 19.2 we mentioned that one of the characteristics of an op amp is high (∞ ?) input resistance. You can see in Figure 19.6 that Q_1 and Q_2 are emitter followers. You may recall from Section 13.3 that the emitter follower can have a relatively high input resistance. This effect is further enhanced by the presence of transistors Q_3 and Q_4 and the fact that current sources are used in both the collector and the emitter circuits. Since current sources ideally have infinite resistance, you should expect the 741 to have high input resistance. The data sheet in Appendix F lists a typical value of $Z_I = 2$ MΩ.

Transistors Q_3 and Q_4 also cause the diff amp to have a fairly high voltage gain.

19.7 Diff Amp Offset Nulling

As you can see from Figure 19.6, the diff amp should be symmetrical. That means that with no input to either pin 2 or pin 3 of the op amp, half of the bias current, I_{C8} should flow through transistors Q_1, Q_3, and Q_5 and the other half should flow through transistors Q_2, Q_4, and Q_6. This should result in exactly zero volts appearing at the output terminal of the op amp. However, as you should expect by now, "exactly" is hard to achieve, even in op amps. An offset of a few millivolts is not uncommon, especially in "C" grade op amps.

Op amp pins 1 and 5 have been provided for the purpose of nulling any offset that may be present in the output in the event that the users application demands such a critical adjustment. The way offset nulling is usually accomplished is to hook the ends of a 10 kΩ pot between pins 1 and 5 and hook the slider of the pot to the V^- power supply. This allows the user to vary the amount of resistance in parallel with resistors R_1 and R_2 in Figure 19.6 and so to adjust the op amp output for exactly zero volts when both inputs are grounded.

EXAMPLE 19.2

Determine the value of the resistor R_x that must be placed between pin 1 of the op amp of Example 19.1 and the negative supply in order to increase the current through the noninverting side of the diff amp by 10 percent.

SOLUTION:

The relevant circuit is shown in Figure 19.7. From Example 19.1, the current source supplying I_{C5} and I_{C6} is $I_{C8} = 18.9 \ \mu A$. Since $I_{C8} = I_{C5} +$

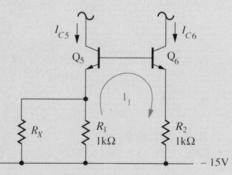

FIGURE 19.7 Example 19.2.

I_{C6}, if I_{C5} is to be increased by 10 percent then I_{C6} must be reduced by 10 percent. Thus the new values for I_{C5} and I_{C6} are

$$I_{C5} = 1.1 * \frac{I_{C8}}{2}$$

$$= 1.1 * \frac{18.9 \ \mu A}{2} = 10.4 \ \mu A$$

and

$$I_{C6} = 0.9 * \frac{I_{C8}}{2}$$

$$= 0.9 * \frac{18.9 \ \mu A}{2} = 8.51 \ \mu A$$

Applying the junction equation

$$V_{BE5} = 26 \text{ mV} * \ln\left(\frac{I_{C5}}{10 \text{ fA}}\right) \qquad (8.4)$$

$$= 26 \text{ mV} * \ln\left(\frac{10.4 \ \mu A}{10 \text{ fA}}\right) = 0.540 \text{ V}$$

and

$$V_{BE6} = 26 \text{ mV} * \ln\left(\frac{I_{C6}}{10 \text{ fA}}\right) \qquad (8.4)$$

$$= 26 \text{ mV} * \ln\left(\frac{8.51 \ \mu A}{10 \text{ fA}}\right) = 0.535 \text{ V}$$

Doing a KVL walk around loop l_1 in Figure 19.7

$$I_{C5}(R_x \parallel R_1) + V_{BE5} - V_{BE6} - I_{C6}R_2 = 0V \qquad (19.5)$$

or

$$R_x \parallel R_1 = \frac{-V_{BE5} + V_{BE6} + I_{C6}R_2}{I_{C5}}$$

$$= \frac{-0.540 \text{ V} + 0.535 \text{ V} + 8.51 \ \mu A * 1 \text{ k}\Omega}{10.4 \ \mu A} = 316 \ \Omega$$

Solving for R_x

$$R_x = 316 \ \Omega \parallel (-R_1)$$
$$= 316 \ \Omega \parallel (-1 \text{ k}\Omega) = 463 \ \Omega$$

19.8 The Voltage Amplifier

We have seen how the diff amp can provide reasonably high input resistances as required by the ideal op amp described in Section 19.2. Another ideal op amp requirement described there is high voltage gain. As we mentioned in Section 19.6, the diff amp has a fairly high voltage gain, due at least in part to the fact that it is driven by current sources. The disadvantage in using current sources is that the output resistance is high. Transistor Q_{16} is an emitter follower which provides a better match between the diff amp's high output resistance and the relatively low input resistance of the common emitter amplifier containing Q_{17}. The relevant circuitry is shown in Figure 19.8.

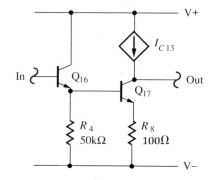

FIGURE 19.8 The voltage amplifier stage.

Since the circuit of Q_{16} will have a voltage gain of approximately one (see Chapter 13), and so will the complementary output stage of Q_{14} and Q_{20} (see Chapter 15), you can see that all of the voltage gain of the 741 must be provided by the diff amp and the circuit of transistor Q_{17}.

EXAMPLE 19.3
Assuming that the voltage gain of the diff amp in Figure 19.2 has a value of 400, estimate the voltage gain of the circuit of transistor Q_{17}.

SOLUTION:
The overall voltage gain for the 741 is listed in Appendix F as typically "$A_{VS} = 200$ V/mV." This means $A_V = 200{,}000$. Assuming that $A_V \approx 1$ for the circuit of Q_{16} and the output circuit, since the overall gain is the product of the stage gains

$$A_{V17} = \frac{A_{VS}}{A_{Vdiff}}$$

$$= \frac{200{,}000}{400} = 500$$

19.9 Biasing the Output Circuit

In Section 19.2 we pointed out that the output portion of the circuit in Figure 19.2 is just the complementary push-pull circuit which we studied in Chapter 15. However, the output biasing circuit presents two unique features. One is the use of the current source, Q_{13}, which we examined in Section 19.5. The other unique feature is the use of transistors Q_{18} and Q_{19} to produce the bias needed to avoid crossover distortion. If you have forgotten this phenomenon and the use of diodes to prevent its occurrence, you should review Section 15.8.

In view of what we did in Section 15.8 and what you know of op amp design practices, you might expect the output biasing circuit in Figure 19.2 to look like the circuit in Figure 19.9. The circuit actually used in Fig-

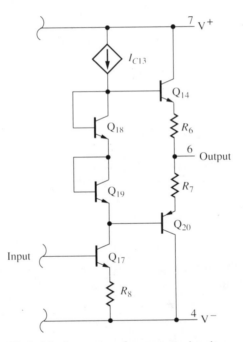

FIGURE 19.9   Diode biasing a complementary circuit.

ure 19.2 is that of Figure 19.10. The reason for using the circuit of Figure 19.10 is that only part of the bias current flows through Q_{18}. The remainder flows through the collector of Q_{19}. Also all of the current flowing through Q_{18} flows through R_{10} (we ignore base currents, remember?). These factors combine to place the bias voltage under the control of the designer.

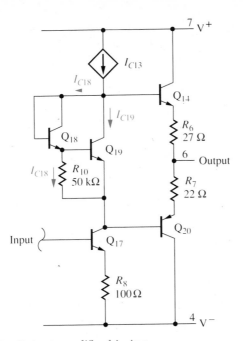

FIGURE 19.10 Output amplifier biasing.

EXAMPLE 19.4

Using ± 15 V power supplies, determine the base bias voltage $V_{B14} - V_{B20}$ for the circuit of Figure 19.10.

SOLUTION:

From Example 19.1 the value of the bias current supply is $I_{C13} = 718$ μA. Assuming that the base currents are negligible

$$I_{C19} = I_{C13} - I_{C18} \tag{19.6}$$

and

$$I_{C18} = \frac{V_{BE19}}{R_{10}} \tag{19.7}$$

Substituting Equation 19.7 into Equation 19.6

$$I_{C19} = I_{C13} - \frac{V_{BE19}}{R_{10}} \tag{19.8}$$

$$= 718 \text{ μA} - \frac{V_{BE19}}{50 \text{ k}\Omega}$$

$$= \frac{35.9 \text{ V} - V_{BE19}}{50 \text{ k}\Omega}$$

Iterating Equation 19.8 with the junction equation

$I_{C19} = \dfrac{35.9\ V - V_{BE19}}{50\ k\Omega} =$	704	705	705	μA
$V_{BE19} = 26\ mV * \ln\dfrac{I_{C19}}{10\ fA} =$	0.649	0.649	0.649	V

Solving Equation 19.6 for I_{C18}

$$I_{C18} = I_{C13} - I_{C19}$$
$$= 718\ \mu A - 705\ \mu A = 13.0\ \mu A$$

Substituting the above result into the junction equation

$$V_{BE15} = 26\ mV * \ln\left(\frac{I_{C15}}{10\ fA}\right) \qquad (8.4)$$

$$= 26\ mV * \ln\left(\frac{13\ \mu A}{10\ fA}\right) = 0.546\ V$$

From Figure 19.10

$$V_{B14} - V_{B20} = V_{BE18} + V_{BE19}$$
$$= 0.546\ V + 0.649\ V = 1.20\ V$$

You should be aware that due to the biasing requirements of the various internal amplifier stages, the maximum output voltage swing, V_{sat} is limited to less than the supply voltage, or "the rail" as it is often called. For the 741C, V_{sat} is limited to approximately 2 V less than the supply voltage. The more expensive versions of the 741 such as the 741E will give much closer to "rail to rail" performance.

19.10   Output Current Limiting

Notice that the output circuit of Figure 19.10 is not protected. If the output terminal became accidentally grounded when a signal was applied, either or both Q_{14} and Q_{20} would probably be damaged. Transistors Q_{15}, Q_{21}, and Q_{23} in Figure 19.2 limit the output current so that shorting the output to ground will not damage a 741. Figure 19.11 shows the relevant portion of

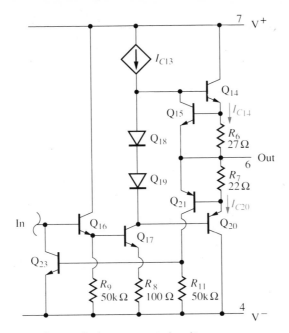

FIGURE 19.11   The equivalent output circuit.

the 741 circuit. To understand how Q_{15} and Q_{23} limit the output current we will examine the circuit behavior both when the output tries to go positive and when it tries to go negative.

First, assume that V_{out} is positive. That means that the input to Q_{16} is negative so that Q_{16} and Q_{17} are shut off and V_{C17} is high. This will reverse bias Q_{20} and so shut it off. Since Q_{17} is shut off, the bias current from Q_{13} flows into the base of Q_{14} turning it on. Current will then flow from V^+ through Q_{14}, through R_6, out of pin 6, and on to whatever load is attached to pin 6. Assuming that the load current is small enough so that the voltage it generates across R_6 is small, transistor Q_{15} will not conduct.

However, if the output current increases to the point that the voltage loss across R_6 reaches the turn-on voltage of transistor Q_{15}, then Q_{15} begins to conduct and the current that would otherwise drive the base of Q_{14} is conducted away by Q_{15} limiting the output of Q_{14}. Thus, the positive output current is limited by the turn-on voltage of Q_{15} and the value of R_6. The essential components of the positive current limiting circuit are shown in Figure 19.12. Remember this rather simple but effective method of current limiting. Its use is not limited to op amps alone. We will encounter another application of it in Chapter 21.

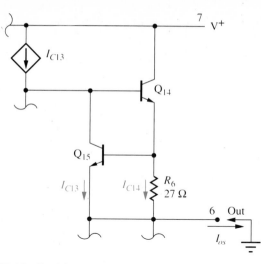

FIGURE 19.12   Positive output current limiting.

EXAMPLE 19.5

If ± 15 V power supplies are used, determine the short circuit positive output current for the circuit of Figure 19.2. Compare with the value listed in Appendix F for the 741.

SOLUTION:

From Example 19.1 when ± 15 V supplies are used, $I_{C13} = 718$ μA. The essential circuit components are shown in Figure 19.12. From the junction equation

$$V_{BE15} = 26 \text{ mV} * \ln\left(\frac{I_{C15}}{10 \text{ fA}}\right) \tag{8.4}$$

$$= 26 \text{ mV} * \ln\left(\frac{718 \text{ μA}}{10 \text{ fA}}\right) = 0.650 \text{ V}$$

The current flowing through R_6 is

$$I_{C14} = \frac{V_{BE15}}{R_6}$$

$$= \frac{0.650 \text{ V}}{27 \text{ Ω}} = 24.1 \text{ mA}$$

The shorted output current, I_{os} is

$$I_{os} = I_{C13} + I_{C14}$$
$$= 718 \ \mu A + 24.1 \ mA = 24.8 \ mA$$

From Appendix F for the 741 the value listed is $I_{os} = 25$ mA so the manufacturer came fairly close (!).

Current-limiting when the output is negative is not quite so simple. To obtain a negative output, assume that the input to Q_{16} in Figure 19.11 is positive. Transistors Q_{16} and Q_{17} will then be turned on causing V_{C17} to go down to near V^-. Transistor Q_{14} will then be reverse biased so it will be turned off. However Q_{20} will be forward biased causing it to conduct and so to pull V_{out} down near V^-.

Assuming that the resulting value of I_{C20} will be small, Q_{21} will not be turned on so no base current will be delivered to Q_{23} and it will also be turned off so Q_{23} will not divert any of the positive signal we originally assumed to be delivered to Q_{16} above.

On the other hand, if the output is shorted so that I_{C20} is so large that the voltage loss across R_7 reaches the turn-on voltage of Q_{21} then it will supply base current to Q_{23}, turning it on and diverting the signal from Q_{16}, thus limiting the current Q_{20} can conduct. The essential parts of the negative current limiting circuit are shown in Figure 19.13.

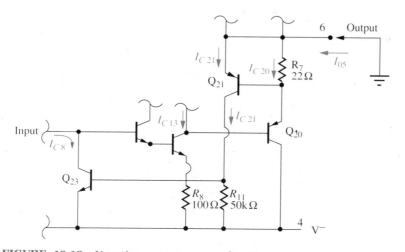

FIGURE 19.13 Negative output current limiting.

EXAMPLE 19.6

Repeat Example 19.5 for negative output current.

SOLUTION:

Assume that when Q_{23} turns on, all of the current delivered to the diff amp by Q_8 is delivered to Q_{23}. From Example 19.1, $I_{C8} = 18.9 \ \mu A$. Then applying the junction equation to Q_{23}

$$V_{BE23} = 26 \ mV * \ln\left(\frac{I_{C23}}{10 \ fA}\right) \tag{8.4}$$

$$= 26 \ mV * \ln\left(\frac{18.9 \ \mu A}{10 \ fA}\right) = 0.555 \ V$$

From Figure 19.13

$$I_{C21} = \frac{V_{BE23}}{R_{11}}$$

$$= \frac{0.555 \ V}{50 \ k\Omega} = 11.1 \ \mu A$$

Applying the junction equation to Q_{21}

$$V_{BE21} = 26 \ mV * \ln\left(\frac{I_{C21}}{10 \ fA}\right) \tag{8.4}$$

$$= 26 \ mV * \ln\left(\frac{11.1 \ \mu A}{10 \ fA}\right) = 0.542 \ V$$

Again from Figure 19.13

$$I_{C20} = \frac{V_{BE21}}{R_7}$$

$$= \frac{0.542 \ V}{22 \ \Omega} = 24.6 \ mA$$

and

$$I_{os} = I_{C20} + I_{C21}$$
$$= 24.6 \ mA + 11.1 \ \mu A = 24.6 \ mA$$

Again the agreement with the data in Appendix F is acceptable.

It is important to recognize that for an op amp to be considered ideal, the current it is called on to deliver must be kept well below the rated maximum value for the op amp.

19.11 Bandwidth Limiting

As we have seen, op amps are direct coupled circuits so they do not have a low frequency cutoff. They will amplify DC. However, if you look again at Figure 19.2, or the more complete circuit diagram for the 741 in Appendix F, you will see that there is a 30 pF capacitor in the circuit. According to Figure 19.2, it will provide feedback between what is essentially the output of the voltage amplifier stage consisting of transistors Q_{16} and Q_{17} and its input. Since the voltage amplifier stage has phase inversion, the feedback will be degenerative. If you need a review of this type of feedback, reread Section 15.14.

Since the capacitor is small, only high frequencies will be fed back. Thus it will reduce the high frequency gain of the op amp. The purpose of this degeneration is to prevent high frequency oscillation. Some op amps require the user to provide this frequency compensation externally. This provides the user the flexibility of using the value that suits the application best, but at the cost of having to hook up additional components.

Figure 19.14 is a copy of the graph labeled "OPEN LOOP FREQUENCY

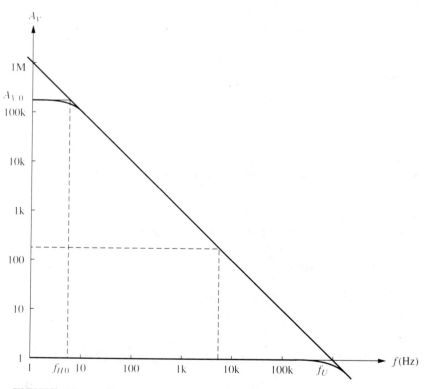

FIGURE 19.14 Frequency response of a 741.

RESPONSE" from the 741 data sheet in Appendix F. You will see that the "open loop high frequency cutoff," f_{HO} for the 741 is just under 10 Hz. Yes, 10 Hz! That is the HIGH frequency cutoff. The graph shows that if you want high gain and high frequency performance, the 741 is not your best choice.

The graph contains no surprises if you recall from Sections 12.9 and 12.10, or from your circuits class, how a single capacitor low pass filter behaves. As you can see on the log log plot of Figure 19.14, the roll off is linear above f_H. The gain drops by one decimal place when the frequency increases by one decimal place. Thus their product remains constant at a value known as the "gain bandwidth product" which we will identify as BW. The definition of BW is

$$BW \equiv A_f f_H \qquad (19.9)$$

where A_f is the gain of the op amp at the frequency f_H.

The gain bandwidth product, BW, is also called the "unity gain bandwidth," because it is obvious from Equation 19.9 that if $A_f = 1$, then $BW = f_H$. This is the value, usually referred to simply as the "bandwidth," BW, which manufacturer's data sheets usually provide to describe an op amp's frequency response.

You will discover in Chapter 20 that the gain of a circuit containing an op amp can be controlled through the use of feedback. It turns out that when the circuit gain is reduced, the frequency response is still described by the graph of Figure 19.14. Thus the high frequency cutoff at any specified gain can be obtained from either Equation 19.9 if a value for BW is known, or from the gain vs frequency graph for the op amp being used.

EXAMPLE 19.7

Find the frequency f_H at which the open loop gain of a 741 op amp is $A_{OL} = 200$.

SOLUTION:

It is a little hard to locate the gain value of 200 on the log scale of Figure 19.14, but since log (200) = 2.30, the desired point is about 3/10 of the distance between $A_V = 100$ and $A_V = 1000$. From Figure 19.14, f_H then appears to be about 7/10 of the distance between $f = 1$ kHz and $f = 10$ kHz. Therefore, to locate f_H

$$f_H = 10^{3.7} = 5.01 \text{ kHz}$$

To solve the problem using Equation 19.9, from Appendix F, for the 741, "BW = 1 MHz." Solving Equation 19.9 for f_H gives

$$f_H = \frac{BW}{A_V} = \frac{1 \text{ MHz}}{200} = 5 \text{ kHz}$$

19.12 Rise Time and Slew Rate

Often manufacturers give the "rise time," t_r, rather than the unity gain bandwidth, BW. It sounds reasonable that they would be related. To determine how they are related, we need to recall how capacitors behave.

Most amplifiers can be considered to behave something like the low-pass circuit of Figure 19.15a. When a "step voltage" such as that of Figure 19.15b is the input, the typical capacitor charging curve of Figure 19.15c is the result. The "rise time," t_r, is defined to be the time required for the output waveform to rise from 10 percent to 90 percent of its final value as shown in Figure 19.15c when the input is a step voltage as shown in Figure 19.15b. Most scopes have 10 percent and 90 percent markings on their screens for this reason.

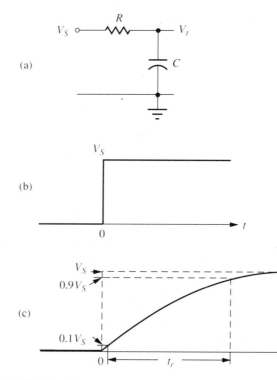

FIGURE 19.15   Rise time.
   (a) Low pass circuit.
   (b) Input voltage waveform.
   (c) Output voltage waveform.

You should recall from your circuits class that the equation that describes the capacitor charging curve can be written

$$V_t = V_S - (V_S - V_0)e^{-t/RC} \qquad (19.10)$$

or

$$RC = \frac{t}{\ln\left(\dfrac{V_S - V_0}{V_S - V_t}\right)} \qquad (19.11)$$

From the charging curve of Figure 19.15c you can see that if $t = t_r$, then $V_0 = 0.1\, V_S$, and $V_t = 0.9\, V_S$. Substituting these values into Equation 19.11 gives

$$RC = \frac{t_r}{\ln (9)} \qquad (19.12)$$

Looking at how the low-pass circuit of Figure 19.15a responds to an AC signal, you remember that it will have a high frequency cutoff, f_H, and at f_H

$$R = X_C = \frac{1}{2\pi f_H C} \qquad (19.13)$$

or

$$f_H = \frac{1}{2\pi RC} \qquad (19.14)$$

If we refer to the unity gain bandwidth then $f_H = BW$. Substituting Equation 19.12 into Equation 19.14 gives the desired relationship

$$BW = \frac{\ln 9}{2\pi t_r} \qquad (19.15)$$

You will often find Equation 19.15 written with the constants in the equation reduced to a single number. That practice tends to hide the basic relationships involved, and the resulting constants are usually not easy to remember.

Another way of indicating the response of a circuit to a step input such as that of Figure 19.15b is the "slew rate." The slew rate is simply the maximum slope of the output waveform. It tells how fast the output voltage can change. The units commonly used for slew rate are volts per microsecond. If you look in Appendix F at the data sheet for the 741 you will see that its slew rate is listed as 0.5 V/μsec.

EXAMPLE 19.8

Using the information from Appendix F, find the gain of a 741C operating at 20 kHz.

SOLUTION:

Appendix F lists the bandwidth for the 741 as $BW = 1.0$ MHz. Thus, from Equation 19.14 we can say that at 20 kHz

$$A_V = \frac{BW}{f} \tag{19.16}$$

$$= \frac{1.0 \text{ MHz}}{20 \text{ kHz}} = 50.0$$

This value also agrees fairly well with the value you can read from the graph in Appendix F.

Again, you should be surprised that the calculated gain is so low. It does not look much like infinity, does it? The 741 is not a high frequency device. If you wish to use an op amp where high frequencies are involved, you should not use a 741.

19.13 Idealizing Real Op Amps

We have pointed out that there is no such thing as an ideal op amp. According to Appendix F, the input impedance of a 741C is about 2 MΩ. That is not ∞ Ω. However, you know that in electronics we usually say that if the input impedance of a device is one decimal place bigger than the resistance of the circuit to which it is connected, the device will not load the circuit significantly and can be ignored. Maybe that is one way of saying that it is effectively an infinite resistance. This places an upper limit on resistances we can use with an op amp at about

$$R_{ckt} < \frac{R_{in}}{10}$$

Using similar reasoning, any resistance that we connect to the output of the op amp should not load it significantly. Thus the circuit resistance should be big compared to the output impedance of the op amp. Therefore we can say

$$10R_{out} < R_{ckt} < \frac{R_{in}}{10} \tag{19.16}$$

Applying the same logic to the infinite gain requirement, the gain of the circuit in which the op amp is used should be much less than the gain of the op amp alone. In order to distinguish between these two different gain values, the gain of the total circuit of which the op amp is a part is usually called the "closed loop gain," A_{CL}, and the gain of the op amp alone is called the "open loop gain," A_{OL}. This is the gain that we can obtain from the data sheets. It is the gain we talked about in Sections 19.11 and 19.12.

Using the above terms and the arguments presented with reference to circuit resistance, we can say

$$A_{OL} > 10A_{CL} \qquad (19.17)$$

Substituting Equation 19.17 into Equation 19.9 tells us what unity gain bandwidth is required for a given circuit gain and frequency if the op amp is to be considered ideal

$$BW > 10A_{CL}f_H \qquad (19.18)$$

19.14 Comparators

We should probably study comparators separately from op amps since their required characteristics are slightly different. However, since the comparator idea is used by op amps, we will look briefly at the idea here.

As the name implies, a comparator just compares. It usually compares an input voltage to a fixed reference voltage. The output voltage is always either as high as it can go, $+V_{sat}$, or as low as it can go, $-V_{sat}$ depending on how the input voltage compares to the reference voltage. Since the comparator output is not intended to go just part way up or part way down, there is no need for its response to be linear. What is required is that it must compare the two voltages precisely (low input offset), and that its output change state from low to high, or from high to low very quickly (high slew rate). We will have more to say about those requirements later.

The differential input feature of op amps together with their ideally infinite gain causes all op amps to function something like special purpose comparators. They compare the voltage present at the two inputs. When one is greater than the other by the smallest amount, the infinite gain feature causes the output to go to V_{sat}.

19.15 Input Offset

As we pointed out in Section 19.7, in the event that the above comparison must be done very precisely, offset nulling may be needed. The circuit usually used is shown in Figure 19.16.

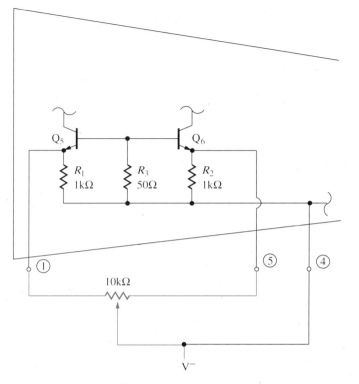

FIGURE 19.16 Offset nulling circuit.

19.16 Zero Crossing Detectors

A common requirement in electronics is to have something happen when a waveform crosses zero. A good example is the trigger circuit of your scope. In order to achieve a stationary pattern on the screen, you would like for the trace to start just as the waveform crosses zero. In other words, you need a zero crossing detector.

A zero crossing detector uses a comparator with one of its inputs connected to the signal of interest and the other input connected to ground. Thus it compares the signal voltage to ground, or to zero volts.

Figure 19.17a contains a zero crossing circuit. The pins are numbered as they are on a typical 8 pin 741 package. For the sake of simplicity, the required power supply connections to the op amp are not shown; however, if you hook up the circuit, you cannot afford that luxury.

Figure 19.17b shows the input and the output waveforms. A noninverting zero crossing detector is shown. We could have designed an inverting zero crossing detector just as easily.

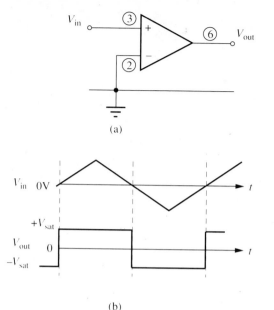

(a)

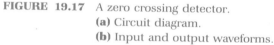

(b)

FIGURE 19.17 A zero crossing detector.
(a) Circuit diagram.
(b) Input and output waveforms.

Notice that the input and the output waveforms are drawn on the same time base. By now you should have developed the practice of drawing carefully coordinated waveforms. It is the best way of demonstrating what is occurring in a circuit and why it is occurring.

You should see that the comparator circuit of Figure 19.17a takes advantage of the infinite gain property of the op amp. If the input is negative by the slightest amount, the output is at $-V_{sat}$. If the input is positive by the slightest amount, the output is at $+V_{sat}$.

You should notice that the output of this circuit is "digital" in the sense that the output is always either "high," at $+V_{sat}$ or "low," at $-V_{sat}$. When it "changes state" by going from one to the other, it goes all the way.

EXAMPLE 19.9

Calculate the minimum gain required for the op amp of Figure 19.17a if its output goes from -13 V to $+13$ V when the input goes from:

(a) -1 μV to $+1$ μV
(b) -1 pV to $+1$ pV
(c) -0 V to $+0$ V

SOLUTION:

$$A_V \equiv \frac{V_{out}}{V_{in}} \qquad (12.17)$$

(a) $A_V = \dfrac{+13\ V - (-13\ V)}{+1\ \mu V - (-1\ \mu V)} = 13{,}000{,}000$

(b) $A_V = \dfrac{+13\ V - (-13\ V)}{+1\ pV - (-1\ pV)} = 13 \times 10^{12}$

(c) $A_V = \dfrac{+13\ V - (-13\ V)}{+0\ V - (-0\ V)} = \infty$

19.17 Voltage Level Detectors

A voltage level detector can be viewed as the more general case of a zero crossing detector. Again, the signal is applied to one of the comparator inputs, but in this case the other comparator input can have a voltage other than zero volts applied to it.

An inverting level detector is shown in Figure 19.18a. Again, for simplicity, the power supply connections to the op amp have been omitted.

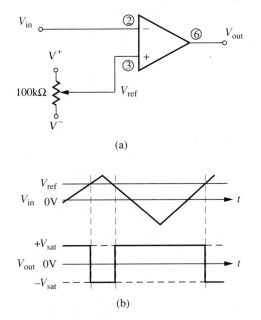

(a)

(b)

FIGURE 19.18 A voltage level detector.
(a) Circuit diagram.
(b) Input and output waveforms.

The reference voltage, V_{ref}, can be obtained from any source, but it is often obtained from the power supplies as shown. This allows the selection of any value for V_{ref} from V^- to V^+ by simply turning the knob on the pot.

Since the circuit shown is an inverting level detector, any time V_{in} is less than V_{ref}, then V_{out} will be high. Any time V_{in} is greater than V_{ref}, then V_{out} will be low. The waveforms are shown in Figure 19.18b.

An obvious application for the level detector is again the trigger circuit of your oscilloscope. Next time you are in the lab, hook up the probe of your scope to your sine wave signal generator. Adjust the controls so that a convenient display is seen. Set the horizontal position control so that you can easily see the beginning of the trace. Observe where on the waveform the trace begins as you turn the trigger level pot first one way and then the other. You will find that you can trigger the trace at any point on the waveform that you desire. The trigger level pot is setting the reference level on a voltage level detector circuit like that of Figure 19.18a.

▶▶▶▶▶▶▶▶ ## 19.18   Troubleshooting

Op amps are quite sturdy and reliable. The most common way they are damaged is by accidentally hooking one of the power supplies to the output, pin 6. The easiest way of determining whether an op amp is damaged is by connecting the inverting input to the output and then checking to see if the voltage is the same at both the inverting and the noninverting inputs. If they are the same, the chances are that the op amp is all right. If they are different, replace the op amp.

19.19   Summary

This chapter provided a brief introduction to op amps, including their internal circuitry, their characteristics, and how those characteristics are used to advantage in circuits containing op amps. Only simple comparators, zero crossing detectors, and level detectors were used in this chapter, but more complex op amp circuits will be considered in future chapters. However, you will find that no matter how complex the circuit, the basic principles set forth in this chapter still apply and are the key to understanding the circuit's behavior.

GLOSSARY

Bandwidth: BW, or f_U The frequency range between the upper cutoff frequency and the lower cutoff frequency. (Section 19.10)

Comparator: An electronic circuit or device that compares an input voltage to a fixed reference voltage. (Section 19.4)

Current mirror: A circuit designed so that its output current is dictated by, or "mirrors" the current in a control circuit. (Section 15.8, 19.5)

Diff amp: A differential amplifier. An amplifier whose output is determined by the difference between two input signals. (Section 19.6)

Inverting input: The input to an electronic circuit that produces an output whose phase is reversed with respect to the input. (Section 19.2)

Noninverting input: The input to an electronic circuit that produces an output whose phase is the same as that of the input (Section 19.2)

op amp: An abbreviation for the term "operational amplifier." An ideal op amp meets these criteria: $R_{in} = \infty \ \Omega$ and $A_V = \infty$. (Section 19.1)

Rail: A slang expression used to indicate the power supply voltage. (Section 19.8)

Rise time: t_r The time required for the output waveform to rise from 10 percent to 90 percent of its final value when the input is a square pulse. (Section 19.11)

Slew rate: A measure of an op amp's switching speed, defined to be the maximum rate of change of its output voltage when a step voltage is applied to its input. It is usually expressed in volts per microsecond. (Section 19.11)

Unity gain bandwidth: BW or f_u The frequency at which the gain of a DC coupled amplifier is reduced to one. (Section 19.10)

PROBLEMS

Section 19.3

1. Record the voltage gain (Large Signal Voltage Gain) and the input impedance for the 741C op amp as listed on the manufacturer's data sheet in Appendix F.

2. Determine the input voltage to a 741C when a source that has an open circuit voltage of 10 mV is connected to one of the 741C's inputs if the source has an internal resistance of (a) 20 kΩ; (b) 2 MΩ.

3. Repeat Problem 2 using a 741E op amp.

4. How much power does the stylus of your record player have to put out if it delivers 100 mV to the 100 kΩ input of your stereo? How much power would it have to put out if the stereo input impedance were $\infty \ \Omega$?

5. How much power would the record player stylus of Problem 4 have to put out if it had to drive an amplifier with the same input impedance as a 741C op amp?

6. What must be the voltage gain of the stereo amplifier of Problem 4 if it delivers 50 W to an 8 Ω speaker?

7. Find the input voltage required to produce an output voltage of 10 V
   (a)   From a 741C op amp.
   (b)   From an ideal op amp.

Section 19.5

8. Draw a load line for the circuit of Figure 19.19. *Hint:* What is the internal resistance of a current source?

9. Determine the value of V_{CE} for the circuit of Figure 19.19.

10. Redraw the circuit of Figure 19.2 replacing all current mirror circuits with the symbol for a constant current source.

11. Repeat Example 19.1 if ±9 V power supplies are used.

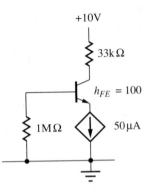

FIGURE 19.19   Problem 19.8.

Section 19.6

12. Determine the value of I_{C3} and I_{C4} in Example 19.1.

13. Determine the value of V_{C5} in Example 19.1 if both inputs are grounded.

14. Repeat Example 19.3 assuming that the emitter follower circuits each have a voltage gain of 0.95.

Section 19.9

15. Sketch the output waveform that would probably result when a sine wave of small amplitude is input to a 741 op amp that has the collector and emitter of transistor Q_{18} shorted together.

16. Determine the Q point current, I_{C14} flowing in the output circuit of a 741C if ±15 V power supplies are used.

17. Using ±9 V power supplies for a 741C op amp, estimate the base bias voltage, $V_{B14} - V_{B20}$ for the output circuit.

Section 19.10

18. If ±9 V power supplies are used, determine the short circuit positive output current for the circuit of Figure 19.2. Compare the result with that of Example 19.5.

Section 19.11

19. Determine the frequency at which a 741C has an open loop voltage gain of 100.
20. Find the open loop gain of a 741C operating at a frequency of 60 Hz.
21. Calculate the unity gain bandwidth for a 741C from the rise time listed in Appendix F.

Section 19.12

22. Suppose that a step voltage is input to a 741C. Determine the time required for the output to increase from 0 V to $0.7 * V^+$.

Section 19.13

23. What is the maximum value of resistance that should be connected to the input of a 741C if it is to be considered ideal?
24. Determine the minimum value of resistance that can be connected to the output of a 741C that is powered by ±15 V supplies if it is to function as an ideal op amp. *Hint:* Equation 19.16 is not the only requirement. See Section 19.10.

Section 19.16

25. Draw the circuit for an inverting zero crossing detector and draw its input and output waveforms on the same time base as was done in Figure 19.17.

Section 19.17

26. Draw the circuit for a noninverting voltage level detector whose reference voltage can be set to any voltage from zero volts to the V^+ supply.
27. Battery-powered smoke detectors typically contain a battery testing circuit that will sound an alarm when the battery voltage drops below about 7 V. Design such a test circuit using a 741C. *Hint:* Use a Zener diode to provide a reference voltage.

Section 19.18

◀ **Troubleshooting**

28. Determine the maximum signal voltage that should exist between Pins 2 and 3 of a 741C for a 10 V output.
29. Describe how using an op amp with some input offset could adversely affect the behavior of a zero-crossing detector.
30. Describe the behavior of the trace on your oscilloscope if an op amp with a low slew rate were used in the level detector that triggers the horizontal trace.

Feedback containing human and electronic components. (Photograph by David Powers/Stock, Boston.)

Chapter 20
Negative Feedback Circuits

SPECIAL TERMS

Open loop gain
Feedback
Closed loop gain
Virtual ground
Differentiator
Integrator
Ramp generator
Buffer

20.1   Introduction and Objectives

In Chapter 19 we looked at the characteristics of ideal op amps and some of the conditions under which real op amps can be considered to be ideal, but we did not use them as amplifiers. In this chapter you should get a better understanding of op amps and how their characteristics can be used to advantage in building both inverting and noninverting amplifiers.

After having studied this chapter you should have a reasonably good idea how the classic ideal op amp model works and how it can be used with real circuits. You should be able to:

- Design, analyze, and troubleshoot simple inverting and noninverting amplifier circuits, voltage followers, summing amplifiers, differentiators and integrators.

20.2   Feedback Defined

Life could not exist without feedback. Your ears provide feedback to let you know if your voice sounds right. That is why totally deaf people usually cannot speak either. When you reach for a glass of whatever you reach for, your eyes provide feedback to tell your brain if your hand is on target or not. You may have seen cases where too much of whatever has damaged the feedback loop.

Feedback is an essential part of most op amp circuits and the reason why they are useful. You will find that feedback changes the overall gain of a circuit. That is why in the last chapter we had to distinguish between closed loop gain, A_{CL}, and open loop gain, A_{OL}.

We first mentioned feedback in Section 15.14. There we defined feedback as the return of output information back to the input. If, as the input signal tends to increase, the signal that is fed back tends to decrease, the result is to reduce the net output. We would say that what is fed back is out of phase with the input, and that the feedback is negative, or degenerative.

On the other hand, if, as the input signal tends to increase, the signal that is fed back also tends to increase, you can see that the result is to increase the net output. The feedback is then said to be in phase with the input, and is positive, or regenerative feedback. As you might expect, this usually is an unstable condition which we usually try to avoid, except for certain kinds of trigger circuits and oscillators which we will discuss in Chapter 22.

20.3 The Classic op Amp Circuit

The circuit of Figure 20.1 is sometimes called the classic op amp circuit because it is the first and most commonly used op amp circuit. Further, most other op amp circuits are just variations of this circuit. Since the input is to the inverting terminal of the op amp, it should come as no surprise that it is an inverting circuit. Resistor R_f is known as the feedback resistor because it feeds part of the output back to the input. It constitutes the "feedback loop." It lets the input "feel" and be influenced by the output. Since the feedback is to the inverting input, it is degenerative. Thus the closed loop gain, A_{CL}, will be less than the open loop gain, A_{OL}, of the op amp alone.

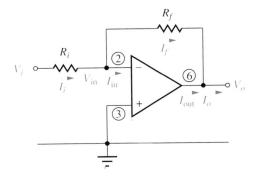

FIGURE 20.1   The classic op amp circuit.

The notations that we will use to describe the circuit and its behavior will be fairly consistent and standard. Most of the variables we will be talking about are indicated on Figure 20.1 and are self-explanatory. The pin numbers are those used for the standard 741 package. Occasionally we will reference voltages or currents to those pin numbers. Obviously, $V_{in} = V_2$. Notice that you will need to make a distinction between V_i and V_{in}. They are NOT equal.

For the most part, we will assume that ideal op amp conditions apply. Thus

$$R_{in} \approx \infty \ \Omega \qquad (19.1)$$

and

$$A_{OL} \approx \infty \qquad (19.2)$$

Looking at Figure 20.1, you should see that Equation 19.1 implies

$$I_{in} \approx 0 \text{ A} \qquad (20.1)$$

Also, if you remember what we said in Chapter 19 about the comparator nature of op amps and how differential amplifiers work, looking at the circuit of Figure 20.1, you can see that

$$V_{in} = V_2 - V_3$$

Then from Equation 12.17 where the definition of voltage gain was given, we can say

$$V_{out} = A_{OL}V_{in}$$

Therefore, you can see that the only way to prevent the output from going to saturation is to keep V_{in} very small. In fact, although the mathematicians and your calculator may not like it, if $A_{OL} = \infty$, then $V_{in} = 0$ V, or $V_2 - V_3 = 0$ V. Of course, in practice, A_{OL} is never truly infinite and so V_{in} will not be exactly zero, but it will be close. We often say that V_{in} is "virtually" zero. Then Equation 19.2 implies that

●
$$V_2 \stackrel{\wedge}{\approx} V_3 \tag{20.2}$$

We will use the "hat" (^) over the equal sign in Approximation 20.2 to remind us that the op amp will "try" to maintain the relationship, but if it cannot due to saturation or for some other reason, then Equation 20.2 and all equations derived from it are invalid.

Almost everything we do with op amps will make use of Approximations 20.1 and 20.2.

Due to Approximation 20.2, we can say that the voltage at pins 2 and 3 are VIRTUALLY equal. Not EXACTLY equal (do not hook a wire between them!) but VIRTUALLY equal. Since pin 3 in Figure 20.1 is grounded, we often say that pin 2 is a "virtual ground." We will return to this idea shortly.

Before doing a formal analysis of the circuit, it might be worth your time to get a feeling for what is "really" happening in Figure 20.1. The instant that you apply a positive voltage, V_i, electric charges begin to flow through R_i. Since $R_{in} = \infty$, charges "begin to pile up" at pin 2, and so to generate a positive voltage, V_{in}. The presence of V_{in} causes V_o to start going very negative very quickly since V_{in} goes to the inverting input and the gain is big. The output continues to go more negative as long as there is a positive voltage at V_{in}. As V_o becomes increasingly negative, it pulls an increasing current through R_f and so it removes charges from pin 2 at an ever-increasing rate. When V_o reaches the place where it is removing every charge from pin 2 that V_i pushes through R_i then V_o no longer increases and equilibrium is established. Obviously, if V_o cannot reach the required voltage due to saturation, or for some other reason, then it cannot prevent a voltage buildup at pin 2. This is when Equation 20.2 becomes invalid.

Our quantitative analysis of the circuit of Figure 20.1 will begin with the KCL equation for pin 2 of the op amp

$$I_i = I_f + I_{in} \qquad (20.3)$$

but

$$I_{in} \approx 0 \text{ A} \qquad (20.1)$$

so

$$I_i \approx I_f \qquad (20.4)$$

Applying Ohm's Law to I_i and I_f in Approximation 20.4 gives

$$\frac{V_i - V_{in}}{R_i} = \frac{V_{in} - V_o}{R_f} \qquad (20.5)$$

Since $V_{in} = V_2$ and since we have just established that for Figure 20.1, V_2 is a virtual ground, then

$$V_{in} \overset{\wedge}{=} 0 \text{ V} \qquad (20.6)$$

Again the "hat" over the equal sign in Equation 20.6 indicates that the op amp will try to make the equation true, but if, due to saturation or for some other reason, it cannot, then everything we are about to do will be invalid.

Substituting Equation 20.6 into Equation 20.5 yields the classic op amp equation

$$\bullet \qquad \frac{V_i}{R_i} \overset{\wedge}{=} -\frac{V_o}{R_f} \qquad (20.7)$$

Or, since voltage gain, A_v is defined as

$$A_v \equiv \frac{V_o}{V_i} \qquad (12.17)$$

we have

$$\bullet \qquad A_{CL} \overset{\wedge}{=} -\frac{R_f}{R_i} \qquad (20.8)$$

Notice that the hat warns us that for the preceding derivation to be valid, two conditions must be met. They are $I_i = I_f$, and $V_{in} = 0$ V.

Equation 20.8 is worth a second look because it says that the gain of the circuit of Figure 20.1 is dependent only on the external resistors used and is independent of the op amp itself. The gain of every other amplifier circuit we have studied has been critically dependent on some device parameter such as h_{fe}, or g_{os}, which was typically not known very accurately.

The gain for this circuit is not dependent on any device parameters because of the infinities in Equations 19.1 and 19.2. As long as R_{in} and A_{OL} are big, their actual values are irrelevant. Another feature worth noting about Equation 20.8 is the minus sign. It reminds us that the circuit of Figure 20.1 is an inverting amplifier.

EXAMPLE 20.1

Determine V_o for the circuit of Figure 20.2 and show that in this case it is reasonable to neglect V_{in} and I_{in}.

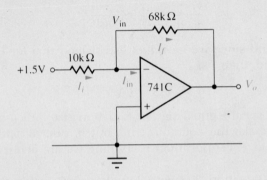

FIGURE 20.2   Example 20.1.

SOLUTION:

Solving Equation 20.7 for V_o gives

$$V_o \hat{=} -V_i \frac{R_f}{R_i} \tag{20.9}$$

$$= -1.5\ \text{V} \frac{68\ \text{k}\Omega}{10\ \text{k}\Omega} = -10.2\ \text{V}$$

According to the data sheet in Appendix F, for the 741C, $R_{in} = 2\ \text{M}\Omega$ and $A_V = 200{,}000$. From the definition of A_V given in Section 12.8 we can write

$$V_{in} = \frac{V_o}{A_V}$$

$$= \frac{-10.2\ \text{V}}{-200{,}000} = 51.0\ \mu\text{V} \qquad (0.000051\ \text{V!!})$$

Since V_{in} is negligible compared to V_i and V_o, it seems reasonable to drop it from Equation 20.5.

If V_{in} is ignored we have

$$I_i = \frac{V_i}{R_i}$$

$$= \frac{1.5 \text{ V}}{10 \text{ k}\Omega} = 150 \text{ μA}$$

Applying Ohm's Law to the op amp input gives

$$I_{in} = \frac{V_{in}}{R_{in}}$$

$$= \frac{51.0 \text{ μV}}{2 \text{ M}\Omega} = 25.5 \text{ pA} \qquad (0.0000254 \text{ μA!!})$$

Since I_{in} is negligible compared to I_i, it seems reasonable to drop it from Equation 20.3.

From Example 20.1 you should see why, as we said above, pin 2 is often called a VIRTUAL ground. From our perspective, its voltage is negligibly small, but it cannot be EXACTLY ground since it is that very small voltage multiplied by the very high (infinite?) gain of the op amp that determines V_o.

20.4 Frequency Characteristics

In Chapter 19 we mentioned the fact that the gain of the 741 is attenuated at high frequencies. Its poor high frequency response is one of the greatest disadvantages of the 741.

EXAMPLE 20.2
Find the high frequency cutoff, f_H, for the circuit of Example 20.1.

SOLUTION:
The closed loop voltage gain of the circuit of Figure 20.2 is

$$A_{CL} = \frac{R_f}{R_i} \qquad (20.8)$$

$$= \frac{68 \text{ k}\Omega}{10 \text{ k}\Omega} = 6.8$$

According to Appendix F, the unity gain bandwidth for the 741C is $BW = 1.0$ MHz. Solving Equation 19.18 for the high frequency cutoff, f_H at the required gain gives

$$f_H = \frac{BW}{A_{CL}}$$

$$= \frac{1.0 \text{ MHz}}{6.8} = 147 \text{ kHz} \qquad (20.10)$$

20.5 Inverting Amplifier Input Impedance

At first you might be surprised that we need to worry about the input impedance of an op amp circuit, since one of the ideal op amp assumptions is that $R_{in} = \infty \ \Omega$. Remember, however that for the classic circuit of Figure 20.3, $V_2 \hat{=} 0$ V. Pin 2 is a virtual ground. That means that the only resistance between V_i and (virtual) ground is R_i. Thus the input impedance is simply R_i. As we saw in Chapter 19, for circuits using a 741, if we are to have a reasonable gain, we are restricted to a value for R_i of no more than around 10 kΩ. Thus the classic op amp circuit has a relatively low input impedance.

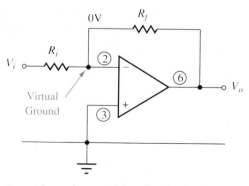

FIGURE 20.3   Input impedance of the classic circuit.

20.6 Inverting Summing Amplifier

Occasionally an amplifier must combine several inputs. Such an amplifier is often called "a mixer." An example is the sound system used by a musical group. It is not uncommon for each sound source to have its own microphone so that the producer can control the various sources to achieve the desired balance. Another example is your car's fuel system. In order to provide the correct amount of fuel, the injector control needs to know such things as throttle position, and engine speed, loading, and temperature.

In both of the above cases several inputs must be combined to produce a single composite output. The summing amplifier shown in Figure 20.4 is a convenient way to achieve this result.

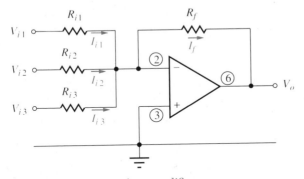

FIGURE 20.4   Inverting summing amplifier.

Notice that no special precaution needs to be taken to isolate the various inputs from one another. Due to the virtual ground at pin 2 there is no tendency for the signal from one input to affect any of the other inputs.

The ideal inverting summing amplifier is a simple extension of the classic op amp circuit. To analyze the circuit we again start by writing the KCL equation for pin 2. Since $I_{in} = 0$ A we have

$$I_{i1} + I_{i2} + \cdots = I_f \tag{20.11}$$

Applying Ohm's Law gives

$$\frac{V_{i1}}{R_{i1}} + \frac{V_{i2}}{R_{i2}} + \cdots \overset{\wedge}{=} -\frac{V_o}{R_f} \tag{20.12}$$

or

$$V_o \overset{\wedge}{=} -\left(V_{i1}\frac{R_f}{R_{i1}} + V_{i2}\frac{R_f}{R_{i2}} + \cdots \right) \tag{20.13}$$

EXAMPLE 20.3

Suppose that a function generator is to be designed that will have a variable AC output of approximately 10 VPP maximum riding on top of a variable DC offset of approximately ±5 VDC maximum. You may have such a function generator in your laboratory. Design a mixer circuit using a 741 op amp that will produce the desired output from a 1 VPP AC source and fixed DC sources of +18 VDC and −18 VDC.

SOLUTION:

The circuit to be designed is shown in Figure 20.5. A 10 kΩ pot is chosen for the offset control, R_{S1}, on the basis that it is small enough to act as a voltage source to the input, and yet large enough that it

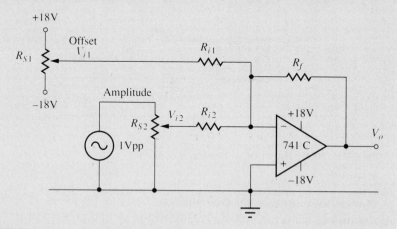

FIGURE 20.5 Example 20.3.

will not load the supplies excessively. A 100 kΩ resistor is chosen for R_{i1} on the basis that it is large compared to R_{S1} and yet it is small compared to R_{in}. Then from Equation 20.7

$$R_f = R_{i1}\frac{V_o}{V_{i1}} \tag{20.14}$$

$$= 100 \text{ k}\Omega \, \frac{5 \text{ V}}{18 \text{ V}} = 27.8 \text{ k}\Omega \qquad (\text{Use } 27 \text{ k}\Omega)$$

Solving Equation 20.7, for R_i gives

$$R_{i2} = R_f \frac{V_{i2}}{V_o} \tag{20.15}$$

$$= 27 \text{ k}\Omega \, \frac{1 \text{ VPP}}{10 \text{ VPP}} = 2.7 \text{ k}\Omega \qquad (\text{Use } 2.7 \text{ k}\Omega)$$

Finally a 1 kΩ pot is chosen for the amplitude control, R_{S2}, on the basis that it is large enough so that it will not place any more of a load on the 1 VPP source than necessary and yet it is reasonably small compared to R_{i2}.

20.7 Differentiator Circuit

Early in your circuits class, you probably learned that electric current is defined as

$$I \equiv \frac{dQ}{dt} \tag{20.16}$$

At that time you probably did not know enough about calculus to do more than recognize that Equation 20.16 meant that current is the rate at which charges pass a given observation point in a circuit. Hopefully, between then and now you have learned a lot. Also from your circuits class you should recall that capacitance is defined as

$$C \equiv \frac{Q}{V} \tag{20.17}$$

and that solving Equation 20.17 for Q and finding the derivative gives the current into a capacitor as

$$I_C = \frac{dQ}{dt} = C\frac{dV}{dt} \tag{20.18}$$

which is just the mathematician's way of saying that current flowing into a capacitor results in an increase of charge on it, and that causes an increase in voltage across it. The actual value of the voltage is not important. What is important is that the current is directly proportional to the rate of change of the voltage. Thus we can use a capacitor to build a circuit that is sensitive to the rate of change of voltage. That is what is known as a "differentiator circuit." A differentiator circuit is shown in Figure 20.6.

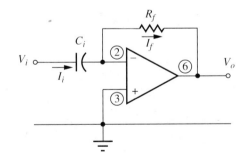

FIGURE 20.6 Differentiator circuit.

As usual, the analysis of the circuit starts with the KCL equation for pin 2. Assuming ideality, we have

$$I_i = I_f \tag{20.4}$$

Substituting I_C from Equation 20.18 for I_i and using Ohm's Law to evaluate I_f gives

$$C_i \frac{dV_i}{dt} = -\frac{V_o}{R_f} \tag{20.19}$$

or

$$V_o = -R_f C_i \frac{dV_i}{dt} \tag{20.20}$$

Although it would be interesting to spend more time on differentiators, they are not used very much in practice. You can see why if you think about how some high frequency noise in V_i would affect the circuit of Figure 20.6. The higher the frequency, the more C_i acts like a coupling capacitor. Thus C_i would couple the high frequency noise directly into the op amp where it would be greatly amplified (∞?).

The output of a differentiator is typically so noisy it is useless. Fortunately most situations where a differentiator might be used can usually be rearranged so that an integrator can be used instead.

20.8   Integrator Circuit

The circuit of Figure 20.7 is an integrator. The only difference between a differentiator and an integrator is that the capacitor and the resistor have exchanged places. Notice that in this case there is no DC feedback in the

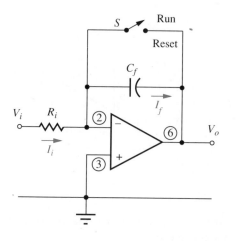

FIGURE 20.7   Integrator circuit.

circuit. This means that if there is any leakage or offset current in the op amp, and there almost always is some, then V_o tends to drift either up or down to saturation. If you work with integrator circuits in the lab, and you should, you should be alert to this tendency. The purpose of the reset switch, S, in Figure 20.7 is to establish the initial conditions by discharging C_f and setting the initial output voltage, V_{oo}, to zero volts.

The principle of operation of the integrator is as follows: if S is opened at $t = 0$ sec, assuming that V_i is positive, charges flowing through R_i begins to produce a positive voltage at pin 2. This causes V_o to go sufficiently negative to attract those charges to C_f where they are stored. As charge builds up on C_f, V_o must go increasingly negative in order to continue removing the charge from pin 2. Obviously the op amp will "give up" when V_o reaches saturation.

The voltage across C_f is proportional to the charge on it. Since the charge on C_f is the summation (the mathematician would say "the integral" of all the charge that has flowed through R_i and since that rate of flow is dependent on V_i, V_o is the integral of V_i over time.

Since differentiator and integrator circuits are similar, the equations that describe their behavior are similar too. Just as before, the analysis of the integrator circuit begins with the KCL equation for pin 2. Assuming ideality we have

$$I_i = I_f \tag{20.4}$$

This time the Ohm's Law equation replaces I_i and Equation 20.18 replaces I_f. This gives

$$\frac{V_i}{R_i} = -C_f \frac{dV_o}{dt} \tag{20.21}$$

or

$$V_{ox} = V_{oo} - \frac{1}{R_i C_f} \int_0^{t_x} V_i \, dt \tag{20.22}$$

When V_i is constant, such as the circuit of Figure 20.8a, the output is linear as shown in Figure 20.8b, and the circuit is known as "a ramp generator." Equation 20.22 can then be solved for $V_{ox} - V_{oo}$, which is the output voltage interval over which the circuit integrates in the integration time t_x. The result is

$$V_{ox} - V_{oo} = -V_i \frac{t_x}{R_i C_f} \tag{20.23}$$

The horizontal sweep generator of your oscilloscope and the horizontal scan circuit of your TV both use this type of circuit.

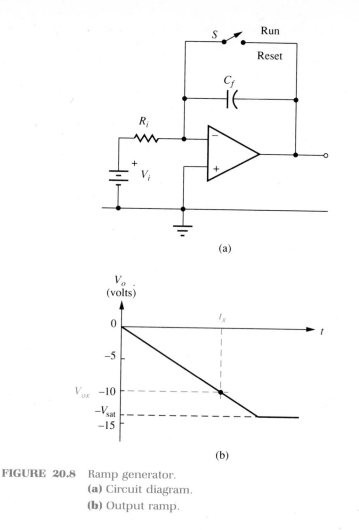

FIGURE 20.8   Ramp generator.
(a) Circuit diagram.
(b) Output ramp.

EXAMPLE 20.4

For the circuit of Figure 20.9a, every time V_o reaches $+5$ V, S is flipped to position 1 and every time V_o reaches -5 V, S is flipped to position 2. Determine the value of R_i that will produce an output frequency of 100 Hz. Draw the waveform.

SOLUTION:

The period of the output waveform at the required frequency is

$$T \equiv \frac{1}{f}$$

$$= \frac{1}{100 \text{ Hz}} = 10 \text{ msec}$$

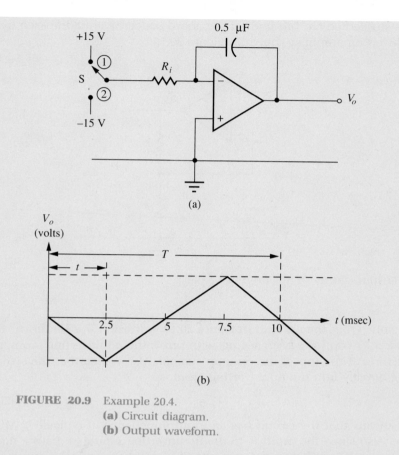

FIGURE 20.9 Example 20.4.
(a) Circuit diagram.
(b) Output waveform.

Thus the time for the waveform to go from $V_o = 0$ V to $V_o = 5$ V is $t = 2.5$ msec. The waveform is shown in Figure 20.9b. From Equation 20.23 we have

$$R_i = -\frac{V_i t}{V_{ot} C_f}$$

$$= -\frac{15 \text{ V} * 2.5 \text{ msec}}{-5 \text{ V} * 0.5 \text{ μF}} = 15 \text{ k}\Omega \qquad \text{(Use 15 k}\Omega\text{)}$$

20.9 Noninverting Amplifier

You may have found Section 20.5 a little frustrating. It seems only reasonable that the circuit designer should be able to capitalize on the very high input impedance of the op amp to construct high input impedance circuits. Also, since the op amp has differential inputs, the designer should be able to build noninverting circuits. Neither are possible with the circuits studied so far, but both of the above concerns are addressed by the noninverting amplifier shown in Figure 20.10.

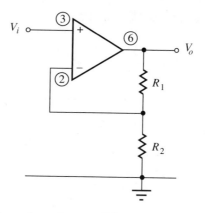

FIGURE 20.10   A noninverting amplifier.

Notice that the op amp in Figure 20.10 is "turned upside down" from the circuits you have been seeing, with pin 3, the noninverting input, now appearing at the top of the op amp symbol. Also notice that the signal is going directly into the noninverting input, so

$$V_i = V_3 \tag{20.24}$$

That means that the circuit has an input impedance of typically 2 MΩ or more. Also since the input is to the noninverting input, we have a noninverting amplifier. That is, the output will be in phase with the input.

You should be warned that the above advantages are not without a price. In the case of the inverting amplifier studied earlier, since the noninverting input is grounded, the gain of that portion of the diff amp inside of the op amp is not too important. Zero volts times any gain is still zero. That is not the case for the noninverting amplifier. Since a nonzero voltage is applied to both inputs of the op amp, in order for it to carry out its comparator function well, the input differential amplifier inside the op amp must be exactly symmetrical. This is one reason why the performance of a noninverting amplifier is typically not quite as good as an inverting amplifier.

The operation of the noninverting circuit depends on the comparator property of the op amp. It "wants" (note the hat) to maintain the condition

$$V_2 \stackrel{\wedge}{=} V_3 \tag{20.2}$$

or

$$V_2 \stackrel{\wedge}{=} V_i \tag{20.25}$$

If V_i increases, the op amp drives V_o toward $+V_{sat}$ ($A_V = \infty$). As V_o increases, V_2 also increases according to the voltage divider relation

$$V_2 = V_o \frac{R_2}{R_1 + R_2} \tag{20.26}$$

When $V_2 = V_3$ again, equilibrium is reestablished and V_o ceases to rise.

Equation 20.26 tells us what V_o will be in terms of V_i. Substituting Equation 20.25 into Equation 20.26 and solving for V_o gives

$$V_o \stackrel{\wedge}{=} V_i \frac{R_1 + R_2}{R_2} \tag{20.27}$$

Thus the voltage gain of the noninverting amplifier is

$$A_v \stackrel{\wedge}{=} \frac{R_1 + R_2}{R_2} \tag{20.28}$$

Again the hat warns you that this is what the op amp "wants" to do.

In contrast with Equation 20.8, there is no negative sign in Equation 20.28, indicating that it is, in fact, a noninverting circuit.

EXAMPLE 20.5

The phono cartridge of a record player has a Thevenin equivalent of $V_{Th} = 0.1$ VPP and $R_{Th} = 100$ kΩ. Design a preamp for it that will have a high frequency cutoff of $f_H = 20$ kHz using a 741C op amp.

SOLUTION:

A noninverting amplifier is the best amplifier to use in order not to load the record player. The circuit connected to the Thevenin equivalent of the record player is shown in Figure 20.11. From Appendix F,

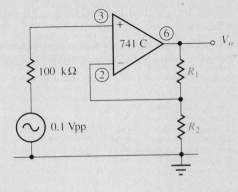

FIGURE 20.11   Example 20.5.

for the 741C $f_U = 1$ MHz. Solving Equation 19.16 for the maximum circuit gain gives

$$A_{CL} = \frac{BW}{10 * f_H}$$ (20.29)

$$= \frac{1 \text{ MHz}}{10 * 20 \text{ kHz}} = 5$$

Again from Appendix F, the input impedance of the 741C is about $R_{in} = 2$ MΩ. Note that this is not the input resistance of the amplifier, which will be much higher due to the circuit configuration, and which would be infinite if we assume an ideal op amp. Rather it is a very conservative estimator of the value of R_2 in order that pin 2 will not load the voltage divider. We will be safe if we choose R_2 thus:

$$R_2 = \frac{R_{in}}{10}$$ (20.30)

$$= \frac{2 \text{ M}\Omega}{10} = 200 \text{ k}\Omega \qquad \text{(Use } 180 \text{ k}\Omega\text{)}$$

Solving Equation 20.28 for R_1

$$R_1 = A_V R_2 - R_2$$ (20.31)

$$= 5 * 180 \text{ k}\Omega - 180 \text{ k}\Omega = 720 \text{ k}\Omega \qquad \text{(Use } 680 \text{ k}\Omega\text{)}$$

You may find the voltage gain in the previous example a little disappointing. As we mentioned in Section 19.11, a 741 is not a high frequency device.

20.10 Voltage Follower Circuit

The circuit of Figure 20.12 is known as a voltage follower, a source follower, unity gain amplifier, a buffer amplifier, or an isolation amplifier, depending on the context in which it is used. It is a noninverting amplifier with 100 percent feedback.

You can see how each of the above names is appropriate if you remember the behavior of a noninverting amplifier. Given the comparator nature of an op amp, it tries to keep $V_2 = V_3$. Since $V_i = V_3$ and $V_o = V_2$ it

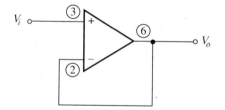

FIGURE 20.12 A voltage follower.

is easy to see that the output voltage will follow the source voltage. It is also obvious that it will have a gain of unity. Also, given the high input impedance of the circuit, it is easy to see how it could be used as a buffer, or an isolator to prevent a load from affecting, or "talking back" to the input circuit.

EXAMPLE 20.6
Construct an "ideal" voltage source of 7.5 V from a 15 V source if the load to be placed on it will vary from open circuit down to 1 kΩ.

SOLUTION:
We will use the same reasoning as we did in Example 20.5 for the choice of R_1. The input impedance of the op amp in the circuit of Figure 20.13 is about 2 MΩ. Thus the voltage divider will not be loaded,

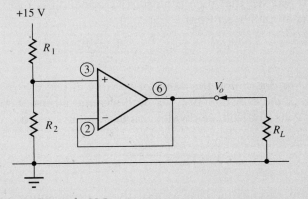

FIGURE 20.13   Example 20.6.

and it will provide an input voltage of 7.5 V to the op amp if $R_1 = R_2 = 100$ kΩ. The voltage follower will act as an effective buffer amplifier so that the voltage divider will be unaffected by the load, R_L. Therefore the output will be constant at $V_o = V_i = 7.5$ V as long as the output current requirement is less than about 25 mA.

20.11 Noninverting Summing Amplifier

In Section 20.6 you learned about the inverting summing amplifier. It is possible to construct a noninverting summing amplifier as shown in Figure 20.14. The analysis of the circuit is somewhat tedious since there are a lot of parallel circuits, but no new concepts are involved.

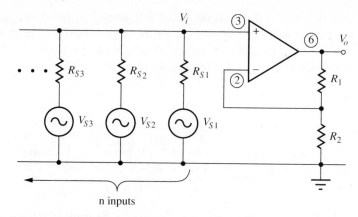

FIGURE 20.14 Noninverting summing amplifier.

You can see that the op amp in Figure 20.14 is part of a simple non-inverting amplifier, so from Section 20.9

$$V_o = V_i \frac{R_1 + R_2}{R_2} \tag{20.27}$$

In order to determine the value of V_i, we will use the principle of super-position. If we consider only V_{S1} and assume all other source voltages to be zero volts, the voltage divider principle allows us to say

$$V_{i1} = V_{S1} \frac{R_{S2} \parallel R_{S3} \parallel \cdots}{R_{S1} + R_{S2} \parallel R_{S3} \parallel \cdots} \tag{20.32}$$

or, in the simplifying case where there are n sources and n identical source resistors each having a value of R,

$$V_{i1} = V_{S1} \left(\frac{\dfrac{R}{n-1}}{R + \dfrac{R}{n-1}} \right)$$

$$= V_{S1} \frac{R}{R(n-1) + R}$$

$$= V_{S1} \frac{1}{n} \tag{20.33}$$

The principle of superposition allows us to determine the value of V_i by adding all of the inputs

$$V_i = (V_{S1} + V_{S2} + \cdots)\frac{1}{n} \tag{20.34}$$

The noninverting summing amplifier is not a particularly good circuit for several reasons. One is that the high input impedance of the noninverting circuit is spoiled due to the fact that all of the inputs are in parallel. Another problem is that since pin 3 of Figure 20.14 is not a virtual ground as pin 2 was for the inverting amplifier of Figure 20.4, the inputs "talk to each other." That is, the signal from one source shows up at the other sources.

Thus, if several high impedance inputs are desired, it is usually better to use a single input noninverting amplifier for each input when high input impedance is required, and then mix them using an inverting summing circuit as shown in Figure 20.15. This allows the added advantage of having separate gain controls on each of the input amplifiers in addition to having a master gain control on the summing amplifier.

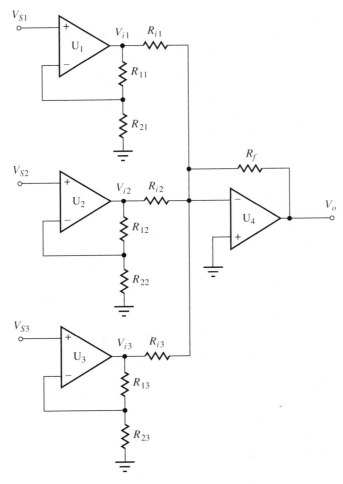

FIGURE 20.15   An improved high impedance mixer.

▶▶▶▶▶▶▶▶   **20.12   Troubleshooting**

As we mentioned in Chapter 19, a good quick in-circuit test of an op amp is to hook a jumper wire around the feedback resistor so that the op amp circuit becomes a voltage follower and then check to see if $V_o = V_3$. If it does, the chances are good that the op amp is functional. If the gain is either much higher than expected or much lower than expected, the feedback loop should be the first place to look for trouble. If R_f is open in the inverting circuit, or if R_1 is open in the noninverting circuit, the gain will be essentially the open loop gain of the op amp. If R_2 in the noninverting circuit is open, the gain will be practically one.

20.13   Summary

In this chapter you were introduced to the most commonly used op amp circuits. In the process your understanding of op amp behavior should have been enhanced. You will find that practically any op amp circuit you will encounter in the future is only an adaptation of the basic ideas presented in this chapter. The following chapters will build on what you should have learned in this chapter.

GLOSSARY

Buffer: An isolating circuit used to prevent undesirable interaction between two parts of a circuit. (Section 20.10)

Closed loop gain: A_{CL} The gain of a circuit or device that contains feedback. (Section 20.2)

Differentiator: A circuit whose output is proportional to the derivative, or the time rate of change of its input signal. (Section 20.7)

Feedback: The return of information from the output back to the input. (Section 2.02)

Integrator: A circuit whose output is the time integral of its input signal. (Section 20.8)

Open loop gain: A_{OL} The gain of a circuit or device that does not contain feedback. (Section 20.2)

Ramp generator: A waveform generator that produces a linear or a triangular-shaped output. (Section 20.8)

Virtual ground: A point in a circuit where the voltage is always so close to ground voltage compared with the other circuit voltages that it can be considered to be grounded. (Section 20.3)

PROBLEMS

Section 20.2

1. When driving on the freeway, feedback helps keep my car traveling at a fairly constant speed. Identify the important components of the feedback loop
   (a) When the cruise control is being used.
   (b) When the cruise control is not being used.
2. As you ride your bicycle, feedback helps to keep you from falling over. Identify the important parts of the feedback loop.

Section 20.3

3. Design a classic op amp circuit that will have a gain of 100 and as high an input impedance as possible. Use a 741C.
4. Determine the voltage gain and the input impedance of the circuit shown in Figure 20.16.

Section 20.4

5. Determine f_H for the circuit of Problem 3.
6. Determine f_H for the circuit shown in Figure 20.16.

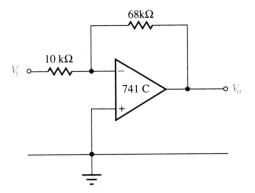

FIGURE 20.16 Problem 20.4.

Section 20.5

7. Design a classic op amp circuit using a 741C. Design for as high an input impedance as possible, as much gain as possible and a high frequency cutoff of 10 kHz.
8. If $+15$ V and -15 V power supplies are used for the circuit of Problem 7, determine the maximum peak-to-peak input signal before clipping will be seen in the output.

Section 20.6

9. Using a 741C op amp in an inverting summing amplifier, design a circuit that will mix the output from a guitar and a microphone and will produce approximately equal volume from each source. Design the circuit for maximum possible input impedance at each input and so that the output will not clip when both sources are operated at full volume if the power supplies for the 741 are $+15$ V and -15 V. Assume that the peak amplitude from the guitar is 1 VP, and the peak amplitude from the microphone is 100 mVP.

10. Show that with each input control in Example 20.3 set for maximum voltage, the output will not reach saturation. Draw the output waveform.

Section 20.7

11. For the differentiator circuit of Figure 20.17a, determine the maximum output voltage for the input shown in Figure 20.17b. Draw the output waveform.

12. Determine the peak value of the output noise voltage for the circuit of Figure 20.17a if there is 10 mV of 5 kHz sine wave noise at the input. *Hint:* Determine X_C and use Equation 20.9.

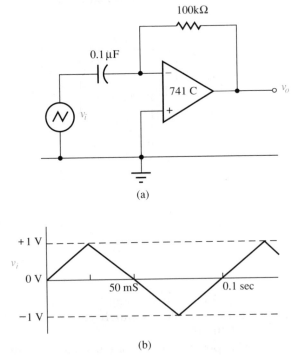

(a)

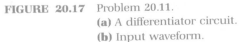

(b)

FIGURE 20.17   Problem 20.11.
(a) A differentiator circuit.
(b) Input waveform.

Section 20.8

13. Determine the output voltage for the circuit of Figure 20.18 one second after switch S is opened and two seconds after switch S is opened.

14. Determine the capacitor size needed in the circuit of Figure 20.18 in order that the output voltage will reach $V_o = +12$ V two seconds after switch S is opened.

15. Determine the peak value of the output noise voltage for the circuit of Figure 20.18 if there is 10 mV of 5 kHz sine wave noise at the input. *Hint:* Determine X_C and use Equation 20.9.

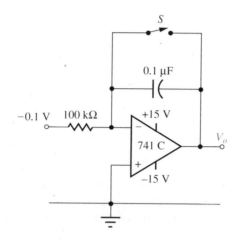

FIGURE 20.18   Problem 20.13.

Section 20.9

16. Using a 741C, design a noninverting amplifier that will have a gain of $A_v = 200$. Estimate f_H for the circuit.

17. Determine the voltage gain for the circuit of Figure 20.19.

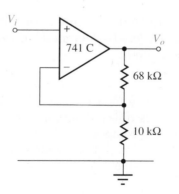

FIGURE 20.19   Problem 20.17.

Section 20.10

18. Determine the voltage gain of the circuit in Figure 20.20 when the pot is set at

   (a)   The top of its travel: $R_1 = 0\ \Omega$ $R_2 = 10\ k\Omega$
   (b)   The center of its travel: $R_1 = 5\ k\Omega$ $R_2 = 5\ k\Omega$
   (c)   The bottom of its travel: $R_1 = 10\ k\Omega$ $R_2 = 0\ \Omega$

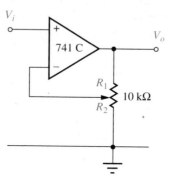

FIGURE 20.20   Problem 20.18.

19. Describe the effect on V_o if, for the circuit of Figure 20.16,
   (a)   The 10 kΩ resistor became an open circuit.
   (b)   The 68 kΩ resistor became an open circuit.
20. Describe the effect on V_o if, for the circuit of Figure 20.19,
   (a)   The 10 kΩ resistor became an open circuit.
   (b)   The 68 kΩ resistor became an open circuit.

Section 20.11

21. Determine the value of V_o for the circuit of Figure 20.14 if $n = 3$, and all of the resistors have a value of 10 kΩ and all of the inputs have a value of 10 mV.

Troubleshooting ▶   ### Section 20.12

22. If $V_i = 10$ mV in Figure 20.16, determine the voltage reading at Pins 2 and 6 of the op amp if a "cold solder joint" made the following resistor appear to be an open circuit:
   (a)   The 10 kΩ resistor.
   (b)   The 68 kΩ resistor.
23. For the circuit of Figure 20.20, if $V_i = 10$ mV and the slider of the pot is set at the center of its travel, determine the effect on V_o if a "dirty pot" condition caused R_1 to appear to become an infinite resistance.
24. Describe the effect on the output of the circuit of Figure 20.18 if the op amp should develop a significant amount of zero offset.

Instrumentation amplifiers in use. (Photograph by Eric Neurath/Stock, Boston.)

Chapter 21
Negative Feedback
Applications

SPECIAL TERMS

Current boost transistor
Series pass transistor
Voltage regulation
Current limiter
Head room
Precision rectifier
Differential signal
Common mode signal
Common mode rejection ratio

21.1 Introduction and Objectives

In the last chapter you should have learned how negative feedback is used in op amp circuits to control the gain of the circuit. In this chapter we will look at some interesting applications.

After having studied this chapter you should have a better understanding of the concept of degenerative feedback and how it can be used to advantage in practical circuit applications. You should be able to:

- Design both a constant current source and a constant voltage source, each of which has current limiting.
- Construct and draw waveforms for a precision rectifier.
- Construct, trace current flow, draw waveforms, calculate voltages, currents and common mode rejection ratio for both a simple differential amplifier and an instrumentation amplifier.

21.2 A Constant Voltage Source

In Chapter 7 you learned that the circuit of Figure 21.1 makes a simple voltage regulator. We rather disparagingly called it a "brute force" regulator. It required a large Zener diode current and a lot of heat sinking.

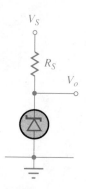

FIGURE 21.1 A Zener regulator.

In Section 13.7 we looked at the regulator circuit of Figure 21.2. The use of the "series pass" transistor, or the "current boost" transistor, as it is often called, greatly reduced the demands on the Zener diode. However, we found that the circuit of Figure 21.2 did not regulate as well as we might wish due to the turn on characteristics of the transistor. The Zener diode might keep the transistor's base voltage constant, but what we needed was to keep the emitter voltage constant. In other words, we

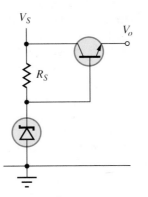

FIGURE 21.2   Zener regulator with a series pass transistor added.

needed to feed information about the output voltage back to the base control circuit. The circuit of Figure 21.3 provides that capability. It is our introduction to the "really high quality regulators" we mentioned at the close of Section 13.7.

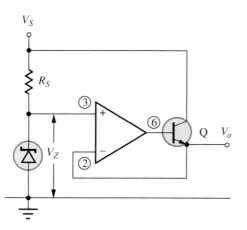

FIGURE 21.3   Regulator circuit with feedback.

If you remember the comparator property of the op amp and how an emitter follower transistor circuit works, you can see immediately how the circuit of Figure 21.3 will operate. From Figure 21.3 you can see that

$$V_3 = V_Z \qquad (21.1)$$

and

$$V_o = V_2 \qquad (21.2)$$

Thus, if V_o is less than V_Z, the Zener diode voltage, the comparator property of the op amp causes V_6 to start to increase. Then, if you remember how emitter followers behave, you know that V_o will follow V_6. When $V_o = V_Z$, then the op amp is "happy," and V_6 no longer increases. Similarly, if V_o is greater than V_Z, the op amp pulls V_6 down, and again V_o follows. Notice that the base voltage is no longer a factor in determining what V_o will be. Since the feedback is coming from the output, the op amp will push the base as hard as it has to in order to obtain $V_o = V_Z$. Thus, assuming that the Zener does its job well, the regulation of the circuit should be essentially zero. We defined regulation in Equation 5.4.

Since the input impedance of the op amp is high, there is very little load on the Zener diode circuit and essentially no load variation. This further reduces the demands on the Zener.

EXAMPLE 21.1

Repeat Example 13.5 using the circuit of Figure 21.3.

SOLUTION:

Example 13.5 was a redesign of Example 7.3 using a series pass transistor. Example 7.3 called for the design of a regulator intended to power a tape recorder ($V_L = 6$ VDC, $I_L = 100$ mADC) from an automobile electrical system ($V_{min} = 13$ VDC, $V_{max} = 15$ VDC). Figure 21.4 contains the relevant circuit. Since the load and the supply are the same as they were in Example 13.5, the transistor will still be a 2N6715.

According to Example 13.5, the required base current for the 2N6715 is $I_B = 1$ mA, which is well within the 741's capability ($I_{omax} = 25$ mA) so the op amp should be able to drive the transistor easily.

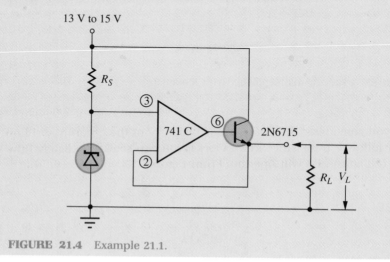

FIGURE 21.4 Example 21.1.

The selection of D_1 is rather simple since pin 3 of the op amp should not load the Zener circuit significantly, and since $V_Z = V_3 \stackrel{\wedge}{=} V_2 = V_o = 6$ V. Remember from Chapter 20 that the meaning of the "hat" above is that the op amp will try to meet the requirements of the equation if it can.

We will select from Appendix F a diode whose Zener voltage is as close to 6 V as possible and whose current consumption is minimum. That will be the 1N5233A. The important data are: $P_{Zmax} = 500$ mW, $V_Z = 6.0$ V, $I_{ZT} = 20$ mA.

The selection of R_S is also simplified by the use of the op amp. In order to avoid the knee of the Zener curve, we will choose

$$I_{Zmin} = \frac{I_{ZT}}{4} \tag{21.3}$$

$$= \frac{20 \text{ mA}}{4} = 5 \text{ mA}$$

The above minimum condition will occur when $V_{Smin} = 13$ V. A KVL walk from V_S through the Zener gives

$$R_S = \frac{V_{Smin} - V_Z}{I_{Zmin}} \tag{21.4}$$

$$= \frac{13 \text{ V} - 6.0 \text{ V}}{5 \text{ mA}} = 1.4 \text{ k}\Omega \qquad \text{(Use 1.2 K}\Omega\text{)}$$

The nearest smaller resistor is specified so that I_Z will never drop below 5 mA.

The maximum power dissipation will occur when $V_{Smax} = 15$ V. The current in the diode circuit will then be

$$I_{Dmax} = \frac{V_{Smax} - V_Z}{R_S} \tag{21.5}$$

$$= \frac{15 \text{ V} - 6 \text{ V}}{1.2 \text{ k}\Omega} = 7.5 \text{ mA}$$

The power rating requirement for R_S is

$$P_{RS} = (I_{Dmax})^2 R_S \tag{21.6}$$

$$= (7.5 \text{ mA})^2 * 1.2 \text{ k}\Omega = 67.5 \text{ mW} \qquad \text{(Use 1/8 W)}$$

The maximum power consumed by the Zener is

$$P_Z = V_Z I_{Dmax} \tag{21.7}$$

$$= 6 \text{ V} * 7.5 \text{ mA} = 45 \text{ mW}$$

which is well below the 500 mW rating of the diode.

You should notice, and be impressed by, how the use of the op amp reduces the requirements on the other circuit components in Example 21.1 above.

21.3   Output Current Limiting—Again

Notice that the feedback in the circuit of Figure 21.3 is taken from the output. Thus any resistance or any nonlinear behavior of components inside the feedback loop, such as the transistor, has essentially no effect on the output. This provides very good regulation and a Thevenin equivalent internal resistance of virtually zero ohms. That means the circuit is almost an ideal voltage source. It also makes for a potentially dangerous situation. If the output accidentally becomes shorted, the resulting zero resistance calls for a rather large current!

Consider the circuit of Figure 21.5. Notice that the feedback is still taken directly from the output. As a result, in spite of the additional circuitry, the output will still be well regulated since the op amp will push V_6 high enough to make $V_o \doteq V_i$, thus overcoming any additional voltage drops we may have added in the circuit. Again the hat warns us that this will only happen if the op amp can make it happen.

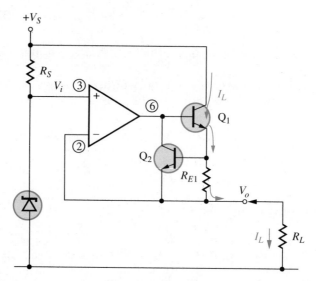

FIGURE 21.5   An output current limited circuit.

You may recognize the circuit of Figure 21.5. It is similar to the output circuit inside of the op amp. We studied that circuit in Section 19.10. If you understood Section 19.10, you know how this circuit operates.

To review what we said in Section 19.10, assume that we start with a fairly large value of R_L. Thus I_L will be small. I_L will flow from V_S down through the series pass transistor, Q_1, and through R_{E1} to R_L as shown in Figure 21.5. Since I_L is small, the voltage generated across R_{E1} will be less than 0.7 V. Thus Q_2 will not conduct and the operation of the circuit will be essentially the same as that of Figure 21.3. As R_L is decreased, I_L increases. When I_L increases to the point that

$$I_{C1}R_{E1} \approx 0.7 \text{ V} \tag{21.8}$$

transistor Q_2 begins to "wake up" and conduct. Part of the output of the 741 is then diverted through Q_2 thus reducing the base drive to Q_1 and limiting its collector current I_{C1max} to

$$I_{C1max} \approx \frac{0.7 \text{ V}}{R_{E1}} \tag{21.9}$$

Since the output of the 741 is limited to about 25 mA, the collector current of Q_2 will never be more than 25 mA, and the output current is limited to about

$$I_{Lmax} \approx \frac{0.7 \text{ V}}{R_{E1}} + 25 \text{ mA} \tag{21.10}$$

In most practical circuits, the 25 mA in Equation 21.10 can be ignored.

EXAMPLE 21.2

For the circuit of Figure 21.6, specify Q_1, Q_2, and R_{E1} if I_L is to be limited to not more than 0.5 A for all possible values of R_L.

SOLUTION:
Specifying Q_1: From Figure 21.6

$$V_{CE1} = 30 \text{ V} - V_{BE2} - V_L \tag{21.11}$$

Thus if $R_L = 0 \ \Omega$ then $V_L = 0 \text{ V}$ and $V_{CE1rate} > 29.3 \text{ V}$. Since $I_{C1} = I_L$ and I_{Lmax} is given as 0.5 A, $I_{C1rate} > 0.5 \text{ A}$. Since both of the preceding maxima will occur at the same time, namely, when $R_L = 0 \ \Omega$, the maximum dissipation for Q_1 will be

$$P_{1max} = 29.3 \text{ V} * 0.5 \text{ A} = 14.7 \text{ W}$$

From the data table in Appendix F, a 2N6413 with a dissipation rating of 15 W would be barely adequate.

Specifying Q_2: From Figure 21.6

$$V_{CE2} = V_{BE1} + V_{BE2} \tag{21.12}$$

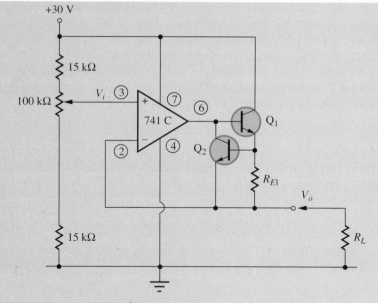

FIGURE 21.6   Example 21.2.

Since neither base to emitter voltage is likely to go as high as 0.8 V, we are probably safe to say $V_{CE2rate} > 1.6$ V. Since I_{C2} can never be more than the maximum current the 741 can deliver, $I_{C2rate} > 25$ mA. Since these maxima will also occur at the same time, again when $R_L = 0$ Ω, the maximum dissipation for Q_2 will be

$$P_{DISS} = 1.6 \text{ V} * 25 \text{ mA} = 40 \text{ mW}$$

You can see that the requirements for Q_2 are not very difficult to meet. Almost any transistor could be used. We will specify a 2N2712.

Specifying R_{E1}: If Q_1 is a 2N6413, then from Appendix F, $h_{FE1} = 150$. Thus, when $R_L = 0$ Ω, $I_{C1max} = 0.5$ A, then

$$I_{B1} = \frac{I_{C1}}{h_{FE}} \tag{9.7}$$

$$= \frac{0.5 \text{ A}}{150} = 3.33 \text{ mA}$$

Since the maximum output current for the 741 is 25 mA, when $R_L = 0$ Ω, then

$$I_{C2} = 25 \text{ mA} - 3.33 \text{ mA} = 21.7 \text{ mA} \tag{21.13}$$

and

$$V_{BE2} = 26 \text{ mV} * \ln\left(\frac{I_{C2}}{10 \text{ fA}}\right) \qquad (8.4)$$

$$= 26 \text{ mV} * \ln\left(\frac{21.7 \text{ mA}}{10 \text{ fA}}\right) = 0.739 \text{ V}$$

Thus

$$R_{E1} = \frac{V_{BE2}}{I_{Lmax}} \qquad (21.14)$$

$$= \frac{0.739 \text{ V}}{0.5 \text{ A}} = 1.48 \ \Omega \qquad (\text{Use } 1.5 \ \Omega)$$

The power rating for R_{E1} is

$$P = I^2 R$$
$$= (0.5 \text{ A})^2 * 1.5 \ \Omega = 0.375 \text{ W} \qquad (\text{Use } 0.5 \text{ W})$$

21.4 Voltage Regulators

As illustrated in Figure 21.7, the entire circuit of any of the regulators we have been discussing could be placed on a single silicon chip and pro-

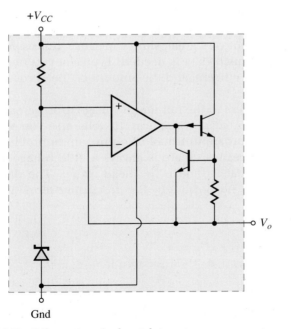

FIGURE 21.7 A three–terminal regulator.

duced as a single integrated circuit. Of course, this is done. As you can see from Figure 21.7, three terminals are required for the simplest such device. Thus they are known as "three-terminal regulators." They are housed in the same package type as those used for transistors.

Some regulators bring the inverting input to the op amp out to a separate "sense" terminal. This allows the designer the additional freedom of controlling the regulator output voltage in terms of some point in the circuit other than the regulator's output. Other designs provide still more flexibility by also bringing out the noninverting op amp terminal, and perhaps the Zener diode terminal as well. This makes it possible for the device to have an adjustable output voltage. You can see that the list of options could grow quite large. Such options are added at the expense of the simple three-terminal design.

One of the most popular three-terminal regulators on the market, the LM340 series is designed for a variety of different output voltages and current ranges. For example, the LM340-5 delivers an output of 5 V and the LM340-24 delivers 24 V. There are other regulators on the market whose output voltage is externally adjustable. Many commercially available regulators have another feature that the regulators we have designed do not have and that is a thermal shutdown feature. For example, when an LM340 is heated up to a temperature of about 175° C it shuts itself off. Such regulators are very forgiving of human mistakes. They are also very easy to use, and reduce the parts count and cost compared to the circuits we talked about in Section 21.3.

You will find a typical manufacturer's data sheet in Appendix F and you will also find some typical data book listings. The amount of data commonly listed is not very great, simply because, unless your application is very critical, not much data is needed. Typically you need to know the required voltage and current, and the amount of "head room" required for the regulator.

"Head room" refers to the minimum voltage drop across the regulator. Obviously, if your application requires 10 volts and you wish to drive it from a 9 V battery, you are in trouble. You may have problems even if you have a 12 V supply because there is always a little voltage lost across the regulator. That is what we mean by "head room." The data books often indicate the required head room by the descriptive term "$V_I - V_O$."

EXAMPLE 21.3
Repeat Example 13.5 using a three-terminal regulator.

SOLUTION:
The circuit is shown in Figure 21.8. In order to select a regulator, the required data are the output voltage (6 V) and current (100 mA). From the data book listing in Appendix F an AN78LO6 with a current rating

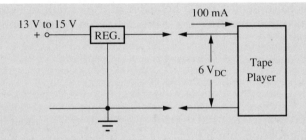

FIGURE 21.8 Example 21.3.

of $I_{Lmax} = 100$ mA would be barely adequate. The AN78N06 would be a safer choice. The regulator voltage rating of $V_{Imax} = 35$ V is more than adequate in either case since the maximum supply voltage is only 15 V. There is adequate head room since the minimum supply voltage of 13 V allows

$$V_I - V_O = 13 \text{ V} - 6 \text{ V} = 7 \text{ V}$$

and according to Appendix F, the AN78N06 requires only 2 V.

As you can see, the "design" above was much easier than Example 13.5. The reason is that the three-terminal regulator contains all of the circuitry of Figure 21.5 on the chip and all you have to do is hook it up.

21.5 A Constant Current Source

Consider the circuit of Figure 21.9. Notice that it is the same circuit as the noninverting amplifiers of Section 20.9 with the exception that R_1 has now been replaced by R_L.

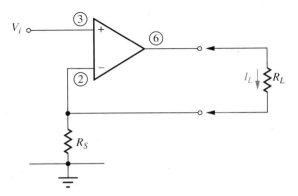

FIGURE 21.9 A constant current source.

In order to understand the behavior of the circuit of Figure 21.9, you only need to remember that the comparator property of the op amp allows us to say

$$I_L \mathrel{\hat{=}} \frac{V_i}{R_S} \tag{21.15}$$

Again the hat warns us that the op amp will try to meet the requirements of the equation, but if it cannot due to saturation, or for some other reason, Equation 21.15 and all other equations derived from it are invalid.

You should notice that the value of R_L does not appear in Equation 21.15. Thus within the ability of the op amp, regardless of the value of R_L, I_L will remain constant. Thus the circuit of Figure 21.9 is a constant current source. A current boost transistor could be added if more current were required.

EXAMPLE 21.4

Determine the maximum value of R_L for which the circuit of Figure 21.10 will act as a constant current source and determine the value of the constant current for the circuit.

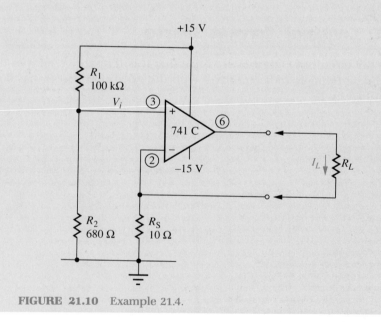

FIGURE 21.10 Example 21.4.

SOLUTION:
The input voltage is

$$V_i = V_{CC}\frac{R_2}{R_1 + R_2} \tag{21.16}$$

$$= 15 \text{ V}\frac{680 \text{ } \Omega}{100 \text{ k}\Omega + 680 \text{ } \Omega} = 101 \text{ mV}$$

Thus

$$I_L \overset{\wedge}{=} \frac{V_i}{R_S} \tag{21.15}$$

$$= \frac{101 \text{ mV}}{10 \text{ } \Omega} = 10.1 \text{ mA}$$

The circuit will no longer be a constant current source when the op amp saturates. According to the data sheet in Appendix H, for a 15 V supply, $V_{\text{sat}} = 13$ V. From the circuit of Figure 21.10

$$R_{L\text{max}} = \frac{V_{\text{sat}} - V_2}{I_L} \tag{21.17}$$

$$= \frac{13 \text{ V} - 101 \text{ mV}}{10.1 \text{ mA}} = 1.28 \text{ K}\Omega$$

21.6 A Precision Rectifier

As you found out in the early chapters of this book, when a diode circuit such as is shown in Figure 21.11a is used to rectify an AC waveform such as that shown in Figure 21.11b, there is always some forward voltage loss across the diode as shown in Figure 21.11c, and it is always a nonlinear loss. That can be a real problem where very small signal amplitudes are involved. Although the Schottky diode which we mentioned in Section 7.3 can reduce that loss to perhaps a tenth of a volt, even that is not enough if you are dealing with signals in the 0.1 V range. The precision rectifier is a rather neat solution to that problem. Consider the circuit in Figure 21.12. Notice that the diodes D_1 and D_2 are inside of the feedback loop. On several previous occasions we have mentioned the advantage of this arrangement.

Figure 21.13a shows the current flow when v_i is positive. Since it is an inverting circuit the voltage at pin 6 goes sufficiently negative to forward bias D_2 and maintain $V_{in} = 0$ V. Thus $V_6 = -V_D$. This reverse biases D_2 (it is circled to indicate reverse bias in Figure 21.13a) so $v_o = 0$ V.

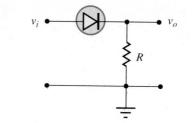

(a)

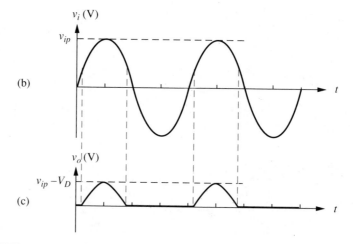

(b)

(c)

FIGURE 21.11 A diode rectifier.
  (a) Circuit diagram.
  (b) Input waveform.
  (c) Output waveform.

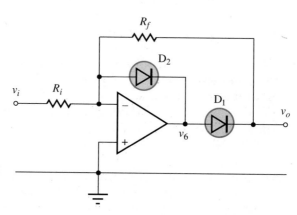

FIGURE 21.12 Inverting precision half-wave rectifier.

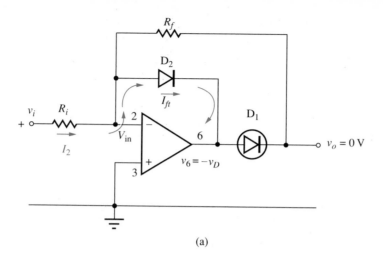

(a)

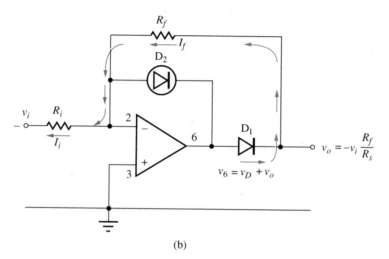

(b)

FIGURE 21.13 Current flow in a precision rectifier.
(a) Current flow on the positive input signal swing.
(b) Current flow on the negative input signal swing.

Figure 21.13b shows the current flow when v_i is negative. Now V_6 goes positive, reverse biasing D_2. It is circled in Figure 21.13b. Pin 6 goes sufficiently positive to force diode D_1 to conduct the current required to meet the op amp requirement

$$I_i = I_f \qquad (20.4)$$

Thus

$$V_o = -V_i \frac{R_f}{R_i} \qquad (20.9)$$

Notice that the diode voltage does not appear in Equation 20.9 so the output voltage is unaffected by the diode's presence in the circuit. The resulting waveforms are shown in Figure 21.14 The "infinite" gain of the op amp compensates for the nonlinear turn-on characteristic of diode D_1 so that, as shown in Figures 21.14b and 21.14c, if the circuit is designed for unity gain, V_6 will be an exact copy of v_i (inverted of course).

As you can see from the waveforms in Figure 21.14 just as v_i crosses zero, ideally, the op amp output must jump from about -0.7 V to $+0.7$ V,

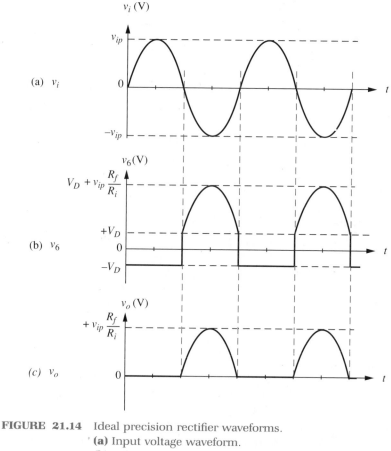

FIGURE 21.14   Ideal precision rectifier waveforms.
(a) Input voltage waveform.
(b) Voltage waveform at pin 6 of the op amp.
(c) Output voltage waveform.

or from $+0.7$ V to -0.7 V in zero time. Given the slew rate of most real op amps, you know that will not happen. The waveform of v_6 and v_o will look more like those shown in Figure 21.15. To the extent that the delay Δt is significant, the rectifier output will be in error.

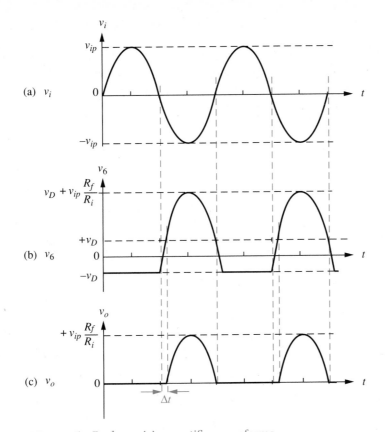

FIGURE 21.15   Real precision rectifier waveforms.
(a) Input voltage waveform.
(b) Voltage waveform at pin 6 of the op amp.
(c) Output voltage waveform.

From Appendix F, the slew rate for a 741 is 0.5 V/μsec. Thus the time it will take for a 741 to slew from −0.7 V to +0.7 V is approximately

$$\Delta t = \frac{0.7 \text{ V} - (-0.7 \text{ V})}{0.5 \text{ V/μsec}} = 2.8 \text{ μsec}$$

Thus, the period of the input waveform must be greater than about 28 μsec for the circuit of Figures 21.12 to act like a "precision" rectifier.

21.7 A Differential Amplifier

Occasionally it is desired to amplify a signal that is "buried" in a lot of noise. Recording the waveform of the nerve impulses from a human heart is a good example. The signal voltage due to the heartbeat is typically less than 1 mV but the normal body noise voltage is usually a hundred times that amount.

Quite often, by careful placement of the ground terminal and the signal probes, it is possible to arrange the signal source so that the signal of interest is a "differential" signal while the noise is "common mode." The circuit of Figure 21.16a is one way of simulating such a situation in the

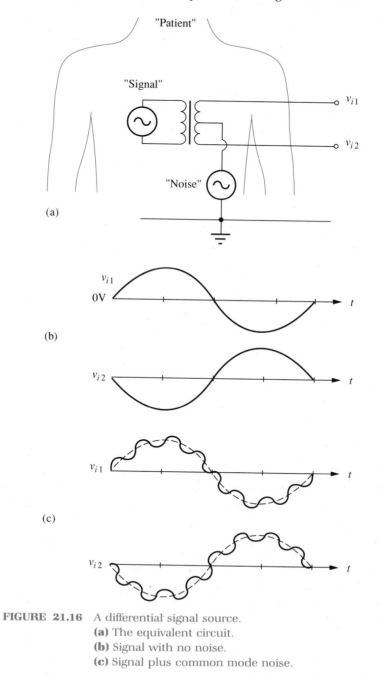

FIGURE 21.16 A differential signal source.
(a) The equivalent circuit.
(b) Signal with no noise.
(c) Signal plus common mode noise.

laboratory. If the amplitude of the signal generator used to simulate the "noise" is set to zero volts, the phase splitting property of the center tapped transformer will cause waveforms at v_{i1} and v_{i2} as shown in Figure 21.16b. We have a "differential signal." Due to their phase relationship, the difference between v_{i1} and v_{i2} is not zero. It is, in fact, their arithmetic sum.

On the other hand, if some "noise" from the noise generator in Figure 21.16a is introduced into the signal, since it is introduced at the center tap of the transformer, the same noise signal in the same phase will be delivered to v_{i1} and v_{i2}. It is a common mode signal. If their amplitudes are equal, since v_{i1} and v_{i2} are in phase, their difference will be zero.

In Figure 21.16 a higher frequency was used for the "noise," but that is not necessarily the case. The relative frequencies is irrelevant. What is important is the phase relationship between v_{i1} and v_{i2}.

Signals of the type described in Figure 21.16 are often encountered in practice.

When you must amplify a differential signal which contains common mode noise such as we have just described, a differential amplifier, or "difference amplifier" as it is often called, is usually the best choice. The circuit of Figure 21.17 is a simple type of differential amplifier. It is important to notice that pin 2 is not a virtual ground in this case since pin 3 is not grounded. The differential amplifier of Figure 21.17 is actually a combination of an inverting amplifier and a noninverting amplifier and it has many of the characteristics of each type.

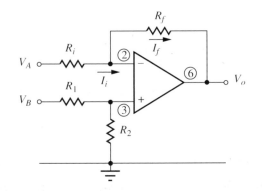

FIGURE 21.17 A simple differential amplifier.

To derive the equation that describes the operation of the circuit of Figure 21.17, we can start, as we have usually done, with the ideal op amp assumption.

$$I_i = I_f \tag{20.4}$$

or

$$\frac{V_A - V_2}{R_i} = \frac{V_2 - V_o}{. \ R_f} \qquad (21.18)$$

$$V_A R_f - V_2 R_f = V_2 R_i - V_o R_i$$

or

$$V_o R_i = V_2 (R_i + R_f) - V_A R_f \qquad (21.19)$$

Applying the voltage divider idea to the noninverting input of Figure 21.17, we can say

$$V_3 = V_B \frac{R_2}{R_1 + R_2} \qquad (21.20)$$

You should remember that

$$V_2 \hat{=} V_3 \qquad (20.2)$$

with the hat indicating that the op amp will try to maintain the equality. Then, using the idea of Equation 20.2, we can substitute V_3 from Equation 21.20 into Equation 21.19. The result is

$$V_o R_i = V_B \frac{R_2}{R_1 + R_2}(R_i + R_f) - V_A R_f \qquad (21.21)$$

Equation 21.21 can be greatly simplified if, as is usually the case for differential amplifiers, we require that

$$R_1 = R_i \qquad (21.22)$$
$$R_2 = R_f \qquad (21.23)$$

Substituting Equations 21.22 and 21.23 into Equation 21.21 gives us the differential op amp equation

$$V_o R_i = V_B R_f - V_A R_f$$

or

$$V_o = (V_B - V_A)\frac{R_f}{R_i} \qquad (21.24)$$

You can easily see that if $V_B = 0$ V, Equation 21.24 reduces to the classic op amp relation of Equation 20.9. We will leave it to you to show that if $V_A = 0$ V, given the circuit of Figure 21.17, and the assumptions of Equations 21.22 and 21.23, then Equation 21.24 reduces to the noninverting relation of Equation 20.27.

21.8 Common Mode Rejection

As you can see from Equation 21.24, for a common mode signal, that is if $V_A = V_B$, then $V_o = 0$ V. However, remember that no equation is any better than the assumptions from which it was derived. For one thing, Equation 21.24 was based on the assumptions of Equations 21.22 and 21.23. You should know by now that the probability of any two resistors being exactly equal is rather small. Also, the gain of the two op amp inputs must be identical. The probability of that occurring is also rather small. In order to compensate for such variations, the differential amplifier circuit is usually modified as shown in Figure 21.18. Resistor R_{2a} is usually chosen to be about 10 percent less than the desired value for R_2, and the pot R_{2b} is chosen to be about 20 percent of R_2. Thus the expected pot setting will be approximately at its midpoint with the possibility of a 10 percent variation either way.

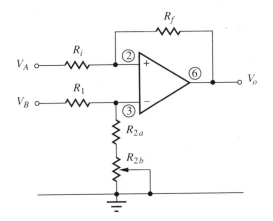

FIGURE 21.18 Differential amplifier with common mode nulling pot.

To adjust pot R_{2b}, the same signal is connected to both inputs and the output is minimized by adjusting the pot setting. This adjustment will allow compensation for some of the previously mentioned practical variations, but probably not all of them. Thus, even if the circuit of Figure 21.18 is perfectly adjusted, if you look carefully enough, you will usually see some common mode signal present in the output. The parameter usually used as a measure of the amplifier's ability to discriminate against common mode signals is the common mode rejection ratio, CMRR. It is defined as the ratio of the amplifier's differential gain, A_{dif} to its common mode gain, A_{com} or

$$CMRR \equiv \frac{A_{dif}}{A_{com}} \qquad (21.25)$$

It is common to express CMRR in dB

$$\text{CMRR} = 20 \log \frac{A_{\text{dif}}}{A_{\text{com}}} \tag{21.26}$$

Although the circuit of Figure 21.18 is an improvement over the basic differential amplifier, it still has some very serious disadvantages. For one thing, its input impedance is low. A KVL walk from V_A into the circuit of Figure 21.17 and back to V_B, remembering that the op amp wants to keep $V_2 \stackrel{\wedge}{=} V_3$, reveals that the differential input impedance is just

$$Z_{AB} = R_i + R_1 \tag{21.27}$$

Since you know about inverting op amp circuits, it should come as no surprise to you that the input impedance at terminal A is

$$Z_A = R_i \tag{21.28}$$

By looking at the circuit of Figure 21.17 or Figure 21.18, you can see that if the op amp is considered ideal, the input impedance at terminal B is

$$Z_B = R_1 + R_2 \tag{21.29}$$

Not only are the above input impedances relatively low, but they are not equal. Therefore the differential circuits seen so far are not very useful in very high impedance circuits.

21.9   Instrumentation Amplifier

Consider the circuit of Figure 21.19. You should recognize the circuit of op amp C as the simple differential amplifier we have been studying. In fact, to improve the common mode rejection, R_2 should be replaced by R_{2a} and R_{2b} of Figure 21.18.

You should recognize that the circuits of op amps A and B are non-inverting amplifiers. Thus the low input impedances and the variations among them that we complained about in Section 21.8 are all taken care of.

Another advantage of the circuit of Figure 21.19 is that the common mode rejection of the circuit is improved over the simple differential amplifier. To understand why, let us first assume that a common mode signal, V_{com} is applied to V_{iA} and V_{iB} of Figure 21.19. Op amp A tries to make $V_{2A} \stackrel{\wedge}{=} V_{\text{com}}$ and op amp B tries to make $V_{2B} \stackrel{\wedge}{=} V_{\text{com}}$. Thus there is no voltage loss across R_4. That means there is no current flowing through R_3 and so $V_A = V_{2A} = V_{\text{com}}$. By the same token, there is no current through R_5 and so $V_B = V_{2B} = V_{\text{com}}$. Thus the common mode voltage gain for op amps A and B is one.

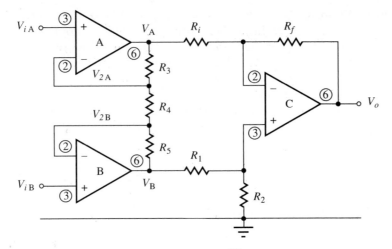

FIGURE 21.19 An instrumentation amplifier.

To determine the differential gain for op amps A and B of Figure 21.19, the voltage divider equation for resistors R_3, R_4, and R_5 can be written as

$$\frac{V_A - V_B}{R_3 + R_4 + R_5} = \frac{V_{2A} - V_{2B}}{R_4} \tag{21.30}$$

Remembering that the op amps will try to maintain $V_2 \stackrel{\wedge}{=} V_3$ we can say

$$\frac{V_A - V_B}{R_3 + R_4 + R_5} = \frac{V_{iA} - V_{iB}}{R_4} \tag{21.31}$$

or

$$V_A - V_B = (V_{iA} - V_{iB})\frac{R_3 + R_4 + R_5}{R_4} \tag{21.32}$$

As you might guess, in order for the circuit to be symmetrical, R_3 and R_5 are always equal. They can be hundreds of times the value of R_4. The result, as you can see from Equation 21.32, is that the differential gain of the circuit can be quite high. We have already shown that the common mode gain of the circuit containing op amps A and B in Figure 21.19 is one. Thus this circuit in combination with the circuit of op amp C with its high common mode rejection property gives the total circuit truly outstanding common mode rejection.

The differential gain of the total circuit can be obtained by substituting Equation 21.31 into Equation 21.24. The result is

$$V_o = -(V_{iA} - V_{iB})\frac{(R_3 + R_4 + R_5)}{R_4}\frac{R_f}{R_i} \tag{21.33}$$

Thus, in addition to the high common mode rejection, the instrumentation amplifier has a very high differential gain, and of course an input impedance limited only by the input impedances of op amps A and B. In short, the total circuit of Figure 21.19 acts like a very high performance op amp. You will find instrumentation amplifiers that have been combined into a single integrated circuit package. The LM725, and the LH0044 are examples. As you might guess an "instrumentation op amp," as it is called, is expensive but it does give outstanding performance.

EXAMPLE 21.5

Find the value required for R_4 in Figure 21.20 in order that $V_o = -1$ V when the differential input is $V_{iA} - V_{iB} = 1$ mV.

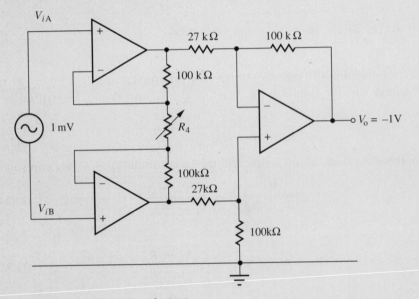

FIGURE 21.20   Example 21.5.

SOLUTION:

From Equation 21.33 we can say

$$V_o R_4 R_i = -(V_{iA} - V_{iB})(R_3 + R_4 + R_5)R_f \qquad (21.34)$$

Substituting numbers from Figure 21.20 into Equation 21.34 gives

$$-1 \text{ V} * R_4 * 27 \text{ k}\Omega = -1 \text{ mV}(100 \text{ k}\Omega + R_4 + 100 \text{ k}\Omega)100 \text{ k}\Omega$$

$$27 \text{ k}R_4 = 20 \text{ M}\Omega + 100 \text{ R}_4$$

or

$$R_4 = \frac{20 \text{ M}\Omega}{26.9 \text{ k}} = 743 \text{ }\Omega$$

21.10   Troubleshooting   ◄◄◄◄◄◄◄◄

The one concept that is common to all negative feedback op amp circuits is that the op amp tries to maintain

$$V_2 \stackrel{\wedge}{=} V_3 \qquad\qquad (20.3)$$

Thus under normal conditions there should be no measurable difference between V_2 and V_3. Since many of the circuits in this chapter have included several interacting components, it can be difficult to determine the cause of a malfunctioning circuit. However, in the event that $V_2 > V_3$, then if the op amp is behaving normally, V_6 should be at negative saturation. If $V_2 < V_3$, then V_6 should be at positive saturation. The circuit defect that most often causes the output of a normal op amp to go to saturation is an open feedback loop. This could result from a defective current boost transistor or other component in the feedback path.

21.11   Summary

In this chapter we reexamined some circuits which we have encountered before such as the voltage regulator circuit and the current limiting circuit and we found that the op amp's characteristics improved the performance of the circuits.

We also encountered the differential amplifier and the instrumentation amplifier, which is an example of how the performance of a circuit can be improved by taking advantage of the characteristics of the op amp.

GLOSSARY

Common mode rejection ratio: CMRR   The ratio of the differential gain of an amplifier to its common mode gain. The ratio is often expressed in dB. (Section 21.8)

Common mode signal: A signal that is the same for two inputs. (Section 21.8)

Current boost transistor: Also known as a "series pass transistor." A transistor placed in series with a load, usually in an emitter follower configuration, for the purpose of increasing the current the circuit can deliver to the load. (Section 21.2)

Current limiter: A circuit which tends to limit the output current of a device or circuit. (Section 21.3)

Differential signal: A signal that is the difference between two input signals. (Section 21.7)

Head room: The amount that the supply voltage must exceed the output voltage of a device. (Section 21.4)

Precision rectifier: A rectifier whose output waveform does not show the customary diode voltage drops but is a very close copy of the input waveform that has been rectified. (Section 21.6)

Series pass transistor: See "current boost transistor." (Section 21.2)

Voltage regulation: The relative amount the output voltage of a circuit changes when a specified load is applied. (Section 21.2)

PROBLEMS

Section 21.2

1. Estimate the value of the base current, I_{BQ1}, the base voltage, V_{BQ1}, and the power, P_{Q1}, dissipated in transistor Q_1 in the circuit of Figure 21.21 if $R_L = 1 \ k\Omega$.

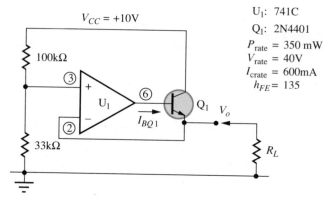

U_1: 741C
Q_1: 2N4401
$P_{rate} = 350 \ mW$
$V_{rate} = 40V$
$I_{crate} = 600mA$
$h_{FE} = 135$

FIGURE 21.21   Problem 21.1.

Section 21.3

2. Determine the minimum allowable value for R_L in Figure 21.21 in order not to exceed the 2N4401 maximum ratings.

Section 21.4

3. Repeat Example 21.1 if the tape recorder requires 9 V at 250 mA. Specify all components.

4. Repeat Example 21.1 if the load is a CB radio that requires 6 V at 2.5 A. *Hint:* A Darlington circuit can increase the current gain. See Section 13.6.

5. Estimate the voltage required at pin 6, the op amp output terminal in Problem 3
   (a) When the load is connected.
   (b) When the load is disconnected.

6. Repeat Problem 5 with reference to the circuit of Problem 4.
7. Redesign the circuit of Example 21.1 to include a current limiting circuit that will prevent damage to the circuit in the event that the output terminals become shorted.
8. Repeat Problem 7 with reference to the circuit of Problem 4.

Section 21.5

9. Use a 741 op amp, ± 15 V power supplies, and a 1 Ω current sensing resistor to design a constant current source that will deliver approximately 50 mA. *Hint:* You may need a current boost transistor.
10. Over what range of load resistor is the circuit of Problem 9 a constant current source?

Section 21.6

11. Determine the maximum frequency for which a 741 can be used to construct a precision rectifier if $R_f = R_i$.

Section 21.7

12. Show that if $V_A = 0$ V in the circuit of Figure 21.17 and if Equations 21.22 and 21.23 are assumed, then Equation 21.24 is equivalent to Equation 20.28.
13. Determine the value of V_o and the value of the differential gain A_{diff} for the circuit in Figure 21.22.

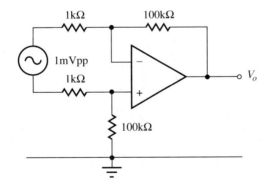

FIGURE 21.22   Problem 21.13.

Section 21.8

14. Determine the value of V_o and the value of the common mode gain, A_{com}, for the circuit in Figure 21.23 if the op amp is assumed to be ideal and

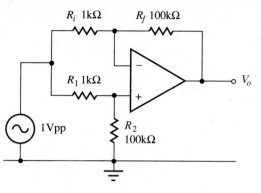

FIGURE 21.23 Problem 21.14.

(a) The resistance of all of the resistors is exactly as shown.
(b) The resistance of all of the resistors is exactly as shown except for R_2 whose true value is 1 percent higher than shown.

Section 21.9

15. Determine the differential input impedance to the circuit of op amp C in the ideal circuit of Figure 21.24. Explain why it is that the low input impedance just calculated is not a great problem in the operation of the overall circuit.

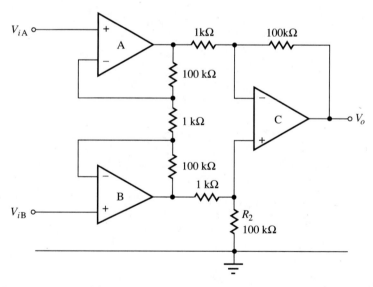

FIGURE 21.24 Problem 21.15.

16. Determine the differential mode voltage gain for the circuit of Figure 21.24.

17. Determine the common mode voltage gain for the circuit of Figure 21.24 if
    (a) The resistances are all exactly as shown.
    (b) The resistance of resistor R_2 is 1 percent high.

Section 21.10   ◀ Troubleshooting

18. Estimate the voltage at op amp pins 2, 3, and 6, and estimate V_o for the circuit of Figure 21.25 if
    (a) The collector of the transistor becomes an open circuit.
    (b) The emitter of the transistor becomes an open circuit.
    (c) The collector becomes shorted to the emitter.

19. Estimate the voltage at op amp pins 2, 3, and 6, and estimate V_o and I_L for the circuit of Figure 21.26 if
    (a) The collector and emitter of transistor Q_2 become shorted together when R_L = (i) 1 kΩ, (ii) 10 Ω, (iii) a dead short.
    (b) The emitter of transistor Q_2 becomes an open circuit when R_L = (i) 1 kΩ, (ii) 10 Ω, (iii) a dead short.

20. Determine the effect on V_o in Example 21.5 if R_4 of Figure 21.20 became:
    (a) A short circuit.
    (b) An open circuit.

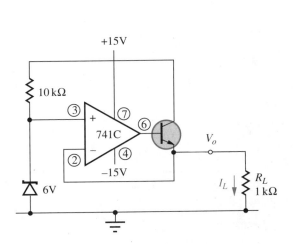

FIGURE 21.25 Problem 21.18.

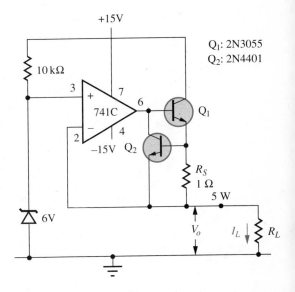

FIGURE 21.26 Problem 21.19.

Positive feedback. (Photograph by Carl Wolinsky/
Stock, Boston.)

Chapter 22
Positive Feedback Circuits

SPECIAL TERMS

Schmitt trigger
Hysteresis
Snap action switch
Trip point
Multivibrator
Astable multivibrator
Monostable multivibrator
One shot

22.1   Introduction and Objectives

In previous chapters we studied negative, or degenerative feedback circuits in which feedback was used to control the gain of the circuit. In this chapter we will introduce the idea of positive feedback. After having studied this chapter you should understand how positive feedback is used in trigger circuits and oscillators and you should be able to:

- Analyze and design Schmitt trigger type circuits using an op amp.
- Analyze and design monostable and astable multivibrators using an op amp.
- Analyze and design square wave, triangle wave, sawtooth wave generators and a Wein bridge sine wave generator using an op amp.

22.2   Inverting Schmitt Trigger

In Section 19.17 we mentioned that the zero crossing detector of Figure 22.1a can be thought of as being a special case of the level detector of Figure 22.1b in which $V_{ref} = 0$ V. It can also be thought of as a special case of the Schmitt trigger shown in Figure 22.1c.

In case you have forgotten how zero crossing detectors and level detectors work, a quick review of Sections 19.16 and 19.17 will show that for the inverting detectors in Figure 22.1, when V_i becomes greater than V_{ref}, V_o will switch to $-V_{sat}$, and when V_i becomes less than V_{ref}, V_o will switch to $+V_{sat}$. In the case of the zero crossing detector and the level detector, the reference voltage that determines the "trip point" for the output is the same whether it is approached from above or below. A Schmitt trigger has the added flexibility of allowing the possibility of one value of V_{ref} for an increasing value of V_{in} and allowing another value of V_{ref} for a decreasing value of V_{in}. The higher value at which the output changes state is known as the "upper trip point," V_{UT}, and the lower value is known as the "lower trip point," V_{LT}. This produces what is sometimes known as a "snap action switch."

In Chapter 19 when we were discussing level detectors, you may have noticed that we refused to talk about a situation that is not uncommon in practice, namely, what happens if $V_{in} = V_{ref}$. What happens to your furnace or air conditioner when the temperature is exactly 70 degrees and the thermostat is set for exactly 70 degrees? In order to avoid this ambiguity, the maker of the thermostat used a snap action switch. That is, some hysteresis was built into the switch. Hysteresis simply means that history has some influence on device behavior. If the input has been low, it will have to go all the way up to the upper trip point before the output will change state. If the input then starts dropping, it will have to go all

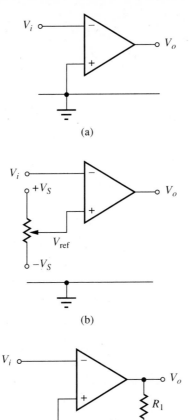

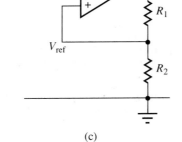

FIGURE 22.1 Some inverting detector circuits
(a) A zero crossing detector.
(b) A level detector.
(c) A Schmitt trigger.

the way down to the lower trip point before the output will change state again. This eliminates the possibility of that ambiguous situation where $V_{in} = V_{ref}$. We no longer have a single V_{ref}, we have two, V_{UT} and V_{LT}.

To understand how a Schmitt trigger works, consider the circuit of Figure 22.2a. First, assume that V_{in} starts out very negative as shown in Figure 22.2b. Since the input goes to the inverting terminal, $V_{out} = +V_{sat}$.

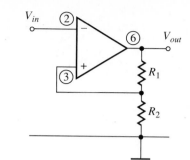

(a)

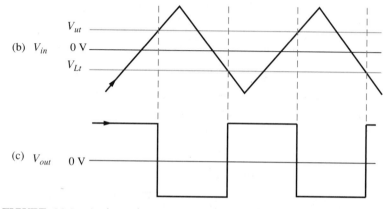

(b) V_{in}

(c) V_{out} 0 V

FIGURE 22.2   An inverting zero symmetric Schmitt trigger.
(a) Circuit diagram.
(b) Input waveform.
(c) Output waveform.

It is the output and the voltage divider, R_1 and R_2 that determine the trip voltage. Thus

$$V_{UT} = V_3 = +V_{sat} \frac{R_2}{R_1 + R_2} \qquad (22.1)$$

As shown in Figure 22.2c, as V_{in} increases, V_{out} remains at $+V_{sat}$ and the op amp compares V_{in} to V_{UT}. When V_{in} exceeds V_{UT}, V_{out} changes state. Then $V_{out} = -V_{sat}$ and the voltage at the noninverting terminal becomes the lower trip point, V_{LT} given by

$$V_{LT} = V_3 = -V_{sat} \frac{R_2}{R_1 + R_2} \qquad (22.2)$$

As you can see either by examining Equations 22.1 and 22.2, or by examining Figure 22.2b, V_{UT} will be positive and V_{LT} will be negative. Further, if you assume that $+V_{sat} = -(-V_{sat})$ then the trip points will be symmetrically spaced about zero.

EXAMPLE 22.1

Using a 741C op amp and power supplies of ±15 V, design a Schmitt trigger circuit with trip points of ±5 V.

SOLUTION:

The circuit is shown in Figure 22.3. According to the data sheet for the 741C in Appendix F, for supplies of ±15 V, $V_{sat} \approx \pm13$ V. Also we know

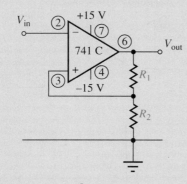

FIGURE 22.3 Example 22.1.

that circuit resistances should be kept below about 200 kΩ. Thus choosing $R_1 = 100$ kΩ and using the voltage divider idea we can say

$$\frac{R_2}{V_{UT}} = \frac{R_1}{V_{sat} - V_{UT}} \tag{22.3}$$

or

$$R_2 = R_1 \frac{V_{UT}}{V_{sat} - V_{UT}}$$

$$= 100 \text{ K}\Omega \frac{5 \text{ V}}{13 \text{ V} - 5 \text{ V}} = 62.5 \text{ k}\Omega \qquad \text{(Use 68 k}\Omega\text{)}$$

The nearest standard resistor is specified.

Notice that the $R_{in} = \infty \, \Omega$ assumption is made in the above solution since Equation 22.3 is only valid for an unloaded voltage divider.

As a practical use for a Schmitt trigger, consider the following situation:

In digital circuits, when signals are transmitted to a destination such as a computer from some remote source such as a user's terminal, the nice square pulses which the terminal generates and which the computer likes to receive become rounded and distorted as they travel over the con-

necting wires. A zero crossing detector is an easy way to restore those
pulses to their nice square shape.

However, in addition to rounding the pulses, the connecting wires
usually pick up a lot of noise as well. The situation is diagrammed, perhaps with some artistic license, in Figure 22.4a. A pulse starts out from

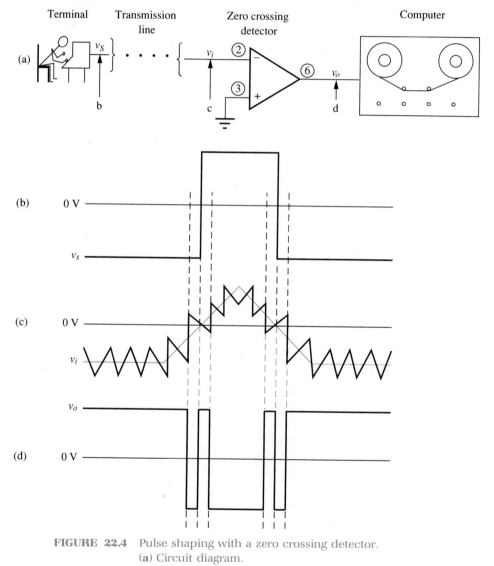

FIGURE 22.4 Pulse shaping with a zero crossing detector.
(a) Circuit diagram.
(b) Waveform leaving user's terminal.
(c) Waveform arriving at the zero crossing detector.
(d) Waveform delivered to the computer.

your terminal as the well-formed pulse of Figure 22.4b. After having traveled by some unknown route, it arrives at the computing center as the rather noisy "pulse" of Figure 22.4c.

If the zero crossing detector shown in Figure 22.4a were used as the pulse-shaping circuit to square up the waveform of Figure 22.4ç, the input to the computer would be the waveform of Figure 22.4d. Instead of seeing the well-defined square pulse that the terminal transmitted, the zero crossing detector would send a pulse to the computer every time the noisy waveform of Figure 22.4c crossed zero. Since the noise causes the incoming waveform to cross zero a couple of times every time the signal crosses zero, the computer would see some false pulses which it would try to interpret. The result could be anything from "garbage" showing up on the screen of the user's terminal to a complete system crash.

A Schmitt trigger, as shown in the circuit of Figure 22.5a is a better solution to the above rather unhappy situation. From your knowledge of Schmitt triggers you know that the output changes state when the input exceeds V_{UT} and it will not change back again until the input goes below V_{LT}. Thus false pulses can be avoided by setting the interval between V_{UT} and V_{LT} greater than the peak-to-peak line noise. As you can see in Figure 22.5d, the result is a nice clean pulse delivered to the computer.

A large part of the job of computer installation and servicing personnel is that of assessing the amount of noise on the signal lines and adjusting the discriminator circuits accordingly.

If you know about digital circuits, you should see several things wrong with the above illustration. Perhaps the most obvious difficulty is that the pulse comes out of the Schmitt trigger inverted. The obvious answer to that problem might be to place an inverter between the Schmitt trigger and the computer.

22.3 Noninverting Schmitt Trigger

Another approach to the above problem might be to design a noninverting Schmitt trigger such as that shown in Figure 22.6a. To analyze the circuit's behavior, assume that the circuit input voltage, V_i, starts out very negative as shown in Figure 22.6b. Since the input is to the noninverting op amp terminal, the output will be at $-V_{sat}$ as shown in Figure 22.6c. The negative voltage at V_o is fed back through R_f and tends to hold pin 3 negative even after V_i becomes positive. When V_i becomes so positive that it overcomes the effect of V_o, it has reached V_{UT}, and the output then switches to $+V_{sat}$ as shown in Figures 22.6b and 22.6c. The positive voltage now at V_o is fed back through R_f so it tends to hold pin 3 positive even after V_i becomes negative. Its effect is overcome by V_i when V_i reaches V_{LT}. Then V_o switches to $-V_{sat}$.

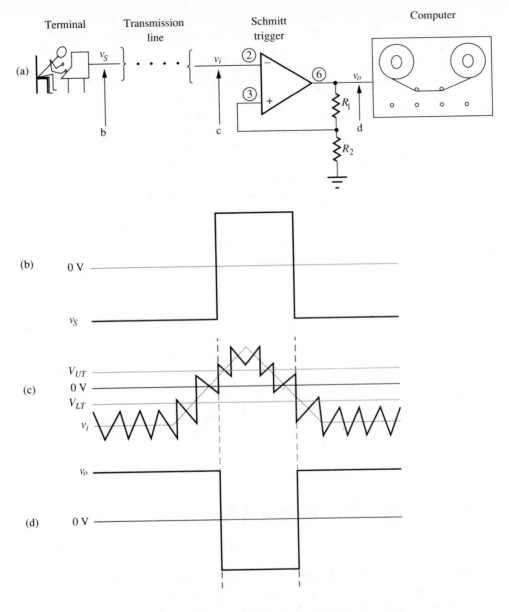

FIGURE 22.5   Pulse shaping with a Schmitt trigger.
(a) Circuit diagram.
(b) Waveform leaving user's terminal.
(c) Waveform arriving at the Schmitt trigger.
(d) Waveform delivered to the computer.

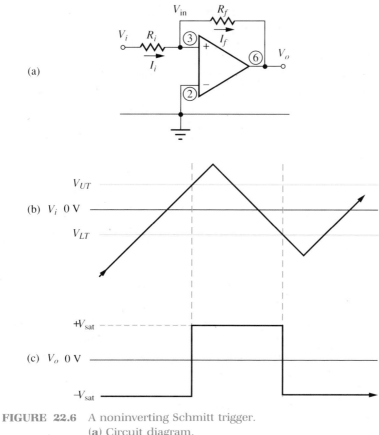

FIGURE 22.6 A noninverting Schmitt trigger.
(a) Circuit diagram.
(b) Input waveform.
(c) Output waveform.

In order to determine the interval between V_{UT} and V_{LT}, the place to begin is where we have begun so many times before

$$I_i = I_f \tag{20.4}$$

If we assume that we have begun with V_i very negative and have moved up the waveform of Figure 22.6b almost, but not quite to V_{UT} so that switching is imminent, then $V_o = -V_{sat}$ and V_3 is almost, but not quite ready to exceed V_2 and cause V_o to switch. Then $V_3 = V_2$. We can then apply Ohm's Law to Equation 20.4. We then have

$$\frac{V_{UT} - V_2}{R_i} = \frac{V_2 - (-V_{sat})}{R_f} \tag{22.4}$$

or

$$V_{UT} - V_2 = (V_2 - (-V_{sat}))\frac{R_i}{R_f} \qquad (22.5)$$

Next, if we assume that switching has occurred so that $V_o = +V_{sat}$ and V_i has passed its maximum value, passed through zero, and is approaching V_{LT}, so that switching is again imminent, Equation 22.4 now becomes

$$\frac{V_{LT} - V_2}{R_i} = \frac{V_2 - (+V_{sat})}{R_f} \qquad (22.6)$$

or

$$V_{LT} - V_2 = (V_2 - (+V_{sat}))\frac{R_i}{R_f} \qquad (22.7)$$

The interval between V_{UT} and V_{LT} can be obtained by subtracting Equation 22.7 from Equation 22.5. The result is

$$V_{UT} - V_{LT} = 2V_{sat}\frac{R_i}{R_f} \qquad (22.8)$$

As you may have guessed, it is not required that the above interval be centered about zero. It can be offset by applying a reference voltage, V_{ref} to pin 2 of the op amp. Adding Equations 22.5 and 22.7 and solving for V_2 gives the required offset

$$V_{ref} = V_2 = \frac{V_{UT} + V_{LT}}{2} * \frac{R_f}{R_i + R_f} \qquad (22.9)$$

EXAMPLE 22.2

Suppose that you have available a transducer that delivers -10 V at $60°F$ and varies linearly up to $+10$ V at $80°F$. Use it and a 741C op amp with ±15 V supplies to design a thermostat that will allow a variation of approximately $\pm1°F$ about any setting within that range.

SOLUTION:

The required circuit is shown in Figure 22.7. If the user selects a thermostat setting of $70°F$ then the control pot will be at the center of its range and $V_{ref} = 0$ V. A setting of $80°F$ gives $V_{ref} = +10$ V, etc. The temperature should then stay within $\pm1°F$ of that setting. If the user has placed the function switch in the "heat" position, when the temperature drops below the user's setting by $1°F$, the op amp output goes to $-V_{sat}$, D_2 conducts, and the furnace comes on.

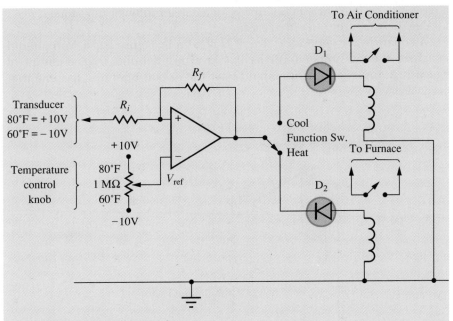

FIGURE 22.7 Example 22.2.

Since the transducer output varies one volt per degree of temperature change, the allowed temperature variation of $\pm 1°F$ gives a trip point range of ± 1 V or $V_{UT} - V_{LT} = 2$ V. Solving Equation 22.8 for R_i, using $R_f = 100$ kΩ gives

$$R_i = R_f \frac{V_{UT} - V_{LT}}{2 V_{sat}} \qquad (22.10)$$

$$= 100 \text{ k}\Omega \, \frac{2 \text{ V}}{2 * 13 \text{ V}} = 7.69 \text{ k}\Omega \qquad (\text{Use } 8.20 \text{ k}\Omega)$$

Since digital circuits demand very square pulses, the need for a Schmitt trigger on the inputs to such circuits is almost universal. Therefore, as you might expect, integrated circuits are available that have the Schmitt trigger right on the chip.

22.4 A Square Wave Generator

A square wave generator is one of a group of circuits known as "multivibrators," perhaps due to the fact that their output "vibrates" between high and low voltage, but ideally spends no time in between. It is a digital type of circuit. A square wave generator is an astable multivibrator since it does not stay permanently in either state.

A square wave generator circuit is shown in Figure 22.8a. You should recognize the op amp circuit as the inverting Schmitt trigger circuit of Figure 22.2. In order to understand how the circuit operates, assume that

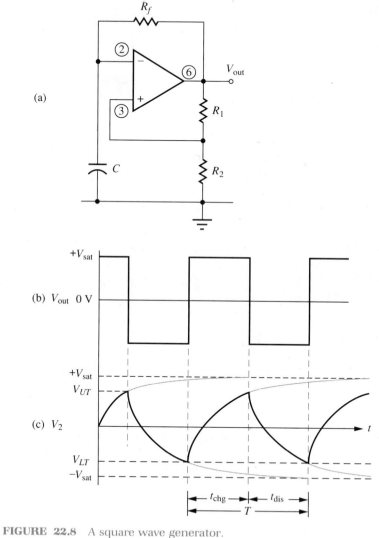

(a)

(b) V_{out} 0 V

(c) V_2

FIGURE 22.8   A square wave generator.
(a) Circuit diagram.
(b) Output waveform.
(c) Waveform at pin 2.

the capacitor is initially discharged. That means that $V_2 = 0$ V. Further, assume that, due to some slight offset, V_{out} is very slightly positive. A fraction of that positive voltage is fed back to pin 3. This positive feedback causes the output to jump to $V_{out} = +V_{sat}$ as shown in Figure 22.8b. The voltage V_3 is then determined by the voltage divider R_1 and R_2 but since it is positive, it holds V_{out} at $+V_{sat}$.

When the output jumps to $+V_{sat}$, it begins to charge the capacitor toward $+V_{sat}$ as shown by the waveform in Figure 22.8c. However, when V_2 exceeds the upper trip point given by

$$V_{UT} = +V_{sat} \frac{R_2}{R_2 + R_1} \tag{22.11}$$

then V_{out} switches to $-V_{sat}$. The capacitor then begins to discharge toward $-V_{sat}$. It discharges until it reaches the lower trip point given by

$$V_{LT} = -V_{sat} \frac{R_2}{R_2 + R_1} \tag{22.12}$$

The process is symmetrical and repeats itself continuously as shown in Figure 22.8b and c.

In order to determine the frequency of oscillation of the circuit of Figure 22.8a, in your circuits class you should have learned that the capacitor charging equation can be written

$$V_t = V_S - (V_S - V_0)e^{-t/(RC)} \tag{22.13}$$

By referring to Figure 22.8c, you can see that in our situation, for the charging curve the variables in Equation 22.13 become: $t = t_{chg}$, $V_0 = V_{LT}$, $V_t = V_{UT}$, and $V_S = +V_{sat}$. Making these substitutions into Equation 22.13 and solving for t_{chg} gives

$$t_{chg} = -R_f C * \ln\left(\frac{V_{sat} - V_{UT}}{V_{sat} - V_{LT}}\right) \tag{22.14}$$

From Figure 22.8 you can see that if the circuit is symmetrical, that is, if $+V_{sat} = -V_{sat}$ then the frequency of oscillation, f, is given by

$$f = \frac{1}{T} = \frac{1}{2t_{chg}} \tag{22.15}$$

EXAMPLE 22.3
Determine the required value of R_f in the circuit in Figure 22.9 in order that the frequency of oscillation will be approximately 100 Hz.

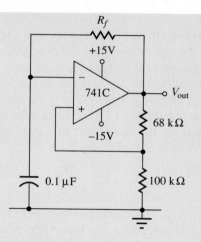

FIGURE 22.9   Example 22.3.

SOLUTION:

The trip points are given by

$$V_{UT} = +V_{sat} \frac{R_2}{R_2 + R_1} \tag{22.11}$$

$$= 13 \text{ V} \frac{100 \text{ k}\Omega}{100 \text{ k}\Omega + 68 \text{ k}\Omega} = 7.74 \text{ V}$$

Assuming that $+V_{sat} = -(-V_{sat})$, then by symmetry

$$V_{LT} = -7.74 \text{ V}$$

Solving Equation 22.15 for the charging time, t_{chg} gives

$$t_{chg} = \frac{1}{2f}$$

$$= \frac{1}{2 * 100 \text{ Hz}} = 5 \text{ msec}$$

Solving Equation 22.14 for R_f gives

$$R_f = -\frac{t_{chg}}{C * \ln\left(\dfrac{V_{sat} - V_{UT}}{V_{sat} - V_{LT}}\right)} \tag{22.16}$$

$$= -\frac{5 \text{ msec}}{0.1 \text{ }\mu\text{F} * \ln\left(\dfrac{13 \text{ V} - 7.74 \text{ V}}{13 \text{ V} - (-7.74 \text{ V})}\right)} = 36.5 \text{ k}\Omega$$

(Use 39 kΩ)

22.5   A Monostable Multivibrator

Early in Section 22.4 we mentioned that the circuit of Figure 22.8 is known as an astable multivibrator because it is, in fact, astable. Its output will not stay permanently either high or low.

Consider the circuit of Figure 22.10a. Switch S could be one of the buttons on your calculator or on the keyboard of your computer. If you look closely at the circuit, you will see that, except for the input circuitry

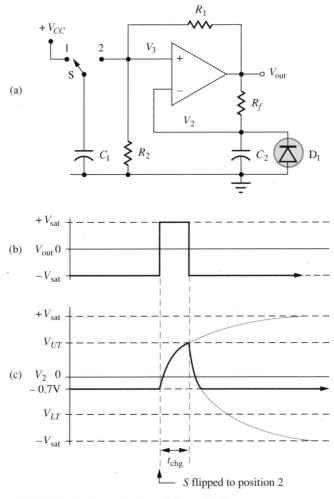

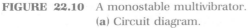

FIGURE 22.10 A monostable multivibrator.
(a) Circuit diagram.
(b) Output waveform.
(c) Waveform at pin 2.

and the diode D_1, it is the same circuit as the astable multivibrator of Figure 22.8.

Upper and lower trip points are determined by the voltage divider R_1 and R_2 in Figure 22.10a just as before, but diode D_1 prevents the circuit from being a square wave generator. Notice that if the output is positive, D_1 is reverse biased so C_2 charges to the upper trip point just as it did in Figure 22.8, but when the output is negative, D_1 is forward biased so C_2 will never charge negatively beyond about -0.7 V. Thus C_2 is never charged to the lower trip point. The output is stable and stays at $-V_{sat}$, held in that state by the fact that D_1 prevents the capacitor from charging to V_{LT}.

Now suppose that switch S is momentarily flipped from position 1 to position 2. Assume that capacitor C_1 is quite small so that it will deliver only a very brief pulse to pin 3 of the op amp. Since it will be a positive pulse, and since, as we have shown the voltage at pin 2 is about -0.7 V, the positive pulse applied to pin 3 will cause the output to go to $+V_{sat}$ where it will stay until C_2 charges up to V_{UT}. This activity is shown in Figures 22.10b and 22.10c.

Thus the width of the pulse in Figure 22.10b, t_{chg}, is independent of how long switch S is held in position 2. This allows the designer to optimize the width of the pulses that the keyboard delivers to your computer. As you would expect, Equation 22.13 applies, and Equation 22.14 now becomes

$$t_{chg} = -R_fC_2 * \ln\left(\frac{V_{sat} - V_{UT}}{V_{sat} - (-0.7 \text{ V})}\right)$$

A monostable multivibrator is also known as a "one shot" because it delivers a single measured pulse every time it is triggered.

22.6 A Triangle Wave Generator

In Section 20.8 we introduced the idea of an integrator circuit, which we also called a ramp generator. If you compare the circuit of U_1 in Figure 22.11a with the circuit in Figure 20.7, you will see that they are essentially the same. U_1 in Figure 22.11a is a ramp generator. If you compare the circuit of U_2 in Figure 22.11a with that of Figure 22.6, you will see that U_2 is a noninverting Schmitt trigger.

In order to understand how the circuit operates, we will begin as we did with the square wave generator by assuming that C_{f1} is discharged and that there is some slight positive voltage at either V_{o1} or V_{o2}. This will cause a very slight positive voltage at the noninverting input to U_2. Due to the

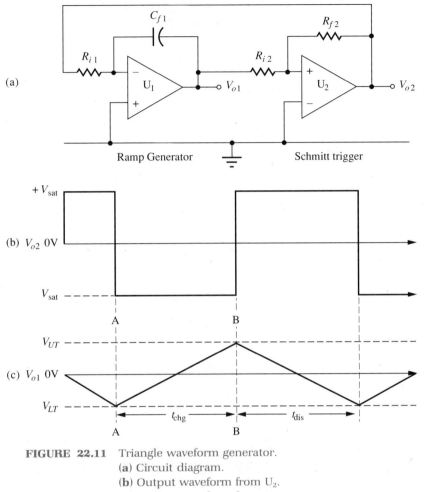

FIGURE 22.11   Triangle waveform generator.
(**a**) Circuit diagram.
(**b**) Output waveform from U_2.
(**c**) Output waveform from U_1.

positive feedback to U_2, V_{o2} will immediately go to $+V_{\text{sat}}$ as shown at the beginning of Figure 22.11b. According to Section 20.8, the ramp generator, U_1 then begins to integrate in a negative direction as shown at the beginning of Figure 22.11c.

According to Section 22.3, when V_{o1} reaches the lower trip point, V_{LT}, then U_2 changes state and V_{o2} goes to $-V_{\text{sat}}$ as shown at Point A of Figure 22.11b. Then U_1 begins to integrate in the positive direction as shown at Point A of Figure 22.11c.

When V_{o1} reaches the upper trip point, V_{UT}, at point B in Figure 22.11c, V_{o2} switches to $+V_{\text{sat}}$ again. The process is symmetrical, assuming that

$+V_{sat} = -(-V_{sat})$ and it repeats itself continuously as shown in Figure 22.11. As you can see, a square wave is available at V_{o2} and a triangle wave is available at V_{o1}.

In order to determine the frequency of the output, the integration voltage interval, $V_{UT} - V_{LT}$ can be determined from the noninverting Schmitt trigger equation developed in Section 22.3

$$V_{UT} - V_{LT} = 2\,V_{sat}\,\frac{R_{i2}}{R_{f2}} \qquad (22.8)$$

Since the input to the integrator is constant at $-V_{sat}$, applying the notation of this chapter to Equation 20.23 gives

$$V_{UT} - V_{LT} = -(-V_{sat})\frac{t_{chg}}{R_{i1}C_{f1}} \qquad (22.17)$$

Combining Equation 22.8 and Equation 22.17 and solving for the integration period t_{chg} gives

$$t_{chg} = 2R_{i1}C_{f1}\,\frac{R_{i2}}{R_{f2}} \qquad (22.18)$$

Again, assuming that the oscillation is symmetric, its frequency is given by

$$f = \frac{1}{2t_{chg}} \qquad (22.15)$$

EXAMPLE 22.4

Determine the value of R_{i1} and R_{i2} for the circuit of Figure 22.12 that will give a triangle wave output of about 20 VPP and 100 Hz.

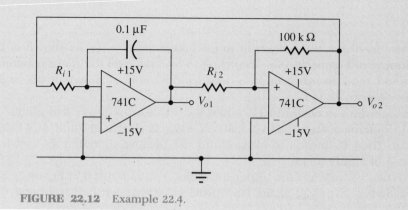

FIGURE 22.12   Example 22.4.

SOLUTION:

Solving Equation 22.8 for R_{i2} gives

$$R_{i2} = R_{f2}\frac{V_{UT} - V_{LT}}{2\,V_{sat}} \qquad (22.19)$$

$$= 100\text{ k}\Omega\,\frac{20\text{ VPP}}{2 * 13\text{ V}} = 76.9\text{ k}\Omega \qquad (\text{Use }82\text{ k}\Omega)$$

The integration time, t_{chg} is

$$t_{chg} = \frac{T}{2} = \frac{1}{2f}$$

$$= \frac{1}{2 * 100\text{ Hz}} = 5\text{ msec} \qquad (22.20)$$

Solving Equation 22.18 for R_{i1} gives

$$R_{i1} = \frac{t_{chg}}{2\,C_{f1}}\frac{R_{f2}}{R_{i2}}$$

$$= \frac{5\text{ msec}}{2 * 0.1\ \mu\text{F}}\frac{100\text{ k}\Omega}{82\text{ k}\Omega} = 30.5\text{ k}\Omega \qquad (\text{Use }33\text{ k}\Omega)$$

22.7 A Sawtooth Oscillator

In Section 20.8 we claimed that a ramp generator is used in the horizontal trace circuit of your oscilloscope or your TV set. If you have ever set the horizontal trace rate of your scope at such a slow rate that your eye could follow the trace, you have probably noticed that the trace travels at a constant rate as it moves from left to right. Then, when it gets to the right side of the screen, it jumps very quickly back to the left side of the screen and begins another trip across the screen. This is an indication that the circuit used is not that of Figure 22.11 or 22.12 because the waveform of the deflection circuit is not symmetrical. The time to charge C_{f1}, t_{chg}, is much different from the discharge time, t_{dis}.

Consider the circuit of Figure 22.13a. Notice that it is the same as Figure 22.11 except for the addition of diode D_1. If we assume ideal components, D_1 will have no effect on the circuit when V_{o2} is negative, since it is then reverse biased. Thus, we can use Equation 22.18 to give the value of t_{chg}. However, when V_{o2} is positive, Equation 22.18 becomes

$$t_{dis} = 2 * 0\ \Omega * C_{f1}\frac{R_{i2}}{R_{f2}} = 0\text{ sec} \qquad (22.21)$$

The above relationships are shown in Figures 22.13b and 22.13c. If you hook up the circuit of Figure 22.13a, and you should, it is doubtful if all of

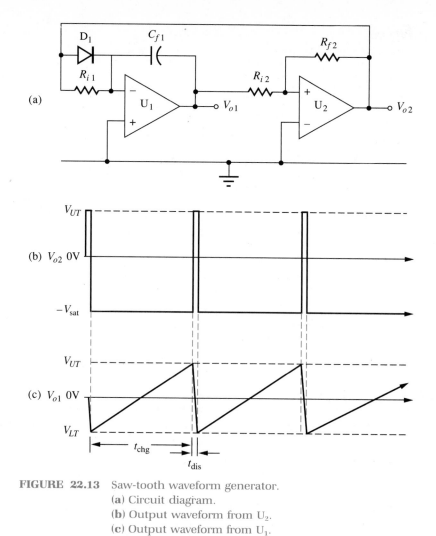

FIGURE 22.13 Saw-tooth waveform generator.
(a) Circuit diagram.
(b) Output waveform from U_2.
(c) Output waveform from U_1.

the components that you use will be ideal (!). Thus it is unlikely that $t_{dis} = 0$ sec. The value actually observed is dependent on the slew rate of the op amps and how much current U_1 can conduct in discharging C_{f1}. For the same reason, if you hook up the circuit, you will find that there is considerable overshoot in the sawtooth waveform unless you put some resistance in series with D_1.

EXAMPLE 22.5
Determine the discharge time, t_{dis}, for Example 22.4 if diode D_1 from Figure 22.13 is added to the circuit of Figure 22.12. Assume that the

slew rate of the op amp is the determining factor in determining the discharge time, t_{dis}.

SOLUTION:

Although the slew rate of U_1 will be a factor in determining t_{dis}, it has to slew only from V_{UT} to V_{LT}. Since U_2 has to slew from $-V_{sat}$ all of the way up to V_{UT} and back to $-V_{sat}$ before the next integration can begin, it will be the controlling factor.

Appendix F lists the slew rate for the 741C as 0.5 V/μsec. Thus the time for U_2 to make the transition from $-V_{sat}$ to V_{UT} and back to $-V_{sat}$ is

$$t_{dis} = 2\frac{V_{UT} - (-V_{sat})}{\text{slew rate}} \tag{22.22}$$

$$= 2\frac{10\text{ V} - (-13\text{ V})}{0.5\text{ V/μsec}} = 92\text{ μsec}$$

22.8 A Wein Bridge Oscillator

The oscillator circuits we have seen so far are relatively stable since at least part of the circuits can be considered to be digital in the sense that the output of the op amp is always at either $+V_{sat}$ or $-V_{sat}$. When a sine wave output waveform is desired, saturation must be avoided since saturation usually produces clipping. Perhaps the most common sine wave oscillator circuit is the Wein bridge circuit shown in Figure 22.14.

In order to understand how the circuit operates, notice that since there is feedback to the noninverting input of the op amp, the possibility

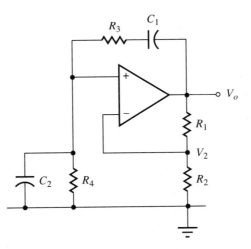

FIGURE 22.14 Basic Wein bridge oscillator circuit.

for oscillation exists. The presence of capacitors in the feedback circuit indicate that the feedback will be frequency-dependent. The higher the frequency, the more signal C_1 will feed back. Thus C_1 and $R_3 + R_4$ form a high pass filter. We talked about the behavior of a high pass filter in Section 12.4. It will produce a positive phase angle in the feedback signal.

On the other hand, you can see that C_2 will tend to conduct the higher frequencies off to ground. Thus C_2 is part of a low pass filter circuit whose behavior we described in Section 12.10. It will tend to produce a negative phase shift in the feedback signal.

Since the feedback is to the noninverting input, in order for oscillation to occur, the phase shift due to C_1 and C_2 must cancel out so that the overall phase shift is zero. This determines the oscillation frequency f_o. In order to determine the value of f_o for the circuit of Figure 22.14, we will begin with the voltage divider equation for the circuit providing feedback to pin 3 of the op amp.

$$\frac{V_o}{R_3 + (-jX_{C1}) + R_4 \| (-jX_{C2})} = \frac{V_3}{R_4 \| (-jX_{C2})} \tag{22.23}$$

or

$$\frac{V_o}{V_3} = \frac{R_3 + (-jX_{C1}) + R_4 \| (-jX_{C2})}{R_4 \| (-jX_{C2})}$$

$$= \frac{R_3 + (-jX_{C1})}{R_4 \| (-jX_{C2})} + 1$$

or, using the "product over the sum" rule for paralleling,

$$\frac{V_o}{V_3} = \frac{(R_3 + (-jX_{C1}))(R_4 + (-jX_{C2}))}{R_4(-jX_{C2})} + 1 \tag{22.24}$$

In order for the Wein bridge circuit to be symmetrical, it is customary that

$$R = R_3 = R_4 \tag{22.25}$$

and

$$C = C_1 = C_2 \tag{22.26}$$

Substituting Equations 22.25 and 22.26 into Equation 22.24 gives

$$\frac{V_o}{V_3} = \frac{(R + (-jX_C))(R + (-jX_C))}{R(-jX_C)} + 1 \tag{22.27}$$

$$= \frac{R^2 + 2R(-jX_C) - X_C^2}{R(-jX_C)} + 1$$

$$= \frac{R}{-jX_C} + 2 - \frac{X_C}{R(-j)} + 1$$

$$= j\frac{R}{X_C} - j\frac{X_C}{R} + 3 \tag{22.28}$$

As we mentioned above, in order for the Wein bridge to oscillate, the net phase shift of the feedback circuit must be zero. This means that the complex terms in Equation 22.28 must sum to zero. Thus

$$j\frac{R}{X_C} - j\frac{X_C}{R} = 0 \tag{22.29}$$

or

$$\frac{R}{X_C} = \frac{X_C}{R}$$

"Cross multiplying" gives

$$R^2 = X_C{}^2$$

or

$$R = X_C = \frac{1}{2\pi f_o C} \tag{22.30}$$

Thus the oscillation frequency for the Wein bridge is given by

$$\bullet \qquad f_o = \frac{1}{2\pi RC} \tag{22.31}$$

As we mentioned at the beginning of this section, in order for an oscillator to produce sine waves, its output must not saturate. In order to prevent saturation, the gain of the amplifier must be carefully controlled. If the gain is too high, the output saturates and clipping results. If the gain is too low, the circuit will not oscillate. The positive feedback makes the gain requirement very critical; a little like trying to balance a pin on its point. Substituting Equation 22.29 into Equation 22.28 shows us that for oscillation, the gain should be

$$A_V = 3.00 \tag{22.32}$$

If you consider only resistors R_1 and R_2 of the circuit in Figure 22.14, you can see that the circuit looks like the simple noninverting amplifier of Section 20.9 which has a gain of

$$A_V = \frac{R_1 + R_2}{R_2} \tag{20.28}$$

It is this part of the circuit that must control the gain. In order to "soften" the rather critical gain requirement mentioned above, and to provide some control over the feedback, the circuit of Figure 22.15, or some variation of it is often used.

In order to understand how the circuit works, note that as long as V_o is small, diodes D_1 and D_2 will not conduct so initially we can ignore them. If we start with the pot at the top of its travel, $R_{1b} = 0\ \Omega$ and $R_2 = 10\ \text{k}\Omega$.

This gives a gain much less than the required $A_V = 3$ so the circuit does not oscillate. As the wiper on the pot is moved toward the bottom of its travel, the required gain is eventually obtained and the circuit begins to oscillate. The use of a 10 kΩ pot allows for a large enough variation in R_1 and R_2 to accommodate any minor circuit variations.

Due to the positive feedback the circuit is unstable. Thus when oscillation begins, the peak amplitude of V_o immediately increases toward V_{sat} which would result in clipping if it were allowed to continue. However, when the voltage across R_{1a} reaches about 0.6 V, diode D_1 begins to conduct on the positive voltage swing and D_2 begins to conduct on the negative swing. This effectively shorts out R_{1a}, thus reducing the gain to a value below the required minimum of $A_V = 3.00$. This limits the resulting amplitude of V_o to whatever value is required to turn on the diodes. Resistor R_{1a} is chosen to be about

$$R_{1a} = \frac{R_1 + R_2}{10} \tag{22.33}$$

so that the change in gain when the diodes begin to conduct will be fairly small. This, in addition to the rather gradual turn on of the diodes, places a "soft" limit on the increase of V_o and so minimizes distortion. In addition, this provides a range in the pot setting between where oscillation begins and where objectionable distortion is first noticed, making its setting less critical.

EXAMPLE 22.6

For the circuit of Figure 22.15, determine the approximate pot setting (the value of R_2) required to produce an output voltage of 20 VPP.

SOLUTION:

Since the circuit is symmetrical, we will look only at the positive peak output. In order to determine the operating point for the diode, we will construct a Thevenin equivalent for the diode circuit as indicated in Figure 22.16. From Figure 22.16a

$$V_{Th} = V_o \frac{R_{1b} + R_2}{R_{1a} + R_{1b} + R_2} \tag{22.34}$$

$$= 10 \text{ V} \frac{10 \text{ k}\Omega}{1.2 \text{ k}\Omega + 10 \text{ k}\Omega} = 8.93 \text{ V}$$

and

$$R_{Th} = R_{1a} \parallel (R_{1b} + R_2) \tag{22.35}$$

$$= 1.2 \text{ k}\Omega \parallel 10 \text{ k}\Omega = 1.07 \text{ k}\Omega$$

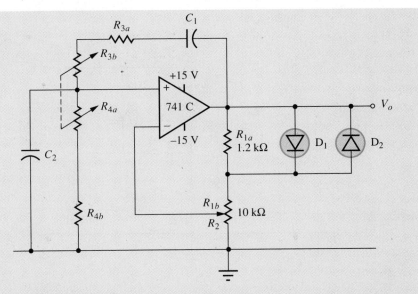

FIGURE 22.15 Wein bridge oscillator with output limiting.

Doing a KVL walk from V_o to V_{Th} in Figure 22.16b gives

$$V_o - V_D - I_D R_{Th} = V_{Th} \tag{22.36}$$

or

$$I_D = \frac{V_o - V_{Th} - V_D}{R_{Th}} \tag{22.37}$$

$$= \frac{10\ V - 8.93\ V - V_D}{1.07\ k\Omega} = \frac{1.07\ V - V_D}{1.07\ k\Omega}$$

Iterating Equation 22.37 with the diode equation gives

$I_D = \dfrac{1.07\ V - V_D}{1.07\ k\Omega} =$	347	411	407	µA
$V_D = 26\ mV * \ln\dfrac{I_D}{10\ fA} =$	0.631	0.635	0.635	V

Now that both V_D and I_D are known, returning to the circuit of Figure 22.16a, the current I_{1a} flowing through R_{1a} is given by

$$I_{1a} = \frac{V_D}{R_{1a}} \tag{22.38}$$

$$= \frac{0.635\ V}{1.2\ k\Omega} = 529\ \mu A$$

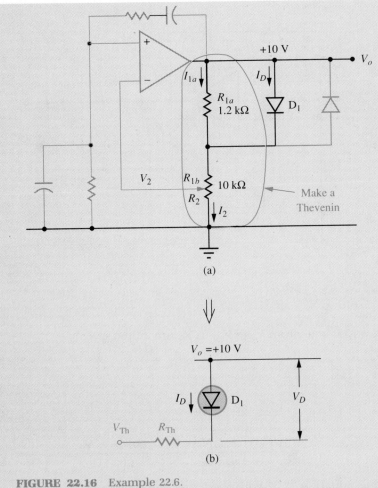

FIGURE 22.16 Example 22.6.
(a) Circuit diagram emphasizing the negative feedback circuit when $v_o > 0V$.
(b) Thevenin equivalent of the negative feedback circuit. Note that $V_{Th} \neq V_2$.

From the circuit of Figure 22.16a you can see that the current flowing through R_2 is given by

$$I_2 = I_{1a} + I_D \qquad (22.39)$$
$$= 529 \ \mu A + 407 \ \mu A = 936 \ \mu A$$

In order to obtain the required gain of $A_V = 3$, the voltage V_2 fed back to pin 2 is

$$V_2 = \frac{V_o}{A_V} \tag{22.40}$$

$$= \frac{10\ \text{V}}{3} = 3.33\ \text{V}$$

Ohm's Law then gives the required value for R_2

$$R_2 = \frac{V_2}{I_2} \tag{22.41}$$

$$= \frac{3.33\ \text{V}}{936\ \mu\text{A}} = 3.56\ \text{k}\Omega$$

22.9 Troubleshooting ◄◄◄◄◄◄◄◄

As we have said before, the most important characteristic you should remember when you are called on to troubleshoot an op amp circuit is the fact that op amps are comparators. Thus the place to begin is to note the voltage at the inverting input and at the noninverting input. If the noninverting input voltage is higher than the inverting input voltage, then the output should be at $+V_{\text{sat}}$. If the noninverting input voltage is lower than the inverting input voltage, the output should be at $-V_{\text{sat}}$. Any other value of V_o indicates either an overloaded output or a defective op amp.

If you are confronted with a Schmitt trigger circuit whose trip points are not where they should be, or an oscillator circuit that either does not oscillate, or oscillates at the wrong frequency, examine the feedback circuit for defective components. As with any other circuit, a thorough understanding of how the circuit should operate is your most valuable troubleshooting tool.

22.10 Summary

This chapter introduced the concept of positive feedback in op amp circuits. We showed how hysteresis can be used in Schmitt trigger type circuits, and how it can be used with timing circuits to construct oscillator circuits. Finally we showed how both positive feedback and negative feedback are combined in the Wein bridge oscillator circuit.

GLOSSARY

Astable multivibrator: A multivibrator whose output cannot be made to stay permanently either high or low, but continually oscillates between states. (Section 22.4)

Hysteresis: That property of a device or circuit which causes a past condition to influence present behavior. (Section 22.2)

Monostable multivibrator: (Also known as a "one shot.") A multivibrator whose output has one stable state. When triggered by an external input its output provides a pulse of fixed duration. (Section 22.5)

Multivibrator: A circuit or device whose output is always either a high voltage or a low voltage, but which ideally spends no time in between. (Section 22.4)

One shot: See MONOSTABLE MULTIVIBRATOR.

Schmitt trigger: An electronic circuit that produces a square wave output by switching at predetermined setpoints on the input signal called the upper trip point, V_{UT} and the lower trip point, V_{LT}. (Section 22.2)

Snap action switch: A switch whose contacts are actuated by an over center spring in such a way that the contacts are always either fully open or fully closed. (Section 22.2)

Trip point: The value of input voltage required to cause the output of a trigger circuit to change state. (Section 22.2)

PROBLEMS

Section 22.2

1. Determine the value of V_o for the circuit of Figure 22.17 when
   - (a) $V_i = -1$ V.
   - (b) $V_i = 0$ V.
   - (c) $V_i = +1$ V.

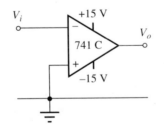

FIGURE 22.17 Problem 22.1.

2. Determine the upper and lower trip points for the Schmitt trigger circuit of Figure 22.18.
3. Draw the input and the output waveforms similar to those in Figure 22.2b and 22.2c if a triangle waveform of 10 VP is input to the circuit in Figure 22.18.

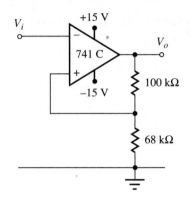

FIGURE 22.18   Problem 22.2.

4. Repeat Problem 3 for the circuit of Figure 22.19.

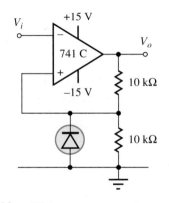

FIGURE 22.19   Problem 22.4.

5. Design an inverting Schmitt trigger using a 741C op amp and ± 15 V supplies that will trip at about ± 1 V.
6. Determine the maximum peak-to-peak noise that can be riding on top of the signal in Problem 2 before the output will begin to show false pulses.

Section 22.3

7. Repeat Problem 5 using a noninverting Schmitt trigger circuit.
8. Determine the upper and lower trip points for the circuit of Figure 22.20.
9. If the input to the circuit of Figure 22.20 varies from $V_i = 0$ V to $V_i = 10$ V, determine the maximum current that the source will have to deliver and the maximum current it will have to sink.

10. For the circuit of Figure 22.20, determine the value of V_{ref} that will give a lower trip voltage of $V_{LT} = 0$ V.

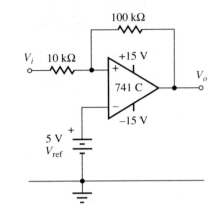

FIGURE 22.20   Problem 22.8.

Section 22.4

11. Calculate the frequency of oscillation of the circuit in Figure 22.21.
12. There will be some leakage current into both the op amp noninverting input and into the capacitor in Figure 22.21. How much total leakage current will it take to stop the circuit from oscillating?

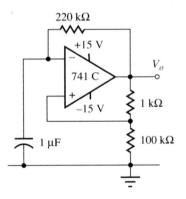

FIGURE 22.21   Problem 22.11.

13. Calculate the size of the capacitor required to cause the circuit of Figure 22.22 to oscillate at a frequency of 100 Hz.

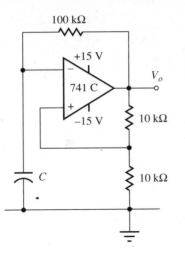

100 kΩ

+15 V

741 C

V_o

-15 V

10 kΩ

C

10 kΩ

FIGURE 22.22   Problem 22.13.

Section 22.5

14. Convert the circuit of Figure 22.22 into a one shot that will produce pulses of approximately 10 msec pulse width.

Section 22.6

15. If the resistor values calculated in Example 22.4 are used but the capacitor in the circuit of Figure 22.12 is changed to $C = 1.0$ μF, determine the effect on both the amplitude and the frequency of
    (a) The triangle wave output, V_{o1}.
    (b) The square wave output, V_{o2}.
16. Repeat Problem 15 if, instead of changing the capacitor value, the 100 kΩ resistor is replaced with a 10 kΩ resistor. *Hint:* If the trip voltage is too high the circuit will not oscillate. How high is "too high"?

Section 22.8

17. Determine the frequency of oscillation of the circuit in Figure 22.14 if $R_3 = R_4 = 100$ kΩ and $C_1 = C_2 = 0.1$ μF.
18. For the circuit of Figure 22.23, determine the value of resistors R_{3a}, R_{3b}, R_{4a}, and R_{4b} if the oscillation frequency is to be adjustable over the range from about $f_o = 100$ Hz to $f_o = 1$ kHz. Note that R_{3b} and R_{4a} are "ganged." The term refers to a single control that varies both pots at the same time.

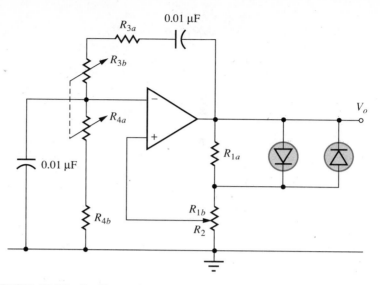

FIGURE 22.23 Problem 22.18.

19. Determine the gain of the amplifier of Example 22.6 before the diodes begin to conduct.
20. Using the pot setting determined in Example 22.6 for the circuit of Figure 22.16, determine what the gain of the amplifier would be if the peak output amplitude were to reach 14 VP.

Troubleshooting ▶   Section 22.9

21. Describe the effect on the output of the circuit of Figure 22.5 if a "cold solder joint" made it appear that:
    (a) $R_1 = \infty \ \Omega$
    (b) $R_2 = \infty \ \Omega$
22. Describe the behavior of the furnace (or air conditioner) in the circuit of Figure 22.7 if:
    (a) Due to a "dirty pot," the slider no longer made contact with the 1 MΩ resistor.
    (b) A "cold solder joint" resulted in $R_f = \infty \ \Omega$.
    (c) Diode D_2 became a short circuit. *Hint:* Relays are not sensitive to the direction of current flow.
23. Determine the voltage reading and waveform at Pins 2, 3, and 6 in the circuit of Figure 22.22 if the 1 µF capacitor became so leaky that it behaved as though there was a 50 kΩ resistor in parallel with it.

A 555 timer is often used to construct a siren. (Photograph by Henley and Savage/The Stock Market.)

Chapter 23
The 555 Timer

SPECIAL TERMS

Flip-flop
High
Low
S
R
Q
$\overline{Q}$
DIP
Current sink
Duty cycle

23.1   Introduction and Objectives

In the last chapter we found that an op amp can be used as a function generator, which is a timing circuit of sorts. In this chapter we will examine the 555 timer which, as its name implies, is specifically intended to be used in timing circuits. It, like the 741 op amp is a very widely used integrated circuit.

After having studied this chapter, you should have an introductory understanding of some of the terminology used in the field of digital electronics. You should also have some idea of what an S-R flip-flop is and how it is used in a 555 timer. You should also know how the 555 timer is constructed in terms of a simplified block diagram. You should be able to:

- Describe the behavior of an S-R flip-flop.
- Describe the behavior of a 555 timer in terms of the simplified block diagram presented in this chapter.
- Use the above block diagram to analyze and design simple circuits containing a 555 timer.

23.2   An S-R Flip-Flop

The 555 timer can be considered to stand at the interface between the analog and the digital branches of electronics. Analog is the field that we have been studying in which voltages and currents vary more or less linearly in response to some stimulus. Digital is the field of "ones" and "zeros," switches that are either on or off, lights that are either lit or not, and never dim. The input to a 555 can be analog in the sense that its voltage may gradually change over time. The output of the 555 is digital in the sense that it is either high or low, on or off.

In order to introduce some of the language of the digital branch of electronics, let us look at what is called an "S-R flip-flop." Consider the circuit of Figure 23.1. You can see where the "S-R" in the name comes from. The noninverting input to op amp U_1 is the S input and the noninverting input to op amp U_2 is the R input.

In order to understand the behavior of the circuit, the value of the resistors in Figure 23.1 is not critical, but we will assume that they all have the same value. We will begin by assuming that a "logical 1" is applied to the S input. That can be accomplished by attaching one end of a jumper wire to the positive power supply and momentarily touching the other end to the S terminal. Since the S terminal is the noninverting input to op amp U_1, this will cause the Q output to "go high." That is, the output of U_1 will go to $+V_{sat}$.

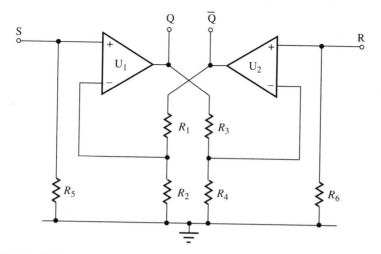

FIGURE 23.1 An op amp S-R flip-flop.

Since $R_3 = R_4$, half of $+V_{sat}$ is then applied to the inverting input of U_2. This causes the $\overline{Q}$ output to "go low." That is, the output of U_2 will go to $-V_{sat}$. The voltage divider R_1 and R_2 return part of $-V_{sat}$ to the inverting input of U_1 effectively locking the Q output high and the $\overline{Q}$ output low.

The jumper wire can now be removed from the S terminal and the output will not change. The S terminal SETS Q high and $\overline{Q}$ low and the circuit "remembers" that it was SET. This is how computers and calculators remember. This condition is shown in Figure 23.2a.

If you now momentarily touch the jumper wire to the R terminal, since the $+15$ V applied to the noninverting input to U_2 is higher than the $+6.5$ V at the inverting input, the $\overline{Q}$ output goes to $+V_{sat}$. This sends $+6.5$ V to the inverting input of U_1, which sends the Q output to $-V_{sat}$. This sends -6.5 V to the inverting input of U_2, locking the $\overline{Q}$ output high and the Q output low.

Again the jumper wire can be removed. The circuit remembers that the R terminal is the last one you touched. These results are summarized in Figure 23.2b. Notice that Q and $\overline{Q}$ are always in opposite states. When one is high, the other is low. In digital language, $\overline{Q}$ is "NOT Q." "NOT" means "in the opposite state." You will find S, R, Q and $\overline{Q}$ terminals on many digital chips, often in combination with other terminals as well. They typically have the meanings that have been attached to them here.

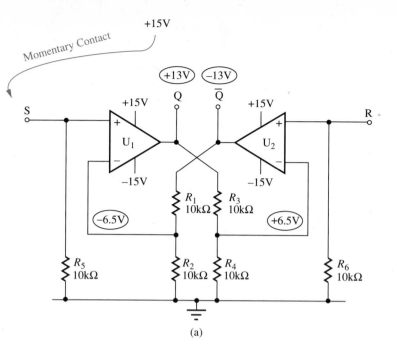

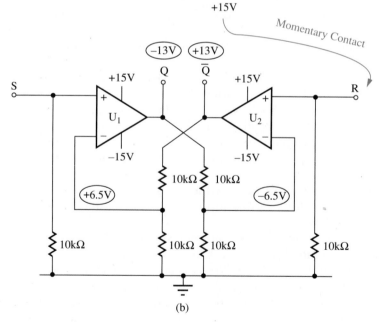

FIGURE 23.2   Operation of an S-R circuit.
(a) Voltage distribution following SET.
(b) Voltage distribution following RESET.

23.3   Introduction to the 555 Timer

Like the 741 op amp, the 555 timer is an integrated circuit that is widely used in electronics. A block diagram of the device is shown in Figure 23.3.

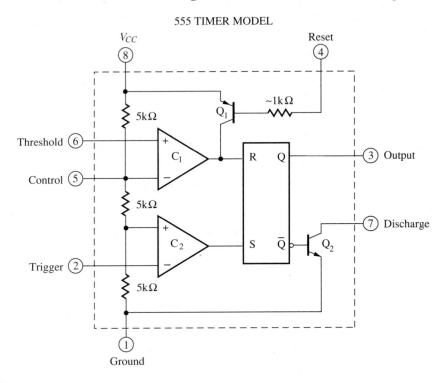

FIGURE 23.3   Block diagram for a 555 timer.

TYPICAL OPERATION:

Supply voltage (V_{CC}):	5 V to 18 V
Supply current (I_{CC}):	10 mA
Output voltage (low):	0.1 V
Output voltage (high):	$V_{CC} - 0.5$ V
Threshold voltage:	2/3 V_{CC}
Control voltage:	2/3 V_{CC} (N/C)
Trigger voltage:	1/3 V_{CC}
Reset voltage:	$< V_{CC} - 0.7$ V
Reset current:	0.1 mA

ABSOLUTE MAXIMUM RATINGS:

Supply voltage (V_{CC}):	18 V
Power dissipation:	600 mW
Input voltages:	V_{CC}
Output current (source or sink):	200 mA
Discharge current (sink)	200 mA

You will find that Figure 23.3 is a simplified version of the diagram on the manufacturer's data sheet. The most important components of the 555 timer are a voltage divider, two comparators, an S-R flip-flop, and a couple of transistors. These are all devices with which you should now be familiar.

In order to understand how the 555 works, first notice that according to Figure 23.3, pin 8 of the standard 8 pin DIP package in which the 555 is usually housed, is labeled "VCC." As you might expect, it is usually connected to the positive power supply which can have any positive voltage between +5 V and +18 V. Pin 1 is usually labeled "GND" and is usually grounded. The 555 has the distinct advantage of requiring only a single power supply.

Most of the internal circuits of the 555 are connected in some way between pins 8 and 1, but for simplicity those connections are not shown in Figure 23.3. One exception is the voltage divider consisting of the three 5 kΩ resistors which normally divide the power supply voltage into three equal parts.

Look next at the "trigger" input which is connected to pin 2. It is usually labeled "TRG." Figure 23.3 shows it going to the inverting input of comparator C_2. The comparator compares the pin 2 input voltage to the voltage at the bottom section of the voltage divider which is normally 1/3 of V_{CC}. Thus, when the input to pin 2 is less than 1/3 of V_{CC}, the output of C_2 SETs the output of the S-R flip-flop high. This output is available at pin 3 which is labeled appropriately enough, "OUT." From your study of Section 23.2 you know that once the output is SET high, it will not change regardless of what voltage is placed on the TRG input. The output will remain SET until it is RESET.

Next let us look at pin 6, the "threshold" input. It is usually labeled "THR." According to Figure 23.3, pin 6 is connected to the noninverting input to comparator C_1. Since the inverting input is connected to the top section of the voltage divider, normally, whenever the input to pin 6 is greater than 2/3 of V_{CC}, the output of comparator C_1 will go high. This RESETs the S-R flip-flop and causes the output at pin 3 to go low.

EXAMPLE 23.1
Draw the output waveform for the circuit of Figure 23.4a if the input waveform is that shown in Figure 23.4b.

SOLUTION:
Since the supply voltage is given on Figure 23.4a as V_{CC} = 6 V, the voltage divider inside the 555 gives a trigger voltage of

$$V_{TRG} = \frac{1}{3} V_{CC} \qquad \text{(23.1)}$$

$$= \frac{1}{3} 6\,V = 2\,V$$

Since the initial value of V_i applied to pin 2 is less than V_{TRG}, V_o is high, typically nearly V_{CC}.

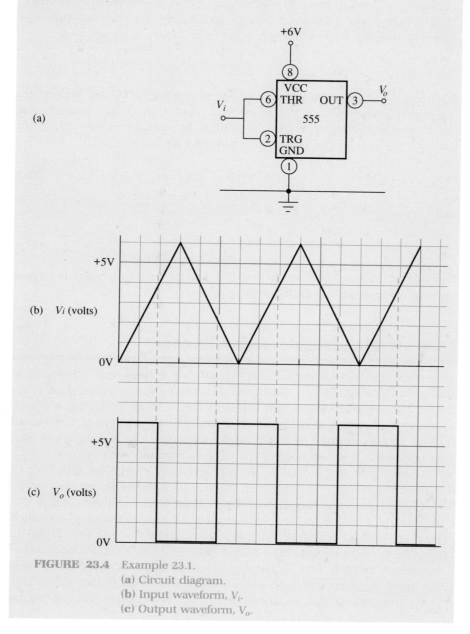

FIGURE 23.4 Example 23.1.
(a) Circuit diagram.
(b) Input waveform, V_i.
(c) Output waveform, V_o.

The voltage divider inside the 555 gives the threshold voltage as

$$V_{THR} = \frac{2}{3} V_{CC} \tag{23.2}$$

$$= \frac{2}{3} 6\,V = 4\,V$$

When V_i applied to pin 6 exceeds V_{THR}, V_o goes low, typically nearly to 0 V, where it stays until V_i becomes less than V_{TRG}, at which time V_o goes high again. The resulting waveform is shown in Figure 23.4c.

Example 23.1 shows that if the language of previous chapters is used, the "upper trip point" voltage, V_{UT}, must be applied to the THR input, terminal 6, and the "lower trip point" voltage must be applied to the TRG input, terminal 2. Thus, for the circuit of Figure 23.4 we have

$$V_{LT} = V_{TRG} = \frac{1}{3} V_{CC} \tag{23.1}$$

and

$$V_{UT} = V_{THR} = \frac{2}{3} V_{CC} \tag{23.2}$$

You should be aware that Equations 23.1 and 23.2 are only valid as long as nothing is connected to pin 5 of the 555.

23.4 A Monostable Multivibrator

In Section 22.5 we described the behavior of a "one shot." The circuit shown in Figure 23.5a is a 555 timer implementation of a one shot. From the user's standpoint the circuit will work the same as the circuit of Figure 22.10a: When switch S is flipped momentarily to position 2, the circuit delivers an output pulse of fixed width independent of how long switch S is in position 2. A description of the circuit operation is as follows:

Assume that the output is initially low. Any charge on capacitor C_2 has been conducted to ground through either diode D_1 or resistor R_2. Thus V_6 is below the threshold voltage. With S in position 1, C_1 is discharged and pin 2 is pulled high through R_1. Such resistors are often known as "pull-up resistors" for that reason. Since the output will not go high until V_2 becomes less than $1/3\ V_{CC}$, $V_{out} = 0\ V$ is a stable condition.

Assume that S is now momentarily flipped to position 2. Since C_1 was discharged, it will briefly pull pin 2 low as shown in Figure 23.5b. This triggers the output high as shown in Figure 23.5d. Capacitor C_2 begins

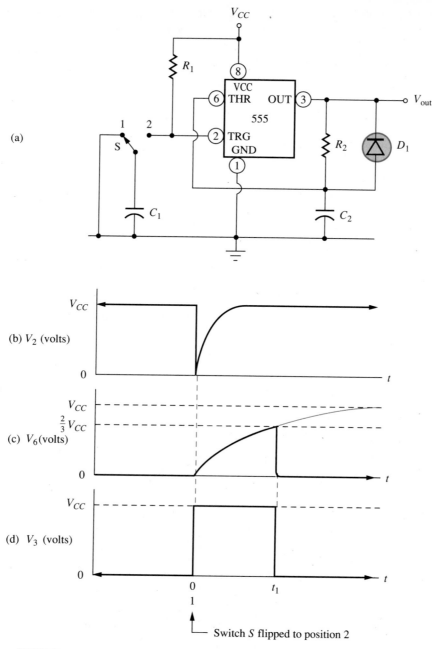

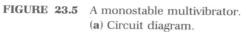

FIGURE 23.5 A monostable multivibrator.
(a) Circuit diagram.
(b) Input voltage waveform.
(c) Voltage waveform at C_2.
(d) Output voltage waveform.

charging through R_2 according to the capacitor charging equation which you should have met in your electric circuits class. It is usually written

$$V_{tc} = V_S - (V_S - V_0)e^{-t_c/(RC)} \qquad (23.3)$$

In this case the above variables have the following meanings:

V_{tc} = capacitor voltage at time t_c

V_S = supply voltage

V_0 = initial capacitor voltage

t_c = capacitor charging time

R = charging resistance

In case your use of logarithms is a little rusty, Equation 23.3 can be written

$$-\frac{t_c}{RC} = \ln\left(\frac{V_S - V_{tc}}{V_S - V_0}\right) \qquad (23.4)$$

When the voltage on the capacitor exceeds V_{tc}, the threshold voltage, which is 2/3 V_{CC} for the circuit of Figure 23.5, then the output drops low as shown in Figure 23.5d. When the output goes low, diode D_1 is forward biased and quickly discharges C_2 as shown in Figure 23.5c. The condition is again stable.

EXAMPLE 23.2

In the circuit of Figure 23.5a, if $C_2 = 0.01$ μF, find the value of R_2 so that the output pulse width will be about $t_c = 1$ msec.

SOLUTION:

Solving Equation 23.4 for R gives

$$R = -\frac{t_1}{C * \ln \dfrac{V_S - V_{t1}}{V_S - V_0}}$$

$$= -\frac{1 \text{ msec}}{0.01 \text{ μF} * \ln \dfrac{V_{CC} - \dfrac{2}{3}V_{CC}}{V_{CC} - 0 \text{ V}}}$$

$$= -\frac{1 \text{ msec}}{0.01 \text{ μF} * \ln \dfrac{1}{3}} = 91.0 \text{ kΩ} \qquad \text{(Use 100 kΩ)}$$

23.5 An Astable Multivibrator

Perhaps the most common use of the 555 is as a tone generator. It beeps at you when you leave the key in your car's ignition, when you hit a wrong key on your computer, and when your microwave is through cooking. It even pulls you over so the police officer can give you a speeding ticket.

Consider the circuit of Figure 23.6a. If we assume that the capacitor is initially discharged, its voltage, and hence the voltage at pin 2 is below V_{TRG} so V_{OUT} is high as shown in part A of Figure 23.6b. The output then begins to charge the capacitor C through resistor R. This is shown in part A of Figure 23.6c.

When the voltage on the capacitor, and hence the voltage at pin 6 exceeds V_{THR}, the output goes low. The capacitor then begins to discharge back through the resistor into pin 3. This is shown in part B of Figures 23.6b and c. When the capacitor voltage becomes less than V_{TRG}, V_{OUT} again goes high and the process repeats.

Since the charging path is the same as the discharging path, following the initial charge from zero, the charging time, t_c, and the discharge time, t_d, should be the same. Thus

$$t_c = t_d = \frac{T}{2} = \frac{1}{2f} \qquad (23.5)$$

Therefore the circuit behavior can be described using the capacitor discharge equation which is just a simplified form of Equation 23.3. It is

$$V_{td} = V_0 e^{-t_d/(RC)} \qquad (23.6)$$

In this case, the above variables have the following meanings:

V_{td} = capacitor voltage after time t_d has elapsed
V_0 = initial capacitor voltage
t_d = capacitor discharging time
R = discharging resistance

Again, if your ability to work with logarithms is a little rusty, Equation 23.3 can be written

$$-\frac{t_d}{RC} = \ln\left(\frac{V_{td}}{V_0}\right) \qquad (23.7)$$

Of course, these relationships can be further simplified, especially for the simple circuit of Figure 23.6, but that does not help you to understand what is really going on in the circuit and will only cause you problems when you try to analyze the more complex circuits later in this chapter. It is better to use the equations as they are at the expense of having to punch another couple of buttons on your calculator.

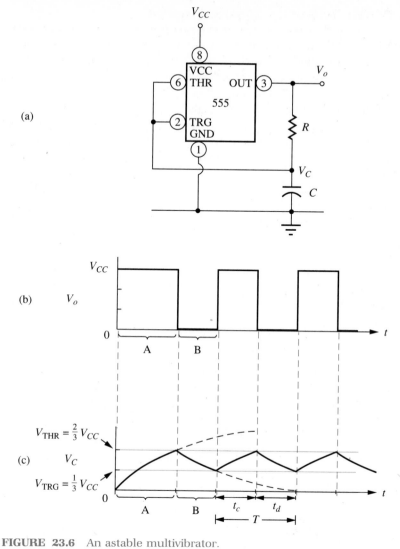

FIGURE 23.6   An astable multivibrator.
(a) Circuit diagram.
(b) Output waveform.
(c) Waveform at capacitor C.

EXAMPLE 23.3

Determine the value of the resistor in Figure 23.6 if the capacitor has a value of 0.01 μF and the frequency of oscillation is required to be about 1 kHz.

SOLUTION:

Due to the symmetry of the circuit, the discharge time is given by

$$t_d = \frac{T}{2} = \frac{1}{2f} \tag{23.5}$$

$$= \frac{1}{2 * 1 \text{ kHz}} = 500 \text{ }\mu\text{sec}$$

From Equation 23.2, V_0, the voltage on the capacitor when it begins to discharge, will be

$$V_0 = \frac{2}{3} V_{CC} \tag{23.8}$$

From Equation 23.1, V_{td}, the voltage on the capacitor when it stops discharging, will be

$$V_{td} = \frac{1}{3} V_{CC} \tag{23.9}$$

Solving Equation 23.7 for the required resistor value gives

$$R = -\frac{t_d}{C * \ln\left(\dfrac{V_{td}}{V_0}\right)} \tag{23.10}$$

Substituting Equations 23.8 and 23.9 into Equation 23.10 gives

$$R = -\frac{t_d}{C * \ln\left(\dfrac{\frac{1}{3} V_{CC}}{\frac{2}{3} V_{CC}}\right)} = -\frac{t_d}{C * \ln(0.5)} \tag{23.11}$$

$$= -\frac{500 \text{ }\mu\text{sec}}{0.01 \text{ }\mu\text{F} * \ln(0.5)} = 72.1 \text{ k}\Omega \qquad (\text{Use } 68 \text{ k}\Omega)$$

23.6 The Discharge Terminal

It would be difficult to devise a simpler oscillator circuit than that of the previous section. One weakness of the circuit is that the output drives the timing circuit. Thus variations in the external load could vary the frequency. The use of the discharge terminal avoids this problem.

You should remember from Section 23.2 that whenever the Q output of the S-R flip-flop inside of the 555 is low, the $\overline{Q}$ output will be high. According to the diagram in Figure 23.3, whenever the $\overline{Q}$ output of the internal S-R flip-flop in a 555 is high, it will supply base current to transistor Q_2, thus turning it on.

Figure 23.3 shows the collector of Q_2 connected to pin 7 and the emitter connected to pin 1. This means that whenever Q_2 is turned on, pin 7 will DIScharge whatever is connected to it to ground. Hence pin 7 is usually labeled "DIS."

To summarize, whenever pin 3 of the 555 is high, $\overline{Q}$ is low so there is no base current to transistor Q_2 and pin 7 is an open circuit. Whenever pin 3 is low, $\overline{Q}$ is high. This turns Q_2 on, effectively grounding pin 7.

You need to understand the difference between pins 3 and 7. Pin 3 can act as either a current source when it is high, or a "current sink" when it is low. Pin 7 can never be a current source since it never goes high. It can only be an open circuit when Q_2 is turned off. It is a current sink when Q_2 is turned on.

EXAMPLE 23.4

For the circuit of Figure 23.7a, indicate where on the waveform of Figure 23.7b LEDs A, B, C, and D are turned on.

SOLUTION:

From Figure 23.7a $V_{CC} = 12$ V so the trigger voltage is

$$V_{TRG} = \frac{V_{CC}}{3} \tag{23.1}$$

$$= \frac{12 \text{ V}}{3} = 4 \text{ V}$$

From Figure 23.7b, the initial value of V_i is less than V_{TRG} so pin 3 is high, i.e., near 12 V. Thus diode A will be forward biased and so turned on. Current will flow from pin 3 into the $+6$ V supply. Diode B will be reverse biased and so, turned off. According to the preceding development, pin 7 will be an open circuit. Thus diodes C and D will feel no bias at all and neither of them will be turned on.

The threshold voltage is

$$V_{THR} = \frac{2}{3} V_{CC} \tag{23.2}$$

$$= \frac{2}{3} 12 \text{ V} = 8 \text{ V}$$

When V_i exceeds V_{THR} then pin 3 goes low ("grounded," if you wish), and, according to the preceding development, pin 7 will also be grounded ("low," if you wish). Thus diodes A and C are reverse biased

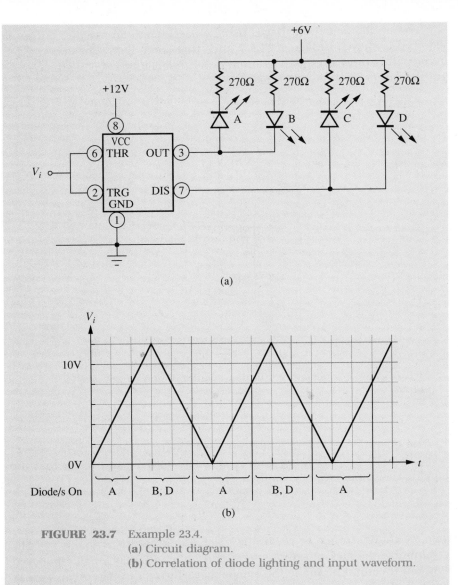

FIGURE 23.7 Example 23.4.
(a) Circuit diagram.
(b) Correlation of diode lighting and input waveform.

so neither of them will conduct. Diodes B and D will both be forward biased so both will be turned on.

Notice that diode C is never turned on since pin 7 never goes high.

The above results are summarized on Figure 23.7b.

Since both pins 3 and 7 become virtually dead shorts to ground you should be careful that neither pin is ever connected directly to V_{CC}.

The oscillator circuit that is usually used when an isolated output is required is shown in Figure 23.8. Notice that some liberty has been taken in relocating pin 7 on the diagram in order to facilitate drawing the circuit, but the device is still a 555 timer.

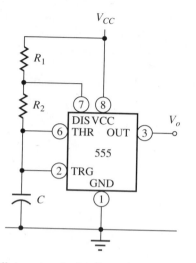

FIGURE 23.8   Oscillator circuit that has the timing circuit isolated from the output.

The operation of the circuit is no longer symmetrical since, as shown in Figure 23.9a, the charging path is through both R_1 and R_2 so that the charging resistance, R_c, is given by

$$R_c = R_1 + R_2 \tag{23.12}$$

while, as shown in Figure 23.9b, the discharging path is through R_2 only. To minimize the lack of symmetry, R_1 can be made fairly small compared to R_2, but it cannot be made too small since, during discharge, pin 7 must absorb both the capacitor discharge current and the current through R_1 as shown in Figure 23.9b.

Equation 23.6 or Equation 23.7 will describe the circuit behavior during capacitor discharge with R_2 being the discharge resistor.

In order to describe the circuit during the charging cycle, we need the more general capacitor equation, Equation 23.3 or 23.4.

As we said in connection with Equation 23.7, for the simple circuit of Figure 23.8, Equation 23.4 could be simplified, but at the expense of your understanding of more complex circuits.

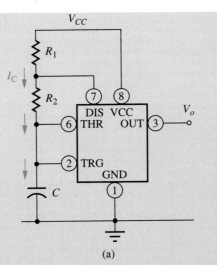

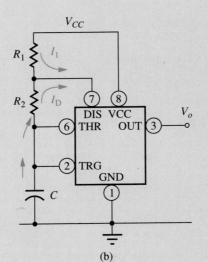

FIGURE 23.9 Oscillator circuit analysis.
(a) Path of the timing capacitor charging current.
(b) Path of the timing capacitor discharging current.

The period of the waveform generated by the circuit of Figure 23.8 is

$$T = t_{chg} + t_{dis}$$ (23.13)

By definition, its frequency is given by

$$f \equiv \frac{1}{T}$$ (23.14)

EXAMPLE 23.5

Determine the frequency of oscillation for the circuit of Figure 23.10.

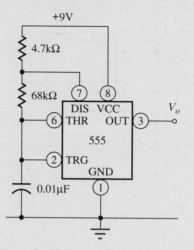

FIGURE 23.10   Example 23.5.

SOLUTION:

For the circuit of Figure 23.10

$$V_{tc} = V_{THR} = \frac{2}{3} V_{CC} \tag{23.15}$$

and

$$V_{td} = V_{TRG} = \frac{1}{3} V_{CC} \tag{23.16}$$

Further, you should recognize that since we are talking about a steady state situation, the discharge cycle will be terminated when the charging cycle begins so that the initial voltage, V_0, for the charging cycle will be the voltage at the end of the discharge cycle, V_{td}. Similarly, V_0 for the discharging cycle will be V_{tc}.

Substituting Equations 23.12, 23.15 and 23.16 into Equation 23.4 and solving for t_c gives

$$t_c = -(R_1 + R_2)C * \ln\left(\frac{V_{CC} - \frac{2}{3} V_{CC}}{V_{CC} - \frac{1}{3} V_{CC}}\right) \tag{23.17}$$

$$= -(R_1 + R_2)C * \ln(0.5)$$

or

$$= -(4.7 \text{ K}\Omega + 68 \text{ k}\Omega) * 0.01 \text{ }\mu\text{F} * \ln(0.5) = 504 \text{ }\mu\text{sec}$$

Substituting Equations 23.15 and 23.16 into Equation 23.7 and solving for the discharge time, t_d, gives

$$t_d = -R_2 C * \ln\left(\frac{\frac{1}{3} V_{CC}}{\frac{2}{3} V_{CC}}\right) = -R_2 C * \ln(0.5) \qquad (23.18)$$

or

$$t_d = -68 \text{ k}\Omega * 0.01 \text{ }\mu\text{F} * \ln(0.5) = 471 \text{ }\mu\text{sec}$$

The period for the waveform is given by

$$T = t_c + t_d \qquad (23.13)$$
$$= 504 \text{ }\mu\text{sec} + 471 \text{ }\mu\text{sec} = 975 \text{ }\mu\text{sec}$$

and the frequency is given by

$$f \equiv \frac{1}{T} \qquad (23.14)$$

$$= \frac{1}{975 \text{ }\mu\text{sec}} = 1.03 \text{ kHz}$$

23.7 A Pulse Generator

As we have seen, the output waveform in the preceding example is not symmetrical. Since the charging path has more resistance than the discharging path, the output is high longer than it is low. In order to describe this situation, the term "duty cycle," D is often used. D is defined to be the ratio of the time a waveform is high, t_{hi} to the period T of the waveform. In equation form

$$D \equiv \frac{t_{hi}}{T} \qquad (23.19)$$

In order to control the duty cycle, the charging path and discharge path must be separately controllable. One way to separate the two paths is shown in Figure 23.11. When the output is high and pin 7 is open, the capacitor charging path is shown in Figure 23.11a. Due to the fact that diode D_1 is part of the charging circuit, V_S in Equation 23.3 and 23.4 now becomes

$$V_{Sc} = V_{CC} - V_D \qquad (23.20)$$

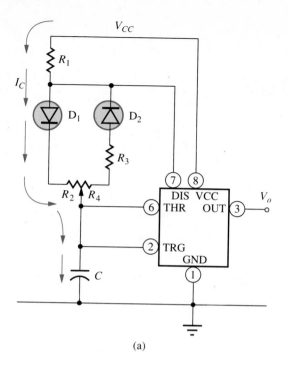

(a)

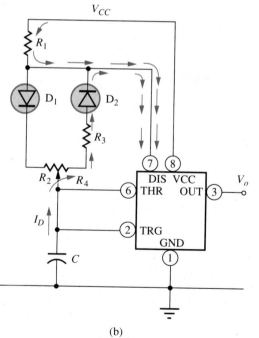

(b)

FIGURE 23.11 A pulse generator.
(a) Path of the timing capacitor charging current.
(b) Path of the timing capacitor discharging current.

The charging resistance, R, in Equation 23.3 or 23.4 becomes

$$R_c = R_1 + R_2 \qquad (23.21)$$

The discharge path is shown in Figure 23.11b. Due to the fact that the discharge path includes diode D_2 the discharge is not directly to ground $(V_S = 0 \text{ V})$. Rather, it is to the diode. This means that Equation 23.3 or 23.4, the general capacitor equation, must be used for the discharge equation as well as for the charge equation. For the discharge path of Figure 23.11b, V_S now becomes

$$V_{Sd} = V_D \qquad (23.22)$$

The discharge resistance becomes

$$R_d = R_3 + R_4 \qquad (23.23)$$

EXAMPLE 23.6

Determine the minimum duty cycle available from the circuit of Figure 23.12.

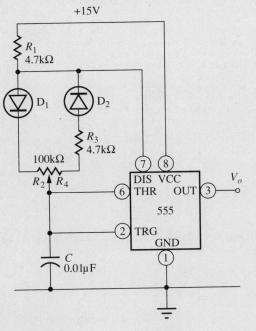

FIGURE 23.12 Example 23.6.

SOLUTION:

For the minimum duty cycle, D_{min}, the pot in Figure 23.12 is set so that $R_2 = 0\ \Omega$ and $R_4 = 100\ k\Omega$. Then the charging resistance is given by

$$R_c = R_1 + R_2 \tag{23.21}$$
$$= 4.7\ k\Omega + 0\ \Omega = 4.7\ k\Omega$$

The charging supply voltage is

$$V_{Sc} = V_{CC} - V_D \tag{23.20}$$
$$\approx 15\ V - 0.7\ V = 14.3\ V$$

The lower and upper trip points will be at 1/3 and 2/3 of V_{CC} as usual. Solving Equation 23.4 for t gives

$$t_c = -R_c C * \ln\left(\frac{V_{Sc} - V_t}{V_{Sc} - V_o}\right)$$

$$= -4.7\ k\Omega * 0.01\ \mu F * \ln\left(\frac{14.3\ V - 10\ V}{14.3\ V - 5\ V}\right) = 36.3\ \mu sec$$

The discharge resistance, R_d, is given by

$$R_d = R_3 + R_4 \tag{23.23}$$
$$= 4.7\ k\Omega + 100\ k\Omega = 105\ k\Omega$$

The discharge supply voltage, V_{Sd}, is given by

$$V_{Sd} = V_D \tag{23.22}$$
$$\approx 0.7\ V$$

The discharge time, t_d, is then given by

$$t_d = -R_d C * \ln\left(\frac{V_{Sd} - V_t}{V_{Sd} - V_o}\right) \tag{23.15}$$

$$= -105\ k\Omega * 0.01\ \mu F * \ln\left(\frac{0.7\ V - 5\ V}{0.7\ V - 10\ V}\right) = 808\ \mu sec$$

The period of the waveform is

$$T = t_c + t_d \tag{23.17}$$
$$= 36.3\ \mu sec + 808\ \mu sec = 844\ \mu sec$$

Since the output is high during the charging time, the duty cycle is now given by

$$D \equiv \frac{t_{hi}}{T} \tag{23.19}$$

$$= \frac{36.3\ \mu sec}{844\ \mu sec} = 0.043\ \text{or } 4.3\%$$

23.8   The Reset Terminal

The previous sections introduced you to the use of the transistor inside the 555 labeled Q_2 in Figure 23.3. While we are thinking about transistors, we should look at pin 4. It is the "ReSeT" terminal, usually labeled "RST." In Figure 23.3 you can see that the base of transistor Q_1 is connected through an appropriate current limiting resistor to pin 4. Transistor Q_1 is a PNP transistor with its emitter connected to V_{CC} and its collector connected to the ReSeT terminal of the S-R flip-flop. Thus whenever Q_1 is turned on, the output will be ReSeT to $V_{out} = 0$ V regardless of what may be happening at pins 2 and 6.

Of course, base current in a transistor can only flow in the direction of the arrow on the emitter. Thus if pin 4 is high, no base current can flow, Q_1 will not be turned on, and pins 2 and 6 will dictate the operation of the circuit. In fact, when pin 4 is not used, it is often connected to pin 8 in order to guarantee that no unwanted circuit resets will occur. On the other hand, if pin 4 is grounded, you can see from Figure 23.3 that base current will flow, turning Q_1 on. This will ReSeT the output to $V_{out} = 0$ V regardless of other circuit conditions.

23.9   The Control Terminal of the 555

The only terminal on the 555 timer that we have not discussed is the "ConTRol" terminal, pin 5, which is usually rather cryptically labeled "CTR." If the 555 is used as we have described it thus far, the trigger point will always be the point where the voltage at pin 2 becomes less than 1/3 of V_{CC} and the threshold will always be the point where the voltage at pin 6 exceeds 2/3 of V_{CC}. Occasionally it is desired to have a different set of trip points. Pin 5 allows for that possibility.

You can see from Figure 23.3 that pin 5 is connected to the upper section of the internal voltage divider composed of three equal resistors of approximately 5 kΩ each. Thus the open circuit voltage at pin 5, V_{CTRo}, is given by

$$V_{CTRo} = \frac{2}{3} V_{CC} \qquad (23.24)$$

The value of V_{CTR} can be pulled down below V_{CTRo} by connecting a resistor between pin 5 and ground. It can also be pulled up above V_{CTRo} by connecting a resistor between pin 5 and V_{CC}.

According to Figure 23.3, comparator C_1 actually compares the voltage at pin 6 to the voltage at pin 5. Therefore the threshold is really reached when the voltage at pin 6 exceeds the voltage at pin 5. Thus a more general equation than Equation 23.2 for finding V_{THR} that is valid regardless of what is connected to pin 5 is

• $$V_{UT} = V_{THR} = V_{CTR} \qquad (23.25)$$

Similarly, you can see from Figure 23.3 that comparator C_2 always compares the voltage at pin 2 to half of voltage at pin 5. Thus a more general equation for finding V_{TRG} than Equation 23.1 is

• $$V_{LT} = V_{TRG} = \frac{1}{2} V_{CTR} \qquad (23.26)$$

EXAMPLE 23.7

Suppose that you have a 50,000 μF 15 V capacitor which you wish to repeatedly charge to near its rated voltage and then discharge as completely as possible into a device which may be assumed to be a dead short to ground. Design the control circuit for the required "zapper."

SOLUTION:

Consider the circuit in Figure 23.13a. We will design the circuit such that when the capacitor voltage reaches about $V_c = 14$ V, the threshold voltage, V_{THR}, will be reached and pin 7 will go low. Current will then flow in the relay coil causing the relay to flip from position 1 to position 2. This will discharge the capacitor. When the capacitor discharges to the point that the capacitor voltage is about $V_d = 1$ V, the trigger voltage, V_{TRG}, will be reached and pin 7 will become an open circuit. The relay will return to position 1, recharging the capacitor. The diode across the relay coil will absorb the "inductive kick" from the relay coil, preventing excessive voltage from damaging the 555.

   Since the 555 trigger voltage cannot be altered directly, it must be set by changing the control voltage. From Equation 23.26 we know that

$$\begin{aligned} V_{CTR} &= 2\, V_{TRG} \\ &= 2 * 1\,V = 2\,V \end{aligned} \qquad (23.27)$$

Figure 23.13b shows the relevant circuitry, including the circuitry internal to the 555. From the circuit you can see that the equivalent resistance identified as R_{eq} is given by

$$R_{eq} = R_3 \,\|\, (R_b + R_c) \qquad (23.28)$$

A voltage divider equation for the circuit can be written as

$$\frac{R_{eq}}{V_{CTR}} = \frac{R_a}{V_{CC} - V_{CTR}} \qquad (23.29)$$

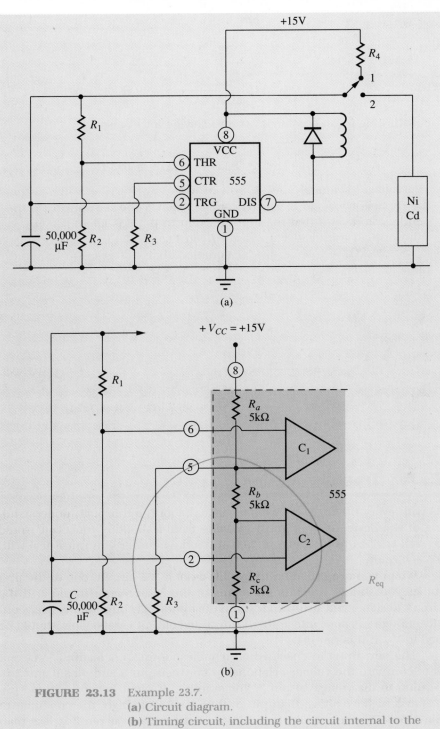

FIGURE 23.13 Example 23.7.
(a) Circuit diagram.
(b) Timing circuit, including the circuit internal to the 555 timer.

or

$$R_{eq} = R_a \frac{V_{CTR}}{V_{CC} - V_{CTR}}$$

$$= 5 \text{ k}\Omega \frac{2 \text{ V}}{15 \text{ V} - 2 \text{ V}} = 769 \text{ }\Omega$$

Substituting into Equation 23.28 and solving for R_3 gives

$$R_3 = R_{eq} \parallel (-(R_b + R_c)) \tag{23.30}$$

$$= 769 \text{ }\Omega \parallel (-(5 \text{ k}\Omega + 5 \text{ k}\Omega)) = 833 \text{ }\Omega \qquad \text{(Use 820 }\Omega\text{)}$$

The voltage divider R_1 and R_2 will provide $V_{THR} = V_{CTR} = 2$ V when the capacitor voltage reaches $V_c = 14$ V. Since the data sheet in Appendix F indicates that the input current to pin 6 is about 0.25 μA, it will not load the divider if resistor R_1 in Figure 23.13 is chosen to be $R_1 = 100$ kΩ. A voltage divider equation that will give the required value for R_2 is

$$\frac{R_2}{V_{THR}} = \frac{R_1}{V_c - V_{THR}} \tag{23.31}$$

or

$$R_2 = R_1 \frac{V_{THR}}{V_c - V_{THR}} \tag{23.32}$$

$$= 100 \text{ k}\Omega \frac{2 \text{ V}}{14 \text{ V} - 2 \text{ V}} = 16.7 \text{ k}\Omega \qquad \text{(Use 18 k}\Omega\text{)}$$

▶▶▶▶▶▶▶ 23.10 Troubleshooting

As with every other device we have talked about, a thorough understanding of the device is your most valuable troubleshooting tool. You would be well advised to have Figure 23.3 at hand whenever you must deal with a 555 circuit.

When you are called on to troubleshoot a 555 circuit, one of the first things you should check is the status of the reset terminal, pin 4. If the circuit designer left it floating, some dust and high humidity is occasionally enough to cause intermittent unwanted circuit resets. Pin 4 should be tied to V_{CC} if it is not used.

The next thing to check when troubleshooting a defective 555 circuit is to compare the voltage relationship between pin 5 and pins 6 and 2 in relation to the voltage at pin 3. For example, if pin 4 is tied high and pin 6 is at a higher voltage than pin 5, then if pin 3 is high, the condition of the 555 is suspect. By the same logic, if the voltage at pin 2 is less than half the voltage at pin 5, pin 3 should be high.

23.11 Summary

We have described the behavior of the 555 timer in terms of a simplified block program and we have illustrated the behavior of each of the input and output terminals with simple examples. If you really understand the block diagram, a knowledge of circuits and careful logic is all that is needed to design or analyze circuits containing a 555.

GLOSSARY

Current sink: A device which ideally acts like a perfect conductor to ground. (Section 23.6)

DIP: An abbreviation for "Dual Inline Package," the most popular type of packaging for integrated circuits. (Section 23.3)

Duty cycle: The ratio of the time a digital pulse is high divided by the period of the waveform. Usually expressed as a percentage. (Section 23.7)

Flip-flop: A bistable multivibrator. That is, a circuit that has two stable states. Its output can be set "high" or "low" by the appropriate input pulse and its output will remain in that state until another input pulse is applied. (Section 23.2)

High: The name applied to the most positive part of a digital signal, assuming that positive logic is used. (Section 23.2)

Low: The name applied to the most negative part of a digital signal, assuming that positive logic is used. (Section 23.2)

Q: The symbol applied to one of the output terminals of a digital circuit having complementary outputs. (Section 23.2)

$\overline{Q}$: The symbol applied to the NOT Q output terminal of a digital circuit having complementary outputs. Its state will always be opposite to the Q output. When Q is high, $\overline{Q}$ is low. When Q is low, $\overline{Q}$ is high. (Section 23.2)

R: The symbol applied to the RESET terminal of a digital circuit. An input pulse applied to the R terminal resets the Q output terminal low. (Section 23.2)

S: The symbol applied to the SET terminal of a digital circuit. An input pulse applied to the S terminal sets the Q output high. (Section 23.2)

PROBLEMS

Section 23.3

1. Describe the effect on the circuit behavior as described in Section 23.2 if the value of resistor R_2 in Figure 23.2
   (a) Jumped to twice its original value.
   (b) Dropped to half its original value.

2. What should be the voltage at pin 5 of the circuit in Figure 23.14?

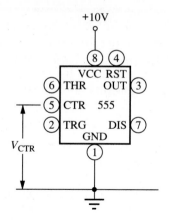

FIGURE 23.14   Problem 23.2.

3. In Chapter 22 what name would we have applied to the circuit of Example 23.1?
4. Using the same time base draw both the input and the output waveforms for the circuit of Figure 23.15a given the input waveform in Figure 23.15b.

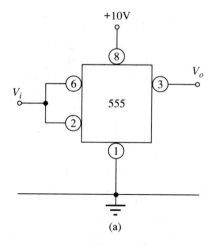

(a)

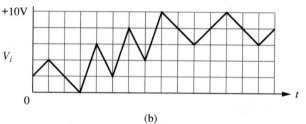

(b)

FIGURE 23.15   Problem 23.4.

Section 23.5

5. Show that for the circuit of Figure 23.16 Equation 23.4 gives the same capacitor charging time as Equation 23.7 gives for the capacitor discharging time.
6. Determine the frequency of oscillation for the circuit of Figure 23.16.
7. For the circuit of Figure 23.17, determine the component values in order to construct a two range variable frequency oscillator. The low range (×1) is to be variable from approximately 10 Hz to 100 Hz and the high range (×10) is to be variable from approximately 100 Hz to 1 kHz.

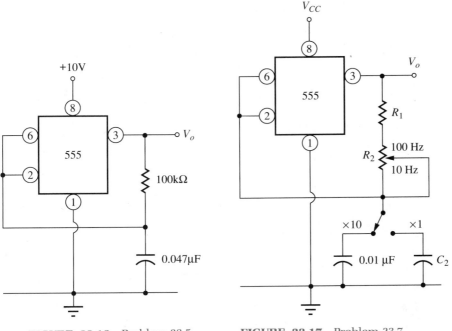

FIGURE 23.16   Problem 23.5.   FIGURE 23.17   Problem 23.7.

Section 23.6

8. Describe the behavior of the circuit of Figure 23.18. Make your description as quantitative as you can.
9. Describe the behavior of the circuit of Figure 23.19. Make your description as quantitative as you can. Compare with Problem 8.
10. Repeat Problem 9 if the polarity of the diode is reversed.
11. Find the maximum value of the total current that flows into pin 7 in Example 23.5.

Section 23.7

12. Determine the duty cycle for both U_1 and U_2 in Figure 23.19.
13. Determine the maximum duty cycle for Example 23.6.

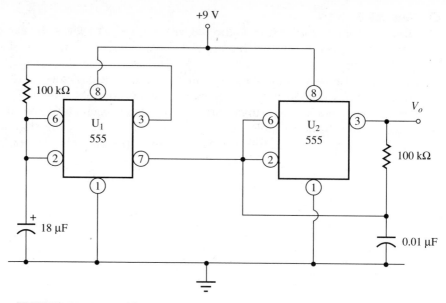

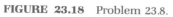

+9 V

FIGURE 23.18 Problem 23.8.

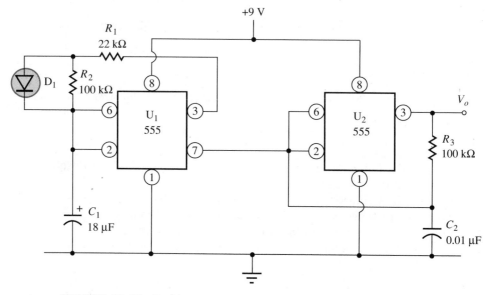

FIGURE 23.19 Problem 23.9.

Section 23.8

14. Describe how the behavior of the circuit in Figure 23.18 would be altered if it were hooked up as shown in Figure 23.20.

15. Describe how the behavior of the circuit in Figure 23.20 would be altered if, instead of hooking pin 7 of U_1 to pin 4 of U_2, pin 3 of U_1 were hooked to pin 4 of U_2. Describe the result if the same exchange of pin 3 for pin 7 of U_1 were made in Figure 23.18.

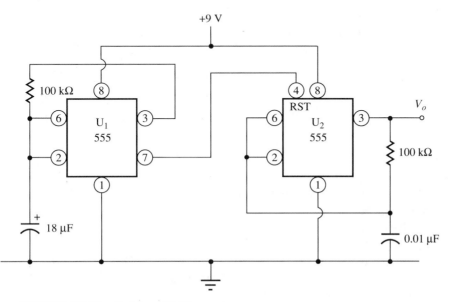

FIGURE 23.20   Problem 23.15.

Section 23.9

16. Repeat Problem 4 if a 10 kΩ resistor were connected between pin 5 of the 555 and ground.

17. What should be the value of resistor R_4 in Example 23.7 if the "zap rate," is to be one discharge every 10 seconds. What will be the maximum current R_4 will have to carry?

18. Ideally what should be the current rating of the relay contacts in Example 23.7? *Hint:* Zap!

19. Describe and explain the behavior of the "zapper" in Figure 23.13 if the discharge "short circuit to ground" in Example 23.7 is replaced by a battery cell whose terminal voltage is 1.2 V.

20. Describe the behavior of the circuit of Figure 23.21. Make your description as quantitative as you can.

21. Repeat Problem 20 if R_2 of Figure 23.21 is connected to pin 3 of U_1 instead of pin 7.

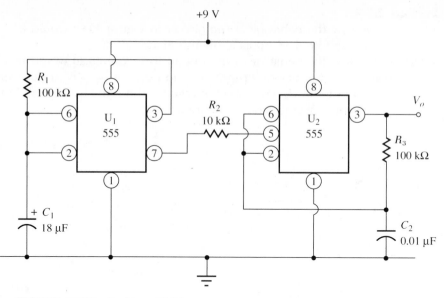

FIGURE 23.21 Problem 23.20.

Troubleshooting ▶ **Section 23.10**

22. Redraw the waveforms in Figure 23.5 if the diode D_1 became:
    (a) An open circuit.
    (b) A short circuit.
23. Describe the effect if, instead of hooking Pin 7 as shown in Figure 23.10, Pin 7 was mistakenly connected to:
    (a) Pin 6.
    (b) Pin 8.
24. For the circuit of Figure 23.20, describe the effect if, instead of hooking Pin 7 of U_1 to Pin 4 of U_2, Pin 7 of U_1 was mistakenly connected to Pin 3 of U_2.

PART V
MISCELLANEOUS SEMICONDUCTOR DEVICES

Glass fibers used in fiberoptic telephone lines.
(Photograph by Ted Horowitz/The Stock Market.)

Applied electronics. (Photograph by Christie Tito.)

Chapter 24
Thyristors and Unijunction Transistors

SPECIAL TERMS

Thyristor
Four-layer diode (FLD)
Holding current (I_H)
Diac
SCR
Crowbar
SCS
Triac
UJT
Valley current
Intrinsic standoff ratio (η)
Relaxation oscillator

24.1   Introduction and Objectives

In this chapter we will begin with a quick overview of four-layer devices in general and then we will look briefly at a few representatives of this class of semiconductor. After studying this chapter, you should understand the basic ideas behind the operation of thyristors and have a reasonably useful model which you can apply to members of this family of devices. In addition you should be able to:

- Describe the behavior of a four-layer diode, a diac, an SCR, an SCS, a triac, and a UJT.
- Draw the transistor equivalent circuit for each of the above devices.
- Analyze circuits containing the above devices.
- Suggest which of the above devices would be best suited to a given circuit requirement.

24.2   Thyristor Properties

Thyristors are digital devices in the sense that they are always either fully turned on (saturated), or fully turned off (cutoff). Thus they waste very little power. When they are turned on, there is only a volt or so lost across them. When they are turned off, they conduct essentially no current. Since $P = VI$, they are efficient. As a result a surprisingly small thyristor can control very heavy loads.

All members of the thyristor family are PNPN devices. That is, they are composed of four semiconductor layers as shown in Figure 24.1a. Thus it

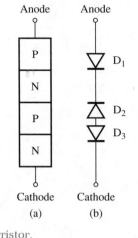

FIGURE 24.1   The thyristor.
(a) Construction.
(b) Diode model.

is tempting to picture them as three diodes in series as shown in Figure 24.1b. However, the diode model is not a very useful way to picture a thyristor.

The behavior of thyristors is better described by the two-transistor model shown in Figure 24.2. Figure 24.2a shows how two transistors could be formed from only three junctions. As you will notice, at least conceptually, one of the junctions shown in Figure 24.1a seems to get split into two pieces and counted twice. Transistor Q_1 in the equivalent circuit of Figure 24.2b is a PNP type that is "upside down" while Q_2 is a normally connected NPN device.

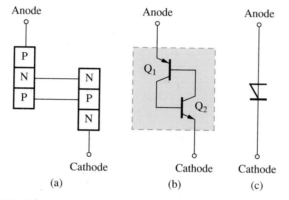

FIGURE 24.2 Thyristor model.
(a) Construction.
(b) Two–transistor model.
(c) Schematic symbol for a four-layer diode.

The various thyristor family members are determined by the way control terminals are attached, or not attached, to the base terminal of the transistors in Figure 24.2b, and how the device is combined with other similar devices.

24.3 The Four-Layer Diode (FLD)

The four-layer diode or FLD is defined to be a *reverse blocking diode thyristor*. It is the simplest member of the thyristor family. You will often hear this device referred to as a "Shockley diode," in honor of one of the pioneers in the field. We will use the more descriptive name, FLD. The FLD is classified as a diode because it is a two-terminal device, having only a cathode terminal and an anode terminal available to the user. The schematic symbol for an FLD is shown in Figure 24.2c.

In order to understand the behavior of a four-layer diode, consider the circuit of Figure 24.3a. The equivalent circuit is shown in Figure 24.3b. Since there is no external connection to either transistor base terminal, you might at first think that the circuit will never conduct. However, on

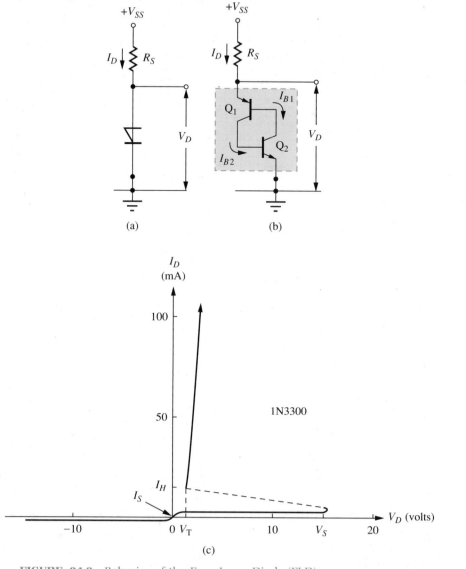

FIGURE 24.3 Behavior of the Four-Layer Diode (FLD).
(**a**) Circuit diagram suitable for constructing an FLD characteristic graph.
(**b**) Transistor equivalent circuit.
(**c**) Characteristic graph for an FLD.

second thought, you know that if you increase V_{SS} far enough, something must happen sooner or later. What happens is something like the Zener breakdown which we introduced in Section 2.11. Here it is known as "breakover." The breakover voltage, or the "switching voltage," V_S as it is more commonly called, is an important FLD parameter. As was the case with the Zener voltage, V_S is determined by the way the device is constructed.

In order to understand how switching occurs, assume that we start with V_{SS} in Figure 24.3 at a low value and slowly increase it. Under ideal conditions, if we assume that Q_2 is initially cut off (no collector current), then $I_{B1} = 0$ A and so Q_1 will also be cut off (no collector current). That means that $I_{B2} = 0$ A thus keeping Q_2 cut off. If you remember Chapter 22, you should recognize this as a positive feedback situation.

Although either transistor could initiate breakover, assume that when V_{SS} reaches the switching voltage, V_S, transistor Q_2 reaches breakover and begins to conduct. This means we have a base current, I_{B1}, flowing from transistor Q_1 which turns it on. Transistor Q_1 then supplies base current to transistor Q_2, locking it on. As you would expect from positive feedback circuits, both transistors saturate very quickly. The typical switching time is less than a microsecond.

Once it is turned on, the voltage across the FLD drops to the on state voltage, V_T. After it has been turned on, the only way to turn off an FLD is to reduce the current to a value below what is known as the "holding current," I_H.

The characteristic graph for a typical FLD, the 1N3300, is shown in Figure 24.3c. Appendix F contains the most important parameters for some representative FLDs as listed in typical data books.

Since maximum and minimum values are usually listed for V_S, the typical value can be approximated by

$$V_S = \sqrt{V_{Smax}V_{Smin}} \tag{24.1}$$

EXAMPLE 24.1
Draw the waveform of V_o in Figure 24.4a. Be as quantitative as you can be.

SOLUTION:
For the 1N3300, Appendix F lists $V_{Smax} = 21$ V; $V_{Smin} = 14$ V. A typical value is approximately

$$V_S = \sqrt{V_{Smax}V_{Smin}} \tag{24.1}$$
$$= \sqrt{21 \text{ V} * 14 \text{ V}} = 17.1 \text{ V}$$

The other required device parameters are: $V_T = 1.5$ V; $I_H = 15$ mA.

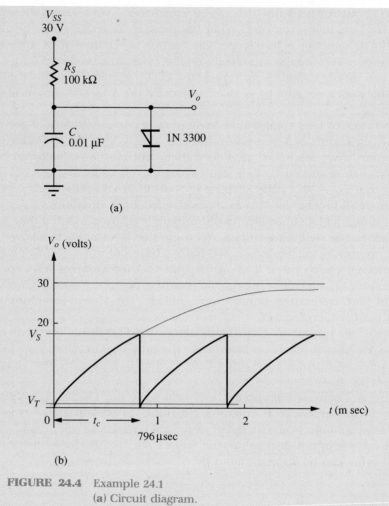

(a)

(b)

FIGURE 24.4 Example 24.1
(a) Circuit diagram.
(b) Output waveform.

The supply, V_{SS} will charge the capacitor through R_S until V_o reaches V_S, the switching voltage of the FLD. When the FLD switches on, the capacitor will be quickly discharged to a voltage of approximately V_T. At that time, the FLD current will be

$$I_{Dmin} = \frac{V_{SS} - V_T}{R_S}$$

$$= \frac{30 \text{ V} - 1.5 \text{ V}}{100 \text{ k}\Omega} = 285 \ \mu\text{A}$$

which is well below the holding current of 15 mA for the 1N3300 so the FLD then switches off and the capacitor-charging cycle repeats.

From your circuits class or from Section 23.4, you know that the capacitor-charging equation can be written

$$-\frac{t_c}{RC} = \ln \frac{V_S - V_{tc}}{V_S - V_0} \tag{23.4}$$

Converting to FLD notation and solving for the capacitor charging time, t_c, gives

$$t_c = -R_S C * \ln \frac{V_{SS} - V_S}{V_{SS} - V_T} \tag{24.2}$$

$$= -100 \text{ k}\Omega * 0.01 \text{ }\mu\text{F} * \ln \frac{30 \text{ V} - 17.1 \text{ V}}{30 \text{ V} - 1.5 \text{ V}} = 796 \text{ }\mu\text{sec}$$

The resulting waveform is shown in Figure 24.4b.

24.4 The Diac

When an FLD is reverse biased, it behaves much the same as an ordinary diode. The diac provides for the capability of having FLD-type switching action on each half of an AC waveform. A diac is functionally the same as hooking two FLDs in parallel with their polarities reversed. The diac circuit symbol shown in Figure 24.5 indicates this property to some extent. Since the device is symmetrical, both terminals are called anodes.

The diac is a logical extension of the capabilities of the FLD. We will find that other members of the thyristor family are related in the same way.

Anode 1

Anode 2

FIGURE 24.5 Schematic symbol for a diac.

24.5   Silicon Controlled Rectifiers (SCR)

Silicon controlled rectifiers are more flexible than FLDs because they are three-terminal devices. They have a control terminal, or "gate," which is connected to the P layer within the thyristor structure as shown in Figure 24.6a. From the equivalent circuit of Figure 24.6b you can see that the gate is really the base of transistor Q_2. The circuit symbol for an SCR is shown in Figure 24.6c.

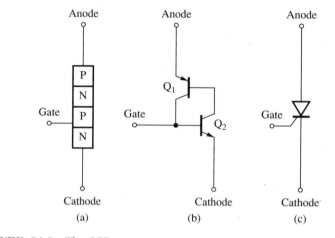

FIGURE 24.6   The SCR.
(a) Structure.
(b) Transistor equivalent circuit.
(c) Schematic symbol.

If you understood our explanation of FLD behavior in terms of the transistor model, you should have no trouble understanding the behavior of SCRs. The FLD was switched into its on state by exceeding the breakover voltage. When an SCR is to be used in a circuit, a device is chosen that has a breakover voltage that is well above the circuit voltage so that breakover will never occur. Rather, the gate terminal controls when the device "fires," i.e., when it switches into its on state. In fact, the breakover voltage, known as V_{DRM} in the language of the SCR, and the reverse blocking voltage, V_{RRM} are usually about the same.

Consider the circuit of Figure 24.7. If we assume that V_{DRM} is greater than V_{SS}, then the SCR will be in its off state and $I_L \approx 0$ A. Both Q_1 and Q_2 of Figure 24.6b will be cut off. The situation is the same as it was when the FLD in Section 24.3 was in its off state. Closing switch S will cause base current to flow into transistor Q_2 in Figure 24.6b. Assuming that this current is above the minimum gate turn-on current, I_{GT}, transistor Q_2 will be turned on. This will mean base current will flow in transistor Q_1. The positive feedback causes both transistors to turn full on, reducing the volt-

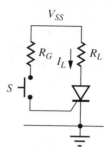

FIGURE 24.7   Circuit containing an SCR.

age across the device to V_T, the on state voltage, which is typically about a volt or two. Provided I_L is above the holding current, I_H, the circuit will latch in the on state. Due to the positive feedback inside the SCR, the gate no longer controls the circuit. Current will continue to flow regardless of whether switch S is open or closed. The only way to turn the SCR off is to reduce I_L to a value below that of I_H.

One last comment needs to be made about real SCRs. Occasionally an SCR will trigger on when power is first applied to the circuit without any gate trigger current being applied. This false triggering typically results when the anode voltage increases too rapidly.

In Section 12.10 we pointed out that the junctions within a transistor have capacitance. Looking at Figure 24.8, you can see that when anode

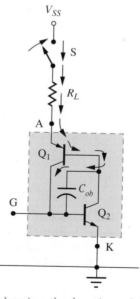

FIGURE 24.8   Circuit showing the location of the internal capacitance C_{ob} inside of an SCR, and how it can produce false triggering.

voltage is applied by closing switch S, even though transistor Q_2 is turned off, a short base current pulse will flow in Q_1 in order to charge the collector junction capacitance C_{ob} of transistor Q_2. If the rate of anode voltage rise, dV/dt, is high enough, the base current pulse will be sufficient to turn on transistor Q_1, triggering the SCR. The maximum allowable value of dV/dt for a given device is usually listed on the data sheet.

To slow the rate of anode voltage rise, a small capacitor, C_A, is often placed in parallel with the anode and cathode terminals as shown in Figure 24.9. The value of the required capacitor can be estimated for a given circuit if the maximum allowable value of dV/dt is known.

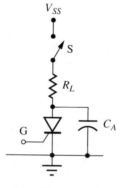

FIGURE 24.9   Placement of the false trigger suppressing capacitor C_A.

Remember that capacitance is defined as

$$C \equiv \frac{Q}{V} \tag{24.3}$$

and current is defined to be

$$I \equiv \frac{dQ}{dt} \tag{24.4}$$

Combining Equation 24.3 into Equation 24.4, the required capacitance, C_A is given by

$$C_A = \frac{I_A}{dV/dt} \tag{24.5}$$

where I_A is the initial current that flows into C_A in Figure 24.9 when switch S is closed. The value of I_A can be found by assuming that C_A is initially discharged, and remembering that the voltage across a capacitor cannot change suddenly. The value of I_A is then given by

$$I_A = \frac{V_{SS}}{R_L} \tag{24.6}$$

EXAMPLE 24.2

Design a circuit such that when a thief tries to steal your tape player, your car's horn will start blowing ($I = 7.5$ A) and will not stop until you turn it off.

SOLUTION:

The circuit is shown in Figure 24.10. When the circuit is armed, current will flow from the car's battery through R_G. As long as the connector that hooks the tape player to the wiring in the car is connected, that current will flow through the grounding wire, GND. No gate current will be supplied to the SCR so the horn will not blow. If the connector is unplugged, the current flowing through R_G will trigger the SCR and the horn will blow until switch S is turned off.

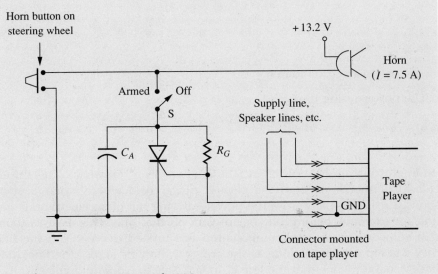

FIGURE 24.10   Example 24.2.

Since the horn requires a current of 7.5 A, a 2N1842 is selected from Appendix F. Both its voltage rating and its current rating are adequate.

For the 2N1842, $dV/dt = 200$ V/μsec. Thus

$$C_A = \frac{I_L}{dV/dt} \tag{24.5}$$

$$= \frac{7.5 \text{ A}}{200 \text{ V/μsec}} = 0.0375 \text{ μF} \qquad \text{(Use 0.039 μF)}$$

From the nominal value of horn current, R_L is

$$R_L = \frac{V_{SS}}{I_L} \tag{24.7}$$

$$= \frac{13.2 \text{ V}}{7.5 \text{ A}} = 1.76 \ \Omega$$

In order to determine the value of R_G, assume that the circuit is armed, the tape player has been unplugged so the required value of $I_{GT} = 150$ mA is flowing from V_{SS} through R_L and R_G into the gate terminal, but triggering has not yet occurred. According to the data in Appendix F, $V_{GT} = 3.5$ V. Doing a KVL walk from V_{SS} to the gate terminal in Figure 24.10 gives

$$V_{SS} - I_{GT}(R_L + R_G) = V_{GT} \tag{24.8}$$

or

$$R_G = \frac{V_{SS} - V_{GT}}{I_{GT}} - R_L \tag{24.9}$$

$$= \frac{13.2 \text{ V} - 3.5 \text{ V}}{150 \text{ mA}} - 1.76 \ \Omega = 62.9 \ \Omega \qquad (\text{Use } 56 \ \Omega)$$

The power rating for R_G is

$$P_{RG} = (150 \text{ mA})^2 * 56 \ \Omega = 1.26 \text{ W} \qquad (\text{Use } 2 \text{ W})$$

An interesting application of an SCR is its use as a "crowbar." This rather colorful term comes from the electric power field where workers often throw a real crowbar across the power carrying rail of an electric railroad line, for example, before they begin work on that line. This assures them that someone in the switching station has indeed deenergized the line they wish to service. As long as the crowbar remains across the line, they have visual evidence that the line is, in fact, safe to work on. In electronics, it is sometimes desirable that a circuit be totally shut down if the voltage or current deviates from some specified value. An SCR is often used for such "crowbar" applications.

24.6 Silicon Controlled Switches (SCS)

The SCS is an obvious extension of the SCR. Figure 24.11a shows the transistor model of the device, and Figure 24.11b shows the circuit symbol for it. As you would expect from looking at Figure 24.11a, Gate 1 of the SCS behaves the same as the gate of an SCR. When current is supplied to G_1, the SCS is triggered on. The SCS provides the added flexibility of an-

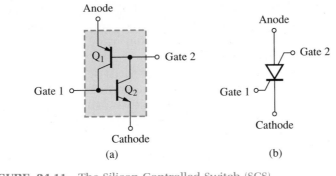

FIGURE 24.11 The Silicon Controlled Switch (SCS).
(a) Transistor equivalent circuit.
(b) The circuit symbol.

other gate, G_2 attached to the base of transistor Q_1. Thus the SCS can be triggered by either supplying current to G_1, or by drawing current from gate G_2.

24.7 Triacs

The SCR has many applications in DC circuits, but it does not switch both halves of an AC waveform since it is a rectifier. The triac has been developed to meet the need for an AC equivalent of the SCR. A triac bears the same relationship to an SCR as a diac does to an FLD.

A triac can be thought of as two SCRs in parallel with polarities reversed and gate terminals tied together. Figure 24.12a shows the SCR equivalent circuit. Figure 24.12b shows the circuit symbol for a triac. Since

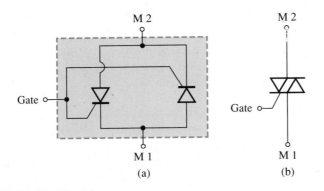

FIGURE 24.12 The triac.
(a) The SCR equivalent circuit.
(b) The circuit symbol.

the device will conduct in either direction, the main terminals are known as M_1 and M_2. The gate voltages and currents are referenced to M_1. The data usually listed for triacs is similar to that listed for SCRs. Appendix F contains some typical triac data.

Triacs are commonly used to control devices operated from the 60 Hz power line in applications such as light dimmers, or motor speed controls. Control of the power delivered to the load is usually accomplished by using some sort of phase shift circuit such as that shown in Figure 24.13a. If both the diac and the triac were removed from the circuit, the source waveform, v_{SS}, and the capacitor waveform, v_C, would appear as shown in Figure 24.13b. Notice the phase shift, ϕ_C.

The circuit in Figure 24.13a is a simplified version of circuits used commercially. One simplification is the use of the 1:1 isolation transformer. Since the load R_L and the control circuit do not share a common point that would be convenient to call "ground," the isolation transformer will allow you to use a scope to examine the waveform across any component in the circuit if you should hook it up, and you should.

Most commercial circuits include a diac in the control circuit as shown in Figure 24.13a. Its function is to provide a sharp trigger pulse to the triac. As the voltage v_C across capacitor C_1 increases from zero, as shown in Figure 24.13b, the diac will prevent trigger current from flowing to the triac until the voltage on C_1 exceeds the trigger voltage, V_{trg} which is the sum of the diac's trigger voltage, V_T, plus the triac's gate voltage, V_{GT}, or

$$V_{trg} = V_T + V_{GT} \tag{24.10}$$

The diac will then trigger, dumping almost all of the charge on C_1 into the gate of the triac. This sharp pulse reduces the triac's switching time. Thus it runs cooler than would otherwise be the case.

In order to do an exact analysis of the circuit of Figure 24.13a, the above very nonlinear components require the solution of a rather awkward differential equation. However, since the error introduced is not large, we will assume that steady state impedance relationships apply. Then, if we ignore the load placed on the capacitor when the diac fires, the waveform of the voltage across the unloaded capacitor in Figure 24.13a would be that of v_{Copen} in Figure 24.13b, and is given by

$$v_C\underline{/\phi_C} = v_S \frac{-jx_C}{R_L + R_1 + (-jx_C)} \tag{24.11}$$

Since the diac discharges the capacitor when it fires, the actual capacitor waveform is more like that shown in Figure 24.13c, but, as we have said, the timing error resulting from assuming the waveform of Figure 24.13b is not unacceptably large. Thus the voltage, v_C, which is applied to the diac in Figure 24.13a will be delayed by a phase angle of ϕ_C which is determined by the phasor v_C in Equation 24.11.

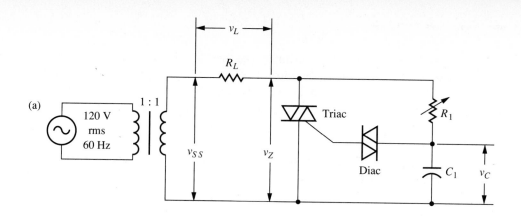

(a)

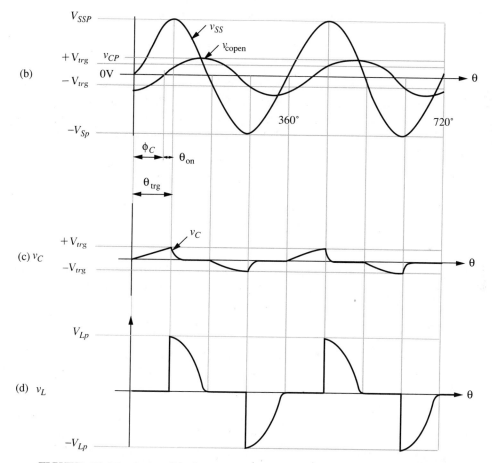

FIGURE 24.13 A simplified triac control circuit.
(a) Circuit diagram.
(b) Supply waveform and waveform that would exist at C_1
if it were not loaded.
(c) Waveform that actually exists across C_1.
(d) Output waveform.

The triac will not fire until v_C reaches a voltage of V_{on} as shown in Figure 24.13b. Looking at the circuit of Figure 24.13a, you can see that the value of V_{on} is determined by the trigger voltage of the diac, V_S, and the gate voltage, V_{GT}, of the triac

$$V_{on} = V_S + V_{GT} \qquad (24.12)$$

As you can see from Figure 24.13b, this results in a further delay, θ_{on}, before triggering can occur. Assuming that v_C is a sine wave, the value of θ_{on} is given by

$$\theta_{on} = \sin^{-1}\left(\frac{V_{trg}}{V_{CP}}\right) \qquad (24.13)$$

Since both θ_{on} and ϕ_C are delay angles as shown in Figure 24.13b, the total angle by which the triac's output is delayed, θ_{trg} is given by finding their numerical sum, or

$$\theta_{trg} = \phi_C + \theta_{on} \qquad (24.14)$$

From Figure 24.13d you can see that the resulting waveform delivered to the load is not a sine wave. Rather, it is essentially the section of the source waveform which begins at θ_{trg} and continues to the end of the half cycle, at which time the triac shuts off. This is repeated on each half cycle as shown in Figure 24.13d.

Since the waveform of v_L is not a sine wave, we cannot use the sine wave equation to determine the rms power delivered to the load, R_L in Figure 24.13a.

$$V_{rms} = \frac{V_P}{\sqrt{2}} \qquad (24.15)$$

Using the definition of rms and a little calculus, we can show that when a sine wave is turned on at θ_{trg} and remains on for the remainder of each half-cycle, V_{rms} is given by

$$V_{rms} = V_P \sqrt{\frac{1}{2} - \frac{\theta^{\circ}_{trg}}{360^{\circ}} + \frac{\sin(2\,\theta_{trg})}{4\pi}} \qquad (24.16)$$

See Appendix D for a derivation of Equation 24.16.

EXAMPLE 24.3
Determine the maximum triggering angle, θ_{trgMax} for the circuit of Figure 24.14.

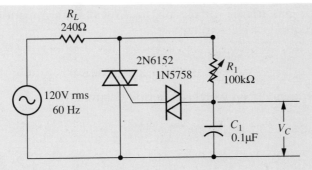

FIGURE 24.14 Example 24.3.

SOLUTION:

Assuming that the source is a sine wave, its peak source voltage is

$$V_{SSP} = \sqrt{2} * V_{SSrms} \tag{24.17}$$
$$= \sqrt{2} * 120 \text{ V} = 170 \text{ VP}$$

The reactance of capacitor C_1 is

$$X_{C1} = -j\frac{1}{2\pi fC} \tag{24.18}$$

$$= -j\frac{1}{2\pi * 60 \text{ Hz} * 0.1 \text{ }\mu\text{F}} = -j26.5 \text{ k}\Omega = 26.5 \text{ k}\Omega\underline{/-90°}$$

The total impedance of the control circuit, $\vec{Z_t}$ is then

$$\vec{Z_t} = \vec{R_L} + \vec{R_1} + \vec{X_C} \tag{24.19}$$
$$= 240 \text{ }\Omega + 100 \text{ k}\Omega - j26.5 \text{ k}\Omega$$
$$= 100 \text{ k}\Omega - j26.5 \text{ k}\Omega = 104 \text{ k}\Omega\underline{/-14.8°}$$

The peak phasor voltage across C_1 is now determined using the voltage divider relation

$$V_{CP} = V_{SSP}\frac{X_C\underline{/-90°}}{Z_t\underline{/\theta}} \tag{24.20}$$

$$= 170 \text{ V}\frac{26.5 \text{ k}\Omega\underline{/-90°}}{104 \text{ k}\Omega\underline{/-14.8°}} = 43.4 \text{ V}\underline{/-75.2°}$$

Thus the phase angle by which v_C lags behind the power line is $\theta_C = 75.2°$.

From Appendix F, the typical turn-on voltage for the 1N5758 diac is

$$V_S = \sqrt{V_{Smax}V_{Smin}} \tag{24.1}$$
$$= \sqrt{24 \text{ V} * 16 \text{ V}} = 19.6 \text{ V}$$

Adding the gate trigger voltage for the 2N6152 from Appendix F gives the total required trigger voltage

$$V_{trg} = V_S + V_{GT} \tag{24.12}$$
$$= 19.6 \text{ V} + 3 \text{ V} = 22.6 \text{ V}$$

The phase angle θ_{on} at which triggering occurs is

$$\theta_{on} = \sin^{-1}\left(\frac{V_{trg}}{V_{CP}}\right) \tag{24.13}$$

$$= \sin^{-1}\left(\frac{22.6 \text{ V}}{43.4 \text{ V}}\right) = 31.4°$$

The maximum total triggering angle, θ_{trgMax}, is given by

$$\theta_{trgMax} = \phi_C + \theta_{on} \tag{24.14}$$
$$= 75.2° + 31.4° = 107°$$

In practice, a larger triggering delay is often required. An increased delay can be achieved by using a circuit such as that in Figure 24.15.

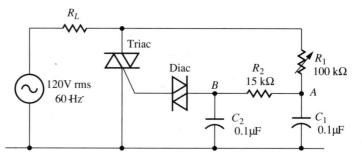

FIGURE 24.15   Triac dimmer circuit.

24.8   Unijunction Transistors (UJT)

A unijunction transistor (UJT) does not fit the definition of a thyristor given in Section 24.2. It is not a PNPN device. Due to its construction it is sometimes referred to as a "double-base diode." It is built from a crystal of silicon containing a rather lightly doped n-type channel with a heavily doped p region on one side of it. Terminals B_1 and B_2 are attached to the ends of the channel and terminal E is attached to the p region as shown in the schematic diagram in Figure 24.16a. As shown by the equivalent circuits in Figures 24.16b and 24.16c, the UJT exhibits thyristor-like behavior, which explains its inclusion here.

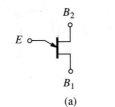

(a)

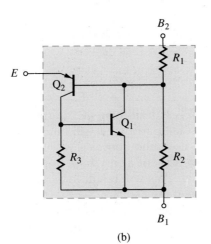

(b)

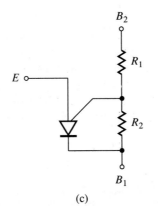

(c)

FIGURE 24.16 Unijunction transistor (UJT).
(a) Schematic symbol.
(b) Transistor equivalent circuit.
(c) SCR equivalent circuit.

In its typical mode of operation, base terminal B_1 can be considered as the common terminal of the UJT. Base terminal B_2 is maintained at a fixed voltage, V_{B2B1}. Assuming that transistor Q_1 in Figure 24.16b is cut off, the base voltage of transistor Q_2 is then determined by V_{B2B1} and the voltage divider ratio for resistors R_1 and R_2. The manufacturers usually list the "interbase resistance," R_{BB} which is defined to be

$$R_{BB} \equiv R_1 + R_2 \qquad (24.21)$$

They also list the voltage divider ratio which they call the "intrinsic stand-off ratio," η (Greek letter eta) which is defined to be

$$\eta \equiv \frac{R_2}{R_1 + R_2} \qquad (24.22)$$

or

$$\eta = \frac{R_2}{R_{BB}} \qquad (24.23)$$

If the emitter voltage V_{EB1} is less than ηV_{B2B1}, then transistor Q_2 in Figure 24.16b will be reverse biased and so it will be cut off. This will keep transistor Q_1 cut off as we assumed above. It will also mean that the emitter terminal of the UJT will present a very high impedance to current flowing into the emitter terminal. This will result in a very small peak emitter current, I_P, in the high impedance state.

As V_{EB1} is increased, the base junction of transistor Q_2 will become forward biased at a peak voltage, V_P, that, as you can see from Figure 24.16b, is approximated by

$$V_P \approx 0.6 \text{ V} + \eta V_{B2B1} \qquad (24.24)$$

When V_{EB1} exceeds V_P, the UJT "fires." That is, it switches to its low impedance state. Base current begins to flow in transistor Q_2 turning it on. This causes sufficient voltage loss across resistor R_3 to turn transistor Q_1 on. Then V_{EB1} drops to its "on" or saturated value, $V_{EB1(sat)}$, where it stays until the emitter current I_E drops below the holding, or "valley" current I_V. Appendix F contains the data listed in most data books for some typical UJTs.

24.9 A UJT Relaxation Oscillator

A common use for a UJT is as the active device in a relaxation oscillator circuit such as that shown in Figure 24.17a. Initially the UJT is "off" and V_{BB} begins charging C through R producing the section of the output waveform in Figure 24.17b labeled "Chg." When V_o exceeds V_P as determined by Equation 24.24, the UJT fires, very quickly discharging C down to the value $V_{EB1(sat)}$, producing the section of the output waveform labeled "Dsc" in Figure 24.17b. Provided the current then flowing through resistor R in Figure 24.17a is less than the valley current I_V for the UJT, it then switches back to its high impedance state and the process repeats.

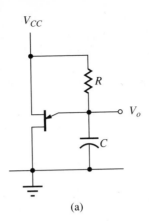

(a)

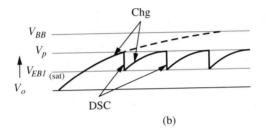

(b)

FIGURE 24.17 A UJT relaxation oscillator.
(a) Circuit diagram.
(b) Output waveform.

The requirement that after C has discharged, the emitter current must be less than I_V if the UJT is to switch back to its high impedance state, places a lower limit on the allowable value of R. Assuming that C is discharged down to $V_E = V_{EB1(sat)}$, doing a KVL walk from V_{BB} through R to the emitter gives

$$V_{BB} - I_V R = V_{EB1(sat)} \tag{24.25}$$

or

$$R_{min} = \frac{V_{BB} - V_{EB1(sat)}}{I_V} \tag{24.26}$$

Since the discharge time is usually negligible, the period of the above oscillation is given by the familiar capacitor charging equation

$$V_t = V_S - (V_S - V_0)e^{-t/RC} \tag{22.13}$$

or

$$T = -RC * \ln\frac{V_{BB} - V_P}{V_{BB} - V_{EB1(sat)}} \tag{24.27}$$

EXAMPLE 24.3

Determine the amplitude and period of the output waveform for the circuit of Figure 24.18 if $R = 10$ kΩ. Also determine the value for R that will produce the maximum frequency of oscillation.

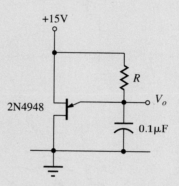

FIGURE 24.18   Example 24.4.

SOLUTION:

From Appendix F, the necessary data for the 2N4948 are: $R_{BB} = 12$ kΩ max.; $\eta = 0.82$ max.; $V_{EB1(sat)} = 3.0$ V max.; $I_v = 4.0$ mA min.

The maximum voltage to which the capacitor will charge is given by

$$V_P \approx 0.6 \text{ V} + \eta V_{B1B2} \tag{24.24}$$
$$= 0.6 \text{ V} + 0.82 * 15 \text{ V} = 12.9 \text{ V}$$

The minimum value to which the capacitor should discharge is given by $V_{EB1(sat)} = 3.0$ V.

Thus the amplitude of the output waveform should be

$$v_o = V_P - V_{EB1(sat)} \tag{24.28}$$
$$= 12.9 \text{ V} - 3 \text{ V} = 9.9 \text{ V}$$

The period of the waveform is given by

$$T = -RC \ln \frac{V_{BB} - V_P}{V_{BB} - V_{EB1(sat)}} \tag{24.27}$$

$$= -10 \text{ k}\Omega * 0.1 \text{ }\mu\text{F} * \ln \frac{15 \text{ V} - 12.9 \text{ V}}{15 \text{ V} - 3 \text{ V}} = 1.74 \text{ msec}$$

Equation 24.27 shows that the smaller R is, the smaller T will be, or the higher the frequency will be. Its minimum value is determined by

$$R_{min} = \frac{V_{BB} - V_{EB1(sat)}}{I_V} \tag{24.26}$$

$$= \frac{15\text{ V} - 3\text{ V}}{4\text{ mA}} = 3\text{ k}\Omega$$

24.10 Troubleshooting ◄◄◄◄◄◄◄◄

As we have said in connection with so many other devices, your best troubleshooting tool is a thorough understanding of how the device is supposed to behave.

The most common failing of most thyristors is early breakover. That is, the device fires before it is designed to, or perhaps without the presence of any gate pulse at all. To check the device for early breakover, isolate the gate so that you are certain that no spurious gate pulses are getting to the device. Then test to see if breakover occurs.

Often, particularly in a noisy industrial environment, high voltage spikes on the power line are responsible for erratic thyristor behavior. In this case the use of an anode capacitor as described in Section 24.5, and perhaps other shielding devices such as a low value inductor in the line is helpful.

24.11 Summary

The family of thyristor devices was introduced and some of the more common members of the family were described in this chapter. The two transistor model for each device was described. The devices examined were the FLD, diac, SCR, SCS, triac, and the UJT. Although not in the strictest sense a thyristor, the UJT was included because of its thyristor-like characteristics.

GLOSSARY

Crowbar: A low resistance shunt which is placed across the terminals of a power source for the purpose of shorting it out. (Section 24.5)

Diac: A bidirectional dipole thyristor. The equivalent of two FLDs in parallel with polarities opposite. (Section 24.4)

Four-layer diode (FLD): Also known as a Shockley diode. Defined as a reverse blocking diode thyristor. (Section 24.3)

Holding current *(I_H)*: The current required to maintain a thyristor in the on state. (Section 24.3)

Intrinsic standoff ratio (η): The voltage divider ratio for the internal voltage divider within a UJT. (Section 24.8)

Relaxation oscillator: An oscillator whose output frequency is determined by the time required to charge or discharge a capacitor or a coil through a resistor. (Section 24.9)

SCR: Silicon controlled rectifier. A thyristor with a control terminal connected to the P-type semiconductor layer within the device. (Section 24.5)

SCS: Silicon controlled switch. A thyristor with one control terminal connected to the N-type semiconductor layer within the device another control terminal connected to the P-type semiconductor layer within the device. (Section 24.6)

Thyristor: The name given to a family of four-layer semiconductor devices that have two stable states of operation; one which exhibits very high internal resistance, and one which exhibits very low internal resistance. (Section 24.2)

Triac: A triode AC semiconductor switch, equivalent to two SCRs in parallel with polarities opposite. (Section 24.7)

UJT: Unijunction transistor. A three-terminal semiconductor device used primarily as a switching device. (Section 24.8)

Valley current *(I_v)*: The holding current for a UJT. (Section 24.8)

PROBLEMS

Section 24.3

1. Describe the most important difference in behavior between a Zener diode whose Zener voltage is 20 V and an FLD whose trigger voltage is 20 V.

2. Determine the most probable value for the trigger voltage of the following FLDs: 1N3301   1N3304.

3. Sketch the turn-on characteristic graph for the FLDs in Problem 2.

4. Determine the period of the waveform for the circuit of Figure 24.4 if the 1N3300 FLD is replaced by a 1N3302.

5. Describe the behavior of the circuit of Problem 4 if the particular 1N3302 you happened to use had a value of V_S within a volt of V_{Smax} as listed in Appendix F.

6. Describe the effect on the circuit of Example 24.1 if the 100 kΩ resistor is replaced by a 1 kΩ resistor. *Hint:* Don't forget I_H.

Section 24.4

7. Describe the effect on the circuit of Example 24.1 if the 1N3300 were replaced with a diac having the same value of V_S as the 1N3300.

8. The circuit of Figure 24.19 is an over-voltage warning circuit. Describe the behavior of the circuit and calculate the circuit's trigger voltage.

9. Design a crowbar circuit whose crowbar voltage is about 35 V. Assume that the circuit is to be fused for 6 A.

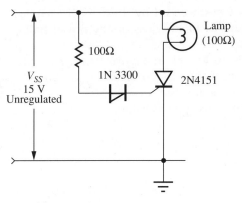

FIGURE 24.19   Problem 24.8.

Section 24.7

10. Derive Equation 24.13

11. Determine the minimum rms power delivered to resistor R_L in Example 24.3.

12. Determine the minimum triggering angle, θ_{trgMin}, for the circuit of Figure 24.14.

Section 24.9

13. Determine the value of the resistor R in the circuit of Figure 24.20 in order that its oscillation frequency will be approximately 1 kHz.

14. Redesign the circuit of Figure 24.20 in order that it will oscillate at approximately 10 kHz.

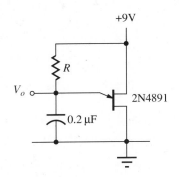

FIGURE 24.20   Problem 24.13.

15. Show how the waveform shown in Figure 24.4b would be affected if breakover occurred at half the value listed in the data books for the 1N3300 in the circuit of Figure 24.4a.

16. Describe the effect on the rest of the circuit if the diac in Figure 24.13 became a short circuit.

17. Describe the effect observed by the user if early breakover caused the 1N5758 in Figure 24.14 to occasionally fire at half of its normal breakover voltage.

18. Assume that $R = 10$ kΩ in Figure 24.18. Describe the effect on the output waveform if the valley current, I_v, for the 2N4948 dropped to:

    (a)   Half of its typical value.

    (b)   One fourth of its typical value.

A photogate tells the computer when a disk is in the
drive.

Chapter 25
Optoelectronic Devices

SPECIAL TERMS

Optoelectronics
Luminous intensity (P_v)
Illumination (E)
Photoconductive device
Photoemissive device
Photovoltaic device

25.1 Introduction and Objectives

In this chapter you will be briefly introduced to some examples of the rapidly expanding family of devices that are concerned with the interaction between light and electricity. Although we will be able to examine only a few of these devices, you should find that the basic understandings which you should have developed in the previous chapters allows you to grasp the principles of operation of these devices rather quickly. At the conclusion of this chapter, you should have some understanding of how optoelectronic devices work and you should be able to:

- Perform rather simple calculations relating to illumination and light intensities.
- Use optoelectronic device characteristic graphs to predict the behavior of circuits containing those devices.

25.2 LEDs

Most of the devices that we have discussed so far have used electrical inputs and electrical outputs. The exception is the LED discussed in Chapter 7. It has an electrical input and a light output. For that reason it should logically be placed in this chapter. However, due to its conceptual simplicity and its wide use, it was included with the other special purpose diodes in Chapter 7. We will look briefly at some other optoelectronic devices in this chapter.

25.3 Some Properties of Light

Since we will be looking at light sensitive devices, let us take time out to talk about some of the relevant properties of light. Because light is a form of energy, the unit commonly used by manufacturers to measure the rate at which light energy is being delivered to their devices is milliwatts of radiant power per square centimeter of exposed surface. Thus we will talk about the luminous intensity P_v of light sources in terms of watts radiated rather than candle power as the physicists would do.

If you touch a lighted incandescent light bulb you quickly realize that its efficiency in converting electrical power into visible light is quite low. Most of the electrical power it consumes is converted to heat. For a typical 10 W bulb, $P_v \approx 0.12$ W of visible light for an efficiency of 1.2 percent. For a 100 W bulb, $P_v \approx 2.4$ W for an efficiency of 2.4 percent.

If a light source is assumed to be isotropic (radiates uniformly in all directions), then geometry will tell us how to determine the resulting illumination striking a photo detector at some arbitrary distance r. All we need to do is imagine that our detector is part of a sphere of radius r surrounding the bulb as shown in Figure 25.1. The output of the bulb will

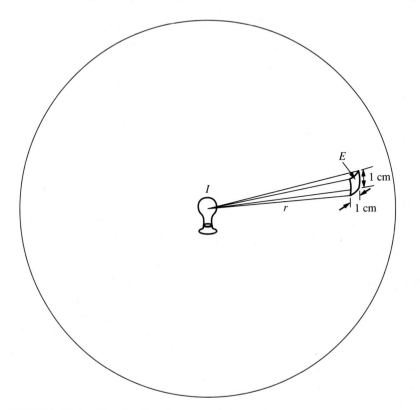

FIGURE 25.1   Illumination from a point source.

be uniformly distributed over the inner surface area of the sphere. The surface area of a sphere is given by

$$A = 4\pi r^2 \qquad (25.1)$$

Thus, the illumination E striking our detector is given by

$$E \equiv \frac{P_v}{A} \qquad (25.2)$$

or, substituting Equation 25.1 into Equation 25.2,

$$E = \frac{P_v}{4\pi r^2} \qquad (25.3)$$

EXAMPLE 25.1

If a 100 W study lamp is suspended 1 meter directly above your desk, what is the illumination on your desk in watts per square centimeter?

SOLUTION:

According to the previous discussion, for a 100 W lamp, $P_v \approx 2.4$ W so

$$E = \frac{P_v}{4\pi r^2} \tag{25.3}$$

$$= \frac{2.4\ \text{W}}{4\pi(100\ \text{cm})^2} = 19.1\ \mu\text{W/cm}^2$$

EXAMPLE 25.2

If a reflector is placed behind the above lamp such that it concentrates all of the light from the lamp in a circle of 50 cm diameter, what is the illumination within the circle?

SOLUTION:

The area of the circle is

$$A = \pi r^2 \tag{25.4}$$

$$= \pi(25\ \text{cm})^2 = 1960\ \text{cm}^2$$

Substituting in Equation 25.2 gives

$$E = \frac{P_v}{A} \tag{25.2}$$

$$= \frac{2.4\ \text{W}}{1960\ \text{cm}^2} = 1.22\ \text{mW/cm}^2$$

In addition to the brightness of light, we will need to refer to the color of light. As you probably know, light can be considered to be a wave, much like the waves on the surface of a lake. The symbol commonly used for the wavelength of light is the Greek letter lambda, λ. Our eyes interpret different wavelengths as different colors. The range of visible light extends from about $\lambda_{\text{red}} = 700$ nm (nanometers) to $\lambda_{\text{violet}} = 400$ nm with all of the colors of the rainbow in between. Our eyes are most sensitive to yellow-green light with a wavelength of about 550 nm. Optoelectronic devices are often most sensitive to infrared light with wavelengths of around 900 nm while others have a spectral response that is close to that of the human eye.

25.4 Photodiodes

In Section 2.4 we introduced the idea of "minority carriers." There we talked mainly about thermally generated minority carriers which can be thought of as resulting when the thermal motion of atoms in a crystal cause electrons to be "knocked loose" from the atoms. Light striking a diode junction can have the same effect.

Diodes specifically designed to be photosensitive are known as "photodiodes." The standard symbol for a photodiode is shown in Figure 25.2a. The construction of a photodiode is shown somewhat schematically in Figure 25.2b. Notice that the battery polarity in Figure 25.2b is such that the junction is reverse biased. That means there will be a fairly wide depletion region and ideally the "dark current" should be zero amps.

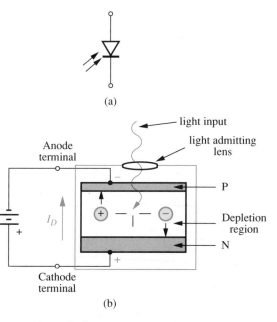

(a)

(b)

FIGURE 25.2   A photodiode.
(a) Schematic symbol.
(b) Construction.

The P region of a photodiode is made so thin that it is transparent to the incoming light. The depletion region is made relatively thick so that the incoming light is absorbed in that region where it produces both positive and negative charges. The positive charges tend to move toward the P region and the negative charges tend to move toward the N region. The result is a photocurrent. Another advantage to making the depletion region quite thick is that the capacitance of the device is reduced, allowing it to be used at relatively high frequencies.

A characteristic graph can be drawn for a photodiode much like the characteristics we drew for transistors in Section 8.10. Instead of the control variable being base current or gate voltage, as it was for transistors, the control variable is now illumination, E. Such a characteristic graph is shown in Figure 25.3. Notice that since the device is reverse biased, the graph looks "upside down" much like the graph for a PNP bipolar transistor.

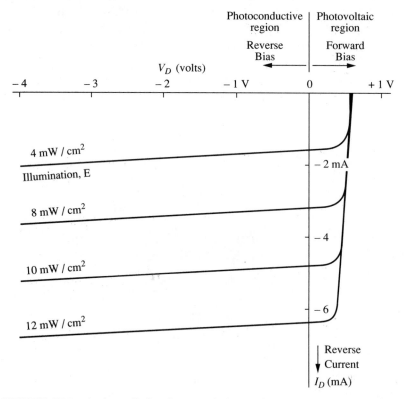

FIGURE 25.3   A photodiode characteristic graph.

Unlike a transistor, the photodiode has a "photovoltaic region." That is, it can be a source of voltage. This is due to the tendency of the positive carriers to move toward the P region and the negative carriers to move toward the N region. In fact, solar cells are simply photodiodes whose photovoltage has been maximized by making their depletion region quite thin. Also, since the current output of a photodiode is proportional to the area of the exposed junction, solar cells typically have relatively large junction areas. Thus, solar cells will usually have too much capacitance to be very useful in AC circuits.

The photodiode in a photodetector circuit usually operates in the photoconductive region of the graph in Figure 25.3. A typical photodetector circuit is shown in Figure 25.4a. A load line can be drawn on the device characteristic graph as shown in Figure 25.4b. The idea is the same as that introduced in Chapter 9 in connection with load lines for transistors.

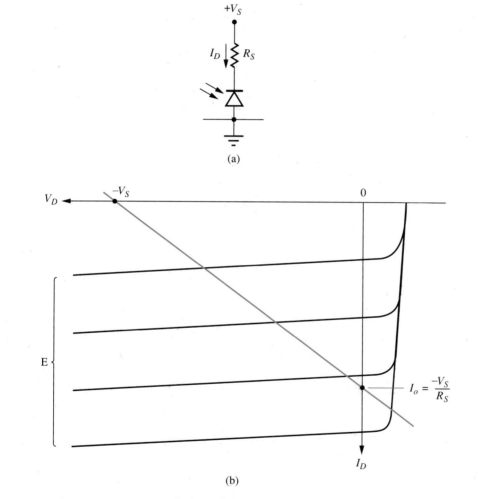

(a)

(b)

FIGURE 25.4 A typical photodiode circuit.
(a) Circuit diagram.
(b) Load line drawn on the characteristic graph.

EXAMPLE 25.3

Determine the current flowing in the circuit of Figure 25.5a if the diode characteristics are given in Figure 25.5b. Assume that the illumination is

(a)   6 mW/cm^2

(b)   12 mW/cm^2

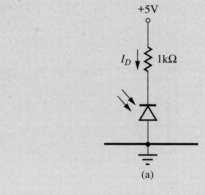

(a)

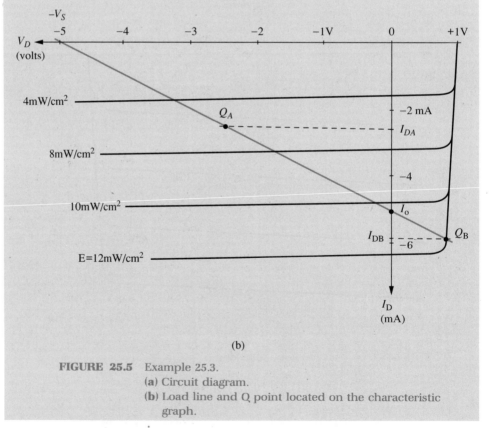

(b)

FIGURE 25.5   Example 25.3.
(a) Circuit diagram.
(b) Load line and Q point located on the characteristic graph.

SOLUTION:

In order to draw the load line, we recognize from Section 9.2 that

$$I_D = 0 \text{ A} \rightarrow V_D = -V_S \qquad (25.5)$$
$$= -5 \text{ V}$$

and

$$V_D = 0 \text{ V} \rightarrow I_0 = \frac{-V_S}{R_S} \qquad (25.6)$$

$$= \frac{-5 \text{ V}}{1 \text{ k}\Omega} = -5 \text{ mA}$$

The resulting load line is drawn on Figure 25.5b. From the load line we read:

(a) $E = 6 \text{ mW/cm}^2 \rightarrow I_{DA} \approx -2.6 \text{ mA}$
(b) $E = 12 \text{ mW/cm}^2 \rightarrow I_{DB} \approx -5.8 \text{ mA}$

Notice that in Part (b) of the above example, the diode is saturated, and is being operated in its photovoltaic region. If the reverse bias on a photodiode is very close to its reverse breakdown value, light generated carriers can precipitate and avalanche much like that produced in a Zener diode. Not surprisingly, photodiodes designed to operate in this mode are known as "avalanche photodiodes."

25.5 Phototransistors

As we pointed out in the last section, a photodiode has some of the characteristics of a transistor. However, its sensitivity is somewhat limited. A phototransistor will provide considerably higher sensitivity, but at some sacrifice in speed.

You should remember that in a typical BJT transistor circuit the collector to base junction is a reverse biased P-N junction. Thus it should come as no surprise to you that, with careful design of this junction and the case in which the device is housed, a phototransistor can be created.

A phototransistor is designed so that light shining through a transparent window in the case of the device will strike the reverse biased C-B junction. The photocurrent produced acts like a current flowing into the base of the transistor. It produces a current flow from collector to emitter which is h_{FE} times as big as the original photocurrent.

Phototransistor symbols are shown in Figure 25.6. Some types do not have a base terminal as shown in Figure 25.6a. Others provide a base terminal as shown in Figure 25.6b.

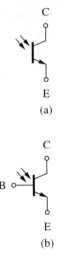

C

E
(a)

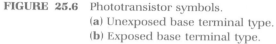

C

B

E
(b)

FIGURE 25.6   Phototransistor symbols.
(a) Unexposed base terminal type.
(b) Exposed base terminal type.

Photodarlington transistors are available for applications where even greater sensitivity is required. In fact, almost any type of device that contains a P-N junction can be designed so that it is photosensitive. It is this characteristic as much as any other that makes possible the widespread use of fiber optics in the electronics industry today.

The device characteristic curves for phototransistors are much the same as the photodiode characteristic curves in Section 25.4 except that there is no photovoltaic region and the sensitivity is greater. Thus many of the ideas we developed beginning in Chapter 8 and onward apply to phototransistors.

25.6   Phototransistor Data

As we found for other devices, typical device data books rarely provide the characteristic curves for phototransistors. Appendix F contains some of the data usually found in typical data books. The first four columns are much the same as those for bipolar transistors. Column 5 contains the "dark current," I_D. That is the leakage current that flows when no light strikes the device. Column 6 contains the "light current," I_L, which is the amount of collector current that flows when the device is exposed to the illumination E_e listed in Column 7. Column 8 contains the wavelength of light to which the device is most sensitive, λ_p. Column 9 indicates the transistor type.

Using I_L and E_e we can define $h_{\lambda e}$, the h parameter for phototransistors which corresponds to h_{fe} for bipolar transistors

$$h_{\lambda e} \equiv \frac{I_L}{E_e} \qquad (25.7)$$

The units for $h_{\lambda e}$ are amp centimeters squared per watt, or Acm^2/W.

EXAMPLE 25.4

Using data from Appendix F, determine the value of $h_{\lambda e}$ for a BPX72 phototransistor, construct a set of characteristic curves for it, draw on the characteristics a load line for the circuit of Figure 25.7a, and determine the level of illumination required to saturate the transistor.

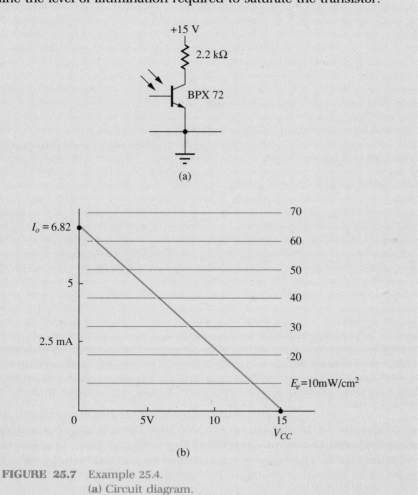

(a)

(b)

FIGURE 25.7 Example 25.4.
(a) Circuit diagram.
(b) Load line drawn on the characteristic graph.

SOLUTION:

The data in Appendix F indicates that for a collector current of $I_L = 500$ μA, the illumination must be $E_e = 4.7$ mW/cm^2. Thus $h_{\lambda e}$ is

$$h_{\lambda e} = \frac{I_L}{E_e} \tag{25.7}$$

$$= \frac{500 \text{ μA}}{4.7 \text{ mW/cm}^2} = 106 \text{ mAcm}^2/\text{W}$$

Using the above value of $h_{\lambda e}$, illumination values are plotted on the graph in Figure 25.7b at 10 mW/cm^2 intervals.

In order to draw the load line, borrowing from Chapter 9, we have

$$I_C = 0 \text{ A} \rightarrow V_{CE} = V_{CC} \tag{9.4}$$

$$= 15 \text{ V}$$

and

$$I_0 = \frac{V_{CC}}{R_C} \tag{9.5}$$

$$= \frac{15 \text{ V}}{2.2 \text{ k}\Omega} = 6.82 \text{ mA}$$

The load line is shown in Figure 25.7b.

From the graph, an illumination of something like 65 mW/cm^2 is required to saturate the transistor. Using $h_{\lambda e}$ and I_0, an analytical solution is given by solving Equation 25.7 for E_e

$$E_{\text{sat}} = \frac{I_0}{h_{\lambda e}} \tag{25.8}$$

$$= \frac{6.82 \text{ mA}}{106 \text{ mAcm}^2/\text{W}} = 64.3 \text{ mW/cm}^2$$

25.7 Optocouplers

An LED and a photodiode or a phototransistor can be combined to form what is called either an "optoisolator" or an "optocoupler." Circuits containing such devices are shown in Figure 25.8. Obviously the phototransistor could be replaced by any other photosensitive transducer such as a photodarlington, a photoSCR, etc. Notice that there is no electrical connection between the input circuit and the output circuit. This can be a great advantage in applications such as the medical field where, for safety reasons it is desirable to electrically isolate a patient from the circuits which monitor his vital signs. It is also a great advantage in sensitive in-

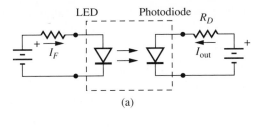

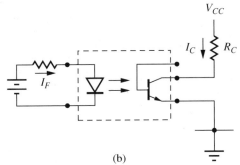

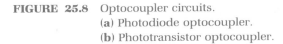

FIGURE 25.8   Optocoupler circuits.
(a) Photodiode optocoupler.
(b) Phototransistor optocoupler.

strumentation circuits where ground loops tend to generate common mode noise which interferes with low level signals.

25.8 Optocoupler Data

Appendix F contains a table of manufacturer's data for optocouplers similar to that contained in typical data books. Since the output circuit contains a transistor, much of the data and symbols should be familiar to you. The first column lists the assigned type number. The next three columns list the transistor's maximum power, voltage, and current ratings. The fifth column, lists the maximum forward current, I_F, that the LED can safely handle. The sixth column lists the maximum isolation voltage, V_{ISO}. That is the maximum voltage that can safely exist between the input LED circuit and the output transistor circuit. The seventh column would be labeled "h_{FE}," if this were a conventional transistor. It lists the forward current transfer ratio, h_F, of the device, which is the ratio of collector current to the forward current of the LED. It is defined to be

$$h_F \equiv \frac{I_C}{I_F} \qquad (25.9)$$

The last two columns give the LED forward voltage, V_F, at a specified forward current, I_F.

EXAMPLE 25.5

Draw a set of characteristics for the 4N31 optocoupler and draw a load line for the circuit of Figure 25.9a. Calculate the value of resistor R_s required to just saturate the output.

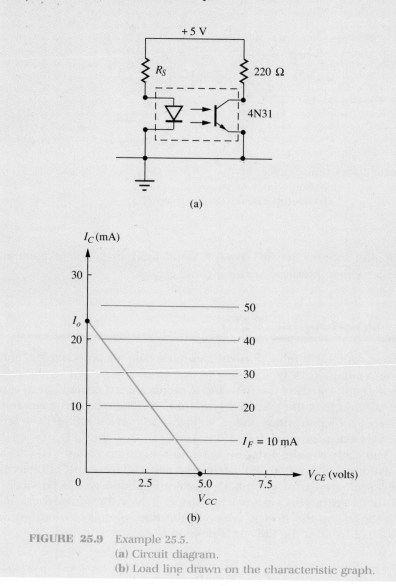

(a)

(b)

FIGURE 25.9 Example 25.5.
(a) Circuit diagram.
(b) Load line drawn on the characteristic graph.

SOLUTION:

From the data in Appendix F, $h_F = 0.5$. From Equation 25.9

$$I_C = h_F I_F \qquad (25.10)$$
$$= 0.5 * I_F$$

Substituting values in Equation 25.10, characteristics are plotted at 10 mA intervals of I_F in Figure 25.9b.

Using the ideas from Section 9.2, according to the circuit of Figure 25.9a, $V_{CC} = 5$ V. This value of V_{CC} is located on Figure 25.9b. Also

$$I_0 = \frac{V_{CC}}{R_{DC}} \qquad (9.5)$$
$$= \frac{5 \text{ V}}{220 \ \Omega} = 22.7 \text{ mA}$$

This value is also located on the characteristic graph in Figure 25.9b and the load line is drawn from V_{CC} to I_0.

Reading from the load line, the value of I_F required to saturate the transistor is about $I_{Fsat} \approx 45$ mA. An analytical solution is given by solving Equation 25.9 for I_F, recognizing that at saturation $I_C \approx I_0$. Then

$$I_{Fsat} = \frac{I_0}{h_F} \qquad (25.11)$$
$$= \frac{22.7 \text{ mA}}{0.5} = 45.5 \text{ mA}$$

According to the data in Appendix F, the forward voltage on the LED will be about $V_F = 1.5$ V when $I_F = 10$ mA. Remembering the very nonlinear characteristic of diodes, it seems reasonable to assume that V_F will not be a great deal more at I_{Fsat}.

A KVL walk from V_{CC} through R_S then gives

$$V_{CC} - I_{Fsat} R_S = V_F \qquad (25.12)$$

or

$$R_S = \frac{V_{CC} - V_F}{I_{Fsat}}$$
$$= \frac{5 \text{ V} - 1.5 \text{ V}}{45.5 \text{ mA}} \approx 77.0 \ \Omega \qquad \text{(Use 68 } \Omega\text{)}$$

The nearest smaller standard size is specified in order to assure that saturation will occur.

▶▶▶▶▶▶▶▶ **25.9   Troubleshooting**

Often problems that occur with optoelectronic devices involve the optics of the devices. Dust can collect on lenses, optical elements can get out of alignment, or aging and deterioration can occur in optical coupling media, degrading the quality of the optical signal. Light leaks or component aging can cause undesirable Q point shifts. All of these possibilities must be considered in addition to the more familiar electrical sources of difficulty when you are called on to troubleshoot optoelectronic devices.

25.10   Summary

In this chapter we surveyed some of the properties of light in order to be able to talk about illumination and luminous intensity. We used the above ideas together with the characteristics of real optoelectronic devices to predict how those devices would behave in electronic circuits.

GLOSSARY

Illumination: E The luminous intensity per unit area of exposed surface. (Section 25.3)

Luminous intensity: P_v The amount of visible light produced by a light source. (Section 25.3)

Optoelectronics: That field of electronics which deals with the interaction of light with electronic devices. (Section 25.2)

Photoconductive device: A device whose electrical resistance is affected by light. (Section 25.4)

Photoemissive device: A device that emits electrons when it is exposed to light.

Photovoltaic device: A device that generates a voltage when it is exposed to light. (Section 25.4)

PROBLEMS

Section 25.3

1. A certain slide projector has a light bulb that is rated at 300 W. It has an efficiency of about 3 percent. The optical system of the projector concentrates about 1/4 of the light produced on your slide when you place it in the projector. If the slide measures 2.4 cm by 3.4 cm, find the illumination on the slide in watts per square centimeter.
2. If 10 percent of the light striking the slide in the projector in Problem 1 is then delivered to a screen that is 35 feet away and measures 3

meters by 4.25 meters, find the illumination on the screen in watts per square centimeter.

3. A spotlight used in a theater is rated at 1000 W. It is 3 percent efficient and half of the resulting light is delivered to the stage. Find the diameter of the circle to which the lamp can deliver an illumination of 80 mW/cm^2 (the illumination delivered by the sun at noonday).

Section 25.4

4. Draw a load line for the circuit of Figure 25.10a.

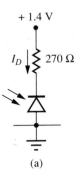

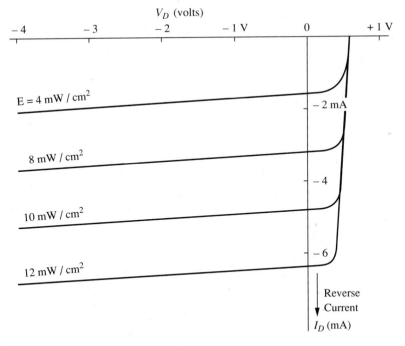

(b)

FIGURE 25.10   Problem 25.4.
    (a) Circuit diagram.
    (b) Photodiode characteristic graph.

5. If the photodiode in the circuit of Figure 25.10a has the characteristics shown in Figure 25.10b, determine the diode current I_D when the illumination is
   (a)   5 mW/cm^2
   (b)   10 mW/cm^2
   (c)   15 mW/cm^2.

Section 25.6

6. Determine the value of $h_{\lambda e}$ for a BPX72E phototransistor from the data in Appendix F and compare with the value of $h_{\lambda e}$ for the BPX72 from Example 25.4.

7. The circuit of Figure 25.11 is the circuit for a photographic exposure meter. What fraction of the light striking the exposure meter must be filtered out if the meter is to read full scale when the illumination reaching the exposure meter is $E_t = 100$ mW/cm^2?

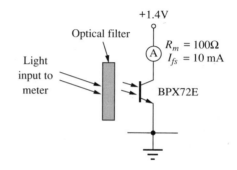

FIGURE 25.11   Problem 25.7.

Section 25.7

8. According to the data in Appendix F for a 4N30 optocoupler, how much input current, I_F, would be required in order to drive the collector current, I_C, to its rated maximum? Is it safe to drive the input this hard?

9. Determine a value for R_S in Example 25.5 that will place the Q point in the center of the load line.

10. How big can the AC input current, i_F, be in Problem 9 before clipping will occur in the AC output waveform?

Section 25.9

11. Describe the effect on the Q point and diode current in Problem 25.5 if dust on the optics reduces the light entering the junction of the photodiode.

12. Describe the effect on the calibration of the exposure meter of Problem 25.7 if the supply voltage drops to 1.2 V, assuming that the characteristic curves for the BPX72E are:

    (a) Perfectly horizontal lines.

    (b) A slope of 1mA/V

13. Describe how light leaking into the 4N31 from some external source would affect the result in Problem 25.9.

Answers to Selected Problems

Chapter 1

1. $I_5 = 12.4$ mA
3. $V_B = 1.2$ V
5. $I_L = 32.1$ μA
7. $R_{2B} = 3.75$ kΩ
9. $V_{TH} = 9.64$ V; $R_{TH} = 643$ Ω
11. $I_N = 15$ mA; $R_N = 643$ Ω
13. $P_L = 3.72$ mW
15. $I_1 = 565$ μA; $I_2 = 7.38$ mA
17. $V_t = 13.1$ V
19. $f_L = 43.5$ Hz
21. (a) An open circuit downstream from point B, or point B shorted directly to V_S.
    (b) An open circuit upstream from point B, or point B shorted directly to V_2.
23. An open circuit downstream from the pot, or both the pot and R_1 shorted (unlikely!)
25. $V_A = 8.50$ V; $V_B = 8.31$ V

Chapter 2

1. (a) 0 A (b) 0 A (c) 22 mA
   (d) 240 mA (e) many amps?
3. $R_D = 16.8$ Ω
5. $I_D \approx -6.5$ nA
7. $P_D = 0.84$ W
9. $V_S = 12.2$ V
11. (a) $I_D = 298$ mA
    (b) $I_D = 0$ A

13. (a) $V_{out} = 0$ V
    (b) $V_{out} = \infty$ V

Chapter 3

1. $V_D = 0.90$ V; $P_D = 90.5$ mW
3. (a) $V_S = 300$ V (b) $V_S = 301$ V
5. $V_S = 301$ V
7. $R_S = 220$ Ω
9. $I_D = 39.4$ mA
11. $V_D \approx 0.85$ V; $I_D \approx 60$ mA
13. $NV_T = 74.2$ mV; $N = 2.85$; $I_S = 664$ μA
15. $V_D = 62.4$ mV $* \ln \dfrac{I_D}{20.2 \text{ nA}}$
17. $V_D = 333$ mV
19. I_D 37.8 mA; $V_D = 0.753$ V
21. $r_D = 6.07$ Ω; 2.6 Ω from Problem 22 (not very good agreement)
23. $v_D = 2.27$ mV (Graphically); $v_D = 0.609$ mV (Theoretically)
25. $V_{out} = 5.12$ V
27. (a) All voltages, currents, and powers are zero.
    (b) $V_{out} = 15$ V; $I_2 = 4.55$ mA; $P_2 = 68.2$ mW; $I_D = 64.7$ mA; $P_D = 49.6$ mW; $I_3 = 64.7$ mA; $P_3 = 921$ mW; $I_S = 69.2$ mA; $P_S = 1.04$ W

Chapter 4

3. $I_{av} = 1.45$ A
5. Transformer: $V_{sec} = 7.58$ V;
$I_{av} = 100$ mA
Diode: $PIV = 19.7$ V; $I_{av} = 100$ mA; $I_{srg} = 33$A
9. Resistor R_S: $R_S = 1$ kΩ;
$P_{RS} = 10$ W
Diode: $PIV = 169$ V; $I_{av} = 50$ mA; $I_{srg} = 168$ mA
13. Transformer: $V_2 = 88.2$ VCT;
$I_{av} = 100$ A
Diode: $PIV = 91.6$ V; $I_{av} = 50$ A;
$I_{srg} = 308$ A; $I_P = 161$ A
Power dissipated by transformer: $P_S = 2$ kW
Power dissipated by each diode: $P_D = 70$ W
15. (a) Blows fuses or overheats and begins to smoke.
(b) Charging current is too low.
17. 1. A diode became an open circuit. Check diodes.
2. Broken hook-up wire. Visual inspection.
3. Open circuit in transformer. Check transformer terminal voltages—with load applied.

Chapter 5

1. $r = 6.15\%$
3. $r = 6.79\%$
5. Blows fuse.
7. Capacitor: $C = 1000$ μF at $V_{rate} = 9.5$ V
Transformer: $V_2 = 8.1$ V;
$I_{av} = 200$ mA
Diodes: $PIV = 10.3$ V; $I_{av} = 100$ mA; $I_{srg} = 32.6$ A; $I_P = 951$ mA
9. $R_F = 100$ Ω; $P_{RF} = 5$ W;
$C_F = 620$ μF
11. Capacitor: $C = 400$ μF at $V_{rate} = 10.6$ V
Transformer: $V_2 = 8.19$ V;
$I_{av} = 80$ mA

Diode: $PIV = 20.6$ V; $I_{av} = 80$ mA; $I_{srg} = 35.8$ A; $I_P = 637$ mA
13. (a) Excessive ripple, low DC voltage output.
(b) High (∞) current, blown fuse.

Chapter 6

3. $v_0 = +0.608$ V to -0.608 V.
I_D is very small.
7. $C = 10$ μF
13.

Inputs			
A	B	C	V_{out}
0	0	0	0
0	0	1	0
0	1	0	0
0	1	1	0
1	0	0	0
1	0	1	0
1	1	0	0
1	1	1	1

15. (a) Positive half of the waveform does not clip.
(b) No signal output.
17. (a) On the negative half cycle, the circuit is shorted, destroying D_1, the transformer, or a fuse.
(b) output voltages will be lower than expected.

Chapter 7

1. $R = 3.3$ kΩ; $P_R = 1$ W
3. 2.5 times
5. (a) $V_{out} = 14.2$ mV   (b) 1.11 V
7. Zener will be destroyed.
9. $\Delta V_Z = 32.0$ mV
11. $\Delta V_Z = 224$ mV
13. $R_S = 270$ Ω
15. $r = 2.62\%$
17. $V_{out} = 0.803$ V; $P_{RS} = 3.17$ W (It will overheat.)

Chapter 8

1. (a) $V_{BE} = 0.711$ V
   (b) $V_{BE} = 0.771$ V
   (c) $V_{BE} = 0.831$ V
3. $I_C = 4.48$ mA
5. (a) $V_{BB} = 0.876$ V
   (b) $V_{BB} = 2.42$ V
   (c) $V_{BB} = 17.3$ V
7. $V_{BB} = 6.12$ V
9. $V_{CE} = 4.31$ V
11. (a) $V_E = 0$ V; $V_B = 0$ V;
    $V_C = 15$ V
    (b) $V_E = 4.21$ V; $V_B = 4.93$ V;
    $V_C = 4.21$ V
13. (a) $V_{BB} = 0.6$ V
    (b) $V_{BB} = 8.31$ V
15. $I_{C1} = 40$ mA; $I_{C2} = 2.0$ mA;
    $I_{C3} = 6$A
17. $V_C = V_E = 4.21$ V; $V_B = 4.93$ V

Chapter 9

1. $V_{CC} = 18$ V; $I_0 = 26.5$ mA
3. (a) $I_C = 0$ A; $V_C = 18$ V
   (b) $I_C = 20$ mA; $V_C = 4.40$ V
   (c) $I_C = 26.5$ mA; $V_C = 0$ V
5. $V_{BB} = 0.924$ V; $V_{CC} = 15$ V;
   $R_C = 1.2$ kΩ; $h_{FE} = 195$
7. $V_{CE} = 10.4$ V
9. $V_{Cmin} = 2.83$ V; $V_{Cmax} = 12.2$ V
11. $v_{CC} = 10.9$ V; $v_{oPP} = 6.76$ VPP
13. $I_B = 41$ μA; $v_{oPP} = 8.2$ VPP;
    $V_{BB} = 5.71$ V
15. $V_C = 5.31$ V; $V_E = 2.07$ V;
    $V_B = 2.75$ V
17. $V_C = V_B = V_E = 2.84$ V

Chapter 10

1. $R_B = 150$ kΩ
3. $I_C = 11$ mA; $I_C^* = 12.1$ mA;
   $S = 99.7\%$
   Theory: $S = 100\%$
5. $R_B = 330$ kΩ
7. $I_C = 2.94$ mA; $I_C^* = 3.06$ mA;
   $S = 39.9\%$
   Theory: $S = 42.9\%$

9. Error $\approx 1\%$
11. $S = 9.09\%$; $V_E = 1.41$ V;
    $V_B = 2.12$ V; $V_C = 5.62$ V
13. $R_1 = 10$ kΩ; $R_2 = 2.7$ kΩ
15. $V_E = 2.61$ V; $V_B = 3.32$ V;
    $V_C = 12.8$ V
17. $V_E = 0$ V; $V_B = 0.722$ V;
    $V_C = 0$ V

Chapter 11

1. (a) Graph: $h_{ie} = 2.58$ kΩ;
   Theory: $h_{ie} = 1.3$ kΩ
   (b) Graph: $h_{ie} = 550$ Ω; Theory:
   $h_{ie} = 260$ Ω
3. $h_{fe} = 240$
5. (a) $1/h_{oe} = 6.0$ kΩ
   (b) $1/h_{oe} = 1.58$ kΩ
7. $h_{fe} = 220$; $h_{ie} = 109$ Ω;
   $1/h_{oe} = 588$ Ω

Chapter 12

1. (a) $R_{in} = 589$ Ω
   (b) $C_i = 6.69$ μF (Use 6.8 μF)
3. $h_{ie} = 1.05$ kΩ
5. (a) $R_{out} = 2.07$ kΩ
   (b) $C_o = 1.96$ μF (Use 2 μF)
7. $C_E = 676$ μF (Use 500 μF)
9. $v_{out} = 312$ mVPP
11. $f_L = 44.4$ Hz
13. $f_H = 155$ kHz
15. $A_v = 167$; $A_i = 2,090,000$;
    $A_P = 85.4$ dB
19. $R_{E1} = 203$ Ω (Use 220 Ω);
    $R_{E2} = 460$ Ω (Use 470 Ω);
    $C_E = 51.2$ μF (Use 50 μF);
    $C_i = 1.44$ μF (Use 1.5 μF)
21. $R_c = 1.57$ kΩ (Use 1.5 kΩ); $R_E = 500$ Ω (Use 470 Ω); $R_1 = 12.9$ kΩ
    (Use 12 kΩ); $R_2 = 4.18$ kΩ (Use
    3.9 kΩ); $C_i = 6.03$ μF (Use 6 μF);
    $C_0 = 1.81$ μF (Use 1.8 μF);
    $C_E = 630$ μF (Use 500 μF)
23. $S = 8.7\%$; $V_E = 2.54$ V; $V_B = 3.25$ V; $V_c = 7.30$ V; $R_{in} = 610$
    Ω; $R_{out} = 980$ Ω; $A_V = 147$;

$v_{outmax} = 3.85$ V_P; $v_{inmax} = 51.7$ mvPP

25. (a) No effect.
    (b) $A_v = -1.62$
27. $V_E = 2.29$ V; $V_B = 2.97$ V; $V_C = 5.73$ V

Chapter 13

1. $R_1 = 45.2$ kΩ (Use 47 kΩ); $R_2 = 185$ kΩ (Use 180 kΩ)
3. $R_{in} = 26.6$ kΩ; $R_{out} = 17.1$ Ω
5. $A_P = 10.6$ dB
7. $f_L = 19.7$ Hz
9. $R_E = 500$ Ω (Use 470 Ω); $R_1 = 1.70$ MΩ (Use 1.8 MΩ) $R_2 = 851$ MΩ (Use 820 MΩ); $C_i = 24.7$ nF (Use 0.027 μF) $C_o = 1.77$ μF (Use 1.8 μF); $C_E = 349$ μF (Use 330 μF)
11. $R_1 = 4.63$ MΩ (Use 4.7 MΩ); $R_2 = 28.9$ MΩ (Use 27 MΩ)
13. (a) $V_{open} = 6.25$ V
    (b) $V_{open} = 6.13$ V
15. (a) $V_E = V_C = 9$ V; $V_B = 7.14$ V
    (b) $P_{RE} = 45$ mW
17. (a) $V_B = 0$ V; $V_E = 0$ V
    (b) $P_{RE} = 0$ W

Chapter 14

3. $P_{DISS} = 104$ mW
5. $I_C = 34.6$ mA
7. (a) $P_{DISS} = 495$ mW
    (b) $P_{DISS} = 500$ mW
    (c) $P_{DISS} = 495$ mW
9. (a) $v_{out} = 3.54$ VP
    (b) Eff = 6.09%
11. $P_{Lmax} = 250$ mW
13. $R_E = 28.3$ Ω (Use 27 Ω); $P_{RE} = 135$ mW (Use 1/4 W); $V_{CC} = 16.8$ V; $R_1 = 1.58$ kΩ (Use 1.5 kΩ); $R_2 = 309$ Ω (Use 330 Ω); Transistor: 2SD1051
15. $R_1 = 14.4$ kΩ (Use 15 kΩ) $R_2 = 3.05$ kΩ (Use 3.3 kΩ)
17. Transistor is cut off.

19. (a) $V_C = 15$ V; $V_B = 0.210$ V
    (b) $V_C = 14.6$ V; $V_B = 3.24$ V

Chapter 15

1. $V_{oav} = 54.0$ V
3. $V_{CE} = 10$ V; $I_C = 0$ A; $I_B = 0$ A
5. Low output impedance.
7. $A_v \approx 1$
9. (a) $P_{Lrms} = 1.28$ Wrms
    (b) $P_{CC} = 1.74$ WDC
    (c) Eff = 73.5%
11. (a) $\theta_{on} = 30°$
    (b) $\theta_{off} = 150°$
    (c) cf = 33.3%
    (d) class C
13. (a) $R_2 = 273$ Ω (Use 270 Ω)
    (b) $R_2 = 309$ Ω (Use 330 Ω)
    (c) $R_2 = 391$ Ω (Use 390 Ω)
15. $i_0 = 0.5$ A; $I_C = 4.3$ mA; $V_{CE} = 5$ V
17. $v_{out} = 7.61$ VPP
19. Use: 2N6715 and 2N6727 $R_E = 0.82$ Ω; $R_1 = 180$ Ω; $R_2 = 15$ Ω; $C_i = 4.7$ μF; $C_o = 330$ μF
21. $P_{DISS} = 1.53$ W; $V_{E1} = 21.3$ V; $V_{B1} = 21.8$ V; $I_C \approx 85$ μA; $V_{E2} = 21.2$ V; $V_{B2} = 20.7$ V
23. (a) No power from the supply.
    (b) $P_2 = 82.6$ W (Q_2 will be destroyed.)

Chapter 16

3. (a) Conduction
    (b) No conduction
5. $\beta = 313$ μs/V
9. $I_D = 5$ mA $\left(1 - \dfrac{V_{GS}}{-4\ V}\right)^2$

11.

V_{GS} (volts)	0	−1	−2	−3
I_D (mA)	10	4.44	1.11	0

13. $V_{GS} = -1.36$ V
15. $V_{DS} < V_{DSknee}$ so $I_D < 1.5$ mA for any value of V_{GS}.
17. $g_{fs} = 3.5$ mS

19. $g_{fs} = 2.5$ ms; $1/g_{os} = 66.7$ kΩ

21. $g_{fs} = 2.68$ ms; $1/g_{os} = 100$ kΩ

23. $I_L = I_{DSS} = 150$ mA;
$R_{Lmax} = 66.7$ Ω

Chapter 17

1. $V_P = -2$ V; $I_{DSS} = 1.5$ mA

3. $R_D = 2.04$ kΩ (Use 2.2 kΩ)

5. $R_{AC} = 1.05$ kΩ

7. $-V_P = 3$ V; $I_{DSS} = 5$ mA;
$v_{DD} = 8.25$ V

9. $R_D = 519$ Ω (Use 560 Ω); $-V_P = 5$ V; $I_{DSS} = 14$ mA; $v_{DD} = 8.7$ V

11. (a) $I_D = 1.77$ mA
(b) $I_D = 3.18$ mA

13. $I_D = 4.95$ mA; $V_{DS} = 7.39$ V;
$V_{GS} = -2.03$ V

15. (a) $V_{DD} = 10.7$ V
(b) $I_D = 389$ μA

17. $R_G = 150$ MΩ

19. $V_{GG} = -2.05$ V; $R_G = 250$ MΩ (Use 10 MΩ ??); $V_{DD} = 10.2$ V; $R_D = 519$ Ω (Use 560 Ω)

21. $C_i = 788$ pF (Use 750 pF);
$C_S = 5.31$ μF (Use 5.6 μF);
$C_o = 1.54$ μF (Use 1.5 μF)

23. $A_v = -\dfrac{g_{fs}R_{AC}}{1 + g_{fs}R_{SQ}}$

25. $R_D = 5.56$ kΩ (Use 5.6 kΩ);
$V_{DD} = 13.9$ V
$R_G = 200$ MΩ (Use 10 MΩ ??);
$R_{SQ} = 764$ Ω (Use 820 Ω);
$C_i = 312$ pF (Use 300 pF);
$C_S = 3.88$ μF (Use 3.9 μF);
$C_o = 0.317$ μF (Use 0.33 μF)

27. $I_D \approx 1.75$ mA; $V_{GS} = -2.15$ V;
$V_D = 10.9$ V; $V_G = -2$ V;
$V_S = 0.15$ V

29. $I_D = 1.73$ mA; $V_{GS} = -2.14$ V;
$V_D = 10.9$ V; $V_G = -2$ V;
$V_S = 0.142$ V

31. $f_L = 97$ kH$_z$; $A_v = -1.17$

33. (a) No effect
(b) $V_S = 0$ V; $V_G = 0$ V;
$V_D = 3.4$ V

Chapter 18

1. Change places with the + and − symbols.

3. Place back to back Zener diodes between gate and source terminals.

5. $R_S = 255$ Ω (Use 270 Ω)

7. $V_D = 13.1$ V

9. $V_D = 11.7$ V

11.

V_D (volts)	−4	−3	−2	−1	0	1
I_D (mA)	0	2.19	8.75	19.7	35.0	54.7
V_{DSknee} (V)	0	1	2	3	4	5

13. $R_D = 950$ Ω (Use 1 kΩ); $V_{DD} = 6.16$ V; $R_G = 2.5$ MΩ (Use 2.2 mΩ); $C_i = 3.32$ nF (Use 3300 pF); $C_o = 4.08$ μF (Use 4 μF)

15. $V_{DSknee} = 3.75$ V; $I_{Dknee} = 3.38$ mA; $v_{DD} = 5.39$ V; $v_o = 1.46$ VPP

17.

V_{GS} (volts)	2.4	2.6	2.8	3.0	3.5	4.0	4.5
I_D (mA)	0	1.25	5.0	11.3	37.8	80.0	138

19. $v_i = 15.2$ VP or 30.4 VPP

21. (a) No change.
(b) Transistors are destroyed, $I_{DP} = 7.45$ A

Chapter 19

1. $A_v = 200,000$; $R_{in} = 2$ MΩ

3. (a) $V_{in} = 9.97$ mV
(b) $V_{in} = 7.5$ mV

5. $P_{out} = 5.0$ nW

7. (a) $V_{in} = 50$ μV
(b) $V_{in} = 0$ V

9. $V_{CE} = 9.43$ V

11. $I_{C13} = I_{C12} = 418$ μA; $I_{C8} = I_{C9} = 16.7$ μA

13. $V_{C5} = -13.9$ V

15. Crossover distortion

17. $V_{B14} - V_{B20} = 1.18$ V

19. $f = 10$ kHz

21. $BW = 1.17$ MHz

23. $R_{ckt} = 200$ kΩ

25. Ground noninverting input, connect V_{in} to inverting input.

29. Offset causes a zero crossing error.

Chapter 20

1. (a) The output from an electric generator mechanically connected to the wheels becomes the reference voltage when the "set" button is pressed. Subsequent output is compared to the reference and the comparator output is fed to the throttle control device.

   (b) Emotion, habit, philosophy, or reason dictates a cruising speed. Subsequent speedometer reading, sound of the engine, vibration of the steering wheel, and passing scenery are compared to the preceding mindset and the result of the comparison is fed to my right foot.

3. $R_f = 180$ kΩ; $R_i = 1.8$ kΩ
5. $f_H = 10$ kHz
7. $R_f = 180$ kΩ; $R_i = 18$ kΩ
9. $R_f = 180$ kΩ; $R_{i(guitar)} = 25.7$ kΩ (Use 27 kΩ)
   $R_{i(mic)} = 2.57$ kΩ (Use 2.7 kΩ)
11. Upward slope: $V_o = -0.4$ V; Downward slope: $V_o = +0.4$ V
13. (a) $V_o = +10$ V
    (b) $V_o = +V_{sat} \approx +14$ V
15. $V_o = -31.8$ μVP
17. $a_v = 7.80$
19. (a) No signal. $V_o = 0$ V
    (b) No feedback. $A_v = A_{OL}$
21. $V_o = 20$ mV
23. $A_v = A_{ol} = 200,000$

Chapter 21

1. $I_B = 18.4$ μA; $V_B = 3.16$ V; $P_Q = 18.6$ mW
3. Q_1: $V_{rate} = 15$ V; $I_{Crate} = 250$ mA; $P_{rate} = 1.5$ W

U_1: $V_{CC} = 15$ V; $I_o = 2.5$ mA
Z_1: $V_Z = 9$ V; $P_Z = $ minimal.
R_S: $R_S = 780$ Ω (Use 680 Ω);
$P_{RS} = 51.2$ mW (Use 1/8 W)
5. (a) $V_6 = 9.80$ V
   (b) $V_6 = 9.52$ V
7. $R_{E1} = 5.3$ Ω (Use 5.6 Ω);
   $P_{RE} = 110$ mW (Use 1/4 W)
   $P_{Q2} = 38.2$ mW (Use any small transistor 2N2712, 2N4401 ?)
9. $R_1 = 200$ kΩ (Use 180 kΩ);
   $R_2 = 602$ Ω (Use 560 Ω)
   $P_{Q1} = 725$ mW (Use 2SD1051)
11. $f = 35.7$ kH$_z$
13. $V_o = 0.1$ VPP; $A_{diff} = 100$
15. $Z_{AB} = 2$ kΩ; R_{oA} and R_{oB} are very low.
17. (a) $V_o = 0$ V; $A_{com} = 0$
    (b) $V_o = 9.8$ mV; $A_{com} = 0.0098$
19. (a) (i) $I_L = 6$ mA; $V_6 = V_o = V_2 = V_3 = 6$ V
        (ii) $I_L = 25$ mA; $V_6 = V_o = V_2 = 250$ mV
        (iii) $I_L = 25$ mA; $V_6 = V_0 = V_2 = 0$ V
    (b) (i) $I_L = 6$ A; $V_o = V_2 = V_3 = 6$ V; $V_6 = 6.71$ V
        (ii) $I_L = 600$ mA; $V_o = V_2 = V_3 = 6$ V; $V_6 = 6.82$ V
        (iii) $I_L = 1$ A; $V_o = V_2 = 0$ V; $V_6 = 1.84$ V

Chapter 22

1. (a) $V_0 = +V_{sat} \approx +13$ V
   (b) $V_o = $ Ambiguous
   (c) $V_0 = -V_{sat} \approx -13$ V
3. $V_o = +V_{sat}$ until $V_i = V_{UT} = +5.26$ V; $V_o = -V_{sat}$ until $V_i = V_{LT} = -5.26$ V.
5. $R_1 = 200$ kΩ (Use 180 kΩ);
   $R_2 = 15$ kΩ (Use 15 kΩ)
7. $R_f = 180$ kΩ; $R_i = 13.8$ kΩ (Use 15 kΩ)
9. Source: $I_i = 180$ μA; Sink: $I_i = -80.0$ μA

11. $f_o = 0.429$ Hz

13. $C = 0.0455$ μF (Use 0.047 μF)

15. (a) $V_{o1} = 10.7$ VP; $f_o = 9.24$ Hz

    (b) $V_{o2} = \pm V_{sat} \approx 13$ VP

17. $f_o = 15.9$ Hz

19. $A_V = 3.15$

21. (a) $V_{UT} = 0$ V; $V_{LT} = 0$ V

    (b) $V_{UT} = +V_{sat}$; $V_{LT} = -V_{sat}$

23. No oscillation. V_o saturates.

Chapter 23

1. No change as long as $V_i > V_2$

3. Inverting Schmitt trigger

5. $t_c = -RC * \ln\left(\dfrac{1}{2}\right)$

   $t_d = -RC * \ln\left(\dfrac{1}{2}\right)$

7. $R_1 = 72.1$ kΩ (Use 68 kΩ); $R_2 = 653$ kΩ (Use 1 MΩ pot?) $C_2 = 0.1$ μF

9. $t_c = 331$ ms $t_d = 1.52$ s

11. $I_7 = 2.00$ mA

13. $D = 95.7\%$

15. (a) No change

    (b) f determined by the maximum switching rate of 555.

17. $R_4 = 104$ Ω (Use 100 Ω); $I_{max} = 140$ mA

19. "Zapping" ceases.

21. V_1 High: $t_c = 916$ μs; $t_d = 693$ μs

    V_1 Low: Operation is the same as problem 20.

23. (a) $t_d = 0$ sec. The output would be a sawtooth waveform

with $f = 1.98$ kHz.

(b) The first time $V_C = V_{UT}$, Pin 7 will attempt to "discharge" the power supply and the 555 will be destroyed.

Chapter 24

1. The voltage of the FLD drops to V_T when it begins to conduct.

3. IN3301: $V_S = 21$ V; $V_T = 1.5$ V; $I_H = 15$ mA; 1N3304: $V_S = 37.8$ V; $V_T = 1.5$ V; $I_H = 20$ mA

5. C charges to V_{SS}. Circuit does not oscillate.

7. No difference in behavior.

9. Use: 2N1843 $C_A = 0.05$ μF; $R_G = 180$ Ω

11. $P_{Lrms} = 19.1$ Wrms

13. $R = 3.14$ kΩ (Use 3.3 kΩ)

15. $V_p = 8.55$ V; $t_C = 284$ μsec

17. The load will receive an increase in power when ever early breakover occurs.

Chapter 25

1. 276 mW/cm^2

3. $d = 15.5$ cm (about 6 inches!)

5. (a) $I_D \approx -2.15$ mA

   (b) $I_D = -4.8$ mA

   (c) $I_D = -6.6$ mA

7. $\%E_f = 66.4\%$

9. $R_S = 154$ Ω (Use 150 Ω)

11. I_D would be reduced.

13. The Q point will move up the load line.

Appendix A
Schematic Symbols

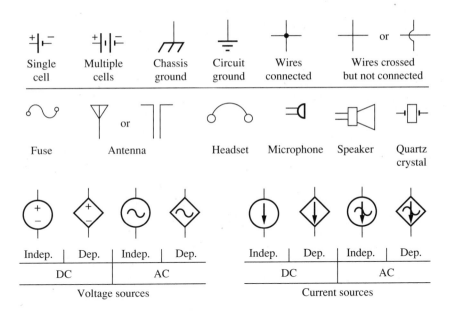

Single cell   Multiple cells   Chassis ground   Circuit ground   Wires connected   Wires crossed but not connected

Fuse   Antenna   Headset   Microphone   Speaker   Quartz crystal

Indep.	Dep.	Indep.	Dep.
DC		AC	

Voltage sources

Indep.	Dep.	Indep.	Dep.
DC		AC	

Current sources

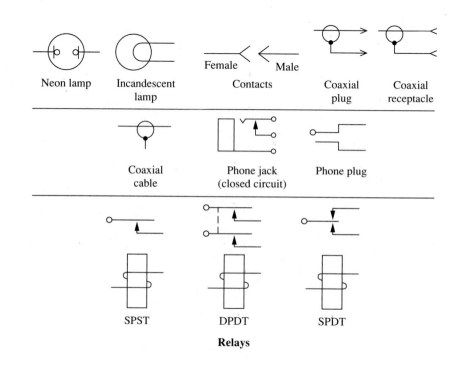

Neon lamp   Incandescent lamp   Female   Male   Contacts   Coaxial plug   Coaxial receptacle

Coaxial cable   Phone jack (closed circuit)   Phone plug

SPST   DPDT   SPDT

Relays

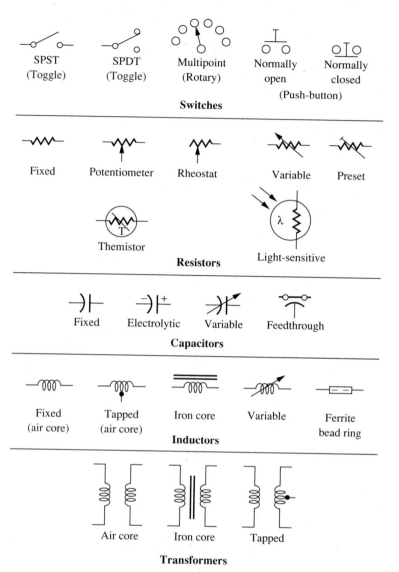

SPST
(Toggle)

SPDT
(Toggle)

Multipoint
(Rotary)

Normally
open

Normally
closed

(Push-button)

Switches

Fixed

Potentiometer

Rheostat

Variable

Preset

Themistor

Light-sensitive

Resistors

Fixed

Electrolytic

Variable

Feedthrough

Capacitors

Fixed
(air core)

Tapped
(air core)

Iron core

Variable

Ferrite
bead ring

Inductors

Air core

Iron core

Tapped

Transformers

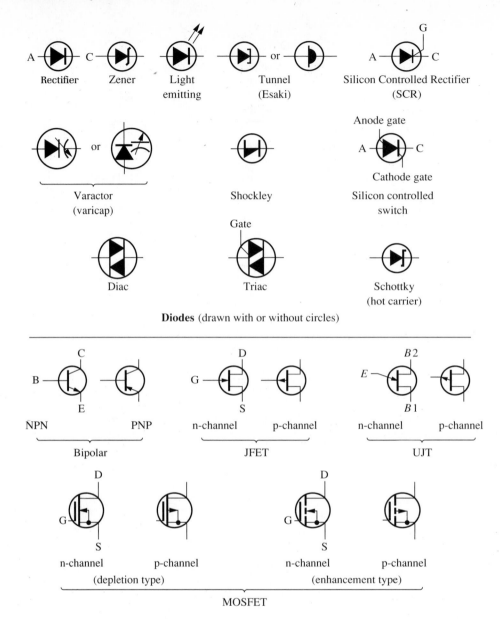

Diodes (drawn with or without circles)

Transistors (drawn with or without circles)

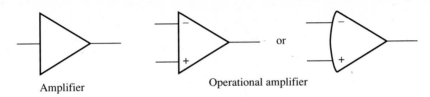

Amplifier

Operational amplifier   or

Amplifiers

Appendix B
Standard Component Values

Color Code

−2 = Silver (multiplier)
−1 = Gold (multiplier)
 0 = Black
 1 = Brown
 2 = Red
 3 = Orange
 4 = Yellow
 5 = Green
 6 = Blue
 7 = Violet
 8 = Grey
 9 = White

First Digit
Second Digit
Multiplier
Tolerance
None 20%
Sil ± 10%
Gol ± 5%

1st
2nd
3rd
Mult.
Tol.
Red = 2%
Brn = 1%

Prefixes

10^{-12} = p (pico)
10^{-9} = n (nano)
10^{-6} = μ (micro)
10^{-3} = m (mili)
10^{+3} = k (kilo)
10^{+6} = M (meg)

Physical Sizes

2 Watt
1 W
1/2 W
1/4 W

Standard Component Values

| Resistors | | | | | Capacitors | | | | | | | | |
| Four-Band | | Five-Band | | | pF | | | μF | | | | | |
10%	5%							0.01	0.1	1	10	100	1000
10	10	100	215	464	10	100	1000	0.01	0.1	1	10	100	1000
	11	102	221	475	11	110	1100						
12	12	105	226	487	12	120	1200	0.012	0.12	1.2			
	13	107	232	499		130	1300						
15	15	110	237	511	15	150	1500	0.015	0.15	1.5	15	150	1500
	16	113	243	523		160	1600						
18	18	115	249	536	18	180	1800	0.018	0.18	1.8	18	180	
	20	118	255	549	20	200	2000	0.020	0.20	2.0	20	200	2000
22	22	121	261	562	22	220	2200		0.22	2.2	22		
	24	124	267	576	24	240	2400					240	
27	27	127	274	590		250	2500		0.25		25	250	2500
	30	130	280	604	27	270	2700	0.027	0.27	2.7	27	270	
33	33	133	287	619	30	300	3000	0.030	0.30	3.0	30	300	3000
	36	137	294	634	33	330	3300	0.033	0.33	3.3	33	330	3300
39	39	140	301	649	36	360	3600						
	43	143	309	665	39	390	3900	0.039	0.39	3.9	39		
47	47	147	316	681			4000	0.040		4.0		400	
	51	150	324	698	43	430	4300						
56	56	154	332	715	47	470	4700	0.047	0.47	4.7	47		
	62	158	340	732	50	500	5000	0.050	0.50	5.0	50	500	5000
68	68	162	348	750	51	510	5100						
	75	165	357	768	56	560	5600	0.056	0.56	5.6	56		5600
82	82	169	365	787			6000	0.060		6.0			6000
	91	174	374	806	62	620	6200						
		178	384	825	68	680	6800	0.068	0.68	6.8			
		182	392	845	75	750	7500				75		
		187	402	866			8000			8.0	80		
		191	412	887	82	820	8200	0.082	0.82	8.2	82		
		196	422	909	91	910	9100						
		200	432	931									
		205	442	953									
		210	453	976									

Appendix C

Standard Unit Prefixes[*]

Prefix	Pronunciation	Multiplication Factor	Symbol
tera	ter′à	10^{12}	T
giga	jĭ′gà	10^{9}	G
<u>mega</u>	<u>mĕg′à</u>	10^{6}	<u>M</u>
<u>kilo</u>	<u>kĭl′ō</u>	10^{3}	<u>k</u>
<u>milli</u>	<u>mĭl′ĭ</u>	10^{-3}	<u>m</u>
<u>micro</u>	<u>mīkrō</u>	10^{-6}	<u>μ</u>
nano	năn′ō	10^{-9}	<u>n</u>
<u>pico</u>	pē′kō	10^{-12}	p
<u>femto</u>	<u>fĕm′tō</u>	10^{-15}	<u>f</u>
atto	ăt′tō	10^{-18}	a

*The most commonly used prefixes are underscored.

Appendix D
Calculus-Based Derivations

D.1 JUNCTION DYNAMIC RESISTANCE (r_D):

To show that

$$r_D = \frac{NV_T}{I_D} \tag{3.21}$$

The junction dynamic resistance, r_D, is defined to be

$$r_D \equiv \frac{dV_D}{dI_D} \tag{D.1}$$

The junction equation is

$$V_D = NV_T * \ln \frac{I_D}{I_S} \tag{3.5}$$

Substituting Equation 3.5 into Equation D.1 gives

$$r_D = \frac{d}{dI} \left(NV_T * \ln \frac{I_D}{I} \right) \tag{D.2}$$

Or, since NV_T is constant

$$r_D = NV_T \frac{d}{dI_D} \left(\ln \frac{I_D}{I_S} \right) \tag{D.3}$$

Since dividing numbers is equivalent to subtracting their logarithms, we have

$$r_D = NV_T \frac{d}{dI_D} (\ln I_D - \ln I_S) \tag{D.4}$$

$$= NV_T \left(\frac{d \ln I_D}{dI_D} - \frac{d \ln I_S}{dI_D} \right)$$

Since $\ln I_S$ is a constant, and the derivative of a constant is zero, we have

$$r_D = NV_T \frac{d \ln I_D}{dI_D} \tag{D.5}$$

Since the derivative of $\ln(x)$ is $\dfrac{1}{x}$, we have

$$r_D = \frac{NV_T}{I_D} \tag{3.21}$$

D.2 RECTIFIER DC AVERAGE CURRENT (I_{av}):

To show that

$$I_{AV} = \frac{I_P}{\theta_{rep}} (\cos \theta_{on} - \cos \theta_{off}) \tag{4.2}$$

An examination of the circuit of Figure D.1a indicates that conduction will begin when V_S exceeds the sum of V_X plus the turn-on voltage of the diode. The electrical angle at that instant is indicated in Figure D.1b as θ_{on}. Similarly, conduction will cease at θ_{off}. The process will be repeated again after an angle of θ_{rep}. In the case of the half-wave rectifier shown, θ_{rep} will be 2π radians. It would be π radians for a full-wave circuit. Conduction will occur during the interval between θ_{on} and θ_{off}.

The calculus expression for averaging the current between those limits is

$$I_{av} = \frac{1}{\theta_{rep}} \int_{\theta_{on}}^{\theta_{off}} I_\theta \, d\theta \tag{D.6}$$

If the circuit is assumed to be linear between θ_{on} and θ_{off} (it will not be, due to nonlinearity of the diode, but hopefully the error will not be too great), the current waveform will be approximated by

$$I_\theta = I_P * \sin \theta \tag{D.7}$$

Substituting Equation D.7 in Equation D.6 gives

$$I_{av} = \frac{1}{\theta_{rep}} \int_{\theta_{on}}^{\theta_{off}} (I_P * \sin \theta) \, d\theta \tag{D.8}$$

Since I_P is constant we have

$$I_{av} = \frac{I_P}{\theta_{rep}} \int_{\theta_{on}}^{\theta_{off}} (\sin \theta) \, d\theta$$

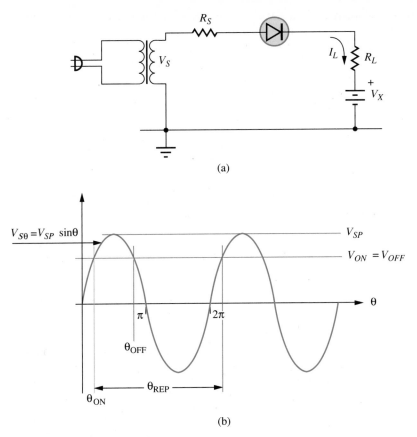

(a)

(b)

FIGURE D.1   Rectifier average current.
(a) Circuit diagram.
(b) Voltage waveform.

Since the integral of the sine function is minus the cosine function, integrating between the indicated limits yields

$$I_{av} = \frac{I_P}{\theta_{rep}} \left((-\cos \theta_{off}) - (-\cos \theta_{on}) \right) \tag{D.9}$$

$$= \frac{I_P}{\theta_{rep}} \left(\cos \theta_{on} - \cos \theta_{off} \right) \tag{4.2}$$

D.3 THE RMS VALUE OF A TRIANGLE WAVE

To show that for a triangle wave

$$v_{rms} = \frac{v_{PP}}{2\sqrt{3}} \tag{5.7}$$

the ripple waveform from a capacitor filtered rectifier approximates a triangle such as that shown in Figure 5.2a. However, the results will be the same if, for simplicity, the waveform is assumed to be a right triangle (a saw-tooth waveform) with its direction reversed. Also, for simplicity we will ignore the DC voltage level. The resulting simplified waveform is shown in Figure D.2.

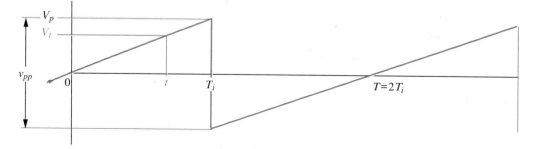

FIGURE D.2 The rms value of a triangular waveform.

To further simplify the work only the positive half of the waveform will be considered. Thus the integration period, T_i, will be only half of the period for the whole waveform. Then from Figure D.2 we have

$$\frac{V_t}{t} = \frac{V_P}{T_i} \tag{D.10}$$

or

$$V_t = \frac{V_P}{T_i} t \tag{D.11}$$

Since "rms" stands for "Root-Mean-Square," we have

$$v_{\text{rms}} \equiv \sqrt{\frac{1}{T_i} \int_0^{T_i} V_t^2 \, dt} \tag{D.12}$$

Substituting Equation D.11 into Equation D.12 gives

$$v_{\text{rms}} = \sqrt{\frac{1}{T_i} \int_0^{T_i} \left(\frac{V_P}{T_i} t\right)^2 dt} \tag{D.13}$$

$$= \sqrt{\frac{1}{T_i} \left(\frac{V_P}{T_i}\right)^2 \int_0^{T_i} t^2 \, dt}$$

$$= V_P \sqrt{\frac{1}{T_i^3} \left[\frac{t^3}{3}\right]_0^{T_i}}$$

$$= V_P \sqrt{\frac{1}{T_i^3}\left[\frac{T_i^3}{3} - 0\right]}$$

$$= \frac{V_P}{\sqrt{3}}$$

Or, since $V_P = \dfrac{V_{PP}}{2}$, we have

$$V_{\text{rms}} = \frac{V_{PP}}{2\sqrt{3}} \tag{5.7}$$

D.4 SLOPE OF CAPACITOR DISCHARGE CURVE

To show that a line that is tangent to the capacitor discharge curve will intersect the time axis at a time, τ given by

$$\tau = RC \tag{5.11}$$

If switch S in the circuit of Figure D.3 is thrown from position 1 to position 2 at time $t = 0$ sec, the capacitor C will discharge into the resistor R. If

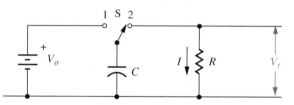

FIGURE D.3 Capacitor discharge circuit.

the voltage V_t is plotted against time, the familiar capacitor discharge curve shown in Figure D.4 results. It is described by the equation

$$V_t = V_o e^{-t/(RC)} \tag{5.9}$$

Equation 5.9 is derived from the definition of capacitance

$$C \equiv \frac{Q}{V} \tag{D.14}$$

Solving Equation D.14 for Q gives

$$Q = CV \tag{D.15}$$

Finding the time derivative of Equation D.15 gives

$$\frac{dQ}{dt} = C\frac{dv}{dt} \tag{D.16}$$

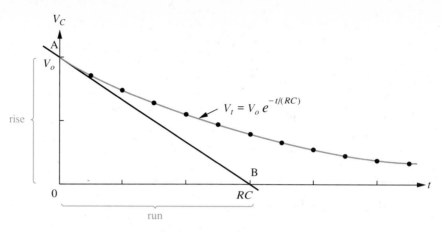

FIGURE D.4   Capacitor discharge graph.

Since current is defined to be

$$I \equiv \frac{dQ}{dt} \tag{D.17}$$

and recognizing that the capacitor will be discharging (negative) as it supplies current to the circuit, we have

$$I = -\frac{dQ}{dt} = -C\frac{dV}{dt} \tag{D.18}$$

Applying Ohm's Law to the circuit of Figure D.3 gives

$$I = \frac{V_t}{R} \tag{D.19}$$

Combining Equations D.18 and D.19 gives

$$C\frac{dV}{dt} = -\frac{V_t}{R}$$

or

$$\frac{dV}{dt} = -\frac{V_t}{RC} \tag{D.20}$$

Remembering that the derivative is the slope of the line, according to Equation D.20, at $t = 0$, the slope is V_0/RC. Since the slope is also the "rise" divided by the "run," if the rise is V_0 as shown in Figure D.4, then the run is RC. Thus the line tangent to the discharge curve at $t = 0$ sec intersects the time axis at $T = RC$.

D.5 STABILITY—FIXED BIAS

To show that for the fixed bias circuit the stability factor S is equal to one.

The stability factor, S, is defined to be

$$S \equiv \frac{\% dI_C}{\% dh_{FE}} \qquad (10.6)$$

where

$$\% dI_C \equiv \frac{dI_C}{I_C} 100\% \qquad (D.21)$$

and

$$\% dh_{FE} \equiv \frac{dh_{FE}}{h_{FE}} 100\% \qquad (D.22)$$

Substituting Equations D.21 and D.22 in Equation 10.6 and doing a little algebra on the result gives

$$S = \frac{h_{FE}}{I_C} * \frac{dI_C}{dh_{FE}} \qquad (D.23)$$

For the fixed bias circuit, from Section 10.2 we have

$$I_C = \frac{V_{CC} - V_{BE}}{\dfrac{R_B}{h_{FE}}} \qquad (10.4)$$

$$= \frac{V_{CC} - V_{BE}}{R_B} * h_{FE} \qquad (D.24)$$

Obviously, V_{CC} and R_B are constants, and if we assume that V_{BE} is essentially constant, then h_{FE} in Equation D.24 is multiplied by a constant. Thus the derivative of I_C with respect to h_{FE} is

$$\frac{dI_C}{dh_{FE}} = \frac{V_{CC} - V_{BE}}{R_B} \qquad (D.25)$$

Substituting Equations D.24 and D.25 in Equation D.23 gives

$$S = \frac{h_{FE} \, R_B}{(V_{CC} - V_{BE}) \, h_{FE}} * \frac{V_{CC} - V_{BE}}{R_B} = 1$$

D.6 STABILITY—SELF BIAS

To show that for the self biased circuit, the stability factor S is given by:

$$S = \frac{R_B}{R_B + h_{FE}R_C} \tag{10.10}$$

From the definition of S we know from Section D.5 that

$$S = \frac{h_{FE}}{I_C} \frac{dI_C}{dh_{FE}} \tag{D.23}$$

In order to evaluate the derivative in Equation D.23 for the self biased circuit, we know from Section 10.4 that

$$I_C = \frac{V_{CC} - V_{BE}}{R_C + \dfrac{R_B}{h_{FE}}} \tag{10.9}$$

or, in order to better fit the form $d(uv) = udv + vdu$

$$I_C = (V_{CC} - V_{BE}) \left(R_C + \frac{R_B}{h_{FE}} \right)^{-1} \tag{D.26}$$

Using the fact that $dx^n = nx^{(n-1)}\,dx$, the derivative of Equation D.26 is

$$dI_C = (V_{CC} - V_{BE})(-1) \left(R_C + \frac{R_B}{h_{FE}} \right)^{-2} d\left(R_C + \frac{R_B}{h_{FE}} \right)$$

$$= -\frac{V_{CC} - V_{BE}}{\left(R_C + \dfrac{R_B}{h_{FE}} \right)^2} d(R_C + R_B h_{FE}^{-1})$$

$$= -\frac{V_{CC} - V_{BE}}{\left(R_C + \dfrac{R_B}{h_{FE}} \right)^2} (-1)\, (R_B h_{FE}^{-2}) dh_{FE}$$

$$= \frac{V_{CC} - V_{BE}}{\left(R_C + \dfrac{R_B}{h_{FE}} \right)^2} \left(\frac{R_B}{h_{FE}^{2}} \right) dh_{FE}$$

or

$$\frac{dI_C}{dh_{FE}} = \frac{(V_{CC} - V_{BE})R_B}{\left(R_C + \dfrac{R_B}{h_{FE}} \right)^2 h_{FE}^{2}} \tag{D.27}$$

Substituting Equations 10.9 and D.27 in Equation D.23 gives

$$S = \frac{h_{FE}}{\dfrac{V_{CC} - V_{BE}}{R_C + \dfrac{R_B}{h_{FE}}}} * \frac{(V_{CC} - V_{BE})R_B}{\left(R_C + \dfrac{R_B}{h_{FE}}\right)^2 h_{FE}{}^2}$$

Simplifying the above equation gives

$$S = \frac{R_B}{R_B + h_{FE}R_C} \qquad (10.10)$$

D.7 STABILITY—EMITTER BIAS

To show that for the emitter biased circuit, the stability factor S is given by

$$S = \frac{R_B}{R_B + h_{FE} R_E} \qquad (10.13)$$

From the definition of S we know from Appendix D.5 that

$$S = \frac{h_{FE}}{I_C} * \frac{dI_C}{dh_{FE}} \qquad (D.23)$$

From Section 10.5 we know that for the emitter biased circuit

$$I_C = \frac{V_{BB} - V_{BE}}{\dfrac{R_B}{h_{FE}} + R_E} \qquad (10.12)$$

or, in order to better fit the form $dx^n = nx^{n-1}dx$,

$$I_C = (V_{BB} - V_{BE}) (R_B h_{FE}{}^{-1} + R_E)^{-1} \qquad (D.28)$$

the derivative of Equation D.28 is

$$
\begin{aligned}
dI_C &= (V_{BB} - V_{BE}) (-1) (R_B h_{FE}{}^{-1} + R_E)^{-2} d(R_B h_{FE}{}^{-1} + R_E) \\
&= -(V_{BB} - V_{BE}) (R_B h_{FE}{}^{-1} + R_E)^{-2} (-1) R_B h_{FE}{}^{-2} dh_{FE}
\end{aligned}
$$

or

$$\frac{dI_C}{dh_{FE}} = \frac{V_{BB} - V_{BE}}{\left(\dfrac{R_B}{h_{FE}} + R_E\right)^2} \frac{R_B}{h_{FE}{}^2} \qquad (D.29)$$

Substituting Equations 10.12 and D.29 into Equation D.23 and simplifying gives

$$S = \frac{R_B}{R_B + h_{FE}R_E}$$

(10.13)

D.8 POWER CURVE TANGENT TO LOAD LINE

To show that if the Q point is correctly chosen, the power curve will be tangent to the load line at the Q point.

From Section 9.7, we will assume that "correctly chosen" means that the Q point is at the center of the AC load line. Thus

$$i_0 = 2I_C$$

(D.30)

$$v_{CC} = 2V_{CE}$$

(D.31)

These relationships are shown in Figure D.5. The slope of the load line, m_{LL} is given by

$$m_{LL} = -\frac{i_0}{v_{CC}}$$

(D.32)

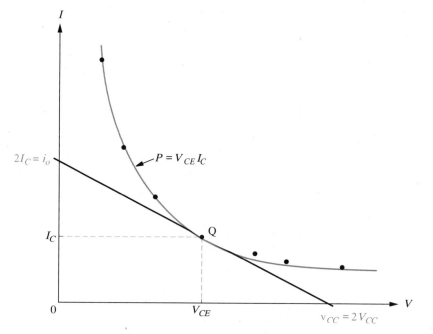

FIGURE D.5 Load line tangent to power curve.

Substituting Equations D.30 and D.31 into Equation D.32 gives

$$m_{LL} = -\frac{2I_C}{2V_{CE}}$$ (D.33)

$$= -\frac{I_C}{V_{CE}}$$

The equation for the power dissipation curve is

$$P_{\text{DISS}} = V_{CE} I_C$$ (14.1)

or, solving for I_C

$$I_C = \frac{P_{\text{DISS}}}{V_{CE}}$$ (D.34)

Since the derivative is the slope of the curve, we have

$$m_{\text{DISS}} = \frac{dI_C}{dV_{CE}}$$ (D.35)

Substituting Equation D.34 into Equation D.35 gives

$$m_{\text{DISS}} = \frac{d}{dV_{CE}}\left(\frac{P_{\text{DISS}}}{V_{CE}}\right)$$

or, in order to better fit the form $dx^n = nx^{n-1}\, dx$,

$$m_{\text{DISS}} = P_{\text{DISS}} \frac{d}{dV_{CE}}(V_{CE})^{-1}$$ (D.36)

$$= P_{\text{DISS}}(-1)V_{CE}^{-2}$$

$$= -\frac{P_{\text{DISS}}}{V_{CE}^{2}}$$

Substituting Equation 14.1 into Equation D.36 gives

$$m_{\text{DISS}} = -\frac{V_{CE} I_C}{V_{CE}^{2}}$$

$$= -\frac{I_C}{V_{CE}}$$ (D.37)

Substituting Equation D.33 into Equation D.37 gives

$$m_{\text{DISS}} = m_{LL}$$

Thus the lines have the same slope at the Q point.

D.9 FET TRANSCONDUCTANCE EQUATION

To show that if g_{os} can be ignored, the transconductance, g_{fs}, for an FET is given by

$$g_{fs} = -\frac{2}{V_P}\sqrt{I_D I_{DSS}}$$

(16.13)

Transconductance is defined to be

$$g_{fs} \equiv \frac{dI_D}{dV_{GS}}$$

(16.11)

In order to evaluate the derivative in Equation 16.11, the transfer equation states

$$I_D = I_{DSS}\left(1 - \frac{V_{GS}}{V_P}\right)^2$$

(16.4)

Substituting Equation 16.4 into Equation 16.11 gives

$$g_{fs} = \frac{d}{dV_{GS}}\left(I_{DSS}\left(1 - \frac{V_{GS}}{V_P}\right)^2\right)$$

(D.38)

$$= I_{DSS}\frac{d}{dV_{GS}}\left(1 - \frac{V_{GS}}{V_P}\right)^2.$$

Differentiating using the standard form $dx^n = nx^{n-1}dx$ gives

$$g_{fs} = I_{DSS} * 2\left(1 - \frac{V_{GS}}{V_P}\right)^1 \frac{d}{dV_{GS}}\left(1 - \frac{V_{GS}}{V_P}\right)$$

(D.39)

$$= 2I_{DSS}\left(1 - \frac{V_{GS}}{V_P}\right)\left(0 - \frac{1}{V_P}\right)$$

$$= -2\frac{I_{DSS}}{V_P}\left(1 - \frac{V_{GS}}{V_P}\right)$$

Solving Equation 16.4 for $\left(1 - \frac{V_{GS}}{V_P}\right)$ gives

$$1 - \frac{V_{GS}}{V_P} = \sqrt{\frac{I_D}{I_{DSS}}}$$

(D.40)

Substituting Equation D.40 into Equation D.39 gives

$$g_{fs} = -2\frac{I_{DSS}}{V_P}\sqrt{\frac{I_D}{I_{DSS}}}$$

(16.16)

or

$$g_{fs} = -\frac{2}{V_P}\sqrt{I_{DSS}I_D}$$

D.10 RMS OUTPUT FROM A TRIAC CIRCUIT

To show that for each half of the AC powerline cycle the rms output from a triac circuit is given by

$$v_{\text{rms}} = V_P\sqrt{\frac{1}{2} - \frac{\theta^\circ{}_{\text{trg}}}{360^\circ} + \frac{\sin(2\theta_{\text{trg}})}{4\pi}} \qquad (24.16)$$

From the definition of rms, for a sine wave that is symmetrical about the voltage axis and begins at θ_{trg} and stops at π as shown in Figure D.6

$$v_{\text{rms}} \equiv \sqrt{\frac{1}{\pi}\int_{\theta_{\text{trg}}}^{\pi}(V_P\sin\theta)^2 d\theta} \qquad (D.41)$$

$$= V_P\sqrt{\frac{1}{\pi}\int_{\theta_{\text{trg}}}^{\pi}\sin^2\theta\, d\theta}$$

$$= V_P\sqrt{\frac{1}{\pi}\left[\frac{\theta}{2} - \frac{\sin 2\theta}{4}\right]_{\theta_{\text{trg}}}^{\pi}}$$

$$= V_P\sqrt{\frac{1}{\pi}\left[\left(\frac{\pi}{2} - 0\right) - \left(\frac{\theta_{\text{trg}}}{2} - \frac{\sin(2\theta_{\text{trg}})}{4}\right)\right]}$$

$$= V_P\sqrt{\frac{1}{2} - \frac{\theta_{\text{trg}}}{2\pi} + \frac{\sin(2\theta_{\text{trg}})}{4\pi}}$$

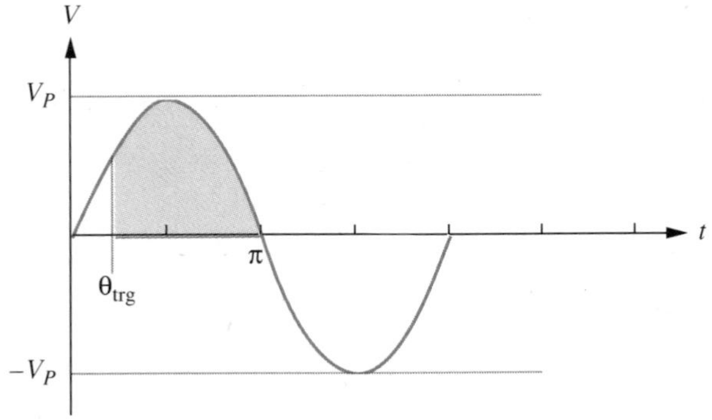

FIGURE D.6   Output from a triac circuit.

Or, if you think in degrees better than in radians

$$v_{\text{rms}} = V_P \sqrt{\frac{1}{2} - \frac{\theta°_{\text{trg}}}{360°} + \frac{\sin(2\theta_{\text{trg}})}{4\pi}} \tag{24.16}$$

Appendix E
Algebraic Derivations

E.1 EMPIRICAL DETERMINATION OF NV_T AND I_S.

To show that given two points on the diode characteristic curve, whose coordinates are (V_{D1}, I_{D1}) and (V_{D2}, I_{D2}), as shown in Figure E.1, the empirical values for the constants in the junction equation are given by

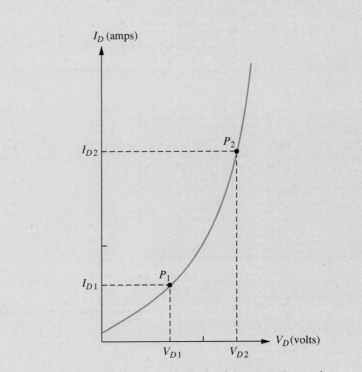

FIGURE E.1 Two points on the diode characteristic graph.

$$NV_T = \frac{V_{D2} - V_{D1}}{\ln\left(\dfrac{I_{D2}}{I_{D1}}\right)} \tag{3.7}$$

and

$$I_S = \frac{I_D}{e^{V_D/(NV_T)}} \tag{3.8}$$

The junction equation is

$$V_D = NV_T * \ln\left(\frac{I_D}{I_S}\right) \tag{3.5}$$

Solving for I_S gives

$$\frac{V_D}{NV_T} = \ln\left(\frac{I_D}{I_S}\right)$$

$$e^{V_D/(NV_T)} = \frac{I_D}{I_S}$$

or

$$I_S = \frac{I_D}{e^{V_D/(NV_T)}} \tag{3.8}$$

Substituting the coordinates for points P_1 and P_2 in Equation 3.8 and solving for NV_T gives

$$I_S = \frac{I_{D2}}{e^{V_{D2}/(NV_T)}} = \frac{I_{D1}}{e^{V_{D1}/(NV_T)}}$$

$$\frac{I_{D2}}{I_{D1}} = \frac{e^{V_{D2}/(NV_T)}}{e^{V_{D1}/(NV_T)}} = e^{\frac{V_{D2} - V_{D1}}{NV_T}}$$

$$\ln\frac{I_{D2}}{I_{D1}} = \frac{V_{D2} - V_{D1}}{NV_T}$$

or

$$NV_T = \frac{V_{D2} - V_{D1}}{\ln\dfrac{I_{D2}}{I_{D1}}} \tag{3.7}$$

Appendix F
Collected Device Data

DIODES (SILICON):

Type No.	MAX. RATINGS			PARAMETERS				Case
	PIV (V)	IF (A)	IFSM (A)	VF (V)	IF (A)	IR (A)	trr (s)	
IN3064	50	0.3	4	0.66 0.8	2m 10m	0.1μ	4n	DO35
IN3595	125	0.5	4	0.6 0.92	1m 0.2	1n	3μ	DO35
1N4001	50							
1N4002	100	Characteristics of all members of the series:						
1N4003	200			0.60	2.3m			
1N4004	400	1	30	0.70	200m	5μ	1μ	DO41
1N4005	600			0.80	245m			
1N4006	800			0.86	1.0			
1N4007	1000							
1N4305	50	0.3	4	0.50 0.70	250μ 10m	0.1μ	2n	DO35
1N4454	50	0.3	4	0.66 0.8	2m 10m	0.1μ	4n	DO35
BA216	10	100m	4	0.56 0.84	0.2m 15m	1.5μ	4n	DO35
BA217	30	100m	4	0.6 1.2	1m 50m	50n	4n	DO35
BA218	50	100m	4	0.6 1.2	1m 50m	50n	4n	DO35
BA219	100	100m	4	0.6 1.1	1m 100m	500n	120n	DO35
BAX13	50	0.30	4	0.6 1.4	2m 75m	0.2μ	4n	DO35
BAY72	100	0.5	4	0.58 0.89	1m 0.1	0.1μ	50n	DO35
BAY82	12	150m	250m	0.59 1.1	0.1m 50m	100n	750p	DO7

Zener Diodes (Silicon):

Type	TOL (±%)	PDISS (W)	VZ (V)	IZT (mA)	ZZT (Ω)	Case
1N5226A	10	500m	3.3	20	28	DO35
1N5227A	10	500m	3.6	20	24	DO35
1N5228A	10	500m	3.9	20	23	DO35
1N5229A	10	500m	4.3	20	22	DO35
1N5230A	10	500m	4.7	20	19	DO35
1N5231A	10	500m	5.1	20	17	DO35
1N5232A	10	500m	5.6	20	11	DO35
1N5233A	10	500m	6.0	20	7.0	DO35
1N5234A	10	500m	6.2	20	7.0	DO35
1N5235A	10	500m	6.8	20	5.0	DO35
1N5236A	10	500m	7.5	20	6.0	DO35
1N5237A	10	500m	8.2	20	8.0	DO35
1N5238A	10	500m	8.7	20	8.0	DO35
1N5239A	10	500m	9.1	20	10	DO35
1N5240A	10	500m	10.0	20	17	DO35
1N5241A	10	500m	11.0	20	22	DO35
1N4728	10	1.0	3.3	76.0	10.0	DO41
1N4729	10	1.0	3.6	69.0	10.0	DO41
1N4730	10	1.0	3.9	64.0	9.0	DO41
1N4731	10	1.0	4.3	58.0	9.0	DO41
1N4732	10	1.0	4.7	53.0	8.0	DO41
1N4733	10	1.0	5.1	49.0	7.0	DO41
1N4734	10	1.0	5.6	45.0	5.0	DO41
1N4735	10	1.0	6.2	41.0	2.0	DO41
1N4736	10	1.0	6.8	37.0	3.5	DO41
1N4737	10	1.0	7.5	34.0	4.0	DO41
1N4738	10	1.0	8.2	31.0	4.5	DO41
1N4739	10	1.0	9.1	28.0	5.0	DO41
1N4740	10	1.0	10.0	25.0	7.0	DO41
1N5913A	10	1.5	3.3	113	10	DO41
1N5914A	10	1.5	3.6	104	9.0	DO41
1N5915A	10	1.5	3.9	96.1	7.5	DO41
1N5916A	10	1.5	4.3	87.2	6.0	DO41
1N5917A	10	1.5	4.7	79.8	5.0	DO41
1N5918A	10	1.5	5.1	73.5	4.0	DO41
1N5919A	10	1.5	5.6	66.9	2.0	DO41
1N5920A	10	1.5	6.2	60.5	2.0	DO41
1N5921A	10	1.5	6.8	55.1	2.5	DO41
1N5922A	10	1.5	7.5	50.0	3.0	DO41
1N5923A	10	1.5	8.2	45.7	3.5	DO41
1N5924A	10	1.5	9.1	41.2	4.0	DO41
1N5925A	10	1.5	10	37.5	4.5	DO41

Zener Diodes (Silicon): (Cont.)

Type	TOL (±%)	PDISS (W)	VZ (V)	IZT (mA)	ZZT (Ω)	Case
1N5926A	10	1.5	11	34.1	5.5	DO41
1N5927A	10	1.5	12	31.2	6.5	DO41
1N5928A	10	1.5	13	28.8	7.0	DO41
1N5929A	10	1.5	15	25.0	9.0	DO41
1N5930A	10	1.5	16	23.4	10	DO41
1N5931A	10	1.5	18	20.8	12	DO41
1N5932A	10	1.5	20	18.7	14	DO41
1N5933A	10	1.5	22	17.0	17.5	DO41
1N5934A	10	1.5	24	15.6	19	DO41
1N5935A	10	1.5	27	13.9	23	DO41
1N5936A	10	1.5	30	12.5	26	DO41
1N5937A	10	1.5	33	11.4	33	DO41
1N5938A	10	1.5	36	10.4	38	DO41
1N5939A	10	1.5	39	9.6	45	DO41
1N5940A	10	1.5	43	8.7	53	DO41
1N5941A	10	1.5	47	8.0	67	DO41
1N5942A	10	1.5	51	7.3	70	DO41
1N5943A	10	1.5	56	6.7	86	DO41
1N5944A	10	1.5	62	6.0	100	DO41
1N4945A	10	1.5	68	5.5	120	DO41
1N5946A	10	1.5	75	5.0	140	DO41
1N5947A	10	1.5	82	4.6	160	DO41
1N5948A	10	1.5	91	4.1	200	DO41
1N5949A	10	1.5	100	3.7	250	DO41
1N5950A	10	1.5	110	3.4	300	DO41
1N5951A	10	1.5	120	3.1	380	DO41
1N5952A	10	1.5	130	2.9	450	DO41
1N5953A	10	1.5	150	2.5	600	DO41
1N5954A	10	1.5	160	2.3	700	DO41
1N5955A	10	1.5	180	2.1	900	DO41
1N5956A	10	1.5	200	1.9	1200	DO41
1N5333A	10	5.0	3.3	380	3.0	17
1N5334A	10	5.0	3.6	350	2.5	17
1N5335A	10	5.0	3.9	320	2.0	17
1N5336A	10	5.0	4.3	290	2.0	17
1N5337A	10	5.0	4.7	260	2.0	17
1N5338A	10	5.0	5.1	240	1.5	17
1N5339A	10	5.0	5.6	220	1.0	17
1N5340A	10	5.0	6.0	200	1.0	17
1N5341A	10	5.0	6.2	200	1.0	17
1N5342A	10	5.0	6.8	175	1.0	17
1N5343A	10	5.0	7.5	175	1.5	17
1N5344A	10	5.0	8.2	150	1.5	17

Zener Diodes (Silicon): (Cont.)

Type	TOL (±%)	PDISS (W)	VZ (V)	IZT (mA)	ZZT (Ω)	Case
1N5345A	10	5.0	8.7	150	2.0	17
1N5346A	10	5.0	9.1	150	2.0	17
1N5347A	10	5.0	10	125	2.0	17
1N5348A	10	5.0	11	125	2.5	17
1N5349A	10	5.0	12	100	2.5	17
1N5350A	10	5.0	13	100	2.5	17
1N5351A	10	5.0	14	100	2.5	17
1N5352A	10	5.0	15	75	2.5	17
1N5353A	10	5.0	16	75	2.5	17
1N5354A	10	5.0	17	70	2.5	17
1N5355A	10	5.0	18	65	2.5	17
1N5356A	10	5.0	19	65	3.0	17
1N5357A	10	5.0	20	65	3.0	17
1N5358A	10	5.0	22	50	3.5	17
1N5359A	10	5.0	24	50	3.5	17
1N5360A	10	5.0	25	50	4.0	17
1N5361A	10	5.0	27	50	5.0	17
1N5362A	10	5.0	28	50	6.0	17
1N5363A	10	5.0	30	40	8.0	17
1N5364A	10	5.0	33	40	10	17
1N5365A	10	5.0	36	30	11	17
1N5366A	10	5.0	39	30	14	17
1N5367A	10	5.0	43	30	20	17
1N5368A	10	5.0	47	25	25	17
1N5369A	10	5.0	51	25	27	17
1N5370A	10	5.0	56	20	35	17
1N5371A	10	5.0	60	20	40	17
1N5372A	10	5.0	62	20	42	17
1N5373A	10	5.0	68	20	44	17
1N5374A	10	5.0	75	20	45	17
1N5375A	10	5.0	82	15	65	17
1N5376A	10	5.0	87	15	75	17
1N5377A	10	5.0	91	15	75	17
1N5378A	10	5.0	100	12	90	17
1N5379A	10	5.0	110	12	125	17
1N5380A	10	5.0	120	10	170	17
1N5381A	10	5.0	130	10	190	17
1N5382A	10	5.0	140	8.0	230	17
1N5383A	10	5.0	150	8.0	330	17
1N5384A	10	5.0	160	8.0	350	17
1N5385A	10	5.0	170	8.0	380	17
1N5386A	10	5.0	180	5.0	430	17
1N5387A	10	5.0	190	5.0	450	17
1N5388A	10	5.0	200	5.0	480	17

Bipolar Junction Transistors (Silicon):

Type No.	Max. Ratings			Parameters				Case
	PDISS (W)	BVCEO (V)	IC (A)	hfe	hoe (mhos)	Cob (F)	Cib (F)	
NPN								
2N2712	200m	18	100m	200	—	12p	—	TO98
2N3904	350m	40	200m	200	40μ	4p	8p	TO92
2N4401	350m	40	600m	135	30μ	6.5p	17p	TO92
2N5810	500m	25	750m	120	—	15p	—	R203a
2N5858	750m	80	1.0	100	—	7p	—	TO105
2SD1051	1.0	40	1.5	220	—	—	—	B37
2N6715	2.0	40	2.0	100	—	—	—	B32
2N6413	15	60	4.0	150	—	—	—	B16
2N3055	117	60	15	40	—	—	—	TO3
PNP								
2N3906	350m	40	200m	200	60μ	4.5p	15p	TO92
2N4403	350m	40	600m	135	60μ	8.5p	20p	TO92
2N5811	500m	25	750m	120	—	—	—	R203a
2N5857	750m	80	1.0	100	—	15p	—	TO105
2SB819	1.0	40	1.5	220	—	45p	—	B37
2N6727	2.0	40	2.0	100	—	—	—	B32
2N6415	15	60	4.0	150	—	—	—	B16
MJ2955	115	60	15	40	—	—	—	TO3

Darlington Transistors (Silicon):

Type No.	Max. Ratings			Parameters		Case
	PDISS (W)	BVCEO (V)	IC (A)	hfe MIN	MAX	
NPN						
2N5306	400m	25	200m	7.0k	70k	TO98
2SD1198	1.0	25	1.0	2.0k	20k	B37
2N6548	2.0	40	2.0	25k	150k	TO202
SDM3300	5.0	40	5.0	1.0k	—	TO33
NSDU45	10	40	5.0	25k	150k	TO202

Junction Field Effect Transistors:

Type No.	Maximum Ratings			Parameters			
	PDISS (W)	BVDSS (V)	IDMAX (A)	VP (V)	IDSS (A)	IGSS (A)	gos (S)
N CHANNEL							
2N4445	400m	25	400m	10	150m	3n	
2N5103	300m	25	30m	4	8m	100p	100μ
2N5391	300m	70	—	2	1.5m	.2n	4μ
2N5457	310m	25	—	3	5m	1.0n	10μ
2N5458	310m	25	—	4	7m	1.0n	10μ
2N5459	310m	25	—	5	9m	1.0n	10μ
2N5668	310m	25	20m	2	3m	2.0n	20μ
2N5669	310m	25	20m	3	7m	2.0n	50μ
2N5670	310m	25	20m	5	14m	2.0n	75μ
P CHANNEL							
2N2609	300m	30	—	4.0	10m	30n	—
2N2844	300m	30	—	1.7	2.2m	30n	—
2N5460	310m	40	—	3.5	3.0m	1μ	75μ
2N5461	310m	40	—	4.3	9.0m	1μ	75μ
2N5462	310m	40	—	5.4	10m	1μ	75μ

Metal–Oxide Semiconductor Field Effect Transistors:

Type No.	Maximum Ratings			Parameters					
	PDISS (W)	BVDSS (V)	IDMAX (A)	VP (V)	IDSS (A)	IGSS (A)	gfs (S)	@ ID (A)	Cis (F)
N CHANNEL	(DEPLETION)								
2SK238	150m	25	10m	2.5	1.5m	100n	—	—	5.0p
3SK121Y	200m	10	50m	4.0	35m	20μ	—	—	2.0p
MFE130	300m	25	30m	4.0	30m	20n	—	—	4.5p
MFE3001	200m	30	20m	8.0	5.0m	10p	—	—	5.0p
N CHANNEL	(ENHANCEMENT)								
VN0106N2	5.0	60	0.8	2.4	1.0μ	100n	300m	0.5	50p
VN0106N3	600m	60	0.5	2.4	1.0μ	100n	300m	0.5	50p
VN0106N4	15	60	2.0	2.4	1.0μ	100n	400m	0.5	50p
VN0106N5	2.2	60	2.0	2.4	10μ	100n	250m	0.5	50p
VN0106N6	300m	60	2.0	2.4	10μ	100n	250m	0.5	50p
VN0106N9	1.0	60	1.0	2.4	1.0μ	100n	400m	0.5	60p
MTD3055E–1	20	60	8.0	3.0	10μ	100n	4m	4.0	500p
P CHANNEL	(ENHANCEMENT)								
VP0106N2	5.0	60	0.5	3.5	10μ	100n	200m	0.5	50p
VP0106N3	600m	60	0.4	3.5	10μ	100n	200m	0.5	50p
VP0106N4	15	60	1.0	3.5	10μ	100n	200m	0.5	50p
VP0106N6	300m	60	0.4	3.5	10μ	100n	200m	0.5	50p
VP0106N9	1.0	60	1.0	3.5	10μ	100n	200m	0.5	60p
MTD2955–1	20	60	8.0	3.0	10μ	100n	3m	6.0	600p

Thyristors:

Type No.	VS MAX (V)	VS MIN (V)	VRRM (V)	VT (V)	IH (A)	IS (A)	ITRM (A)	ITSM (A)
DIACS:								
1N5758	24	16	4.0	5	—	100μ	2.0	—
1N5759	28	19	4.0	5	—	100μ	2.0	—
1N5760	32	24	4.0	7	—	100μ	2.0	—
1N5761	36	28	4.0	7	—	100μ	2.0	—
1N5762	40	32	4.0	7	—	100μ	2.0	—
FOUR-LAYER DIODES:								
1N3300	21	14	30	1.5	15m	1.0μ	10	50
1N3301	26	17	30	1.5	15m	1.0μ	10	50
1N3302	32	21	40	1.5	20m	1.0μ	10	50
1N3303	39	26	50	1.5	20m	1.0μ	10	50
1N3304	46	31	50	1.5	20m	1.0μ	10	50

Silicon Controlled Rectiers:

Type No.	VDRM (V)	IT (A)	VTM (V)	IH (A)	IDoff (A)	VGT (V)	IGT (A)	dV/dt (V/μS)
2N4151	25	5.0	2.0	60m	2.0m	2.5	60m	100
2N4152	50	5.0	2.0	60m	2.0m	2.5	60m	100
2N4153	100	5.0	2.0	60m	2.0m	2.5	60m	100
2N4154	200	5.0	2.0	60m	2.0m	2.5	60m	100
2N4156	400	5.0	2.0	60m	2.0m	2.5	60m	100
2N4157	500	5.0	2.0	60m	2.0m	2.5	60m	100
2N4158	600	5.0	2.0	60m	2.0m	2.5	60m	100
2N1842	25	10	2.5	—	22m	3.5	150m	200
2N1843	50	10	2.5	—	19m	3.5	150m	200
2N1844	100	10	2.5	—	12m	3.5	150m	200
2N1845	150	10	2.5	—	6.5m	3.5	150m	200
2N1846	200	10	2.5	—	6.0m	3.5	150m	200
2N1847	250	10	2.5	—	5.5m	3.5	150m	200
2N1848	300	10	2.5	—	5.0m	3.5	150m	200
2N1849	400	10	2.5	—	4.0m	3.5	150m	200
2N1850	500	10	2.5	—	3.0m	3.5	150m	200
TRIACS:								
2N6151	200	10	1.8	75m	2.0m	3.0	125m	100
2N6152	400	10	1.8	75m	2.0m	3.0	125m	100
2N6153	600	10	1.8	75m	2.0m	3.0	125m	100
2N6157	200	30	2.0	200m	2.0m	3.4	250m	100
2N6158	400	30	2.0	200m	2.0m	3.4	250m	100
2N6159	600	30	2.0	200m	2.0m	3.4	250m	100

Unijunction Transistors:

Type No.	Maximum Values					Min.
	Pt (mW)	RBB (kΩ)	η	VEB1 (sat) (V)	IP (μa)	Iv (mA)
2N2417A	300	6.8	0.62	5.0	—	8.0
2N2421A	300	6.8	0.75	5.0	—	8.0
2N2418A	300	9.1	0.62	5.0	—	8.0
2N2422A	300	9.1	0.75	5.0	—	8.0
2N4891	360	9.1	0.82	4.0	5.0	6.0
2N4948	360	12	0.82	3.0	2.0	4.0

Phototransistors:

Type No.	Max. Ratings							Type
	PD (W)	VCEO (V)	IC (A)	ID (A)	IL (A)	Ee W/cm^2	λp (m)	
BPX71	100m	50	20m	25n	500μ	4.7m	800n	NPN
BPX72	180m	50	25m	100n	500μ	4.7m	800n	NPN
BPX72D	180m	50	25m	100n	850μ	4.7m	800n	NPN
BPX72E	180m	50	25m	100n	1.4m	4.7m	800n	NPN
BPX95C1	180m	30	25m	100n	3.0m	1.0m	800n	NPN
BPX96C2	100m	30	25m	100n	10m	1.0m	800n	NPN

Optocouplers: Phototransistor Output–NPN

Type No.	Maximum Ratings					hF (min) IC/IF	Diode Char.	
	PD (W)	VCE (V)	IC (A)	IF (A)	VISO (V)		VF (V)	IF (A)
4N31	250m	30	100m	80m	1.5k	500m	1.5	10m
4N30	250m	30	100m	80m	1.5k	1.0	1.5	10m
4N28	250m	30	100m	80m	2.0k	100m	1.5	10m
4N26	250m	30	100m	80m	2.0k	200m	1.5	10m
4N29	250m	30	100m	80m	2.5k	1.0	1.5	10m
4N32	250m	30	125m	80m	6.0k	500m	1.5	10m
4N37	500m	30	100m	60m	1.5k	1.0	1.5	10m
4N36	500m	30	100m	60m	2.5k	1.0	1.5	10m
4N35	500m	30	100m	60m	3.5k	1.0	1.5	10m

Fixed Positive Voltage Regulators:

Type no.	VO NOM (V)						ILmax (A)	VImax (V)	VI–VO (V)	Case
AN78L04 . . . AN78L24	4 5 6 7 8 9 10 12 15 20 24						100m	35	1.7	MS47
AN78N04 . . . AN78N24	4 5 6 7 8 9 10 12 15 20 24						300m	35	2.0	MS46
AN78M05 . . . AN78M24	5 6 7 8 9 10 12 15 20 24						500m	35	2.0	TO–220
IP7805CT	5						1.0	10	2.3	TO–220
IP7812CT	12						1.0	19	2.5	TO–220
IP7815CT	15						1.0	23	2.5	TO–220
LAS1505 . . . LAS1524	5 6 8 10 12 15 18 20 24						1.5	35	2.4	TO–3
LAS1605 . . . LAS1615	5 8 12 15						2.0	30	2.6	TO–3

Fixed Negative Voltage Regulators:

Type no.	VO NOM (V)						ILmax (A)	VImax (V)	VI–VO (V)	Case
AN79L06 . . . AN79L24	6 7 8 9 10 12 15 18 20 24						100m	−35	0.8	MS47

BAY82/1N4244
1N4376
Ultra Fast Switching Diodes

- t_{rr}...750 ps (MAX)
- C...0.8 pF (MAX) 1N4244

PACKAGES

BAY82	DO-7
1N4244	DO-7
1N4376	DO-7

ABSOLUTE MAXIMUM RATINGS (Note 1)

Temperatures

Storage Temperature Range	−65°C to +200°C
Maximum Junction Operating Temperature	+175°C
Lead Temperature	+260°C

Power Dissipation (Note 2)

Maximum Total Power Dissipation at 25°C Ambient	250 mW
Linear Power Derating Factor (from 25°C)	1.67 mW/°C

Maximum Voltage and Currents

WIV	Working Inverse Voltage	10 V (12 V BAY82)
I_O	Average Rectified Current	50 mA
I_F	Continuous Forward Current	150 mA
i_f	Peak Repetitive Forward Current	150 mA
$i_{f(surge)}$	Peak Forward Surge Current	
	Pulse Width = 1 s	250 mA

ELECTRICAL CHARACTERISTICS (25°C Ambient Temperature unless otherwise noted)

SYMBOL	CHARACTERISTIC	BAY82		1N4244		1N4376		UNITS	TEST CONDITIONS
		MIN	MAX	MIN	MAX	MIN	MAX		
V_F	Forward Voltage	0.90	1.35			0.89	1.10	V	I_F = 50 mA
		0.80	1.00		1.00	0.81	0.95	V	I_F = 20 mA
		0.77	0.94			0.76	0.88	V	I_F = 10 mA
		0.64	0.79			0.64	0.74	V	I_F = 1.0 mA
		0.53	0.66			0.52	0.61	V	I_F = 0.1 mA
		0.41	0.53			0.42	0.50	V	I_F = 10 μA
I_R	Reverse Current				100		100	nA	V_R = 10 V
					100		100	μA	V_R = 10 V, T_A = 150°C
			100					nA	V_R = 12 V
			50					μA	V_R = 12 V, T_A = 100°C
					250			nA	V_R = 15 V
BV	Breakdown Voltage	15		20		20		V	I_R = 5.0 μA
C	Capacitance		1.3		0.8		1.0	pF	V_R = 0, f = 1 MHz
t_{rr}	Reverse Recovery Time (Note 3)		750		750		750	ps	I_f = I_r = 10 mA, R_L = 100 Ω

NOTES:
1. These ratings are limiting values above which the serviceability of the diode may be impaired.
2. These are steady state limits. The factory should be consulted on applications involving pulsed or low duty-cycle operation.
3. Recovery to I_r = 1.0 mA.
4. For product family characteristic curves, refer to Chapter 4, D3.

*Datasheets reprinted with permission of National Semiconductor.

FAIRCHILD
A Schlumberger Company

2N4400/FTSO4400
2N4401/FTSO4401

Small Signal General Purpose Amplifiers & Switches

- V_{CEO} ... 40 V (Min)
- h_{FE} ... 100-300 @ 150 mA (2N/FTSO4401); 40 (Min) @ 500 mA (2N/FTSO4401)
- t_{on} ... 35 ns (Max) @ 150 mA
- t_{off} ... 255 ns (Max) @ 150 mA
- Complements ... 2N4402, 2N4403

PACKAGE

2N4400	TO-92
2N4401	TO-92
FTSO4400	TO-236AA/AB
FTSO4401	TO-236AA/AB

ABSOLUTE MAXIMUM RATINGS (Note 1)

Temperatures

Storage Temperature	–55° C to 150° C
Operating Junction Temperature	150° C

Power Dissipation (Notes 2 & 3)

	2N	FTSO
Total Dissipation at 25° C Ambient Temperature	0.625 W	0.350 W*
25° C Case Temperature	1.0 W	

Voltages & Currents

V_{CEO}	Collector to Emitter Voltage (Note 4)	40 V
V_{CBO}	Collector to Base Voltage	60 V
V_{EBO}	Emitter to Base Voltage	6.0 V
I_C	Collector Current	600 mA

ELECTRICAL CHARACTERISTICS (25° C Ambient Temperature unless otherwise noted) (Note 6)

SYMBOL	CHARACTERISTIC	4400 MIN	4400 MAX	4401 MIN	4401 MAX	UNITS	TEST CONDITIONS
$BV_{CEO(sus)}$	Collector to Emitter Sustaining Voltage (Note 5)	40		40		V	$I_C = 1.0$ mA, $I_B = 0$
BV_{CBO}	Collector to Base Breakdown Voltage	60		60		V	$I_C = 100 \mu A$, $I_E = 0$
BV_{EBO}	Emitter to Base Breakdown Voltage	6.0		6.0		V	$I_E = 100 \mu A$, $I_C = 0$
I_{CEX}	Collector Cutoff Current		100		100	nA	$V_{CE} = 35$ V, $V_{EB} = 0.4$ V
I_{BL}	Base Reverse Current		100		100	nA	$V_{CE} = 35$ V, $V_{EB} = 0.4$ V

NOTES:
1. These ratings are limiting values above which the serviceability of any individual semiconductor device may be impaired.
2. These are steady state limits. The factory should be consulted on applications involving pulsed or low duty cycle operations.
3. These ratings give a maximum junction temperature of 150° C and (TO-92) junction-to-case thermal resistance of 125° C/W (derating factor of 8.0 mW/° C); junction-to-ambient thermal resistance of 200° C/W (derating factor of 5.0 mW/° C); (TO-236) junction-to-ambient thermal resistance of 357° C/W (derating factor of 2.8 mW/° C).
4. Rating refers to a high current point where collector to emitter voltage is lowest.
5. Pulse conditions: length = 300 μs; duty cycle ≤ 2%.
6. For product family characteristic curves, refer to Curve Set T145.
* Package mounted on 99.5% alumina 8 mm x 8 mm x 0.6 mm.

ELECTRICAL CHARACTERISTICS (25° C Ambient Temperature unless otherwise noted) (Note 6)

SYMBOL	CHARACTERISTIC	4400		4401		UNITS	TEST CONDITIONS
		MIN	MAX	MIN	MAX		
h_{FE}	DC Current Gain	20		20			$I_C = 100\ \mu A$, $V_{CE} = 1.0$ V
		40		40			$I_C = 1.0$ mA, $V_{CE} = 1.0$ V
				80			$I_C = 10$ mA, $V_{CE} = 1.0$ V
h_{FE}	DC Pulse Current Gain (Note 5)	50	150	100	300		$I_C = 150$ mA, $V_{CE} = 1.0$ V
		20		40			$I_C = 500$ mA, $V_{CE} = 2.0$ V
$V_{CE(sat)}$	Collector to Emitter Saturation Voltage (Note 5)		0.4		0.4	V	$I_C = 150$ mA, $I_B = 15$ mA
			0.75		0.75	V	$I_C = 500$ mA, $I_B = 50$ mA
$V_{BE(sat)}$	Base to Emitter Saturation Voltage (Note 5)	0.75	0.95	0.75	0.95	V	$I_C = 150$ mA, $I_B = 15$ mA
			1.2		1.2	V	$I_C = 500$ mA, $I_B = 50$ mA
C_{cb}	Collector to Base Capacitance		6.5		6.5	pF	$V_{CB} = 5.0$ V, $I_E = 0$, $f = 100$ kHz
C_{eb}	Emitter to Base Capacitance		30		30	pF	$V_{BE} = 0.5$ V, $I_C = 0$, $f = 100$ kHz
h_{fe}	Small Signal Current Gain	20	250	40	500		$I_C = 1.0$ mA, $V_{CE} = 10$ V, $f = 1.0$ kHz
h_{ie}	Input Impedance	0.5	7.5	1.0	15	$k\Omega$	$I_C = 1.0$ mA, $V_{CE} = 10$ V, $f = 1.0$ kHz
h_{oe}	Output Admittance	1.0	30	1.0	30	μmhos	$I_C = 1.0$ mA, $V_{CE} = 10$ V, $f = 1.0$ kHz
h_{re}	Voltage Feedback Ratio	0.1	8.0	0.1	8.0	$\times 10^{-4}$	$I_C = 1.0$ mA, $V_{CE} = 10$ V, $f = 1.0$ kHz
f_T	Current Gain Bandwidth Product	200		250		MHz	$I_C = 20$ mA, $V_{CE} = 10$ V, $f = 100$ MHz
t_d	Turn On Delay Time (test circuit no. 559)		15		15	ns	$I_C = 150$ mA, $V_{CC} = 30$ V, $I_{B1} = 15$ mA
t_r	Rise Time (test circuit no. 559)		20		20	ns	$I_C = 150$ mA, $V_{CC} = 30$ V, $I_{B1} = 15$ mA
t_s	Storage Time (test circuit no. 560)		225		225	ns	$I_C = 150$ mA, $V_{CC} = 30$ V, $I_{B1} = I_{B2} = 15$ mA
t_f	Fall Time (test circuit no. 560)		30		30	ns	$I_C = 150$ mA, $V_{CC} = 30$ V, $I_{B1} = I_{B2} = 15$ mA

μA741
Operational Amplifier

Linear Division Operational Amplifiers

Description

The μA741 is a high performance monolithic operational amplifier constructed using the Fairchild Planar Epitaxial process. It is intended for a wide range of analog applications. High common mode voltage range and absence of latch up tendencies make the μA741 ideal for use as a voltage follower. The high gain and wide range of operating voltage provide superior performance in integrator, summing amplifier, and general feedback applications.

- **No Frequency Compensation Required**
- **Short Circuit Protection**
- **Offset Voltage Null Capability**
- **Large Common Mode And Differential Voltage Ranges**
- **Low Power Consumption**
- **No Latch Up**

Absolute Maximum Ratings

Storage Temperature Range
Metal Can and Ceramic DIP   −65°C to +175°C
Molded DIP and SO-8   −65°C to +150°C
Operating Temperature Range
Extended (μA741AM, μA741M)   −55°C to +125°C
Commercial (μA741EC, μA741C)   0°C to +70°C
Lead Temperature
Metal Can and Ceramic DIP
(soldering, 60 s)   300°C
Molded DIP and SO-8
(soldering, 10 s)   265°C
Internal Power Dissipation[1, 2]
8L-Metal Can   1.00 W
8L-Molded DIP   0.93 W
8L-Ceramic DIP   1.30 W
SO-8   0.81 W
Supply Voltage
μA741A, μA741, μA741E   ± 22 V
μA741C   ± 18 V
Differential Input Voltage   ± 30 V
Input Voltage[3]   ± 15 V
Output Short Circuit Duration[4]   Indefinite

Notes

1. $T_{J\ Max} = 150°C$ for the Molded DIP and SO-8, and 175°C for the Metal Can and Ceramic DIP.
2. Ratings apply to ambient temperature at 25°C. Above this temperature, derate the 8L-Metal Can at 6.7 mW/°C, the 8L-Molded DIP at 7.5 mW/°C, the 8L-Ceramic DIP at 8.7 mW/°C, and the SO-8 at 6.5 mW/°C.
3. For supply voltages less than ± 15 V, the absolute maximum input voltage is equal to the supply voltage.
4. Short circuit may be to ground or either supply. Rating applies to 125°C case temperature or 75°C ambient temperature.

Connection Diagram
8-Lead Metal Package
(Top View)

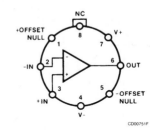

CD00751F

Lead 4 connected to case.

Order Information

Device Code	Package Code	Package Description
μA741HM	5W	Metal
μA741HC	5W	Metal
μA741AHM	5W	Metal
μA741EHC	5W	Metal

Connection Diagram
8-Lead DIP and SO-8 Package
(Top View)

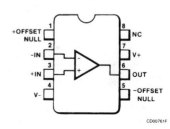

CD00761F

Order Information

Device Code	Package Code	Package Description
μA741RM	6T	Ceramic DIP
μA741RC	6T	Ceramic DIP
μA741SC	KC	Molded Surface Mount
μA741TC	9T	Molded DIP
μA741ARM	6T	Ceramic DIP
μA741ERC	6T	Ceramic DIP
μA741ETC	9T	Molded DIP

Equivalent Circuit

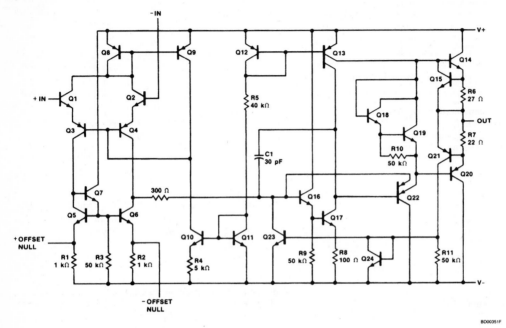

μA741

μA741 and μA741C

Electrical Characteristics $T_A = 25°C$, $V_{CC} = \pm 15$ V, unless otherwise specified.

Symbol	Characteristic		Condition	μA741			μA741C			Unit
				Min	Typ	Max	Min	Typ	Max	
V_{IO}	Input Offset Voltage		$R_S \leqslant 10$ kΩ		1.0	5.0		2.0	6.0	mV
$V_{IO\ adj}$	Input Offset Voltage Adjustment Range				±15			±15		mV
I_{IO}	Input Offset Current				20	200		20	200	nA
I_{IB}	Input Bias Current				80	500		80	500	nA
Z_I	Input Impedance			0.3	2.0		0.3	2.0		MΩ
I_{CC}	Supply Current				1.7	2.8		1.7	2.8	mA
P_c	Power Consumption				50	85		50	85	mW
CMR	Common Mode Rejection			70			70	90		dB
V_{IR}	Input Voltage Range			±12	±13		±12	±13		V
PSRR	Power Supply Rejection Ratio				30	150				μV/V
			$V_{CC} = \pm 5.0$ V to ±18 V					30	150	
I_{OS}	Output Short Circuit Current				25			25		mA
A_{VS}	Large Signal Voltage Gain		$R_L \geqslant 2.0$ kΩ, $V_O = \pm 10$ V	50	200		20	200		V/mV
V_{OP}	Output Voltage Swing		$R_L = 10$ kΩ	±12			±12	±14		V
			$R_L = 2.0$ kΩ	±10			±10	±13		
TR	Transient Response	Rise time	$V_I = 20$ mV, $R_L = 2.0$ kΩ, $C_L = 100$ pF, $A_V = 1.0$		0.3			0.3		μs
		Overshoot			5.0			5.0		%
BW	Bandwidth				1.0			1.0		MHz
SR	Slew Rate		$R_L \geqslant 2.0$ kΩ, $A_V = 1.0$		0.5			0.5		V/μs

μA741

μA741 and μA741C (Cont.)

Electrical Characteristics Over the range of $-55°C \leqslant T_A \leqslant +125°C$ for μA741, $0°C \leqslant T_A \leqslant +70°C$ for μA741C, unless otherwise specified.

Symbol	Characteristic	Condition	μA741			μA741C			Unit
			Min	Typ	Max	Min	Typ	Max	
V_{IO}	Input Offset Voltage							7.5	mV
		$R_S \leqslant 10 \ k\Omega$		1.0	6.0				
$V_{IO \ adj}$	Input Offset Voltage Adjustment Range			±15			±15		mV
I_{IO}	Input Offset Current							300	nA
		$T_A = +125°C$		7.0	200				
		$T_A = -55°C$		85	500				
I_{IB}	Input Bias Current							800	nA
		$T_A = +125°C$		0.03	0.5				μA
		$T_A = -55°C$		0.3	1.5				
I_{CC}	Supply Current	$T_A = +125°C$		1.5	2.5				mA
		$T_A = -55°C$		2.0	3.3				
P_c	Power Consumption	$T_A = +125°C$		45	75				mW
		$T_A = -55°C$		60	100				
CMR	Common Mode Rejection	$R_S \leqslant 10 \ k\Omega$	70	90					dB
V_{IR}	Input Voltage Range		±12	±13					V
PSRR	Power Supply Rejection Ratio			30	150				μV/V
A_{VS}	Large Signal Voltage Gain	$R_L \geqslant 2.0 \ k\Omega, \ V_O = ±10 \ V$	25			15			V/mV
V_{OP}	Output Voltage Swing	$R_L = 10 \ k\Omega$	±12	±14					V
		$R_L = 2.0 \ k\Omega$	±10	±13		±10	±13		

7

μA741

μA741A and μA741E
Electrical Characteristics $T_A = 25°C$, $V_{CC} = \pm 15$ V, unless otherwise specified.

Symbol	Characteristic	Condition			Min	Typ	Max	Unit
V_{IO}	Input Offset Voltage	$R_S \leqslant 50$ Ω				0.8	3.0	mV
I_{IO}	Input Offset Current					3.0	30	nA
I_{IB}	Input Bias Current					30	80	nA
Z_I	Input Impedance	$V_{CC} = \pm 20$ V			1.0	6.0		MΩ
P_C	Power Consumption	$V_{CC} = \pm 20$ V				80	150	mW
PSRR	Power Supply Rejection Ratio	$V_{CC} = +10$ V, -20 V to $V_{CC} = +20$ V, -10 V, $R_S = 50$ Ω				15	50	μV/V
I_{OS}	Output Short Circuit Current				10	25	40	mA
A_{VS}	Large Signal Voltage Gain	$V_{CC} = \pm 20$ V, $R_L \geqslant 2.0$ kΩ, $V_O = \pm 15$ V			50	200		V/mV
TR	Transient Response	Rise time	$A_V = 1.0$, $V_{CC} = \pm 20$ V, $V_I = 50$ mV, $R_L = 2.0$ kΩ, $C_L = 100$ pF			0.25	0.8	μs
		Overshoot				6.0	20	%
BW	Bandwidth				0.437	1.5		MHz
SR	Slew Rate	$V_I = \pm 10$ V, $A_V = 1.0$			0.3	0.7		V/μs

The following specifications apply over the range of $-55°C \leqslant T_A \leqslant +125°C$ for the μA741A, and $0°C \leqslant T_A \leqslant +70°C$ for the μA741E.

Symbol	Characteristic	Condition			Min	Typ	Max	Unit
V_{IO}	Input Offset Voltage						4.0	mV
$\Delta V_{IO}/\Delta T$	Input Offset Voltage Temperature Sensitivity						15	μV/°C
$V_{IO\ adj}$	Input Offset Voltage Adjustment Range	$V_{CC} = \pm 20$ V			10			mV
I_{IO}	Input Offset Current						70	nA
$\Delta I_{IO}/\Delta T$	Input Offset Current Temperature Sensitivity						0.5	nA/°C
I_{IB}	Input Bias Current						210	nA
Z_I	Input Impedance				0.5			MΩ
P_C	Power Consumption	$V_{CC} = \pm 20$ V	μA741A	$-55°C$			165	mW
				$+125°C$			135	
			μA741E				150	
CMR	Common Mode Rejection	$V_{CC} = \pm 20$ V, $V_I = \pm 15$ V, $R_S = 50$ Ω			80	95		dB
I_{OS}	Output Short Circuit Current				10		40	mA
A_{VS}	Large Signal Voltage Gain	$V_{CC} = \pm 20$ V, $R_L \geqslant 2.0$ kΩ, $V_O = \pm 15$ V			32			V/mV
		$V_{CC} = \pm 5.0$ V, $R_L \geqslant 2.0$ kΩ, $V_O = \pm 2.0$ V			10			
V_{OP}	Output Voltage Swing	$V_{CC} = \pm 20$ V	$R_L = 10$ kΩ		± 16			V
			$R_L = 2.0$ kΩ		± 15			

μA741

Typical Performance Curves

Voltage Gain vs Supply Voltage for μA741/A

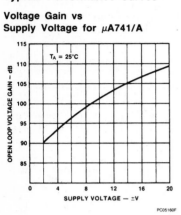

PC05160F

Output Voltage Swing vs Supply Voltage for μA741/A

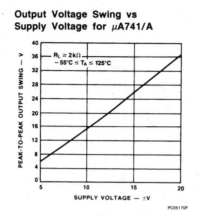

PC05170F

Input Common Mode Voltage vs Supply Voltage for μA741/A

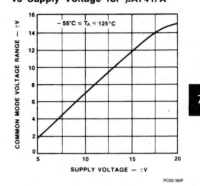

PC05180F

7

Voltage Gain vs Supply Voltage for μA741C/E

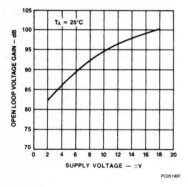

PC05190F

Output Voltage Swing vs Supply Voltage for μA741C/E

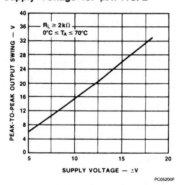

PC05200F

Input Common Mode Voltage Range vs Supply Voltage for μA741C/E

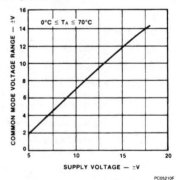

PC05210F

Transient Response for μA741C/E

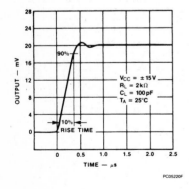

PC05220F

Transient Response Test Circuit for μA741C/E

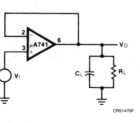

CR01470F

Lead numbers are shown for metal package only

Common Mode Rejection Ratio vs Frequency for μA741C/E

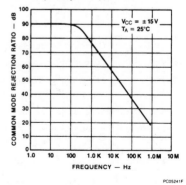

PC05241F

μA741

Typical Performance Curves (Cont.)

Frequency Characteristics vs Supply Voltage for μA741C/E

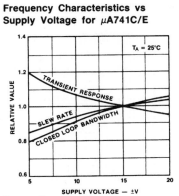

PC05251F

Voltage Offset Null Circuit for μA741C/E

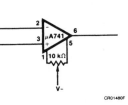

CR01480F

Lead numbers are shown for metal package only

Voltage Follower Large Signal Pulse Response for μA741C/E

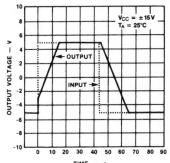

PC05260F

Power Consumption vs Supply Voltage

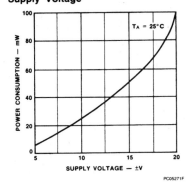

PC05271F

Open Loop Frequency Response

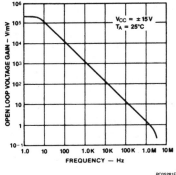

PC05281F

Open Loop Phase Response vs Frequency

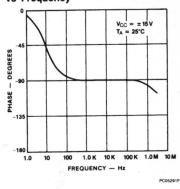

PC05291F

Input Offset Current vs Supply Voltage

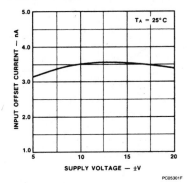

PC05301F

Input Impedance and Input Capacitance vs Frequency

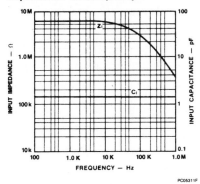

PC05311F

Output Resistance vs Frequency

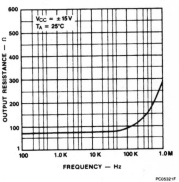

PC05321F

Typical Performance Curves (Cont.)

Output Voltage Swing vs Load Resistance

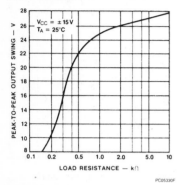

PC05330F

Output Voltage Swing vs Frequency

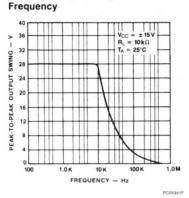

PC05341F

Input Noise Voltage vs Frequency

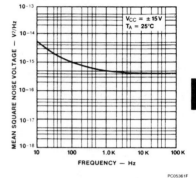

PC05361F

7

Input Noise Current vs Frequency

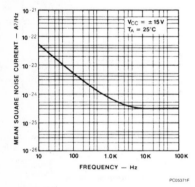

PC05371F

Broadband Noise for Various Bandwidths

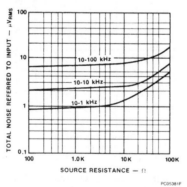

PC05381F

Input Bias Current vs Temperature for μA741/A

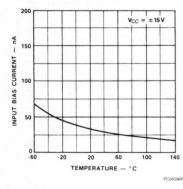

PC05390F

Input Impedance vs Temperature for μA741/A

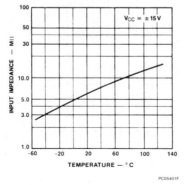

PC05401F

Short Circuit Current vs Temperature for μA741/A

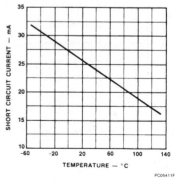

PC05411F

µA741

Typical Performance Curves (Cont.)

Input Offset Current vs Temperature for µA741/A

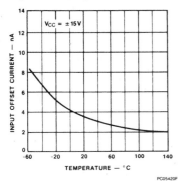

PC05420F

Power Consumption vs Temperature for µA741/A

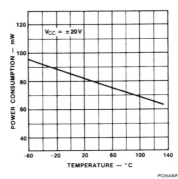

PC05430F

Frequency Characteristics vs Temperature for µA741/A

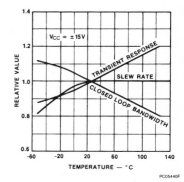

PC05440F

Input Bias Current vs Temperature for µA741C/E

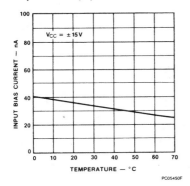

PC05450F

Input Impedance vs Temperature for µA741C/E

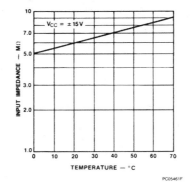

PC05461F

Input Offset Current vs Temperature for µA741C/E

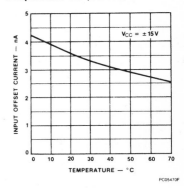

PC05470F

Power Consumption vs Temperature for µA741C/E

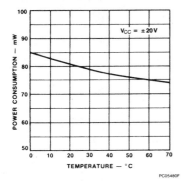

PC05480F

Short Circuit Current vs Temperature for µA741C/E

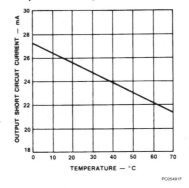

PC05491F

Frequency Characteristics vs Temperature for µA741C/E

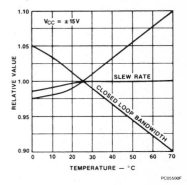

PC05500F

Appendix G
SuperCalc Solutions

In the following examples the format usually used is:

HEADING:

The first few SuperCalc rows are used to identify and specify the problem.
The remaining rows are used as follows:

DATA INPUT:

Column A: Text identifying the input data.
Column B: Numerical value of the input data.
Column C: Units for the input data.
Column D: Text descriptive of the input data.

CALCULATIONS:

Column A: Text identifying the displayed result.
Column B: Formula which calculates the displayed result.
Column C: Units for the displayed result.
Column D: Copy of the SuperCalc formula in Column B.

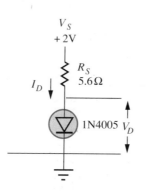

FIGURE G.1   Example 3.5

EXAMPLE 3.5

```
   !   A   !!   B   !!   C   !!   D   !!   E   !!   F   !!   G
1  EXAMPLE 3.5:  SOLVING THE JUNCTION EQUATION
2
3  Find the diode voltage,  VD, and current, ID in Figure G.1
4
5  DATA INPUT:
6    VS     =           2 V      = SOURCE VOLTAGE
7    RS     =         5.6 OHM    = SOURCE RESISTANCE
8    VT     =        .0428 V     = DIODE VOLTAGE PARAMETER
9    IS     =       1.9e-9 A     = DIODE CURRENT PARAMETER
10
11 CALCULATIONS:
12   ID     =     .2321429 A     = (B6-.7)/B7
13   VD1    =     .7969792 V     = B8*LN(B12/B9)
14
15   ID2    =     .2148251 A     = (B6-B13)/B7
16   VD2    =     .7936610 V     = B8*LN(B15/B9)
17
18   ID3    =     .2154177 A     = (B6-B16)/B7
19   VD3    =     .7937789 V     = B8*LN(B18/B9)
20
21   ID4    =     .2153966 A     = (B6-B19)/B7
22   VD4    =     .7937747 V     = B8*LN(B21/B9)
```

A more elegant procedure which uses the automatic iteration feature of
SuperCalc, but which hides the details of the iteration is:

```
   !   A   !!   B   !!   C   !!   D   !!   E   !!   F   !!   G
1  EXAMPLE 3.5
2
3  Find the diode voltage,  VD, and current, ID in Figure G.1
4
5  DATA INPUT:
6    VS     =           2 V      = SOURCE VOLTAGE
7    RS     =         5.6 OHM    = SOURCE RESISTANCE
8    NVT    =        .0428 V     = DIODE VOLTAGE PARAMETER
9    IS     =       1.9e-9 A     = DIODE CURRENT PARAMETER
10
11 CALCULATIONS:
12   ID1    =     .2153973 A     = (B6-B13)/B7
13   VD1    =     .7937748 V     = B8*LN(B12/B9)
```

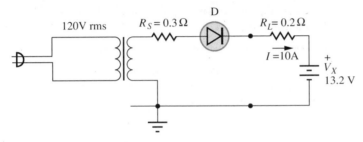

FIGURE G.2   Example 4.2

EXAMPLE 4.2

```
    :    A    ::    B    ::    C    ::    D    ::    E    ::    F    ::    G
1   EXAMPLE 4.2
2
3   Specify transformer and diode for the circuit of Figure G.2
4
5   DATA INPUT:
6     VX     =        13.2 VDC    = NOMINAL BATTERY VOLTAGE
7     RX     =          .2 OHM    = ESTIMATED BATTERY RESISTANCE
8     RS     =          .3 OHM    = TRANSFORMER EQUIVALENT RESISTANCE
9     IAV    =          10 ADC    = DC CHARGING CURRENT
10
11  CALCULATIONS:
12    IP     =    34.92395 AP     = PI*B9/COS(B14*PI/180)
13    VSP    =    31.59250 VP     = B12*(B7+B8)+.026*LN(B12/1E-14)+B6
14    OON    =    25.90069 DEG    = 180/PI*ASIN((B6+.6)/B13)
15    VSRMS  =    22.33927 VRMS   = B13/SQRT(2)
16    PIV    =    44.79250 VP     = B13+B6
17    ISURG  =    102.1140 AP     = (B13-B18)/B8
18    VDSRG  =    .9584192 VP     = .026*LN(B17/1E-14)
19
20  TRANSFORMER SPECS:
21    VSRMS  =    22.33927 VRMS   = B15
22    IAV    =          10 ARMS   = B9
23
24  DIODE SPECS:
25    PIV    =    44.79250 VP     = B16
26    IAV    =          10 ADC    = B9
27    IP     =    34.92395 AP     = B12
28    ISURG  =    102.1137 AP     = B17
```

Note that even though the above solution uses iteration to obtain the value
of both the diode voltage and current, and the conduction angle, the result

does not vary appreciably from the values obtained using the simpler approximate method of Example 4.2.

The following are rather generalized design recipes based on the power supply design recipes of Table 5.1. They all use the same design specs which were taken from Example 5.3.

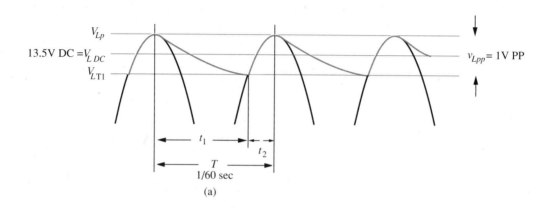

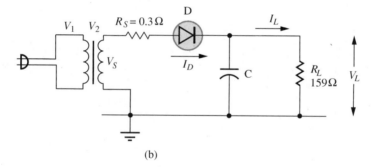

FIGURE G.3 Half-wave power supply design.
(a) Output waveform.
(b) Circuit diagram.

HALF-WAVE POWER SUPPLY DESIGN

```
   ┊   A   ┊┊   B   ┊┊   C   ┊┊   D   ┊┊   E   ┊┊   F   ┊┊   G
1  HALF WAVE POWER SUPPLY DESIGN
2
3  Design the power supply in Figure G.3
4
5  DATA INPUT:
6    VLDC  =        13.5 VDC
7    VLPP  =         1 VPP
```

```
 8   RL    =           159 OHM
 9   RS    =            .3 OHM
10   FS    =            60 HZ
11
12 CALCULATIONS:
13   VLT1  =            13 V      = B6-B7/2
14   VLP   =            14 V      = B6+B7/2
15   T2    =      .0010086 SEC    = ACOS(B13/B14)/(2*PI*B10)
16   T     =      .0166667 SEC    = 1/B10
17   T1    =      .0156580 SEC    = B16-B15
18   C     =      .0013288 FRD    = B17/(B8*LN(B14/B13))
19   ILDC  =      .0849057 ADC    = B6/B8
20   IDAV  =      .0849057 AAV    = B19
21   IDP   =      1.402962 AP     = B19*B16/B15
22   VDP   =      .8469442 VP     = .026*LN(B21/1E-14)
23   VSP   =      15.26783 VP     = B21*B9+B22+B14
24   V2RMS =      10.79599 VRMS   = B23/SQRT(2)
25   ISURG =      48.06963 AP     = (B23-B22)/B9
26   PIV   =      28.76783 VP     = B23+B6
27
28 COMPONENT SPECIFICATIONS:
29
30 TRANSFORMER:
31   V2RMS =      10.79599 VRMS
32   I2AV  =      .0849057 AAV
33
34 DIODE:
35   PIV   =      28.76783 VP
36   IDAV  =      .0849057 AAV
37   IDP   =      1.402962 AP
38   ISURG =      48.06963 AP
39
40 CAPACITOR:
41   C     =      .0013288 FRD
42   VLP   =            14 VP
```

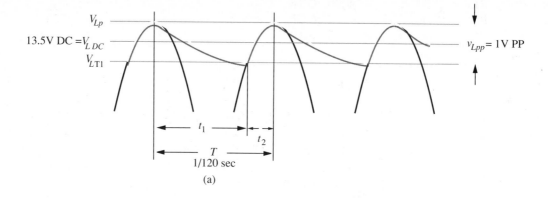

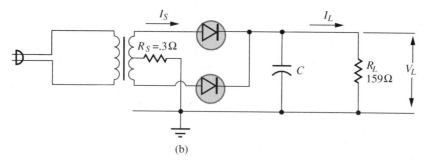

FIGURE G.4 Center tapped power supply design.
(**a**) Output waveform.
(**b**) Circuit diagram.

```
CT POWER SUPPLY DESIGN
   ¦   A   ¦¦   B   ¦¦   C   ¦¦   D   ¦¦   E   ¦¦   F   ¦¦   G
1  CENTER TAPPED POWER SUPPLY DESIGN
2
3  Design the power supply in Figure G.4 given:
4
5  DATA INPUT:
6    VLDC   =        13.5 VDC
7    vLPP   =           1 VPP
8    RL     =         159 OHM
9    RS     =          .3 OHM
10   FS     =          60 HZ
11
12 CALCULATIONS:
13   VLT1   =          13 V      = B6-B7/2
14   VLP    =          14 V      = B6+B7/2
15   T2     =     .0010086 SEC   = ACOS(B13/B14)/(2*PI*B10)
```

```
16    T      =    .0083333 SEC     = 1/(2*B10)
17    T1     =    .0073247 SEC     = B16-B15
18    C      =    .0006216 FRD     = B17/(B8*LN(B14/B13))
19    ILDC   =    .0849057 ADC     = B6/B8
20    IDAV   =    .0424528 AAV     = B19/2
21    IDP    =    .7014810 AP      = B19*B16/B15
22    VDP    =    .8289224 VP      = .026*LN(B21/1E-14)
23    VSP    =   15.03937  VP      = B21*B9+B22+B14
24    V2RMS  =   21.26888  VRMS    = 2*B23/SQRT(2)
25    ISURG  =   47.36815  AP      = (B23-B22)/B9
26    PIV    =   29.03937  VP      = B23+B14
27
28 COMPONENT SPECIFICATIONS:
29
30 TRANSFORMER:
31   V2RMS =   21.26888  VRMS
32   I2AV  =    .0849057 AAV
33
34 DIODES (2 REQUIRED):
35   PIV   =   29.03937  VP
36   IDAV  =    .0424528 AAV
37   IDP   =    .7014810 AP
38   ISURG =   47.36815  AP
39
40 CAPACITOR:
41   C     =    .0006216 FRD
42   VLP   =          14 VP
```

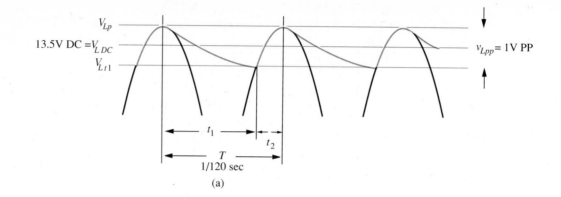

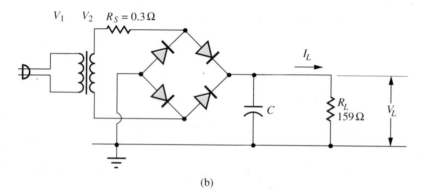

FIGURE G.5   Bridge power supply design.
(a) Output waveform
(b) Circuit diagram.

BRIDGE POWER SUPPLY

```
      :   A   ::   B   ::   C   ::   D   ::   E   ::   F   ::   G
1   BRIDGE POWER SUPPLY DESIGN
2
3   Design the power supply in Figure G.5 given:
4
5   DATA INPUT:
6     VLDC   =        13.5 VDC
7     vLPP   =           1 VPP
8     RL     =         159 OHM
9     RS     =          .3 OHM
10    FS     =          60 HZ
11
12  CALCULATIONS:
13    VLT1   =          13 V       = B6-B7/2
```

```
14   VLP   =         14 V     = B6+B7/2
15   T2    =    .0010086 SEC   = ACOS(B13/B14)/(2*PI*B10)
16   T     =    .0083333 SEC   = 1/(2*B10)
17   T1    =    .0073247 SEC   = B16-B15
18   C     =    .0006216 FRD   = B17/(B8*LN(B14/B13))
19   ILDC  =    .0849057 ADC   = B6/B8
20   IDAV  =    .0424528 AAV   = B19/2
21   IDP   =    .7014810 AP    = B19*B16/B15
22   VDP   =    .8289224 VP    = .026*LN(B21/1E-14)
23   VSP   =   15.86829 VP    = B21*B9+2*B22+B14
24   V2RMS =   11.22057 VRMS  = B23/SQRT(2)
25   ISURG =   47.36815 AP    = (B23-2*B22)/B9
26   PIV   =   14.82892 VP    = B14+B22
27
28  COMPONENT SPECIFICATIONS:
29
30  TRANSFORMER:
31   V2RMS =   11.22057 VRMS
32   I2AV  =    .0849057 AAV
33
34  DIODES (4 REQUIRED):
35   PIV   =   14.82892 VP
36   IDAV  =    .0424528 AAV
37   IDP   =    .7014810 AP
38   ISURG =   47.36815 AP
39
40  CAPACITOR:
41   C     =    .0006216 FRD
42   VLP   =         14 VP
```

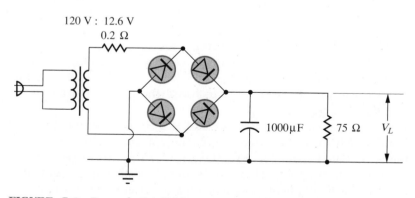

FIGURE G.6   Example 5.4: Bridge power supply analysis.

EXAMPLE 5.4

```
 :   A   ::    B   ::    C    ::    D    ::    E    ::    F    ::    G
1  EXAMPLE 5.4
2
3  Find VLDC, %r, %vr for the circuit of Figure G.6
4
5  DATA INPUT:
6   VS    =        12.6 VRMS  = SOURCE VOLTAGE
7   RS    =          .2 OHM   = SOURCE RESISTANCE
8   F     =          60 HZ    = LINE FREQUENCY
9   C     =        .001 FRD   = FILTER CAPACITANCE
10  RL    =          75 OHM   = LOAD RESISTANCE
11
12 CALCULATIONS:
13  T     =     .0083333 SEC   = 1/(2*B8)
14  VSP   =     17.81909 VP    = SQRT(2)*B6
15  VLP   =     16.07907 VP    = (B14-2*B20)*B10/(B7+B10)
16  VLT1  =     14.61261 V     = B15*EXP(-B17/((B7+B10)*B9))
17  T1    =     .0071916 SEC   = B13-180/PI*(ACOS(B16/B15))/(360*B8)
18  VLDC  =     15.34584 VDC   = (B15+B16)/2
19  ILP   =     1.493490 AP    = B18/B10*B13/(B13-B17)
20  VDP   =     .8485700 VP    = .026*LN(B19/1E-14)
21  vLPP  =     1.466461 VPP   = B15-B16
22  VLRMS =     .4233309 VRMS  = B21/(2*SQRT(3))
23  VLOPN =     16.61909 V     = B14-1.2
24  %R    =     2.758603 %     = B22/B18*100
25  %VR   =     8.297023 %     = (B23-B18)/B18*100
```

Even though the above procedure uses iteration to obtain both the capacitor discharge time t_1, (Cell B17) and also the peak diode voltage V_{DP}, (Cell B20), the results are nearly the same as those obtained in the text assuming a peak diode voltage of 0.8 V.

TRANSISTOR AMPLIFIER DESIGN

The following four examples are cumulative in the sense that each one is an extension of those that have gone before. If you save each of them on a disk, you will find that it is easy to revise the previous example to include the new material.

Whenever the value of a resistor that is a circuit component is calculated, the user is expected to enter the value of the resistor to be actually used in SuperCalc Column B of the next row. This entry is the value used in subsequent calculations. This avoids tolerance stackup, permits the exercise of designer judgment, and keeps the programming simple.

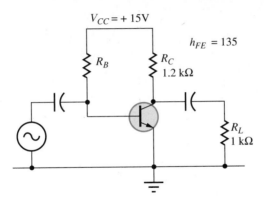

FIGURE G.7   Example 10.1: Fixed biased circuit.

EXAMPLE 10.1: FIXED BIAS

```
  ::  A  ::    B    ::    C    ::    D    ::    E    ::    F    ::    G
1   EXAMPLE 10.1: FIXED BIAS
2
3   Design the amplier of Figure G.7
4
5   DATA INPUT:
6    VCC   =         15 V        = DC  SUPPLY VOLTAGE
7    RC    =       1200 OHM      = COLLECTOR RESISTOR
8    RL    =       1000 OHM      = EXTERNAL LOAD RESISTANCE
9    HFE   =        135          = TRANSISTOR CURRENT GAIN (BETA)
10
11  CALCULATIONS:
12   RDC   =       1200 OHM      = B7
13   RAC   = 545.4545 OHM        = 1/(1/B7+1/B8)
14   IC    = .0085938 A          = B6/(B12+B13)
15   VBE   = .7144663 V          = .026*LN(B14/1E-14)
16   IB    = .0000637 A          = B14/B9
17   RB    = 224412.7 OHM        = (B6-B15)/B16
18       USE    220000 OHM       = NEAREST STANDARD RESISTOR
```

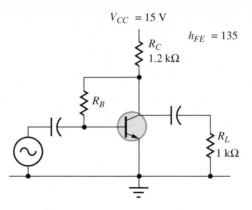

$V_{CC} = 15$ V

R_C
1.2 kΩ

$h_{FE} = 135$

R_B

R_L
1 kΩ

FIGURE G.8   Example 10.4: Self biased circuit.

EXAMPLE 10.4: SELF BIAS

```
    !  A   !!   B    !!   C    !!    D    !!    E    !!    F    !!   G
1   EXAMPLE 10.4: SELF BIAS
2
3   Design the amplier of Figure G.8
4
5   DATA INPUT:
6   VCC   =        15 V           = DC  SUPPLY VOLTAGE
7   RC    =      1200 OHM         = COLLECTOR RESISTOR
8   RL    =      1000 OHM         = EXTERNAL LOAD RESISTANCE
9   HFE   =       135             = TRANSISTOR CURRENT GAIN (BETA)
10
11  CALCULATIONS:
12  RDC   =      1200 OHM         = B7
13  RAC   =  545.4545 OHM         = 1/(1/B7+1/B8)
14  IC    =  .0085938 A           = B6/(B12+B13)
15  VBE   =  .7144663 V           = .026*LN(B14/1E-14)
16  IB    =  .0000637 A           = B14/B9
17  RB    =  62412.75 OHM         = (B6-B14*B7-B15)/B16
18      USE     68000 OHM         = NEAREST STANDARD RESISTOR
```

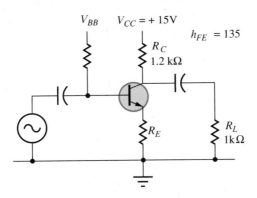

FIGURE G.9   Example 10.6: Emitter biased circuit.

EXAMPLE 10.6: EMITTER BIAS

```
    ! A  !!   B   !!   C   !!   D   !!   E   !!   F   !!   G
1   EXAMPLE 10.6: EMITTER BIAS
2
3   Design the amplier of Figure G.9
4
5   DATA INPUT:
6   VCC   =         15 V          = DC SUPPLY VOLTAGE
7   RC    =       1200 OHM        = COLLECTOR RESISTOR
8   RL    =       1000 OHM        = EXTERNAL LOAD RESISTANCE
9   HFE   =        135            = TRANSISTOR CURRENT GAIN (BETA)
10
11  CALCULATIONS:
12  RE    =        400 OHM        = B7/3
13     USE         390 OHM        = NEAREST STANDARD RESISTOR
14  RDC   =       1590 OHM        = B7+B13
15  RAC   =   935.4545 OHM        = 1/(1/B7+1/B8)+B13
16  IC    =  .0059395 A           = B6/(B14+B15)
17  VE    =  2.316415 V           = B16*B13
18  VBE   =  .7048617 V           = .026*LN(B16/1E-14)
19  VB    =  3.021276 V           = B18+B17
20  IB    =  .0000440 A           = B16/B9
21  RB    =       5265 OHM        = B9*B13/10
22     USE        5600 OHM        = NEAREST STANDARD RESISTOR
23  VBB   =  3.267657 V           = B19+B20*B22
```

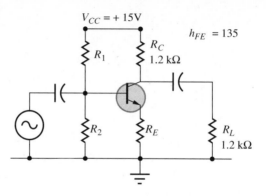

FIGURE G.10   Example 10.8: Divider biased circuit.

EXAMPLE 10.8 DIVIDER BIAS

```
   !   A   ::   B   ::    C    ::    D    ::    E    ::   F   ::   G
1   EXAMPLE 10.8: DIVIDER BIAS
2
3   Design the amplifier of Figure G.10
4
5   DATA INPUT:
6     VCC    =          15 VOLT   = DC SUPPLY VOLTAGE
7     RC     =        1200 OHM    = COLLECTOR RESISTOR
8     RL     =        1000 OHM    = EXTERNAL LOAD RESISTANCE
9     HFE    =         135        = TRANSISTOR CURRENT GAIN (BETA)
10
11  CALCULATIONS:
12    RE     =         400 OHM    = B7/3
13       USE         390 OHM    = NEAREST STANDARD RESISTOR
14    RDC    =        1590 OHM    = B7+B13
15    RAC    =    935.4545 OHM    = 1/(1/B7+1/B8)+B13
16    IC     =     .0059395 AMP   = B6/(B14+B15)
17    VE     =    2.316415 VOLT   = B16*B13
18    VBE    =    .7048617 VOLT   = .026*LN(B16/1E-14)
19    VB     =    3.021276 VOLT   = B18+B17
20    IB     =    .0000440 AMP    = B16/B9
21    RB     =        5265 OHM    = B9*B13/10
22    VBB    =    3.252918 VOLT   = B19+B20*B21
23    R1     =    24278.20 OHM    = B21*B6/B22
24       USE       27000 OHM    = NEAREST STANDARD RESISTOR
25    R2     =    7476.647 OHM    = B24*B22/(B6-B22)
26       USE        6800 OHM    = NEAREST STANDARD RESISTOR
```

TRANSISTOR AMPLIFIER DESIGN—MAXIMUM GAIN

The following bipolar junction transistor circuit design procedure was taken from the design procedure presented in Table 12.1. The circuit used is that of Example 12.13. The circuit is shown in Figure G.11. Since all of the emitter resistance is bypassed, rather than delete the voltage gain term throughout the recipe, a large number is entered for AV in Cell B9. This produces a negative value for R_{E1} in Cell B25. R_{E1} is then set to zero ohms in Cell B26.

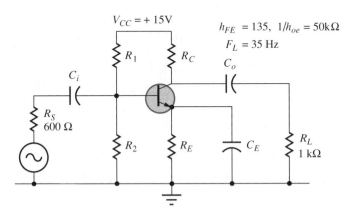

$$V_{CC} = +15\text{V}$$

$$h_{FE} = 135, \quad 1/h_{oe} = 50\text{k}\Omega$$

$$F_L = 35 \text{ Hz}$$

FIGURE G.11   Example 12.13

EXAMPLE 12.13

```
     !  A  !!   B   !!   C   !!   D   !!   E   !!   F   !!   G
 1  DIVIDER BIASED BJT SMALL SIGNAL AMPLIFIER DESIGN
 2
 3  Design the amplifier of Figure G.11
 4
 5  DATA INPUT:
 6   VCC   =          15 VOLT  = DC  SUPPLY VOLTAGE
 7   RS    =         600 OHM   = SIGNAL SOURCE RESISTANCE
 8   RL    =        1000 OHM   = EXTERNAL LOAD RESISTANCE
 9   AV    =     1000000       = MAXIMUM GAIN
10   FL    =          35 HZ    = REQUIRED LOW FREQUENCY CUTOFF
11   HFE   =         135       = TRANSISTOR CURRENT GAIN  (BETA)
12   1/HOE =       50000 OHM   = TRANSISTOR OUTPUT RESISTANCE
13
14  CALCULATIONS:
15   RC    =    1020.408 OHM   = 1/(1/B6-1/B10)
16      USE       1000 OHM   = NEAREST STANDARD RESISTOR
17   RE    =    333.3333 OHM   = B14/3
18   IC    =    .0081818 AMP   = B6/(B16+B17+1/(1/B16+1/B8))
```

```
19  VBE   =   .7131891 VOLT  =  .026*LN(B19/1E-14)
20  VE    =  2.727273 VOLT  =  B18*B17
21  VB    =  3.440462 VOLT  =  B19+B20
22  VC    =  6.818182 VOLT  =  B6-B18*B16
23  IB    =   .0000606 AMP   =  B18/B11
24  HIE   =       429 OHM   =  .026/B23
25  RE1   =  -3.17728 OHM   =  1/(1/B16+1/B8)/B9-B24/B11
26     USE         0 OHM   =  NEAREST STANDARD RESISTOR
27  RE2   =  333.3333 OHM   =  B17-B26
28     USE       330 OHM   =  NEAREST STANDARD RESISTOR
29  RB    =      4455 OHM   =  B11*(B26+B28)/10
30  VBB   =  3.710462 VOLT  =  B21+B23*B29
31  R1    =  18009.89 OHM   =  B29*B6/B30
32     USE     18000 OHM   =  NEAREST STANDARD RESISTOR
33  R2       5915.947 OHM   =  B32*B30/(B6-B30)
34     USE      5600 OHM   =  NEAREST STANDARD RESISTOR
35  RIN   =  389.8439 OHM   =  1/(1/B32+1/B34+1/(B24+B11*B26))
36  CI    =   .0000046 FRD   =  1/(2*PI*B10*(B7+B35))
37  ROUT  =  980.3922 OHM   =  1/(1/B16+1/(B12+B26))
38  CO    =   .0000023 FRD   =  1/(2*PI*B10*(B37+B8))
39  CE    =   .0006565 FRD   =
40  = 1/(2*PI*B10*(1/(1/B28+1/(B26+(B24+1/(1/B32+1/B34+1/B7))/B11))))
```

TRANSISTOR AMP. DESIGN—SPECIFIED GAIN

The following design example is the same as the preceding one, with the exception that the circuit gain is required to be 10. The circuit is that of Figure G.12.

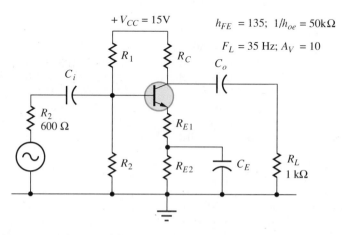

FIGURE G.12 BJT amplifier design—limited gain.

```
        :   A   ::    B    ::    C    ::    D    ::   E    ::   F   ::    G
1   DIVIDER BIASED BJT SMALL SIGNAL AMPLIFIER DESIGN
2
3   Design the amplifier of Figure G.12
4
5   DATA INPUT:
6     VCC    =           15 VOLT   = DC SUPPLY VOLTAGE
7     RS     =          600 OHM    = SIGNAL SOURCE RESISTANCE
8     RL     =         1000 OHM    = EXTERNAL LOAD RESISTANCE
9     AV     =           10        = REQUIRED VOLTAGE GAIN
10    FL     =           35 HZ     = REQUIRED LOW FREQUENCY CUTOFF
11    HFE    =          135        = TRANSISTOR CURRENT GAIN (BETA)
12    1/HOE  =        50000 OHM    = TRANSISTOR OUTPUT RESISTANCE
13
14  CALCULATIONS:
15    RC     =     1020.408 OHM    = 1/(1/B8-1/B12)
16       USE         1000 OHM    = NEAREST STANDARD RESISTOR
17    RE     =     333.3333 OHM    = B15/3
18    IC     =     .0081818 AMP    = B6/(B16 +B17+1/(1/B16+1/B8))
19    VBE    =     .7131891 VOLT   = .026*LN(B18/1E-14)
20    VE     =     2.727273 VOLT   = B18*B17
21    VB     =     3.440462 VOLT   = B19+B20
22    VC     =     6.818182 VOLT   = B6-B18*B16
23    IB     =     .0000606 AMP    = B18/B11
24    HIE    =          429 OHM    = .026/B23
25    RE1    =     46.82222 OHM    = 1/(1/B16+1/B8)/B9-B24/B11
26       USE           47 OHM    = NEAREST STANDARD RESISTOR
27    RE2    =     286.3333 OHM    = B17-B26
28       USE          270 OHM    = NEAREST STANDARD RESISTOR
29    RB     =       4279.5 OHM    = B11*(B26+B28)/10
30    VBB    =     3.699825 VOLT   = B21+B23*B29
31    R1     =     17350.14 OHM    = B29*B6/B30
32       USE        18000 OHM    = NEAREST STANDARD RESISTOR
33    R2     =     5893.436 OHM    = B32*B30/(B6-B30)
34       USE         5600 OHM    = NEAREST STANDARD RESISTOR
35    RIN    =     2619.514 OHM    = 1/(1/B32+1/B34+1/(B24+B11*B26))
36    CI     =     .0000014 FRD    = 1/(2*PI*B10*(B7+B35))
37    ROUT   =     980.4102 OHM    = 1/(1/B16+1/(B12+B26))
38    CO     =     .0000023 FRD    = 1/(2*PI*B10*(B37+B8))
39    CE     =     .0001009 FRD    =
40     = 1/(2*PI*B10*(1/(1/B28+1/(B26+(B24+1/(1/B32+1/B34+1/B7))/B11))))
```

TRANSISTOR VOLTAGE AMPLIFIER ANALYSIS

The following example is adapted from the transistor amplifier circuit
analysis procedure of Table 12.2. The circuit designed in the preceding
example is used.

EXAMPLE 12.13

```
:  A  ::  B  ::  C  ::  D  ::  E  ::  F  ::  G
1  DIVIDER BIASED BJT SMALL SIGNAL AMPLIFIER ANALYSIS
2
3  Analyze the circuit designed in Example 12.13
4
5  DATA INPUT:
6   VCC   =          15 VOLT   = DC SUPPLY VOLTAGE
7   R1    =       18000 OHM    = BASE BIAS RESISTOR
8   R2    =        5600 OHM    = BASE BIAS RESISTOR
9   RC    =        1000 OHM    = COLLECTOR RESISTOR
10  RE1   =          47 OHM    = EMITTER RESISTOR
11  RE2   =         270 OHM    = EMITTER RESISTOR
12  RS    =         600 OHM    = SIGNAL SOURCE RESISTANCE
13  RL    =        1000 OHM    = EXTERNAL LOAD RESISTOR
14  HFE   =         135        = TRANSISTOR CURENT GAIN (BETA)
15  1/HOE =       50000 OHM    = TRANSISTOR COLLECTOR RESISTANCE
16
17 CALCULATIONS:
18  VBB   =    3.559322 VOLT   = B6*B8=(B8+B7)
19  RB    =    4271.186 OHM    = 1/(1/B7+1/B8)
20  S     =    .0907485        = B19/(B19+B14*(B10+B11))
21  IC    =    .0081637 AMP    = (B18-B22)/(B10+B11+B19/B14)
22  VBE   =    .7131316 VOLT   = .026*LN(B21/1E-14)
23  VE    =    2.587903 VOLT   = B21*(B10+B11)
24  VB    =    3.301034 VOLT   = B22+B23
25  VC    =    6.836268 VOLT   = B6-B21*B9
26  IB    =    .0000605 AMP    = B21/B14
27  HIE   =    429.9504 OHM    = .026/B26
28  RIN   =    2619.656 OHM    = 1/(1/B7+1/B8+1/(B27+B14*B10))
29  ROUT  =    980.4102 OHM    = 1/(1/B9+1/(B15+B10))
30  AV    =    9.864527        = B14*1/(1/B15+1/B9+1/B13)/(B27+B14*B10)
31  AI    =    25.84167        = B30*B28/B13
```

COMPLEMENTARY AMPLIFIER DESIGN

The following complementary symmetry design procedure is an implementation of the recipe presented in Table 15.1. Example 15.6 is used. The circuit is shown in Figure G.13.

Note that user intervention is required to standardize resistor sizes and also to set N, the number of biasing diodes in Cell B26. If N is too large, a negative value will be calculated for $R2$.

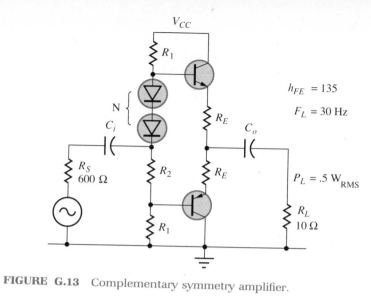

V_{CC}

R_1

N

C_i

R_E

C_o

R_S
600 Ω

R_2

R_E

$P_L = .5\ W_{RMS}$

R_1

R_L
10 Ω

$h_{FE} = 135$

$F_L = 30\ Hz$

FIGURE G.13 Complementary symmetry amplifier.

EXAMPLE 15.6

	A		B		C		D		E		F		G

```
 1  COMPLEMENTARY SYMMETRY AMPLIFIER DESIGN
 2
 3  DATA INPUT:
 4    PL    =              .5 WRMS = RMS POWER TO THE LOAD
 5    RL    =              10 OHM  = RESISTANCE OF THE LOAD
 6    RS    =             600 OHM  = RESISTANCE OF THE SOURCE
 7    FL    =              30 HZ   = LOW FREQUENCY CUTOFF
 8    hFE   =             135      = TRANSISTOR GAIN
 9
10  CALCULATIONS:
11    RE    =               1 OHM  = B5/10
12        USE               1 OHM  = NEAREST STANDARD RESISTOR
13    R1    =             297 OHM  = B8*(B12+B5)/5
14        USE             270 OHM  = NEAREST STANDARD RESISTOR
15    iLP   =         .3162278 AMP = SQRT(2*B4/B5)
16    VBEP  =         .8082074 VOLT = .026*LN(B15/1E-14)
17    VCC   =         9.838337 VOLT = 2*(B15*(B14/B8+B12+B5)+B16)
18    IC    =         .0031623 AMP = B15/100
19    VE1   =         4.922331 VOLT = B17/2+B18*B12
20    VE2   =         4.916006 VOLT = B17/2-B18*B12
21    VBE   =         .6884729 VOLT = .026*LN(B18/1E-14)
22    VB1   =         5.610804 VOLT = B19+B21
23    VB2   =         4.227533 OHM  = B20-B21
24    ID    =         .0156575 AMP = B23/B14
25    VD    =         .7300641 VOLT = .026*LN(B24/1E-14)
26    N     =               1      = USE THIS NUMBER OF BIASING
                                     DIODES
27    R2    =        41.71835 OHM  = (B22-B23-B26*B25)/B24
```

```
28      USE           39 OHM    = NEAREST STANDARD RESISTOR
29   PDISS  =   .1730574 WATT   = B20*(B18+B15/PI^2)
30   HIE    =  34.87040 OHM     = .026*PI*B8/B15
31   RIN    = 123.9871 OHM      = 1/(2/B14+1/(B30+B8*(B12+B5)))
32   ROUT   =  2.074626 OHM     = B12+(B30+1/(2/B14+1/B6))/B8
33   CI     =  .0000073 FRD     = 1/(2*PI*B7*(B6+B31))
34   CO     =  .0004394 FRD     = 1/(2*PI*B7*(B32+B5))
35   PSPY   =  1.175467 WATT    = B17*(B24+B18+B15/PI)
36   EFF    =  42.53628 %       = B4/B35*100
```

JFET AMPLIFIER DESIGN

The following JFET examples are cumulative in the sense that the solution of Examples 17.2 and 17.3 are subsets of the solution of Example 17.4. If you save these procedures on a disk your work will be facilitated. As was the case in previous design procedures, user intervention is required for the selection of standard resistor values.

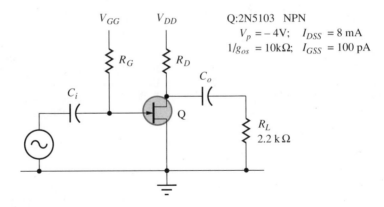

V_{GG} V_{DD} Q:2N5103 NPN

$V_p = -4V; \quad I_{DSS} = 8 \text{ mA}$

$1/g_{os} = 10k\Omega; \quad I_{GSS} = 100 \text{ pA}$

R_G R_D C_o

C_i Q R_L 2.2 kΩ

FIGURE G.14   Example 17.2

EXAMPLE 17.2

```
 !   A   !!   B   !!   C   !!   D   !!   E   !!   F   !!   G
1  JFET TWO SUPPLY AMPLIFIER DESIGN
2  DESIGN THE CIRCUIT SHOWN IN FIGURE G.14
3  DATA INPUT:
4    RL    =      2200 OHM    = RESISTANCE OF THE LOAD
5    VP    =        -4 VOLT   = PINCHOFF VOLTAGE
6    IDSS  =      .008 AMP    = SATURATION CURRENT
7    IGSS  =    -1e-10 AMP    = GATE LEAKAGE CURRENT
8    1/GOS =     10000 OHM    = JFET OUTPUT RESISTANCE
9
```

```
10  CALCULATIONS:
11  RD      =     2820.513  OHM   = 1/(1/B4-1/B8)
12     USE          2700  OHM   = NEAREST STANDARD RESISTOR
13  RAC     =     1212.245  OHM   = 1/(1/B12+1/B4)
14  ID      =      .0028284  AMP   = B6/SQRT(8)
15  VGS     =     -1.62159  VOLT  = B5*(1-SQRT(B14/B6))
16  VDS     =     10.26921  VOLT  = (B6-B14)*B13-B5
17  VDD     =     17.90597  VOLT  = B16+B14*B12
18  RG      =      4.0000e9  OHM   = B5/(10*B7)
19     USE      10000000  OHM   = NEAREST STANDARD RESISTOR
20  VGG     =     -1.62259  VOLT  = B15+B7*B19
```

EXAMPLE 17.3

```
    !  A   !!  B   !!  C   !!  D   !!  E   !!  F   !!  G
1   JFET TWO SUPPLY AMPLIFIER DESIGN
2
3   DATA INPUT:
4   RL      =          100  OHM   = RESISTANCE OF THE LOAD
5   VP      =          -10  VOLTS = PINCHOFF VOLTAGE
6   IDSS    =          .15  AMPS  = SATURATION CURRENT
7   IGSS    =         -3e-9  AMPS  = GATE LEAKAGE CURRENT
8   GOS     =        1e-50  SIEM  = OUTPUT CONDUCTANCE
9
10  CALCULATIONS:
11  RD      =          100  OHM   = 1/(1/B4-B8)
12     USE           100  OHM   = NEAREST STANDARD RESISTOR
13  RAC     =           50  OHM   = 1/(1/B12+1/B4)
14  ID      =      .0530330  AMP   = B6/SQRT(8)
15  VGS     =     -4.05396  VOLT  = B5*(1-SQRT(B14//B6))
16  VDS     =     14.84835  VOLT  = (B6-B14)*B13-B5
17  VDD     =     20.15165  VOLT  = B16+B14*B12
18  RG      =      3.3333e8  OHM   = B5=(10*B7)
19     USE      10000000  OHM   = NEAREST STANDARD RESISTOR
20  VGG     =     -4.08396  VOLT  = B15+B7*B19
```

SOURCE BIASED JFET AMPLIFIER DESIGN

The following JFET circuit design procedure was derived from the design recipe of Table 17.1. As was the case in the design procedures previously presented, user intervention is required for the standardization of resistors. The circuit is shown in Figure G.15.

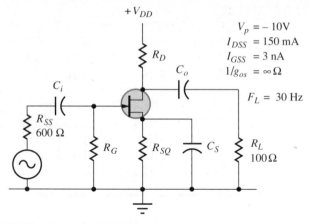

$V_p = -10\text{V}$
$I_{DSS} = 150\text{ mA}$
$I_{GSS} = 3\text{ nA}$
$1/g_{os} = \infty\,\Omega$

$F_L = 30\text{ Hz}$

FIGURE G.15 Example 17.4

EXAMPLE 17.4

```
  !   A   !!    B    !!    C    !!    D    !!    E    !!    F    !!    G
1  JFET AMPLIFIER DESIGN
2
3  Design the amplier in Figure G.15
4
5  DATA INPUT:
6    VP    =      -10 VOLT    = FET PINCHOFF SPEC
7    IDSS =      .15 AMP     = FET DRAIN SATURATION SPEC
8    IGSS =     -3e-9 AMP    = FET GATE CURRENT SPEC
9   1/GOS =     1e50 OHM     = FET DRAIN RESISTANCE SPEC
10   RSS  =      600 OHM     = SOURCE RESISTANCE
11   RL   =      100 OHM     = LOAD RESISTANCE
12   FL   =       30 HZ      = LOW FREQUENCY CUTOFF
13
14 CALCULATIONS:
15   RD   =      100 OHM     = 1/(1/B11-1/B9)
16     USE       100 OHM     = NEAREST STANDARD RESISTOR
17   RAC  =       50 OHM     = 1/(1/B16+1/B11)
18   ID   = .0530330 A       = B7/SQRT(8)
19   VGS  = -4.05396 V       = B6*(1-SQRT(B18/B7))
20   VS   = 4.053964 V       = -B19
21   VDS  = 14.84835 V       = (B7-B18)*B17-B6
22   VD   = 18.90231 V       = B21+B20
23   VDD  = 24.20561 V       = B22+B18*B16
24   RG   = 3.3333e8 OHM     = B6/(10*B8)
25     USE   10000000 OHM    = NEAREST STANDARD RESISTOR
26   RSQ  = 76.44229 OHM     = B20/B18
27     USE       82 OHM      = NEAREST STANDARD RESISTOR
```

```
28    RIN  = 10000000 OHM     = B25
29    ROUT =      100 OHM     = 1/(1/B16+1/B9)
30    CI   = 5.30e-10 FRD     = 1/(2*PI*B12*(B10+B28))
31    CS   = .0000647 FRD     = 1/(2*PI*B12*B27)
32    CO   = .0000265 FRD     = 1/(2*PI*B12*(B29+B11))
33    GFS  = .0178381 SIEM    = -2/B6*SQRT(B7*B18)
34    AV   = -.891905         = -B33*(1/(1/B9+1/B16+1/B11))
```

EXAMPLE 17.6

```
   :  A  ::  B   ::   C   ::   D   ::   E   ::   F   ::   G
1  JFET AMPLIFIER DESIGN
2
3  Design the amplier in Figure 17.15
4
5  DATA INPUT:
6     VP   =          -3 VOLT   = FET PINCHOFF SPEC
7     IDSS =        .005 AMP    = FET DRAIN SATURATION SPEC
8     IGSS =       -1e-9 AMP    = FET GATE CURRENT SPEC
9   1/GOS =       100000 OHM    = FET DRAIN RESISTANCE SPEC
10    RSS  =       100000 OHM    = SOURCE RESISTANCE
11    RL   =         1000 OHM    = LOAD RESISTANCE
12    FL   =           30 HZ     = LOW FREQUENCY CUTOFF
13
14 CALCULATIONS:
15    RD   =     1010.101 OHM    = 1/(1/B11-1/B9)
16      USE         1000 OHM    = NEAREST STANDARD RESISTOR
17    RAC  =          500 OHM    = 1/(1/B16+1/B11)
18    ID   =     .0017678 AMP    = B7/SQRT(8)
19    VGS  =     -1.21619 VOLT   = B6*(1-SQRT(B18/B7))
20    VS   =     1.216189 VOLT   = -B19
21    VDS  =     4.616117 VOLT   = (B7-B18)*B17-B6
22    VD   =     5.832306 VOLT   = B21+B20
23    VDD  =     7.600073 VOLT   = B22+B18*B16
24    RG   =          3e8 OHM    = B6/(10*B8)
25      USE     10000000 OHM    = NEAREST STANDARD RESISTOR
26    RSQ  =     687.9806 OHM    = B20/B18
27      USE          680 OHM    = NEAREST STANDARD RESISTOR
28    RIN  =     10000000 OHM    = B25
29    ROUT =     990.0990 OHM    = 1/(1/B16+1/B9)
30    CI   =     5.25e-10 FRD    = 1/(2*PI*B12*(B10+B28))
31    CS   =     .0000078 FRD    = 1/(2*PI*B12*B27)
32    CO   =     .0000027 FRD    = 1/(2*PI*B12*(B29+B11))
33    GFS  =     .0019820 SIEM   = -2/B6*SQRT(B7*B18)
34    AV   =     -.986076        = -B33*(1/(1/B9+1/B16+1/B11))
```

The following is an example of the iterating feature of SuperCalc:

EXAMPLE 17.7

```
   :   A   ::   B   ::   C   ::   D   ::   E   ::   F   ::   G
1  JFET AMPLIFIER CIRCUIT ANALYSIS
2
3  ANALYZE THE CIRCUIT IN EXAMPLE 17.6
4
5  DATA INPUT:
6     VP   =          -3 VOLT  = FET PINCHOFF SPEC
7     IDSS =        .005 AMP   = FET DRAIN SATURATION SPEC
8     IGSS =       -1e-9 AMP   = FET GATE CURRENT SPEC
9    1/GOS =      100000 OHM   = FET DRAIN RESISTANCE SPEC
10    RSS  =      100000 OHM   = SIGNAL SOURCE RESISTANCE
11    RL   =        1000 OHM   = LOAD RESISTANCE
12    RSQ  =         680 OHM   = TRANSISTOR SOURCE RESISTOR
13    RG   =    10000000 OHM   = GATE RESISTOR
14    RD   =        1000 OHM   = DRAIN RESISTOR
15    CI   =     5.6e-10 FRD   = INPUT CAPACITOR
16    CS   =     .000008 FRD   = SOURCE BYPASS CAPACITOR
17    CO   =    .0000027 FRD   = OUTPUT CAPACITOR
18    VDD  =         7.6 VOLT  = POWER SUPPLY VOLTAGE
19
20 CALCULATIONS:
21    VGS  =    -1.21019 VOLT  = B6*(1-SQRT(B22/37))
22    ID   =    .0017797 AMP   = -B21/B12
23    VD   =    5.820306 VOLT  = B18-B22*B14
24    VS   =    1.210192 VOLT  = B22*B12
25    VDS  =    4.610113 VOLT  = B23-B24
26    RIN  =    10000000 OHM   = B13
27    ROUT =    990.0990 OHM   = 1/(1/B14+1/B9)
28    FL   =    28.13913 HZ    = 1/(2*PI*B15*(B10+B26))
29    FL   =    29.25642 HZ    = 1/(2*PI*B16*B12)
30    FL   =    29.61977 HZ    = 1/(2*PI*B17*(B27+B11))
31    GFS  =    .0019887 SIEM  = -2/B6*SQRT(B7*B22)
32    AV   =    -.989397       = -B31*(1/(1/B9+1/B14+1/B11))
```

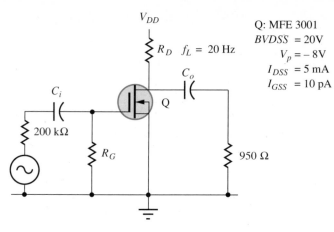

V_{DD}

R_D $f_L = 20$ Hz

C_o

C_i

200 kΩ

R_G

950 Ω

Q: MFE 3001
$BVDSS = 20$V
$V_p = -8$V
$I_{DSS} = 5$ mA
$I_{GSS} = 10$ pA

FIGURE G.16   Example 18.2

EXAMPLE 18.2
The circuit is copied here as Figure G.16

```
     ::  A   ::   B   ::   C   ::   D   ::   E   ::   F   ::   G
 1   MOSFET AMPLIFIER CIRCUIT DESIGN
 2
 3   DESIGN THE CIRCUIT SHOWN IN FIGURE G.16
 4
 5   DATA INPUT:
 6       VP    =          -8 VOLT  = FET PINCHOFF SPEC
 7       IDSS  =        .005 AMP   = FET DRAIN SATURATION SPEC
 8       IGSS  =       1e-11 AMP   = FET GATE CURRENT SPEC
 9     1/GOS   =        1e50 OHM   = FET DRAIN RESISTANCE SPEC
10       RSS   =      200000 OHM   = SOURCE RESISTANCE
11       RL    =         950 OHM   = LOAD RESISTANCE
12       FL    =          20 HZ    = LOW FREQUENCY CUTOFF
13
14   CALCULATIONS:
15       RD    =         950 OHM   = 1/(1/B11-1/B9)
16       USE          1000 OHM   = NEAREST STANDARD RESISTOR
17       RAC   =    487.1795 OHM   = 1/(1/B16+1/B11)
18       ID    =        .005 AMP   = B7
19       VGS   =           0 VOLT  = MIDPOINT BIAS
20       VGSK  =           4 VOLT  = -B6/2
21       VDSK  =          12 VOLT  = B20-B6
22       IDK   =      .01125 AMP   = B7*(1-B20/B6)^2
23       VDS   =    15.04487 VOLT  = (B22-B18)*B17+B21
24       VDD   =    20.04487 VOLT  = B23+B18*B16
25       RG    =       -8e10 OHM   = B6/(10*B8)
```

```
26      USE    10000000 OHM    = NEAREST STANDARD RESISTOR
27   RIN  =    10000000 OHM    = B25
28   ROUT =        1000 OHM    = 1/(1/B16+1/B9)
29   CI   =    7.80e-10 FRD    = 1/(2*PI*B12*(B10+B27))
30   CO   =     .0000041 FRD   = 1/(2*PI*B12*(B28+B11))
31   GFS  =      .00125 SIEM   = -2/B6*SQRT(B7*B18)
32   AV   =     -.608974       = -B33*(1/(1/B9+1/B16+1/B11))
```

EXAMPLE 18.6

```
  :   A   ::   B   ::   C   ::   D   ::   E   ::   F   ::   G
1  MOSFET COMPLEMENTARY SYMMETRY AMPLIFIER DESIGN
2
3  DATA INPUT:
4     PL   =           .5 WRMS  = RMS POWER TO THE LOAD
5     RL   =           10 OHM   = RESISTANCE OF THE LOAD
6     RSS  =          600 OHM   = RESISTANCE OF THE SOURCE
7     FL   =           30 HZ    = LOW FREQUENCY CUTOFF
8     hFE  =          135       = TRANSISTOR GAIN
9
10 TRANSISTOR SELECTION:
11    PDISS =         .25 WATT  = B4/2
12    BVDSS =    13.41641 VOLT  = 6*SQRT(B4*B5)
13
14 SELECT TRANSISTORS BASED ON ABOVE APPROXIMATIONS
15    VP   =          3.5 VOLT  = PINCHOFF VOLTAGE RATING
16    IGSS =     .0000001 AMP   = GATE LEAKAGE CURRENT RATING
17    GFSX =           .2 SIEM  = TRANSCONDUCTANCE AT IDX
18    IDX  =           .5 AMP   = DRAIN TEST CURRENT
19
20 CALCULATIONS:
21    RSQ  =            1 OHM   = B5/10
22     USE              1 OHM   = NEAREST STANDARD RESISTOR
23    iLP  =     .3162278 AMP   = SQRT(2*B4/B5)
24    IDO  =         .245 AMP   = (B15*B17/2)^2/B18
25    VGSP =     7.476354 VOLT  = B15*(1+SQRT(B23/B24))
26    VDSK =     3.976354 VOLT  = B25-B15
27    VDD  =     14.90972 VOLT  = 2*(B23*(B21+B5)+B26)
28    ID   =     .0031623 AMP   = B23/100
29    VGS  =     3.897635 VOLT  = B15*(1+SQRT(B28/B24))
30    VS1  =     7.458021 VOLT  = B27/2+B28*B21
31    VS2  =     7.451697 VOLT  = B27/2-B28*B21
32    VG1  =     11.35566 VOLT  = B30+B29
33    VG2  =     3.554061 VOLT  = B31-B29
```

```
34    R1   =   3554061. OHM    = B33/(10*B16)
35      USE    3300000 OHM    = NEAREST STANDARD RESISTOR
36    R2   =   7243900. OHM    = B35*(B32-B33)/B33
37      USE    6800000 OHM    = NEAREST STANDARD RESISTOR
38    RIN  =   1650000 OHM    = B35/2
39    ROUT =         1 OHM    = B22
40    CI   =   1.607e-9 FRD    = 1/(4*PI*B7*(B6+B38))
41    CO   =   .0004823 FRD    = 1/(2*PI*B7*(B39+B5))
```

Appendix H
PSpice Solutions

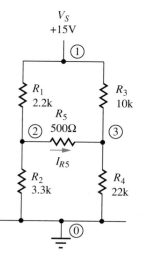

V_S
+15V

①

R_1
2.2k

R_5
R_3
10k

② 500Ω ③

I_{R5}

R_2
3.3k

R_4
22k

⓪

FIGURE H.1   Problem 1.12

PROBLEM 1.12 Unbalanced Wheatstone bridge.
****   CIRCUIT DESCRIPTION

```
*FIND  I(R5) IN FIGURE H.1
.DC VS 15 15.5 1
VS 1 0 DC
R1 1 2 2.2K
R2 2 0 3.3K
R3 1 3 10K
R4 3 0 22K
R5 2 3 500
.PRINT DC I(R5)
.END
```

  VS   I(R5)
 1.500E+01   −1.509E−04

PROGRAM EXPLANATION:

Line 1.   Comment line ignored by PSpice.

Line 2.   "DC" defined: PSpice requires that DC be a sweep of voltages. DC will sweep from 15 V to 15.5 V in 1-V increments. Thus the first data point will be taken at VS = 15 V and the second data point will be out of range so only one data point will be recorded; the one we want. This is one way of circumventing PSpice's demand for a sweep of DC voltages.

Line 3.   VS is connected between nodes 1 (+) and 0 (−). VS is defined by the term "DC" (see above). For node identification see Figure H.1.

Line 4.   R1 is connected between nodes 1 (+) and 2 (−) and has a value of 2.2 KΩ.

. . .

Line 9.   For each data point of the DC sweep (the definition in line 2 allows only one data point) the current in R5 is printed. The negative value indicates that current is actually flowing in the opposite direction to that assumed in line 8 when R5 was defined to be connected between nodes 2 (+) and 3 (−).

Line 10. For historical reasons an "end card" must terminate the program.

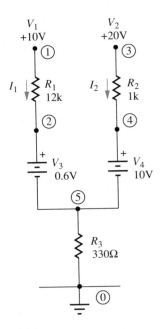

FIGURE H.2   Problem 1.15

PROBLEM 1.15 Kirchhoff circuit.
****   CIRCUIT DESCRIPTION

```
*FIND CURRENTS  I(R1)  AND I(R2) IN FIGURE H.2
*I(R1) = CURRENT FROM (−) V(1).
*I(R2) = CURRENT FROM (−) V(2).
V1 1 0 10V
V2 3 0 20V
V3 2 5 .6V
V4 4 5 10V
R1 1 2 12K
R2 3 4 1K
R3 5 0 330
.OP
.END
```

NODE	VOLTAGE	NODE	VOLTAGE	NODE	VOLTAGE	NODE	VOLTAGE
(1)	10.0000	(2)	3.2214	(3)	20.0000	(4)	12.6214
(5)	2.6214						

      VOLTAGE SOURCE CURRENTS

NAME	CURRENT
V1	−5.649D−04
V2	−7.379D−03
V3	5.649D−04
V4	7.379D−03

Thus $I_{R1} = \underline{564.9\mu A}$ and $I_{R2} = \underline{7.379mA}$

PROGRAM EXPLANATION:

Lines 1–3.   Comment lines ignored by PSpice.

Lines 3–7.   Voltage sources defined. The voltage of each source is included with its definition and no DC sweep is defined. Thus a ".PRINT DC" statement cannot be used.

Lines 8–10. Resistors defined.

Line 11.   Operating point data printed out: All node voltages and all supply currents are listed. Currents from (−) V1 and V2 are the currents sought. Thus

$$IR1 = - (-564.9 \ \mu A) = 564.9 \ \mu A$$
$$IR2 = - (-7.379 \ mA) = 7.379 \ mA$$

Line 12.   End of the program.

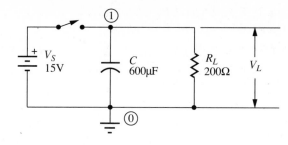

FIGURE H.3 Problem 1.17

PROBLEM 1.17 Capacitor discharge.
**** CIRCUIT DESCRIPTION

```
*FIND  VL  1/60 SEC AFTER S OPENS IN FIGURE H.3
*VL = VC
C 1 0 600E-6 IC=15V
R 1 0 200
.TRAN 16.7MS 16.7MS UIC
.PRINT TRAN V(C)
.END
```

**** TRANSIENT ANALYSIS

TIME	V(C)
.000E+00	1.500E+01
1.670E-02	1.305E+01

Thus V_L = <u>13.05V</u>

PROGRAM EXPLANATION:

Lines 1–2. Comment statements ignored by PSpice.

Line 3. Capacitor connected between nodes 1 and 0. C = 600 µF. Initially charged to 15 V.

Line 4. Resistor connected between nodes 1 and 0. R = 200Ω.

Line 5. Transient analysis defined to have an increment of 16.7 ms and to terminate after 16.7 ms. This produces two data points, the initial and the final values.

Line 6. Print the transient data point values of VC.

Line 7. End of program.

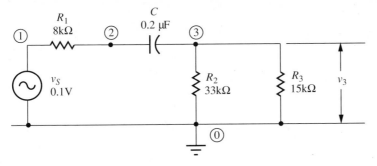

FIGURE H.4   Problem 1.19: Circuit

PROBLEM 1.19 High pass circuit.
```
****        CIRCUIT DESCRIPTION

*FIND  FL  FOR  FIGURE H.4
.AC DEC 5 .1 10KHZ
VS 1 0 AC .1V
R1 1 2 8K
C 2 3 .2UF
R2 3 0 33K
R3 3 0 15K
.PROBE V(3)
*GRAPH IS  FIGURE H.5
*FL  IS WHERE EXTRAPOLATION OF DECREMENT LINE INTERSECTS 0 DB.
.END
```

Thus from Figure H.5, F_L = <u>44.4Hz</u>

PROGRAM EXPLANATION:

Line 1.   Comment line ignored by PSpice.

Line 2.   "AC" defined. PSpice requires that AC be a sweep of frequencies. AC will sweep by decades with 5 data points per decade starting at F = 0.1 Hz and ending at F = 10 kHz.

Line 3.   VS is a voltage source connected between nodes 1 and 0 as shown in Figure H.4. It is defined by "AC" in line 2 and has an amplitude of 0.1 VP.

Lines 4–7.   Circuit component specification. See previous examples.

Line 8.   PROBE makes the node 3 voltage data available for graphing.

Lines 9–10. Comment statements.

Line 11.   End of program.

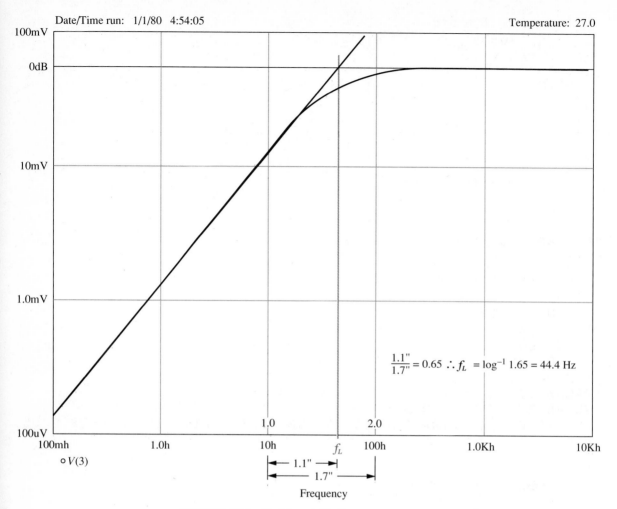

$$\frac{1.1''}{1.7''} = 0.65 \therefore f_L = \log^{-1} 1.65 = 44.4 \text{ Hz}$$

○ V(3)

Frequency

FIGURE H.5   Problem 1.19: Frequency response graph

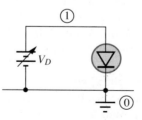

FIGURE H.6   Section 2.6: Circuit for drawing diode characteristic graph

SECTION 2.6 PSPICE DEFAULT DIODE CHARACTERISTIC
****   CIRCUIT DESCRIPTION

```
*DRAW DIODE CHARACTERISTIC. USE FIGURE H.6
.DC VD .4 .9V 10MV
VD 1 0 DC
D 1 0 D
.MODEL D D
.PROBE
*GRAPH IS FIGURE H.7
.END
```

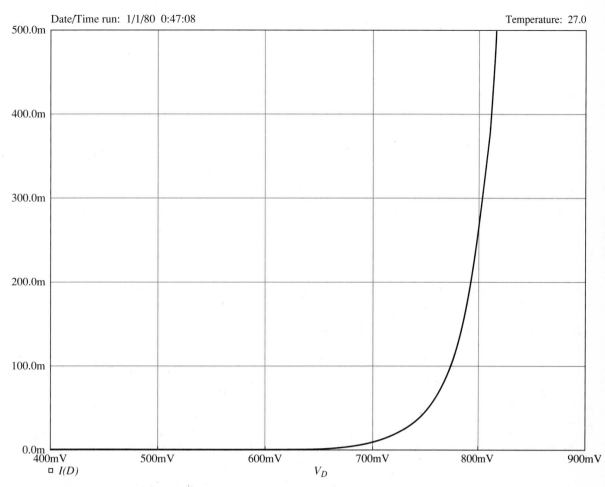

FIGURE H.7   PSpice default diode characteristic graph.

PROGRAM EXPLANATION:

Line 1. Comment line.

Line 2. "DC" defined: PSpice requires that DC be a sweep of voltages. DC will sweep from 0.4 V to 0.9 V in 10 mV increments.

Line 3. VD is a voltage source connected between nodes 1 and 0 and defined by DC as defined in line 2.

Line 4. "D" is a diode connected between nodes 1 (anode) and 0 (cathode) and defined by the model D in line 5.

Line 5. "MODEL" defines D in line 4 to be the default diode model which is approximately:

$$V_D = 26 \text{ mV} * \ln \frac{I_D}{10 \text{ fA}} \tag{3.10}$$

Line 6. PROBE makes the node 3 voltage data available for graphing.

Line 7. Comment statement.

Line 8. End of program.

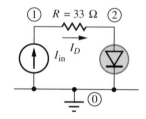

FIGURE H.8   Example 2.1

EXAMPLE 2.1 A. Supply voltage needed for a given current.

```
****       CIRCUIT DESCRIPTION

*FIND  VS  TO GIVE  ID = 20MA IN FIGURE H.8
*IIN = ID = 20MA
IIN 0 1 20MA
R 1 2 33
D 2 0 D
.MODEL D D
.END
```

```
 NODE     VOLTAGE      NODE     VOLTAGE
(  1)      1.3926     (  2)      .7326
```

Thus V_S = <u>1.3926V</u>

PROGRAM EXPLANATION:

Lines 1–2. Comment statements.

Line 3. IIN is an independent current source with current flowing from node 0 (+) to node 1 (−). It has a magnitude of 20 mA. In PSpice, M stands for E − 3; MEG stands for E + 6.

Line 4. R is a resistor connected between nodes 1 and 2 and having a resistance of 33Ω.

Line 5. D is a diode connected between nodes 2 (anode) and O (cathode) and having characteristics defined by MODEL D in line 6.

Line 6. MODEL D defines diode D to have the default parameters.

Line 7. END terminates the program.

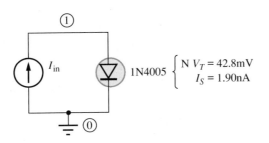

FIGURE H.9 Example 3.4: Specifying diode parameters.

EXAMPLE 3.4 Using diode specifying parameters.
```
****        CIRCUIT DESCRIPTION

*FIND   VD   FOR   ID = 20MA IN FIGURE H.9
*IIN =  ID = 20MA
*  N =  N*VT/VT = 42.8MV/25.9MV = 1.65
*  IS = 1.90 NA
IIN 0 1 20MA
D 1 0 D
.MODEL D D (N=1.65 IS=1.9NA)
.END

****        DIODE MODEL PARAMETERS

IS         1.90D-09
N              1.650

  NODE    VOLTAGE
(  1)      .6900
```

Thus V_D = <u>0.6900V</u>

PROGRAM EXPLANATION:

Lines 1–4. Comment lines.

Line 5.   IIN is a current source with current flowing from node 0 to node 1. The current has a value of 20 mA.

Line 6.   D is a diode connected between nodes 1 (anode) and O (cathode) and having characteristics defined by MODEL D.

Line 7.   MODEL D defines diode D to have an emission coefficient of N = 1.65 and a reserve saturation current of IS = 1.9 nA.

Line 8.   END terminates the program.

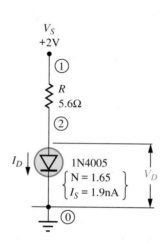

FIGURE H.10 Example 3.5: Current flow through a real diode.

EXAMPLE 3.5 Current flowing in a real diode.
```
****      CIRCUIT DESCRIPTION

*FIND  ID  IN  FIGURE H.10
*N = 1.65   IS = 1.9NA
VS 1 0 2V
R 1 2 5.6
D 2 0 D
.MODEL D D (N=1.65 IS=1.9NA)
.OP
.END
****      DIODE MODEL PARAMETERS

IS          1.90D-09
N                1.650
```

```
NODE    VOLTAGE    NODE    VOLTAGE
( 1)     2.0000    ( 2)     .7916
    VOLTAGE SOURCE CURRENTS
    NAME          CURRENT
    VS          -2.158D-01
```

Thus $I_D = $ <u>215.8mA</u>

PROGRAM EXPLANATION:

Since the required diode current is the current flowing out of $(-)$ the voltage supply, V_S, and since the operating point statement, .OP, causes all voltage source currents to be printed, it is used to obtain the required diode current. Otherwise, the program is very similar to previous circuits.

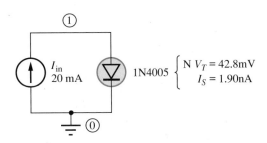

FIGURE H.11   Example 3.7: Diode dynamic resistance, r_D.

EXAMPLE 3.7 DYNAMIC RESISTANCE RD
```
****      CIRCUIT DESCRIPTION

*FIND  RD  FOR  FIGURE H.11
* N = 1.65  IS = 1.90NA
IIN 0 1 20MA
D 1 0 D
.MODEL D D (N=1.65 IS=1.9NA)
.OP
.END
****      DIODE MODEL PARAMETERS

IS        1.90D-09
N             1.650

NODE    VOLTAGE
( 1)     .6900

****      OPERATING POINT INFORMATION
```

**** DIODES

```
MODEL        D
ID           2.00E-02
VD           .690
REQ          2.13E+00
CAP          .00E+00
```

Thus $r_D = \underline{2.13\Omega}$

PROGRAM EXPLANATION:

Since the operating point information contains the required equivalent resistance, REQ, the program is essentially the same as others previously described.

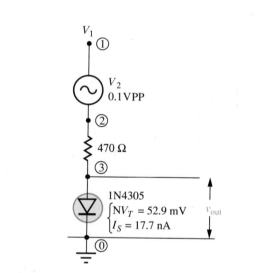

V_1 ①

V_2 0.1VPP

②

470 Ω

③

1N4305

$\begin{cases} NV_T = 52.9 \text{ mV} \\ I_S = 17.7 \text{ nA} \end{cases}$ v_{out}

⓪

FIGURE H.12 Example 3.8: Response to an AC signal.

EXAMPLE 3.8A SIGNAL SWITCHING : VDC = 0V
****   CIRCUIT DESCRIPTION

```
*FIND  VOAC WHEN  V1 = 0V IN FIGURE H.12
*N = N*VT/VT = 52.9MV/25.9MV = 2.04
*IS = 17.7NA
V2 2 0 AC .1V
.AC LIN 1 1KHZ 1KHZ
R 2 3 470
D 3 0 D
```

```
.MODEL D D (N=2.04 IS=17.7NA)
.PRINT AC V(D)
.END
```

```
****        DIODE MODEL PARAMETERS

IS          1.77D-08
N               2.040

****        SMALL SIGNAL BIAS SOLUTION

 NODE    VOLTAGE    NODE    VOLTAGE
(  2)      .0000    (  3)      .0000

****        AC ANALYSIS

     FREQ         V(D)
  1.000E+03     9.998E-02
```

Thus v_{out} = <u>99.98mV</u>

PROGRAM EXPLANATION:

Lines 1–3. Comment lines.

Line 4.   V2 is a voltage source connected between nodes 2 and 0. Since V1 = 0 V, node 1 = node 0, V2 is defined by AC which is defined in line 5. It has an amplitude of 0.1 V.

Line 5.   AC is defined to vary linearly with one data point being taken between the starting frequency of 1 kHz and the end frequency of 1 kHz. This is one way of circumventing the PSpice requirement that AC sweep a range of frequencies.

Lines 6–8. Similar to previous examples.

Line 9.   PRINT specifies that the AC analysis is to be used and that the diode voltage is to be printed.

Line 10. Terminates the program.

EXAMPLE 3.8B SIGNAL SWITCHING : VDS = 5V
```
****        CIRCUIT DESCRIPTION

*FIND  VOAC WHEN  V1 = 5V  IN  FIGURE H.12
*N = N*VT/VT = 52.9MV/25.9MV = 2.04
*IS = 17.7NA
V1 1 0 5V
V2 2 1 AC .1V
.AC LIN 1 1KHZ 1KHZ
R 2 3 470
D 3 0 D
```

```
.MODEL D D (N=2.04 IS=17.7NA)
.PRINT AC V(D)
.END
```

****   DIODE MODEL PARAMETERS

```
IS          1.77D-08
N              2.040
```

****   SMALL SIGNAL BIAS SOLUTION

NODE	VOLTAGE	NODE	VOLTAGE	NODE	VOLTAGE
(1)	5.0000	(2)	5.0000	(3)	.6942

****   AC ANALYSIS

```
    FREQ        V(D)
1.000E+03    1.210E-03
```

Thus v_{out} = __1.210mV__

PROGRAM EXPLANATION:

Except for the addition of the DC voltage source $V1$, this is the same as the program for Example 3.8A.

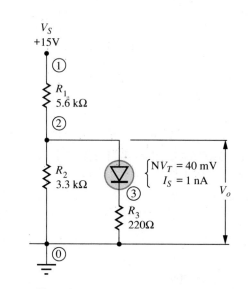

FIGURE H.13   Problem 3.21

PROBLEM 3.21 OUTPUT VOLTAGE
****   CIRCUIT DESCRIPTION

```
*FIND THE VALUE OF VO  IN  FIGURE H.13
* N = NVT/VT = 40MV/25.9MV = 1.55
* IS = 1NA
VS 1 0 15V
R1 1 2 5.6K
R2 2 0 3.3K
R3 3 0 220
D 2 3 D
.MODEL D D (N=1.55 IS=1NA)
.END
```

****   DIODE MODEL PARAMETERS

```
IS         1.00D-09
N             1.550
```

****   SMALL SIGNAL BIAS SOLUTION

NODE	VOLTAGE	NODE	VOLTAGE	NODE	VOLTAGE
(1)	15.0000	(2)	1.0617	(3)	.4768

Thus V_0 = <u>1.0617V</u>

PROGRAM EXPLANATION:

This program is essentially the same as others previously described.

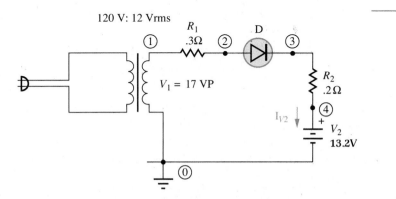

FIGURE H.14 Example 4.1: A half-wave battery charger.

EXAMPLE 4.1: BATTERY CHARGER
****   CIRCUIT DESCRIPTION

```
*FIND EFFECTIVE  DC  CURRENT IN FIGURE H.14.
.TRAN .1MS 16MS
V1 1 0 SIN(0 17V 60HZ)
R1 1 2 .3
D 2 3 D
.MODEL D D
R2 3 4 .2
V2 4 0 12.6V
.PROBE
*OUTPUT WAVEFORMS ARE IN FIGURES H.15 AND H.16.
.END

****      DIODE MODEL PARAMETERS

IS       1.00D-14
```

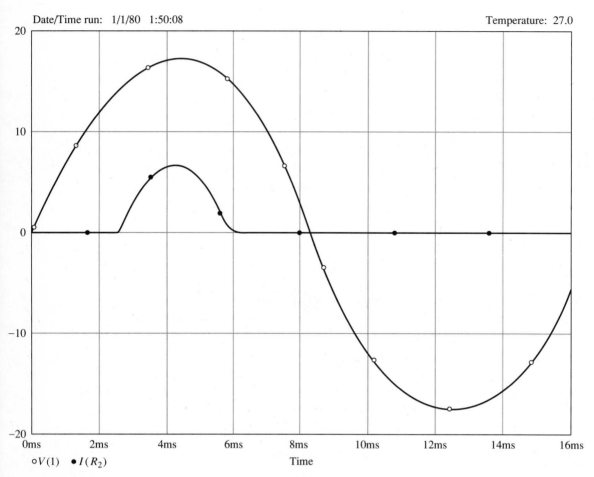

Date/Time run: 1/1/80 1:50:08   Temperature: 27.0

$\circ V(1)$ $\bullet I(R_2)$ Time

FIGURE H.15   Example 4.1: Battery charger waveforms.

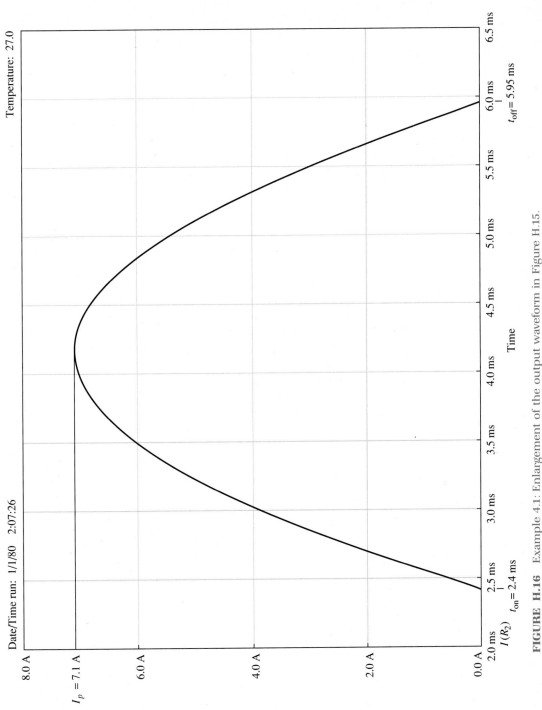

FIGURE H.16 Example 4.1: Enlargement of the output waveform in Figure H.15.

The current waveform in Figure H.15 is greatly expanded in Figure H.16. It indicates that conduction begins at about $t_{on} = 2.4$ ms and ends at about $t_{off} = 5.95$ ms. It also indicates the peak current to be about $I_P = 7.1$ A. Since $f = 60$ Hz, we have

$$\theta_{on} = 360°ft_{on} = 360° * 60 \text{ Hz} * 2.4 \text{ ms} = 51.8°$$

and

$$\theta_{off} = 360°ft_{off} = 360° * 60 \text{ Hz} * 5.95 \text{ ms} = 129°$$

Since $\theta_{rep} = 2\pi$, we have

$$I_{DC} = \frac{I_P}{\theta_{rep}} (\cos \theta_{on} - \cos \theta_{off}) \tag{4.1}$$

$$= \frac{7.1 \text{ A}}{2\pi} (\cos 51.8° - \cos 129°) = \underline{\underline{1.40 \text{ A}}}$$

PROGRAM EXPLANATION:

Line 1.   Comment line.

Line 2.   Transient analysis. The circuit is evaluated every 0.1 ms for 16 ms.

Line 3.   V1 is a voltage source connected between nodes 1 and 0. It is a sine wave source with 0 V DC offset, an amplitude of 17 V, and a frequency of 60 Hz.

Lines 4–8. Circuit components specification as explained in previous examples. Note that the MODEL statement in line 6 uses the default parameters to define the diode.

Line 9.   PROBE stores all voltage and current data for use in the Probe graphics program used to generate Figures H.15 and H.16.

Line 10. Comment line.

Line 11. END terminates the program.

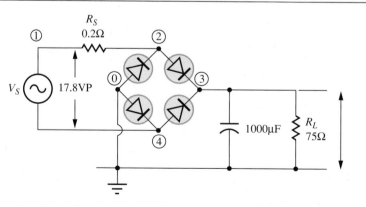

FIGURE H.17   Example 5.4: Bridge power supply circuit.

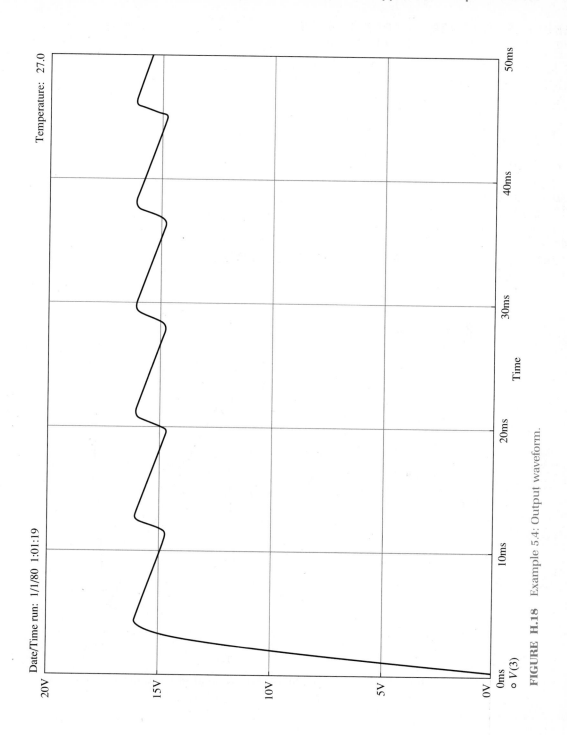

FIGURE H.18   Example 5.4: Output waveform.

```
EXAMPLE 5.4 POWER SUPPLY RIPPLE
****      CIRCUIT DESCRIPTION
*FIND RIPPLE FOR  FIGURE H.17 (page 878)
.TRAN 10MS 50MS 0 .1MS
VS 1 4 SIN(0 17.8VP 60HZ)
RS 1 2 .2OHM
D1 0 2 D
D2 0 4 D
D3 4 3 D
D4 2 3 D
.MODEL D D
C 30 1000UF
RL 3 0 75OHM
.PROBE
*FIGURES H.18 AND H.19 CONTAIN THE OUTPUT WAVEFORMS.
.END
```

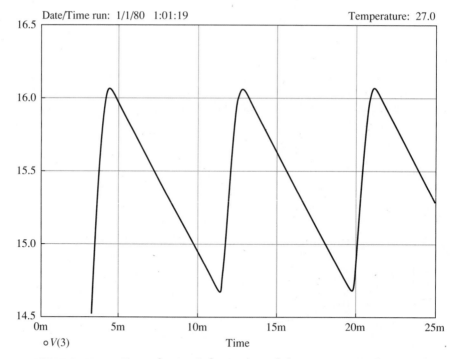

Date/Time run: 1/1/80 1:01:19
Temperature: 27.0
16.5
16.0
15.5
15.0
14.5
0m
5m
10m
15m
20m
25m
o V(3)
Time

FIGURE H.19 Example 5.4: Enlargement of the output waveform in Figure H.18.

PROGRAM EXPLANATION:

Line 1. Comment line

Line 2. Transient analysis. The circuit is evaluated every 10 ms for 50 ms beginning at 0 seconds with plotting points every 0.1 ms. This gives a slightly smoother plot than is given by the default plotting interval.

The remainder of the program is essentially the same as that of previous examples.

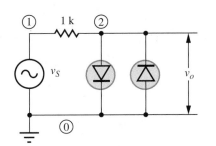

FIGURE H.20   Section 6.2: Clipping circuit.

```
SECTION 6.2: CLIPPING CIRCUIT BEHAVIOR
****      CIRCUIT DESCRIPTION

*PLOT INPUT AND OUTPUT WAVEFORMS FOR FIGURE H.20
.TRAN 10MS 50MS 0 .1MS
VS 1 0 SIN(0 .5V 100HZ 0 -20)
R1 1 2 1KOHM
D1 2 0 D
D2 0 2 D
.MODEL D D
.PROBE
*OUTPUT WAVEFORMS ARE IN FIGURES 6.2 AND H.21
.END

****      DIODE MODEL PARAMETERS

IS        1.00D-14
```

PROGRAM EXPLANATION:

Line 1. Comment line

Line 2. Transient analaysis identical to the previous example.

Line 3. VS is a voltage source connected between nodes 1 and 0. It is a sine wave source with 0 V DC offset, an initial amplitude of 0.5 V,

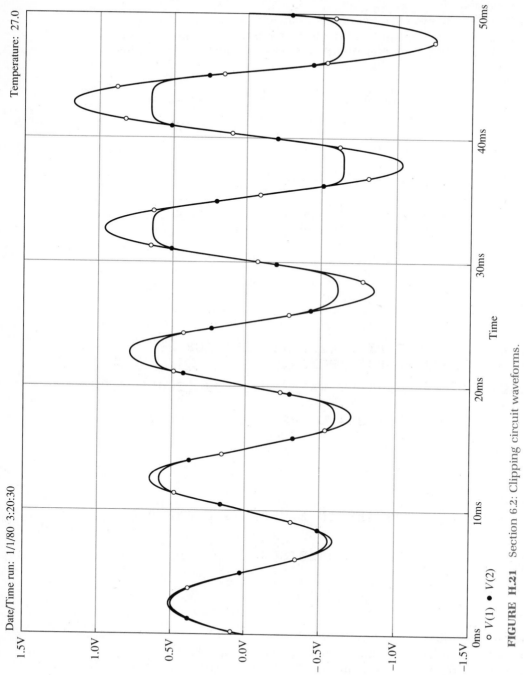

Date/Time run: 1/1/80 3:20:30

Temperature: 27.0

○ V(1) ● V(2)

FIGURE H.21 Section 6.2: Clipping circuit waveforms.

a frequency of 100 Hz, no delay in starting the wave, and a damping factor of -20. The negative damping factor is what causes the amplitude in Figure H.21 to slowly increase.

The remainder of the program is similar to those previously described.

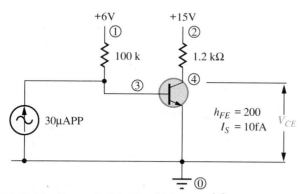

+6V   +15V

① | ②

100 k   1.2 kΩ

③   ④

30μAPP

$h_{FE} = 200$
$I_S = 10fA$ V_{CE}

⓪

FIGURE H.22   Example 9.3: Fixed bias amplifier.

EXAMPLE 9.3: LOCATION OF Q POINT
****   CIRCUIT DESCRIPTION

```
*Q  POINT AND OUTPUT WAVEFORM FOR FIGURE H.22
VBB 1 0 6VDC
VCC 2 0 15VDC
RB 1 3 100KOHM
RC 2 4 1.2KOHM
Q 4 3 0 Q
.MODEL Q NPN (IS=10FA BF=200)
.OP
IIN 0 3 SIN(0 15UAP 100HZ)
.TRAN 30MS 30MS 0 60US
.PROBE
*OUTPUT VOLTAGE WAVEFORM IS IN FIGURE H.23
.END
```

****   BJT MODEL PARAMETERS

TYPE	NPN
IS	1.00D-14
BF	200.000

Date/Time run: 2/17/89 1:49:59   Temperature: 27.0

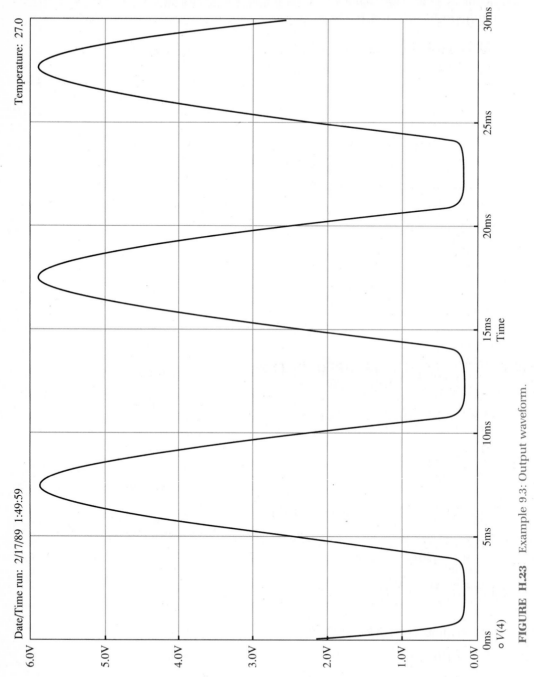

o V(4)

FIGURE H.23 Example 9.3: Output waveform.

```
****        SMALL SIGNAL BIAS SOLUTION

 NODE  VOLTAGE      NODE  VOLTAGE      NODE  VOLTAGE      NODE   VOLTAGE
(  1)   6.0000     (  2)  15.0000     (  3)   .7161     (  4)   2.3186
```

**** OPERATING POINT INFORMATION

**** BIPOLAR JUNCTION TRANSISTORS

```
MODEL      Q
IB         5.28E-05
IC         1.06E-02
VBE         .716
VBC       -1.60
VCE        2.32
BETADC    200.
BETAAC    200.
```

PROGRAM EXPLANATION:

Line 1. Comment line.

Line 2. VBB is the base supply connected beteen nodes 1(+) and 0(−) in Figure H.22. *VBB* has a voltage of 6 VDC.

Line 3. VCC is the collector supply connected between nodes 2(+) and 0(−) having a voltage of 15 VDC.

Line 4. RB is the base resistor connected between nodes 1 and 3. RB has a resistance of 100kΩ.

Line 5. RC is the collector resistor connected between nodes 2 and 4. RC has a resistance of 1.2kΩ.

Line 6. Q is the transistor connected to nodes 4(collector), 3(base), and 0(emitter). Its parameters are specified as model Q.

Line 7. Model Q is defined to be an NPN transistor with the default model changed to IS = 10 fA and BF (h_{FE}) = 200.

Line 8. .OP is a command to list the operating point information.

Line 9. IIN is the input current signal source connected between nodes 0(−) and 3(+). It is defined to be a sine wave with no offset, an amplitude of 15 μAP, and a frequency of 100 Hz.

Line 10. .TRAN is a command to perform a transient analysis of the circuit's response to the input signal. Data is to be reported every 30 ms (irrelevant here) for a total of 30 ms starting at 0 s and calculated every 60 μs.

Line 11. .PROBE is a command to store data for the graphics program.

Line 12. A comment line.

Line 13. .END terminates the program.

Running the Probe program produced the plot shown in Figure H.23.

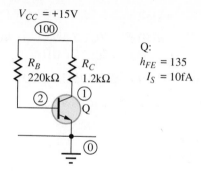

FIGURE H.24 Example 10.2: Fixed bias amplifier for comparison with Example 10.3.

EXAMPLE 10.2: FIXED BIAS CIRCUIT ANALYSIS
```
****          CIRCUIT DESCRIPTION

*ANALYZE THE CIRCUIT OF FIGURE H.24
VCC 100 0 15VDC
RB 100 2 220KOHM
RC 100 1 1.2KOHM
Q 1 2 0 Q
.MODEL Q NPN (IS=1E-14 BF=135)
.OP
.END

****          BJT MODEL PARAMETERS

TYPE          NPN
IS            1.00D-14
BF            135.000

****          OPERATING POINT INFORMATION

MODEL         Q
IB            6.50E-05
IC            8.77E-03
VBE           .711
VBC           -3.77
VCE           4.48
BETADC        135.
BETAAC        135.
```

PROGRAM EXPLANATION:

Line 1. Comment line.
Line 2. VCC is the power supply connected between nodes 100(+) and
  0(−). It has a value of 15 VDC.

Line 3. RB is the base bias resistor connected between nodes 100 and 2. It has a value of 220 kΩ.

Line 4. RC is the collector resistor connected between nodes 100 and 1. It has a value of 1.2 kΩ.

Line 5. Q is the transistor connected to nodes 1(C), 2(B) and 0(E). It has the parameters defined by MODEL Q.

Line 6. .MODEL Q is an NPN transistor with IS = 10 fA and BF (h_{FE}) = 135.

Line 7. .OP is the command to display the operating point information.

Line 8. .END terminates the program.

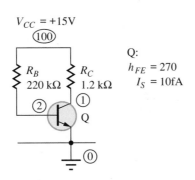

FIGURE H.25 Example 10.3: Fixed bias amplifier for comparison with Example 10.2.

EXAMPLE 10.3: STABILITY (COMPARE EXAMPLE 10.2)
```
****       CIRCUIT DESCRIPTION

*ANALYZE THE CIRCUIT OF FIGURE H .25
VCC 100 0 15VDC
RB 100 2 220KOHM
RC 100 1 1.2KOHM
Q 1 2 0 Q
.MODEL Q NPN (IS=1E-14 BF=270)
.OP
.END

****       BJT MODEL PARAMETERS
TYPE          NPN
IS         1.00D-14
BF          270.000

****       OPERATING POINT INFORMATION
IB          6.50E-05
IC          1.24E-02
VBE          .720
```

VBC	.553
VCE	.168
BETADC	190.
BETAAC	270.

PROGRAM EXPLANATION:

This is the same as the previous program with h_{FE} changed.

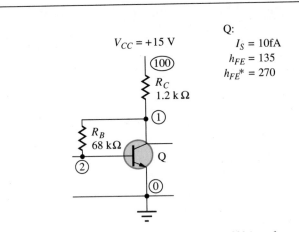

$$V_{CC} = +15 \text{ V}$$

Q:
$I_S = 10\text{fA}$
$h_{FE} = 135$
$h_{FE}* = 270$

R_C
$1.2 \text{ k}\Omega$

R_B
$68 \text{ k}\Omega$

Q

FIGURE H.26   Example 10.5: Stability of a self biased amplifier.

EXAMPLE 10.5A: SELF BIAS CIRCUIT ANALYSIS (HFE = 135)
```
****      CIRCUIT DESCRIPTION

*ANALYZE THE CIRCUIT OF FIGURE H.26
VCC 100 0 15VDC
RC 100 1 1.2K
RB 1 2 68K
Q 1 2 0 Q
.MODEL Q NPN (IS = 10FA BF = 135)
.OP
.END

****      BJT MODEL PARAMETERS

TYPE         NPN
IS           1.00D-14
BF           135.000

****      OPERATING POINT INFORMATION

MODEL        Q
IB           6.18E-05
IC           8.35E-03
```

```
VBE          .710
VBC         -4.20
VCE          4.91
BETADC      135.
BETAAC      135.
```

PROGRAM EXPLANATION:

This program is similar to those previously described.

EXAMPLE 10.5B: SELF BIAS CIRCUIT ANALYSIS (HFE = 270)
```
****        CIRCUIT DESCRIPTION

*ANALYZE THE CIRCUIT OF FIGURE H.26*
VCC 100 0 15VDC
RC 100 1 1.2K
RB 1 2 68K
Q 1 2 0 Q
.MODEL Q NPN (IS = 10FA BF = 270)
.OP
.END

****        BJT MODEL PARAMETERS

             Q
TYPE         NPN
IS           1.00D-14
BF           270.000

****        OPERATING POINT INFORMATION

MODEL        Q
IB           3.63E-05
IC           9.81E-03
VBE          .714
VBC         -2.47
VCE          3.18
BETADC      270.
BETAAC      270.
```

PROGRAM EXPLANATION:

This program is similar to those previously described.

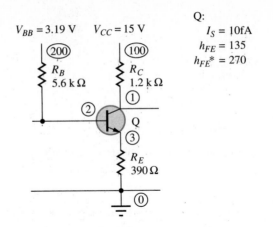

$V_{BB} = 3.19\ \text{V}$ $V_{CC} = 15\ \text{V}$

Q:
$I_S = 10\text{fA}$
$h_{FE} = 135$
$h_{FE}^* = 270$

FIGURE H.27   Example 10.7: Stability of an emitter biased amplifier.

EXAMPLE 10.7A: EMITTER BIAS (HFE =135)
```
****       CIRCUIT DESCRIPTION

*ANALYZE THE CIRCUIT OF FIGURE H.27
VCC 100 0 15VDC
VBB 200 0 3.19VDC
RC 100 1 1.2K
RB 200 2 5.6K
RE 3 0 390
Q 1 2 3 Q
.MODEL Q NPN (IS=10FA BF=135)
.OP
.END

****       BJT MODEL PARAMETERS

           Q
TYPE       NPN
IS         1.00D-14
BF         135.000

****       OPERATING POINT INFORMATION

MODEL      Q
IB         4.25E-05
IC         5.73E-03
VBE        .700
VBC        -5.17
VCE        5.87
BETADC     135.
BETAAC     135.
```

PROGRAM EXPLANATION:

This program is similar to those previously described.

EXAMPLE 10.7B: EMITTER BIAS (HFE = 270)
```
****        CIRCUIT DESCRIPTION

*ANALYZE THE CIRCUIT OF  FIGURE H.27*
VCC 100 0 15VDC
VBB 200 0 3.19VDC
RC 100 1 1.2K
RB 200 2 5.6K
RE 3 0 390
Q 1 2 3 Q
.MODEL Q NPN (IS=10FA BF=270)
.OP
.END

****        BJT MODEL PARAMETERS

              Q
TYPE         NPN
IS          1.00D-14
BF           270.000

****        OPERATING POINT INFORMATION

MODEL        Q
IB           2.24E-05
IC           6.04E-03
VBE          .702
VBC          -4.69
VCE          5.39
BETADC       270.
BETAAC       270.
```

PROGRAM EXPLANATION:

This program is similar to those previously described.

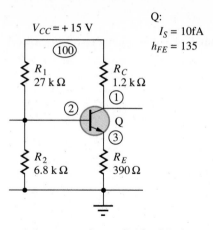

FIGURE H.28 Example 10.9: Voltage divider biasing.

EXAMPLE 10.9: DIVIDER BIAS ANALYSIS (HFE = 135)
```
****      CIRCUIT DESCRIPTION
*ANALYZE THE CIRCUIT OF FIGURE H.28
VCC 100 0 15VDC
RC 100 1 1.2K
RE 3 0 390OHM
R1 100 2 27K
R2 2 0 6.8K
Q 1 2 3 Q
.MODEL Q NPN (IS=10FA BF=135)
.OP
.END
```

```
****      BJT MODEL PARAMETERS

           Q
TYPE       NPN
IS         1.00D-14
BF         135.000
```

```
****      OPERATING POINT INFORMATION

MODEL      Q
IB         3.97E-05
IC         5.35E-03
VBE        .698
VBC        -5.77
VCE        6.47
BETADC     135.
BETAAC     135.
```

PROGRAM EXPLANATION:

This program is similar to those previously described.

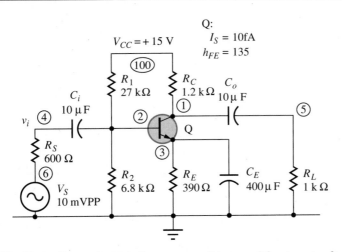

FIGURE H.29   Example 12.1: Common emitter amplifier input voltage.

```
EXAMPLE 12.1
****       CIRCUIT DESCRIPTION
*FIND   VI   FOR THE CIRCUIT OF   FIGURE H.29
VCC 100 0 15VDC
RC 100 1 1.2K
RE 3 0 390OHM
R1 100 2 27K
R2 2 0 6.8K
RS 6 4 600OHM
RL 5 0 1K
CI 4 2 10UF
CO 1 5 10UF
CE 3 0 400UF
VS 6 0 AC 10MV
.AC LIN 1 10KHZ 10KHZ
Q 1 2 3 Q
.MODEL Q NPN (IS=10FA BF=135)
.OP
.PRINT AC V(4) I(RS) V(5) I(RL)
.END

****       BJT MODEL PARAMETERS
TYPE         NPN
IS          1.00D-14
BF           135.000
```

```
****          SMALL SIGNAL BIAS SOLUTION
  NODE     VOLTAGE    NODE · VOLTAGE      NODE     VOLTAGE     NODE   VOLTAGE
(   1)     8.5744   (   2)    2.8023    (   3)     2.1038   (   4)     .0000
(   5)      .0000   (   6)     .0000    (100)    15.0000

****          OPERATING POINT INFORMATION
IB           3.97E-05
IC           5.35E-03
VBE           .698
VBC          -5.77
VCE          6.47
BETADC       135.
****          AC ANALYSIS
    FREQ          V(4)          I(RS)          V(5)          I(RL)
  1.000E+04       4.925E-03     8.459E-06      5.561E-01      5.561E-04
```

PROGRAM EXPLANATION:

In order to obtain the required v_{in}, ".AC" is used. .AC requires a range of frequencies. Since only a single frequency is needed, .AC is instructed to vary linearly from 10 kHz to 10 kHz with one data point being taken. Note that a fairly high frequency and fairly large capacitors were used in order that they would act as good conductors for the signal.

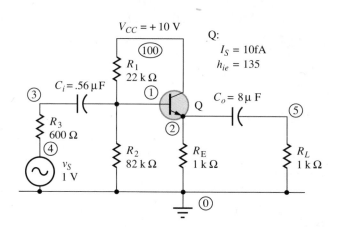

FIGURE H.30   Example 13.2: Emitter follower amplifier input voltage.

EXAMPLE 13.2
```
****          CIRCUIT DESCRIPTION.
*FIND   VI   FOR THE CIRCUIT OF FIGURE H.30
VCC 100 0 10VDC
RE 2 0 1KOHM
```

```
R1 100 1 22K
R2 1 0 82K
RS 4 3 600OHM
RL 5 0 1K
CI 3 1 .56UF
CO 2 5 8UF
VS 4 0 AC 1V
.AC LIN 1 10KHZ 10KHZ
Q 100 1 2 Q
.MODEL Q NPN (IS=10FA BF=170)
.OP
.PRINT AC V(3) I(RS) V(5) I(RL)
.END
```

****   BJT MODEL PARAMETERS

	Q
TYPE	NPN
IS	1.00D-14
BF	170.000

****   OPERATING POINT INFORMATION

MODEL	Q
IB	3.81E-05
IC	6.48E-03
VBE	.703
VBC	-2.78
VCE	3.48
BETADC	170.

****   AC ANALYSIS

FREQ	V(3)	I(RS)	V(5)	I(RL)
1.000E+04	9.601E-01	6.649E-05	9.525E-01	9.525E-04

The voltage gain is

$$A_V \equiv \frac{v_{\text{out}}}{v_{\text{in}}} \qquad\qquad (11.17)$$

$$= \frac{V(5)}{V(3)}$$

$$= \frac{0.09525 \text{ V}}{0.9601 \text{ V}}$$

$$= \underline{0.992} \qquad\qquad (0.992 \text{ in Example 13.2})$$

The input resistance is

$$R_{in} = \frac{V(3)}{I(RS)}$$

$$= \frac{0.9601 \text{ V}}{66.49 \text{ μA}}$$

$$= \underline{14.4 \text{ kOhm}} \qquad (14.4 \text{ kOhm in Example 13.2})$$

To determine the output resistance the open circuit voltage, v_{open}, is needed. Deleting the "CO" line and the "RL" line from the circuit description and inserting a "V(2)" instruction in the ".PRINT" line gives the following AC analysis:

****   AC ANALYSIS

FREQ	V(3)	I(RS)	V(2)
1.000E+04	9.633E-01	6.115E-05	9.595E-01

The open circuit voltage, v_{open}, is this new value of V(2). The output resistance can then be calculated from

$$R_{out} = \frac{v_{open} - v_L}{i_L}$$

$$= \frac{0.9595 \text{ V} - 0.9525 \text{ V}}{952.5 \text{ μA}}$$

$$= 7.35 \text{ Ω} \qquad (8.19 \text{ Ω in Example 13.2})$$

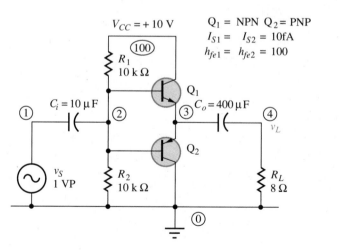

FIGURE H.31   Example 15.1: Unbiased complementary symmetry amplifier.

EXAMPLE 15.1

```
****        CIRCUIT DESCRIPTION

*FIND   vLP   FOR THE CIRCUIT OF FIGURE H.31
VCC 100 0 10VDC
VIN 1 0 SIN(0 1V 10KHZ)
.TRAN 25US 200US
R1 100 2 10K
R2 2 0 10K
RL 4 0 8OHM
CI 1 2 10UF
CO 3 4 400UF
Q1 100 2 3 Q1
.MODEL Q1 NPN (IS=10FA BF=100)
Q2 0 2 3 Q2
.MODEL Q2 PNP (IS=10FA BF=100)
.PRINT TRAN V(2) V(3)
.PROBE
*WAVEFORMS ARE SHOWN IN FIGURE H.32
.END

****        BJT MODEL PARAMETERS
            Q1          Q2
TYPE        NPN         PNP
IS          1.00D-14  1.00D-14
BF          100.000    100.000

****        INITIAL TRANSIENT SOLUTION
 NODE   VOLTAGE      NODE  VOLTAGE      NODE  VOLTAGE      NODE  VOLTAGE
(  1)    .0000     (  2)  5.0000     (  3)  5.0000     (  4)    .0000
(100)  10.0000

****        TRANSIENT ANALYSIS
    TIME          V(2)          V(3)
  .000E+00      5.000E+00      5.000E+00
 2.500E-05      5.999E+00      5.254E+00
 5.000E-05      4.999E+00      5.001E+00
 7.500E-05      4.007E+00      4.751E+00
 1.000E-04      5.000E+00      5.000E+00
 1.250E-04      5.999E+00      5.254E+00
 1.500E-04      4.999E+00      5.001E+00
 1.750E-04      4.007E+00      4.751E+00
 2.000E-04      5.000E+00      5.000E+00
```

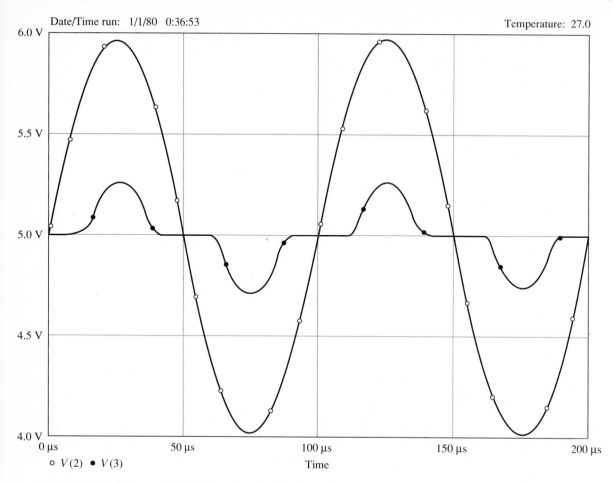

Date/Time run: 1/1/80 0:36:53   Temperature: 27.0

○ V(2) ● V(3)   Time

FIGURE H.32 Example 15.1: Input and output waveforms.

The input and output waveforms are shown in Figure H.32. Hopefully this is the most severe case of crossover distortion you will ever see.

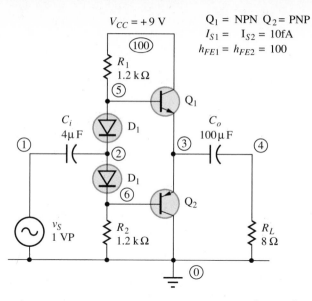

$V_{CC} = +9$ V

$Q_1 = $ NPN $Q_2 = $ PNP
$I_{S1} = I_{S2} = 10$fA
$h_{FE1} = h_{FE2} = 100$

(100)

R_1
1.2 k Ω

(5)

Q_1

C_i
4μF

D_1

(1)

(2)

(3)

C_o
100μF

(4)

D_1

(6)

Q_2

v_S
1 VP

R_2
1.2 kΩ

R_L
8 Ω

(0)

FIGURE H.33   Example 15.2: Diode biased complementary symmetry amplifier.

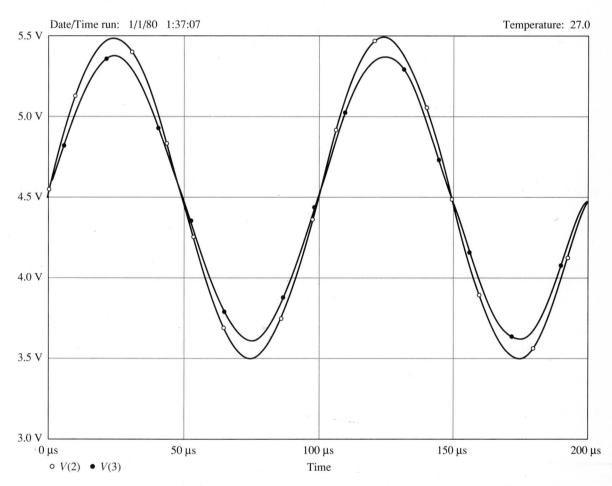

Date/Time run: 1/1/80 1:37:07

Temperature: 27.0

5.5 V

5.0 V

4.5 V

4.0 V

3.5 V

3.0 V

0 μs   50 μs   100 μs   150 μs   200 μs

∘ V(2)   • V(3)

Time

FIGURE H.34   Example 15.2: Input and output waveforms.

EXAMPLE 15.2
```
****        CIRCUIT DESCRIPTION
*FIND   Q  POINT,   vLP  FOR THE CIRCUIT OF FIGURE H.33
VCC 100 0 9VDC
VIN 1 0 SIN(0 1V 10KHZ)
.TRAN 25US 200US
R1 100 5 1.2K
R2 6 0 1.2K
RL 4 0 8OHM
CI 1 2 4UF
CO 3 4 100UF
Q1 100 5 3 Q1
.MODEL Q1 NPN (IS=10FA BF=100)
Q2 0 6 3 Q2
.MODEL Q2 PNP (IS=10FA BF=100)
D1 5 2 D
D2 2 6 D
.MODEL D D
.OP
.PRINT TRAN V(2) V(3)
.PROBE
*WAVEFORMS ARE SHOWN IN FIGURE H.34
.END


****        DIODE MODEL PARAMETERS
IS          1.00D-14


****        BJT MODEL PARAMETERS
            Q1          Q2
TYPE        NPN         PNP
IS          1.00D-14    1.00D-14
BF          100.000     100.000


****        SMALL SIGNAL BIAS SOLUTION
  NODE   VOLTAGE     NODE   VOLTAGE     NODE   VOLTAGE     NODE   VOLTAGE
(  1)     .0000    (  2)    4.5000    (  3)    4.5000    (  4)     .0000
(  5)    5.1848    (  6)    3.8152    (100)    9.0000


****        OPERATING POINT INFORMATION
**** DIODES
            D1          D2
MODEL       D           D
ID          3.15E-03    3.15E-03
VD          .685        .685
```

****   BIPOLAR JUNCTION TRANSISTORS

	Q1	Q2
MODEL	Q1	Q2
IB	3.15E-05	-3.15E-05
IC	3.15E-03	-3.15E-03
VBE	.685	-.685
VBC	-3.82	3.82
VCE	4.50	-4.50
BETADC	100.	100.

****   TRANSIENT ANALYSIS

TIME	V(2)	V(3)
.000E+00	4.500E+00	4.500E+00
2.500E-05	5.488E+00	5.375E+00
5.000E-05	4.479E+00	4.491E+00
7.500E-05	3.498E+00	3.614E+00
1.000E-04	4.501E+00	4.497E+00
1.250E-04	5.490E+00	5.376E+00
1.500E-04	4.480E+00	4.491E+00
1.750E-04	3.499E+00	3.615E+00
2.000E-04	4.502E+00	4.500E+00

The input and the output waveforms are shown in Figure H.34. Note that, true to the current mirror idea, there is no observable crossover distortion. It would be instructive for you to change the ".TRAN" statement in the above circuit description to something like ".TRAN 1US 55US 45US 50NS" and take a really close look at the crossover point. You would find no observable crossover distortion. Real diodes and transistors usually do not match this well.

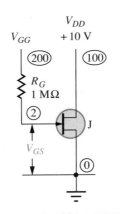

FIGURE H.35   Example 16.1: Fixed bias JFET circuit.

EXAMPLE 16.1A
```
****      CIRCUIT DESCRIPTION
*FIND  VGS  FOR THE CIRCUIT OF FIGURE H.35 (VGG = -2V)
VDD 100 0 10VDC
VGG 200 0 -2VDC
RG 200 2 1MEGOHM
J 100 2 0 J
.MODEL J NJF (VTO=-3V BETA=6E-4)
.OP
.END
```

```
****      JFET MODEL PARAMETERS
TYPE        NJF
VTO         -3.000
BETA      6.00D-04
```

```
****      OPERATING POINT INFORMATION
ID        6.00E-04
VGS       -2.00
VDS       10.0
GM        1.20E-03
```

EXAMPLE 16.1B
```
****      CIRCUIT DESCRIPTION
*FIND  VGS  FOR THE CIRCUIT OF FIGURE H.35 (VGG = +2V)
VDD 100 0 10VDC
VGG 200 0 +2VDC
RG 200 2 1MEGOHM
J 100 2 0 J
.MODEL J NJF (VTO=-3V BETA=6E-4)
.OP
.END
```

```
****      JFET MODEL PARAMETERS
TYPE        NJF
VTO         -3.000
BETA      6.00D-04
```

```
****      OPERATING POINT INFORMATION
ID        6.65E-03
VGS       .329
VDS       10.0
GM        3.99E-03
```

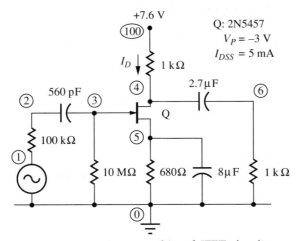

FIGURE H.36 Example 17.8: Source biased JFET circuit.

*EXAMPLE 17.8

The important JFET parameters listed in Appendix F are: $V_P = -3\ V$; $I_{DSS} = 5\ mA$. The "BETA" parameter required by PSpice is

$$\beta = \frac{I_{DSS}}{V_P^2} \tag{16.2}$$

$$= \frac{5\ mA}{(-3\ V)^2} = 556\ \mu S/V$$

```
****      CIRCUIT DESCRIPTION
*DETERMINE DRAIN CURRENT, ID, FOR FIGURE H.36
VCC 100 0 7.6VDC
RG 3 0 10MEGOHM
RD 100 4 1KOHM
J 4 3 5 J
.MODEL J NJF (VTO=-3V BETA=556US/V)
RSQ 5 0 680OHM
.OP
.END

****      JFET MODEL PARAMETERS
TYPE          NJF
VTO          -3.000
BETA          5.56D-04

****      SMALL SIGNAL BIAS SOLUTION
 NODE   VOLTAGE    NODE   VOLTAGE    NODE   VOLTAGE    NODE   VOLTAGE
(  3)     .0001  (  4)    5.8196  (  5)    1.2106  (100)    7.6000
```

```
****        OPERATING POINT INFORMATION
ID          1.78E-03
VGS         -1.21
VDS          4.61
GM          1.99E-03
```

---

*EXAMPLE 17.8

From Appendix F, for the 2N5103: $V_P = -4$ V; $I_{DSS} = 8$ mA. Thus BETA required by PSpice is

$$\beta = \frac{I_{DSS}}{V_P^2} \tag{16.2}$$

$$= \frac{8 \text{ mA}}{(-4)^2} = 500 \text{ }\mu\text{S/V}$$

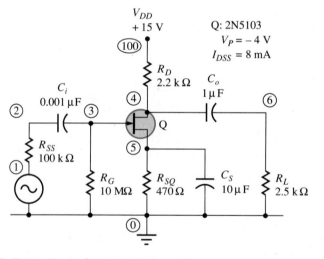

FIGURE H.37   Example 17.9: JFET amplifier.

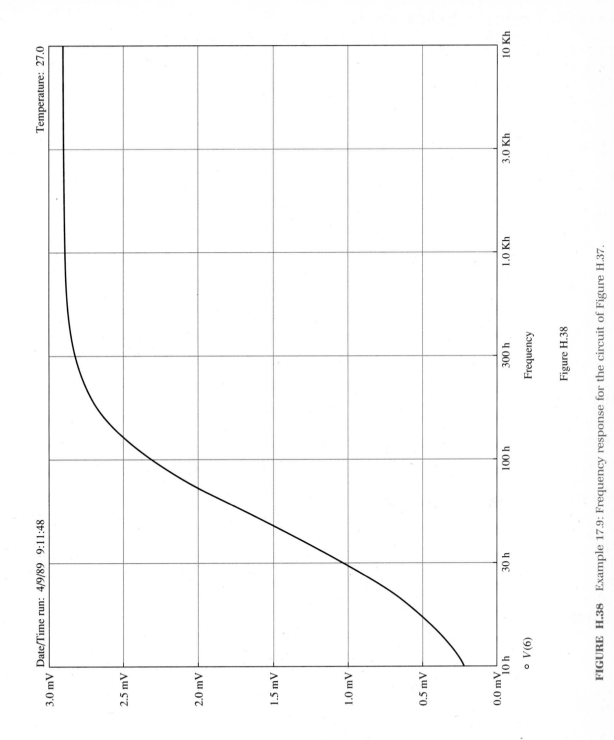

FIGURE H.38   Example 17.9: Frequency response for the circuit of Figure H.37.

```
****          CIRCUIT DESCRIPTION
*FIND  Q  POINT,  AV,   AND  FL  FOR   FIGURE H.37
VDD 100 0 15VDC
VSS 1 0 AC 1MV
.AC DEC 11 10HZ 10KHZ
RSS 1 2 100KOHM
CI 2 3 .001UF
RG 3 0 10MEGOHM
JQ 4 3 5 Q
.MODEL Q NJF (VTO=-4V BETA=500US/V)
RD 100 4 2.2KOHM
RSQ 5 0 470OHM
CS 5 0 10UF
CO 4 6 1UF
RL 6 0 2.5KOHM
.OP
.PRINT AC V(6)
.PROBE
*THE FREQUENCY RESPONSE GRAPH IS IN FIGURE H.38
.END

****      JFET MODEL PARAMETERS
TYPE          NJF
VTO          -4.000
BETA      5.00D-04

****      SMALL SIGNAL BIAS SOLUTION
  NODE   VOLTAGE     NODE   VOLTAGE     NODE   VOLTAGE     NODE   VOLTAGE
(  1)     .0000    (  2)     .0000    (  3)     .0001    (  4)   8.0456
(  5)    1.4857    (  6)     .0000    (100)   15.0000

****      OPERATING POINT INFORMATION
ID        3.16E-03
VGS      -1.49
VDS       6.56
GM        2.52E-03

****      AC ANALYSIS
   FREQ         V(6)
 1.000E+01    2.094E-04
 1.233E+01    2.955E-04

   . . .         . . .

 5.337E+01    1.636E-03
 6.579E+01    1.885E-03
```

8.111E+01   2.115E-03
1.000E+02   2.315E-03

. . .   . . .

3.511E+03   2.914E-03
1.000E+04   2.914E-03

The signal source, V_{SS}, connected between node 1 and ground is an AC source putting out 1 mV. The .AC statement causes the frequency to vary by decades with 11 points plotted per decade from 10 Hz to 10 kHz. Many of the data points have been deleted from the resulting AC analysis. The last two data points indicate that at midband the output voltage is 2.914 mV. Thus

$$A_V = \frac{v_{out}}{v_{in}}$$

$$= \frac{2.914 \text{ mV}}{1 \text{ mV}} = \underline{\underline{2.914}}$$

The output at f_L is

$$v_{fL} = \frac{v_{mid}}{\sqrt{2}}$$

$$= \frac{2.914 \text{ mV}}{\sqrt{2}} = 2.06 \text{ V}$$

From the AC analysis, f_L is between 65.79 Hz and 81.11 Hz.

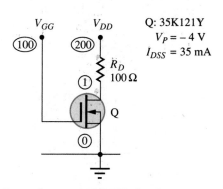

FIGURE H.39.   Example 18.1: MOSFET circuit.

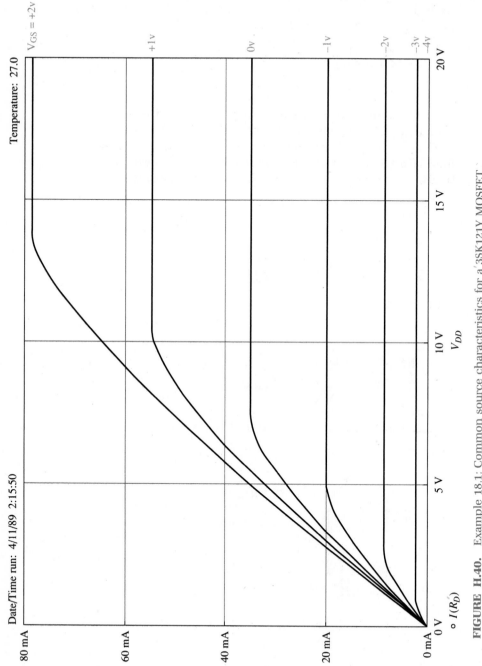

*EXAMPLE 18.1

From Appendix H for the 3SK121Y $V_P = -4$ V; $I_{DSS} = 35$ mA. Converting to PSpice terminology, VTO $= -4$ V and

$$KP = 2\beta = 2\frac{I_{DSS}}{V_P^2} \qquad\qquad (18.2)$$

$$= 2\frac{35 \text{ mA}}{(-4 \text{ V})^2} = 4.38 \text{ m S/V}$$

****   CIRCUIT DESCRIPTION

```
*PLOT DRAIN CHARACTERISTICS FOR FIGURE H.39
VGG 100 0 DC
VDD 200 0 DC
.DC VDD 0V 20V .2V VGG -4V +2V 1V
.OPTIONS LIMPTS=800
RD 200 1 100
M 1 100 0 0 Q
.MODEL Q NMOS (VTO=-4V KP=4.38MS/V)
.PROBE
*CHARACTERISTICS ARE DISPLAYED IN FIGURE H.40
.END
```

****   MOSFET MODEL PARAMETERS

TYPE	NMOS
LEVEL	1.000
VTO	-4.000
KP	4.38D-03

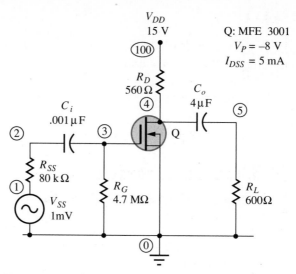

FIGURE H.41. Example 18.3: MOSFET amplifier circuit.

```
*EXAMPLE 18.3

****      CIRCUIT DESCRIPTION
*FIND THE  DB  GAIN IN FIGURE H.41
VDD 100 0 15VDC
VSS 1 0 AC 1MV
.AC LIN 1 10KHZ 10KHZ
RSS 1 2 80KOHM
CI 2 3 .001UF
RG 3 0 4.7MEGOHM
RD 100 4 560OHM
M 4 3 0 0 Q
.MODEL Q NMOS (VTO=-8V KP=156US/V)
CO 4 5 4UF
RL 5 0 600OHM
.PRINT AC V(2) I(RSS) V(5) I(RL)
.END


****      MOSFET MODEL PARAMETERS

TYPE          NMOS
LEVEL          1.000
VTO           -8.000
KP         1.56D-04
```

```
****        SMALL SIGNAL BIAS SOLUTION
```

NODE	VOLTAGE	NODE	VOLTAGE	NODE	VOLTAGE	NODE	VOLTAG
(1)	.0000	(2)	.0000	(3)	.0000	(4)	12.204
(5)	.0000	(100)	15.0000				

```
****        AC ANALYSIS
```

FREQ	V(2)	I(RSS)	V(5)	I(RL)
1.000E+04	9.833E-04	2.092E-10	3.554E-04	5.924E-07

Interpreting the above output:

$$v_{in} = V(2) = 983.3 \ \mu V \qquad i_{in} = I(RSS) = 209.2 \ pA$$

$$v_{out} = V(5) = 355.4 \ \mu V \qquad i_{out} = I(RL) = 592.4 \ nA$$

Thus

$$A_v \equiv \frac{v_{out}}{v_{in}}$$

$$= \frac{355.4 \ \mu V}{983.3 \ \mu V} = 0.361$$

$$A_i \equiv \frac{i_{out}}{i_{in}}$$

$$= \frac{592.4 \ nA}{209.2 \ pA} = 2830$$

$$dB_A = 10 * \log(A_v A_i)$$

$$= 10 * \log(0.361 * 2830) = \underline{\underline{30.1}}$$

Appendix I
r Parameters vs h Parameters

Rather than using h_{FE} or h_{fe} to represent the current gain of a transistor, you will often see the Greek letter β used. That is the source of the term BETA used by PSpice. It is the term used in the r parameter transistor model.

To complete the r parameter model, obviously the current that enters the base of an NPN transistor must flow through the crystal for some distance before it reaches the junction. Thus you would expect some resistance, r_b, to be present in the base region. Similarly, the emitter current must flow through some semiconductor material before it can leave the transistor at the emitter terminal. Thus you would expect some resistance, r_e, to be present in the emitter region. This leads to the r parameter model for a transistor that is shown in Figure I.1.

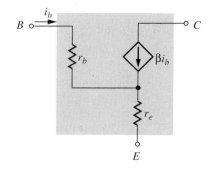

FIGURE I.1.

As you might expect, due to the small distances involved, r_b is usually small. It is under 10Ω for most transistors. Since it carries only the base current which is also comparatively small, r_b is usually ignored. It is probably more accurate to say that the effect of r_b is usually lumped together with r_e since they are both inside the transistor and their effects are hard

913

to separate. This simplification leads to the equivalent circuit shown in Figure I.2.

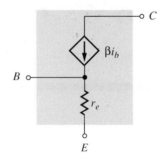

FIGURE I.2

In order to simplify circuit analysis, it is convenient to reference r_e to the base circuit rather than the emitter circuit. This can be done by taking advantage of the Miller Effect. The value of r_e is simply multiplied by the current gain of the transistor. The result is the r parameter equivalent circuit shown in Figure I.3(a).

For comparison, the simplified h parameter model proposed in Section 11.8 is shown in Figure I.3(b). As you can see, aside from differences in notation, the only real difference between the two is that the h parameter model allows for the inclusion of $1/h_{oe}$ in the event that its effect is significant. To convert from one notation to the other, we have:

$$\beta = h_{fe}$$
$$\beta r_e = h_{ie}$$

The only real advantage one model has over the other is that the manufacturers and the data books have usually chosen to list the h parameters. In order to better prepare you to use those resources, this book uses the h parameter model.

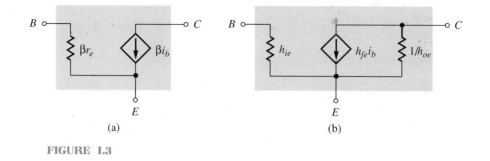

(a)   (b)

FIGURE I.3

INDEX

TABLE 12.1 Common Emitter Amplifier Design Procedure

STD. Choose nearest standard resistor value.

1. $R_C \overset{*}{=} R_L \parallel (-1/h_{oe})$

2. $R_E \overset{*}{=} \dfrac{R_{DC}}{4} = \dfrac{R_C}{3}$ (STD. if step 10 is omitted)

3. $I_C \overset{*}{=} \dfrac{V_{CC}}{R_{DC} + R_{AC}} = \dfrac{V_{CC}}{R_C + R_E + R_C \parallel R_L}$

4. $V_{BE} \approx 26 \text{ mV} * \ln\dfrac{I_C}{10 \text{ fA}}$

5. $V_E = I_C R_E$

6. $V_B = V_{BE} + V_E$

7. $V_C = V_{CC} - I_C R_C$

8. $I_B = \dfrac{I_C}{h_{FE}}$

9. $h_{ie} = \dfrac{26 \text{ mV}}{I_B}$

10. $R_{E1} = \dfrac{R_C \parallel R_L}{A_v} - \dfrac{h_{ie}}{h_{fe}}$

11. $R_{E2} = R_E - R_{E1}$

12. $R_B = \dfrac{S}{1-S} h_{FE} R_E \overset{*}{\approx} \dfrac{1}{10} h_{FE} R_E$

13. $V_{BB} = V_B + I_B R_B$

14. $R_1 = R_B \dfrac{V_{CC}}{V_{BB}}$

15. $R_2 = R_1 \dfrac{V_{BB}}{V_{CC} - V_{BB}}$

16. $R_{in} = R_1 \parallel R_2 \parallel (h_{ie} + h_{fe} R_{E1})$

17. $R_{out} = R_C \parallel (1/h_{oe} + R_{E1})$

18. $C_i = \dfrac{1}{2\pi f_L (R_S + R_{in})}$

19. $C_o = \dfrac{1}{2\pi f_L (R_{out} + R_L)}$

20. $C_E = \dfrac{1}{2\pi f_L \left(R_{E2} \parallel \left(R_{E1} + \dfrac{h_{ie} + R_1 \parallel R_2 \parallel R_S}{h_{fe}} \right) \right)}$

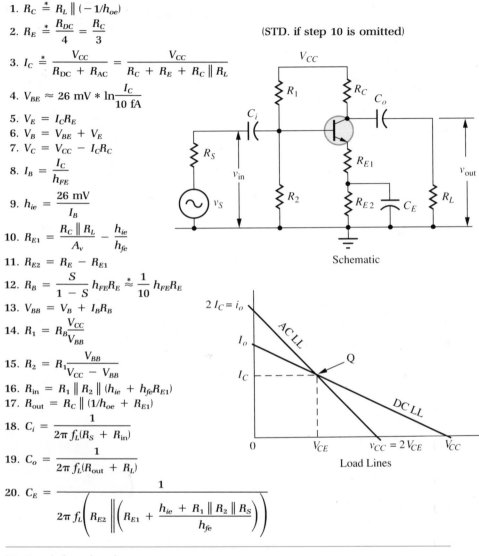

Schematic

Load Lines

*Designer's formula only.